STRATIGRAPHIC EVOLUTION OF FORELAND BASINS

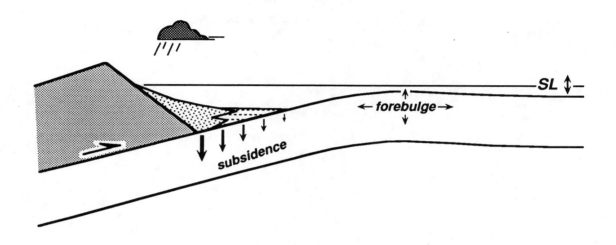

Edited by

Steven L. Dorobek, *Texas A&M University, College Station, TX*
and
Gerald M. Ross, *Institute of Sedimentary and Petroleum Geology, Calgary, Alberta, Canada*

Copyright 1995 by
SEPM (Society for Sedimentary Geology)

Peter A. Scholle, *Editor of Special Publications*
Special Publication No. 52

Tulsa, Oklahoma, U.S.A.

March, 1995

A Publication of
SEPM (Society for Sedimentary Geology)

ISBN 1-56576-016-6

© 1995 by
SEPM (Society for Sedimentary Geology)
P.O. Box 4756
Tulsa, Oklahoma 74131

Printed in the United States of America

STRATIGRAPHIC EVOLUTION OF FORELAND BASINS
INTRODUCTION

A strong case can be made that foreland basins are where the causal links between sedimentation and tectonic events were first recognized, as evidenced by the interpretations of geologists working in classic foreland areas such as the Alpine foreland and the Appalachian Basin (Dana, 1873; Stille, 1913; Kay, 1947, 1951; Trümpy, 1960). The development and acceptance of plate tectonic theory provided a mechanism that coupled hinterland deformation, foreland basin subsidence, and sedimentation (Dickinson, 1973; Price, 1973; Bally and Snelson, 1980). Since then, foreland basins have been the focus of many studies, as indicated by classic papers on their geodynamic evolution and sedimentary fill (Beaumont, 1981; Jordan, 1981; Quinlan and Beaumont, 1984; Stockmal and others, 1986; Flemings and Jordan, 1989; Sinclair and others, 1991), paleohydrology (Bethke, 1985; Garven, 1989), thermal history (Hitchon, 1984; Kominz and Bond, 1986), and several recent compendia on foreland basins and their sedimentary fill (Allen and Homewood, 1986; Macqueen and Leckie, 1992).

This Special Publication was derived from a Research Symposium entitled "Stratigraphic Sequences in Foreland Basins," held at the annual joint meeting of the American Association of Petroleum Geologists and SEPM on June, 1992 in Calgary, Alberta, Canada. Approximately half of the papers in this volume were given as presentations at the Calgary meeting, while the other half were contributed after the meeting. This volume does not provide a comprehensive synthesis of foreland basin stratigraphy because every foreland basin has a unique kinematic history and tectonic setting which, in turn, results in unique stratigraphic sequences. We hope instead that this volume provides a well-balanced perspective of current research on foreland basin stratigraphy and also serves as another element in the evolving framework that comprises our understanding of foreland basins. Given that so many of earth's resources (potable ground water, hydrocarbons, and base metal deposits) are found in foreland basins and that foreland basin strata often provide the only preserved record of the tectonic events that led to basin development, the impetus for continued studies of foreland basin strata should remain for many generations of geologists to come.

QUANTITATIVE MODELS OF FORELAND BASIN EVOLUTION

Early quantitative models for the geodynamic evolution and subsidence history of foreland basins examined the flexural effects of vertical loading on a lithospheric plate. These models attempted to find the best-fit between observed or reconstructed basin profiles and model profiles that were constructed by varying the rheological properties of the plate and the dimensions and density of the orogenic wedge that served as the vertical load. Erosion of the orogenic wedge and redistribution of the vertical load was added to subsequent models. Recent modeling efforts and stratigraphic studies have examined how rheological and mechanical heterogeneities in the loaded plate can influence subsidence patterns and basin geometries. Intraplate stresses also are gaining wider scrutiny as an important component of foreland basin subsidence and its consequent effects on sedimentation patterns.

Several papers in this volume have incorporated additional factors that affect the stratigraphic fill in foreland basins, even though they may have only subtle stratigraphic expressions. Johnson and Beaumont present an elegant model that relates climate, orographically-controlled precipitation, drainage patterns and river power, balance between tectonic and erosional mass fluxes, basin type, and basin subsidence to stratigraphic fill in several hypothetical foreland basins. Peper and others present models which suggest that variations in intraplate stress can be as significant as eustasy in its relative effect on the stratigraphic evolution of foreland basins, although rate of orogenic wedge growth is also important. The Permian Basin of West Texas and southeastern New Mexico is used by Yang and Dorobek as an example of how intraforeland, structural discontinuities act to partition strain during basin development and to demonstrate how these structural elements might influence regional subsidence and sedimentation patterns.

PROVENANCE STUDIES

The provenance of siliciclastic sediment in foreland basins has been used for over a century to provide important constraints on the sources of siliciclastic sediment in foreland basins and on the relative timing of deformation and depositional events.

Provenance studies, however, have evolved significantly beyond classical petrographic approaches which focus on the composition and relative proportions of framework components. Two papers in this volume use U-Pb isotopic analysis of detrital zircons to place additional constraints on sediment provenance in foreland basins. Mustard and others demonstrate that the Cretaceous–Paleogene Nanaimo Group along the Cordilleran margin of southern Canada received detritus from source areas that were hundreds of kilometers inboard of the basin and imply long-distance transport across tectonic strike. The Devonian clastic wedge of the Canadian Arctic is an order of magnitude larger than the classic Catskill delta of the Appalachian Basin, yet fundamental questions regarding sediment provenance persist, in part reflecting the very quartzose framework composition of the Arctic sediments. McNicoll and others use ages of detrital zircons to document a Caledonide-Greenland source area for the clastic wedge in the Canadian Arctic, effectively constraining the foreland fill as being transported longitudinally along the basin axis, parallel to the Ellesmerian orogenic front.

Mudrocks often comprise a large part of the total volume of foreland basin sediment, yet their provenance has rarely been examined. Andersen uses whole-rock chemical composition and clay mineralogy in Ordovician mudstones of the Appalachian Basin in order to determine if the mudstones record first-order trends in provenance that are related to tectonic events.

REGIONAL STUDIES OF FORELAND BASINS

The remainder of this volume consists of regional studies that illustrate how interactions between tectonics, climate, sea-level change, and sedimentation affect foreland basin stratigraphy.

In their study of conglomeratic units in Lower Cretaceous, nonmarine strata from central Wyoming, May and others document partitioning of the Wind River Basin by a series of horsts and grabens, which controlled the distribution and geometry of fluvial networks. Upper Paleozoic strata in the Black Warrior Basin, described in a paper by Thomas, illustrate how the location of the basin next to two, nearly orthogonal orogenic fronts (i.e., the Ouachita and Appalachian thrust belts) influenced subsidence and sedimentation patterns in the basin. Carbonate strata are not generally considered to be signature lithologies of foreland basins, with much greater emphasis placed on siliciclastic facies. Dorobek presents an overview of modern and ancient marine carbonate platforms in foreland basin settings and suggests that these strata may provide a more sensitive record of basin evolution than siliciclastic facies. In this context, Yang and Dorobek examine the stratigraphic and structural evolution of the classic Permian Basin of West Texas and New Mexico as they relate to compressional tectonics of the Marathon-Ouachita fold belt and associated intraforeland deformation.

A number of papers in this volume focus on the Late Devonian-Mississippian Antler orogeny and its effects on foreland basin development along the western margin of North America. Mississippian carbonate platform sedimentation in southern Nevada and eastern California, a structurally complex and poorly understood part of the Antler puzzle, are described in a paper by Stevens and others. Giles and Dickinson discuss stratigraphic relationships in the classical Antler foreland basin of Nevada and Utah, but emphasize interactions between eustasy and flexure on both the basinward and cratonward side of the inferred peripheral bulge and the consequent stratigraphic signature. Savoy and Mountjoy analyze the Devono-Mississippian succession of the southern Canadian Rocky Mountains in the context of Antler-age plate convergence, a view that was somewhat heretical previously but which is gaining wider acceptance amongst Canadian geologists.

Li and others provide a departure from the Antler theme in their study of the stratigraphic evolution of the thick, nonmarine fill of the Triassic-Jurassic Ordos Basin of China. They suggest that the Ordos Basin formed as a response to oblique collision and closure of a complex and irregular Tethyan margin.

The volume is completed by two papers by Gardner on mid-Cretaceous strata from the Western Interior of North America. The first paper provides an overview of controls on regional base-level changes which, in turn, affected the hierarchy of time-stratigraphic units across the Western Interior Foreland Basin. In his second paper, Gardner focuses on central Utah, where detailed stratigraphic and facies analyses provide evidence for eustasy, not tectonism, as the dominant control on sedimentation.

ACKNOWLEDGMENTS

First and foremost, we would like to thank the contributors to this volume for entrusting us with their manuscripts and for their generally timely responses to our requests. Part of the expense for the color cross section that is included in a pocket at the back of this volume was defrayed by Mobil. We thank them for making production of the cross section a reality. The following colleagues served as efficient and thorough reviewers of papers, an often thankless but absolutely essential job:

R. Debruin
T. E. Ewing
P. B. Flemings
K. A. Giles
J. P. Grotzinger
M. T. Harris
P. L. Heller
T. E. Jordan
C. Kerans
T. Lawton
W. C. McClelland
D. Nummedal
C. Paola
G. Quinlan
R. Rainbird
S. K. Reid
B. Richards
T. Ryor
P. Schwans
M. Smith
R. J. Weimer
and four anonymous reviewers

REFERENCES

ALLEN, P. A. AND HOMEWOOD, P., eds., 1986, Foreland Basins: Oxford, International Association Sedimentologists Special Publication 8, 453 p.

BALLY, A. W. AND SNELSON, S., 1980, Realms of subsidence: Canadian Society of Petroleum Geology Memoir 6, p. 1–94.

BEAUMONT, C., 1981, Foreland basins: Geophysical Journal of the Royal Astronomical Society, v. 65, p. 291–329.

BETHKE, C. M., 1985, A numerical model of compaction-driven groundwater flow and heat transfer and its application to the paleohydrology of intracratonic basins: Journal of Geophysical Research, v. 90, p. 6817–6828.

DANA, J. D., 1873, On some results of the earth's contraction from cooling, including a discussion of the origin of mountains and the nature of the earth's interior: American Journal of Science, v. 3, 5, p. 423–443.

DICKINSON, W. R., 1974, Plate tectonics and sedimentation, in Dickinson, W. R., ed., Tectonics and Sedimentation: Society of Economic Paleontologists and Mineralogists Special Publication 22, p. 1–27.

FLEMINGS, P. B. AND JORDAN, T. E., 1989, A synthetic stratigraphic model of foreland basin development: Journal of Geophysical Research, v. 94, p. 3851–3866.

GARVEN, G., 1989, A hydrologic model for the formation of the giant oil sands deposits of the western Canada sedimentary basin: American Journal of Science, v. 289, p. 105–166.

HITCHON, B., 1984, Geothermal gradients, hydrodynamics, and hydrocarbon occurrences, Alberta, Canada: American Association of Petroleum Geologists Bulletin, v. 68, p. 713–743.

JORDAN, T. E., 1981, Thrust loads and foreland basin evolution, Cretaceous, Western United States: American Association of Petroleum Geologists Bulletin, v. 65, p. 2506–2520.

KAY, M., 1947, Geosynclinal nomenclature and the craton: American Association of Petroleum Geologists Bulletin, v. 31, p. 1289–1293.

KAY, M., 1951, North American geosynclines: Boulder, Geological Society of America Memoir 48, 143 p.

KOMINZ, M. A. AND BOND, G. C., 1986, Geophysical modelling of the thermal history of foreland basins: Nature, v. 320, p. 252–256.

MACQUEEN, R. W. AND LECKIE, D. A., 1992, Foreland Basins and Fold Belts: Tulsa, American Association of Petroleum Geologists Memoir 55, 460 p.

PRICE, R. A., 1973, Large-scale gravitational flow of supracrustal rocks, Southern Canadian Rockies, in de Jong, K.A., and Scholten, R., eds., Gravity and Tectonics: New York, John Wiley, p. 491–502.

QUINLAN, G. M. AND BEAUMONT, C., 1984, Appalachian thrusting, lithospheric flexure and the Paleozoic stratigraphy of the eastern interior of North America: Canadian Journal of Earth Sciences, v. 21, p. 973–996.

SINCLAIR, H. D., COAKLEY, B. J., ALLEN, P. A., AND WATTS, A. B., 1991, Simulation of foreland basin stratigraphy using a diffusion model of mountain belt uplift and erosion: an example from the Central Alps, Switzerland: Tectonics, v. 10, p. 599–620.

STILLE, H., 1913, Evolution und Revolutionen in der Erdgeschichte: Berlin, Borntaeger, 32 p.

STOCKMAL, G. S. AND BEAUMONT, C., 1987, Geodynamic models of convergent margin tectonics: the southern Canadian Cordillera and the Swiss Alps, *in* Beaumont, C., and Tankard, A. J., eds., Sedimentary Basins and Basin-Forming Mechanisms: Calgary, Canadian Society of Petroleum Geologists Memoir 12, p. 393–411.

TRÜMPY, R., 1960, Paleotectonic evolution of the Central and Western Alps: Geological Society of America Bulletin, v. 71, p. 843–908.

Steven L. Dorobek and Gerald M. Ross, Editors

CONTENTS

INTRODUCTION
STRATIGRAPHIC EVOLUTION OF FORELAND BASINS . . . . *Steven L. Dorobek and Gerald M. Ross* iii

I. QUANTITATIVE MODELS OF FORELAND BASIN EVOLUTION AND STRATIGRAPHIC DEVELOPMENT

PRELIMINARY RESULTS FROM A PLANFORM KINEMATIC MODEL OF OROGEN EVOLUTION, SURFACE PROCESSES AND THE DEVELOPMENT OF CLASTIC FORELAND BASIN STRATIGRAPHY . . . . *David D. Johnson and Christopher Beaumont* 3

IMPLICATIONS OF OROGENIC WEDGE GROWTH, INTRAPLATE STRESS VARIATIONS, AND EUSTATIC SEA-LEVEL CHANGE FOR FORELAND BASIN STRATIGRAPHY—INFERENCES FROM NUMERICAL MODELING . . . . . *Tim Peper, Ronald van Balen, and Sierd Cloetingh* 25

THE PERMIAN BASIN OF WEST TEXAS AND NEW MEXICO: FLEXURAL MODELING AND EVIDENCE FOR LITHOSPHERIC HETEROGENEITY ACROSS THE MARATHON FORELAND . . . . *Kenn-Ming Yang and Steven L. Dorobek* 37

II. PROVENANCE STUDIES IN FORELAND BASINS

PROVENANCE OF MUDSTONES FROM TWO ORDOVICIAN FORELAND BASINS IN THE APPALACHIANS
. . . . *C. Brannon Andersen* 53

PROVENANCE OF THE UPPER CRETACEOUS NANAIMO GROUP, BRITISH COLUMBIA: EVIDENCE FROM U-Pb ANALYSES OF DETRITAL ZIRCONS . . . . *Peter S. Mustard, Randall R. Parrish, and Vicki McNicoll* 65

PROVENANCE OF THE DEVONIAN CLASTIC WEDGE OF ARCTIC CANADA: EVIDENCE PROVIDED BY DETRITAL ZIRCON AGES . . . . *Vickie J. McNicoll, J. Chris Harrison, Hans P. Trettin, and Ray Thorsteinsson* 77

III. REGIONAL STUDIES OF FORELAND BASINS

CHRONOSTRATIGRAPHY AND TECTONIC SIGNIFICANCE OF LOWER CRETACEOUS CONGLOMERATES IN THE FORELAND OF CENTRAL WYOMING . . . . *Michael T. May, Lloyd C. Furer, Erik P. Kvale, Lee J. Suttner, Gary D. Johnson, and James H. Meyers* 97

DIACHRONOUS THRUST LOADING AND FAULT PARTITIONING OF THE BLACK WARRIOR FORELAND BASIN WITHIN THE ALABAMA RECESS OF THE LATE PALEOZOIC APPALACHIAN-OUACHITA THRUST BELT . . *William A. Thomas* 111

SYNOROGENIC CARBONATE PLATFORMS AND REEFS IN FORELAND BASINS: CONTROLS ON STRATIGRAPHIC EVOLUTION AND PLATFORM/REEF MORPHOLOGY . . . . *Steven L. Dorobek* 127

THE PERMIAN BASIN OF WEST TEXAS AND NEW MEXICO: TECTONIC HISTORY OF A "COMPOSITE" FORELAND BASIN AND ITS EFFECTS ON STRATIGRAPHIC DEVELOPMENT . . . . *Kenn-Ming Yang and Steven Dorobek* 149

DEVELOPMENT OF THE MISSISSIPPIAN CARBONATE PLATFORM IN SOUTHERN NEVADA AND EASTERN CALIFORNIA ON THE EASTERN MARGIN OF THE ANTLER FORELAND BASIN . . . . *Calvin H. Stevens, Darrel Klingman, and Paul Belasky* 175

THE INTERPLAY OF EUSTASY AND LITHOSPHERIC FLEXURE IN FORMING STRATIGRAPHIC SEQUENCES IN FORELAND SETTINGS: AN EXAMPLE FROM THE ANTLER FORELAND, NEVADA AND UTAH . . . . *Katherine A. Giles and William R. Dickinson* 187

CRATONIC-MARGIN AND ANTLER-AGE FORELAND BASIN STRATA (MIDDLE DEVONIAN TO LOWER CARBONIFEROUS) OF THE SOUTHERN CANADIAN ROCKY MOUNTAINS AND ADJACENT PLAINS
. . . . *Lauret E. Savoy and Eric W. Mountjoy* 213

UPPER TRIASSIC-JURASSIC FORELAND SEQUENCES OF THE ORDOS BASIN IN CHINA . . . . . *Li Sitian, Yang Shigong, and Tom Jerzykiewicz* 233

TECTONIC AND EUSTATIC CONTROLS ON THE STRATAL ARCHITECTURE OF MID-CRETACEOUS STRATIGRAPHIC SEQUENCES, CENTRAL WESTERN INTERIOR FORELAND BASIN OF NORTH AMERICA . . . . . . *Michael H. Gardner* 243

THE STRATIGRAPHIC HIERARCHY AND TECTONIC HISTORY OF THE MID-CRETACEOUS FORELAND BASIN OF CENTRAL UTAH . . . . *Michael H. Gardner* 283

PART I
QUANTITATIVE MODELS OF FORELAND BASIN EVOLUTION AND STRATIGRAPHIC DEVELOPMENT

PRELIMINARY RESULTS FROM A PLANFORM KINEMATIC MODEL OF OROGEN EVOLUTION, SURFACE PROCESSES AND THE DEVELOPMENT OF CLASTIC FORELAND BASIN STRATIGRAPHY

DAVID D. JOHNSON AND CHRISTOPHER BEAUMONT
Oceanography Department, Dalhousie University, Halifax, Nova Scotia, B3H-4J1, Canada

ABSTRACT: Clastic foreland basin stratigraphy is primarily determined by the relative rates of first-order basin controlling processes; the rate of mass accretion to an orogen by thrust tectonics, the rate of mass redistribution by surface processes, the rate of flexural isostatic compensation, and the rate of absolute sea level change. We have developed a composite planform foreland basin model to look for model stratigraphic signatures which reflect either the dominant influence of one of these basin controlling processes or interaction among several processes.

The foreland basin model links component models of orogen tectonics, surface processes, lithospheric flexure and eustasy in an internally consistent manner. The tectonic model uses critical wedge principles to construct a doubly-vergent wedge-shaped orogen. The flexural isostasy model uses either an elastic or a thermally-activated linear visco-elastic lithospheric rheology. The surface processes model couples hillslope (mass diffusion) and climate-mediated fluvial (mass advection) transports to erode, redistribute and deposit mass across the orogen, foreland basins and peripheral bulges.

Preliminary results are presented from two models which illustrate the terminal stages of ocean closure, the ensuing continent-continent collision, and the kinematic growth of an orogen with two flanking foreland basins.

In the first model, there is no significant strike variation in model processes, therefore, cross-sections of any model are sufficient to analyze the basins and compare them with previously published results. The contrasting stratigraphic architecture of the basins is controlled by the inherent tectonic asymmetry and by the erosion and sediment flux which become progressively asymmetric as a consequence of the relative positions of the basins on the windward and leeward sides of the growing orogen.

The second model demonstrates the complexities that result when there is a significant strike variation in tectonic processes. This model takes the form of a diachronous continent-continent collision between two continental margins inclined at an angle of ~25°. The model collision zone evolves in time and along strike from accretionary prism to orogen. Sediment flux into the windward foreland basin is greatest adjacent to the largest part of the orogen. This region of the basin becomes subaerial first and the drainage network develops a longitudinal trunk river system, similar to those common to many foreland basins. The combination of lateral and longitudinal fluvial transport results in diachronous filling of the marine basin by an assemblage of fluvial and marine facies which prograde down the basin axis.

INTRODUCTION

A foreland basin forms by flexural isostatic compensation of the lithosphere in response to tectonic loading in a convergent orogen (Fig. 1). Clastic sedimentary filling of the basin is a consequence of surface processes which erode, transport and deposit material. The rate of transport of surface material from the orogen to the basin is a function of climate, sea level, and surface topography. Surface topography is a function of the rate of orogen growth, isostatic compensation and the material properties limiting the rates of surface processes. Clastic foreland basin stratigraphy is therefore the product of linked processes and contains a synoptic record of the competition among rates of orogen growth, flexural isostatic compensation, mass redistribution by surface processes and the response to eustatic sea level changes, among other processes.

Quantitative foreland basin models described by Jordan (1981), Beaumont (1981), Karner and Watts (1983), Quinlan and Beaumont (1984), Stockmal and others (1986), and Beaumont and others (1988), for example, produced chronological records of the flexural isostatic response of the lithosphere to tectonic and sedimentary loads. No attempt was made to model sediment transport in these models, and basins were filled to a specified level without regard to process.

Flemings and Jordan (1989) made a significant improvement over previous models by introducing linear diffusive mass transport to represent the redistribution of mass from the orogen to the basin by surface processes. Their one-dimensional model also included a wedge shaped orogenic load which advances over the autochthon during periods of tectonic convergence (Flemings and Jordan, 1989, 1990; Jordan and Flemings, 1991). The result is a cross-sectional representation of synthetic foreland basin stratigraphy which contains a chronostratigraphic record of the geometric development of the basin and a model lithostratigraphic record of depositional environments based on surface slope.

Other authors, for example Sinclair and others (1991) and Paola and others (1992), have also used diffusive transport models to represent mass redistribution by surface processes. This approach is based on models of linear subaerial hillslope diffusion (e.g., Culling, 1960; Coleman and Watson, 1983; Hanks and others, 1984) and models of the delta development (Kenyon and Turcotte, 1985; Syvitski and others, 1988).

We continue in the spirit of these simple process models for clastic foreland basin stratigraphy. In this paper, we present some preliminary results from a kinematic planform foreland basin model designed to explore the relationships among the tectonic, surface mass transport and isostatic components and the relative importance of each component to the stratigraphy.

The first part of the paper contains a brief description of the component parts of the model and explains the differences in our approach from that of previous models. The second part contains results from two planform models. Both models illustrate the terminal stages of ocean closure, the ensuing continent-continent collision and the kinematic evolution of an orogen with two flanking foreland basins.

In Model 1, there is no significant strike variation in model processes, therefore, a single cross-section of the model is sufficient to analyze the basins and compare them with previously published models. The model was chosen to demonstrate the effects of tectonic and climatic asymmetry on the stratigraphic architecture of the two basins.

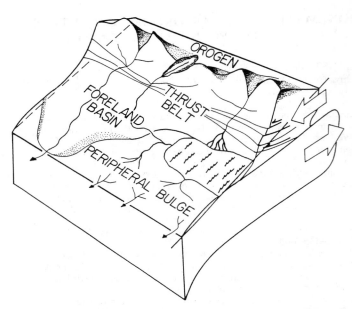

Fig. 1—Schematic drawing showing the fundamental elements of an orogen-foreland basin system: a compressional orogen; erosion, transportation and deposition of sediment in the foreland basin; and flexural isostatic compensation of the lithosphere in response to tectonic and sedimentary loads. The basin may be filled to different degrees along strike depending of the relative rates of mass flux into the orogen, denudation and sedimentation by surface processes, isostatic compensation and eustatic changes in sea level, among other processes.

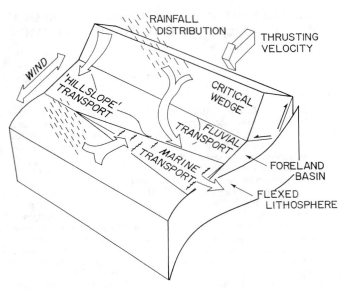

Fig. 2—Schematic drawing of the kinematic foreland basin model. The compressional orogen is modelled as a doubly-vergent wedge, based on critical Coulomb wedge theory. The surface processes model couples long range "fluvial" advective mass transport and short range linear diffusive "hillslope" mass transport to erode, transport and deposit mass in the subaerial environment. Redistribution of material in the marine environment is by linear diffusion. Tectonic and sedimentary loads are flexurally compensated using an elastic lithosphere.

Model 2 examines the complexities that arise when Model 1 is modified to include a significant strike variation in tectonic processes. By inclining the convergent margins we model a diachronous collision zone. It represents, in simplified form, the collision of linear and sawtooth margins in which promontories and recesses determine the sedimentary facies distribution in the foreland basins.

COMPONENTS OF THE PLANFORM KINEMATIC MODEL

The foreland basin model links a kinematic model of orogen tectonics, mass redistribution by surface processes, lithospheric flexure and eustasy (Fig. 2). The tectonic model uses critical wedge principles to construct the orogenic load as a doubly-vergent wedge. The surface processes model incorporates marine and hillslope transport by mass diffusion and fluvial transport. The planform flexural isostasy model uses either an elastic or a thermally-activated linear visco-elastic rheology to calculate the displacement of the lithosphere in response to a vertical load. Eustasy is introduced by varying the model sea level.

The model, which is described in greater detail by Beaumont and Johnson (in prep.) is summarized below to provide the reader with an understanding of the principal interactions among the model components. The overall model is characterized by the mass fluxes associated with the component models, the tectonic flux $Q_T(x,y,t)$, the surface processes flux $Q_S(x,y,t)$, and the isostatic flux $Q_I(x,y,t)$.

Tectonic Model

A simple model of orogen tectonics is required which conserves mass and approximates the planform kinematics and tectonic load distribution in an orogenic belt. It is more important that the model correctly predict the orogen geometry than the internal deformation because it is only the mass distribution that is needed in order to calculate lithospheric flexure. Although the mechanics of orogens remains poorly understood, critical wedge theory (Chapple, 1978; Stockmal, 1983; Davis and others, 1983; Dahlen, 1984; Dahlen and others, 1984; Dahlen and Suppe, 1988) provides a basic working model. In particular, numerical models of the growth of doubly-vergent critical wedges can be used to approximate the evolution of complete sections of plane-strain compressional orogens (Beaumont and others, 1992b; Willett and others, 1993), and we base our kinematic tectonic model on their numerical model results. In this regard, the model is different from the one-sided orogen of Flemings and Jordan (1989), their crustal-scale fault-bend-fold model (Flemings and Jordan, 1990; Jordan and Flemings, 1991) or the geometrical model of the Alps (Sinclair and others, 1991). The geometry of the orogen is based on kinematic rules and its development is linked to other model processes in a manner that provides a natural evolution and not solely by loading to an externally specified shape.

We assume that beneath a convergent orogen part of the lithosphere detaches and is asymmetrically underthrust and subducted. The overlying crust undergoes compressional deformation under shear stress applied to its base by the subducting lithosphere. We use the analogy between the geometry of convergent orogenic belts and the mechanics of doubly-vergent wedges which is based on sandbox experiments by Malavieille (1984), and numerical models of doubly-vergent Coulomb wedges (Beaumont and others, 1992b; Willett and others, 1993). Their model results show that subduction creates two oppositely vergent back-to-back tectonic wedges in an overlying Coulomb crust, because the basal shear stress reverses sign at the point where the lithosphere detaches and subducts. The wedges are asymmetric, but at large growth or when their base is weak, the geometrical asymmetry is small. We assume here that the asymmetry may be neglected and that in section, a model orogen takes the form of symmetric back-to-back Coulomb wedges which grow by accretion at their toes or by transfer of material across their common vertical boundary (cvb) (Fig. 3). The geometry of both wedges is calculated using the small angle, uniform taper, critical Coulomb wedge theory (Dahlen, 1984) with the added constraint that the wedges have the same height along their common vertical boundary. The surface slope, α, of a critical wedge is related to the effective internal and basal coefficients of friction, μ and μ_b, and the basal slope, β, by:

$$\alpha = ((\beta + \mu_b)/(1 + 2\mu)) - \beta. \quad (1)$$

The effective internal angle of friction $\phi = \arctan \mu$ and $\phi_b = \arctan \mu_b$. Initially, $\beta(x,y,t) = 0°$ because the base of the wedges is taken to be the bottom of a uniformly thick layer that detaches above the subducting lithosphere. $\beta(x,y,t)$ evolves both in space and with time as the detachment surface is flexed during isostatic compensation of the model (Fig. 3). We assume that the spatial gradient of $\beta(x,y,t)$ is sufficiently small that the local critical surface slope $\alpha(x,y,t)$ may still be calculated using the uniform taper equation with the current local value of $\beta(x,y,t)$ (Fig. 3).

Current terminology for orogens and foreland basins is particularly problematic when orogens are double sided. To distinguish the two wedges and the two foreland basins, we follow the terminology recently proposed by Willett and others (1993), which uses the prefix "pro" to denote elements of the foreland and orogen associated with the converging lithosphere, and "retro" for those elements on the stationary lithosphere. This partly conforms with Dickinson's (1974) foreland basin terminology in that his retro-arc foreland basin becomes our retro-foreland basin, thereby avoiding the "arc" which is commonly absent. "Pro," as the opposite of "retro" (cf. prograde/retrograde) is used to rename Dickinson's (1974) peripheral foreland basin as a pro-foreland basin. Similarly, the oppositely vergent tectonic wedges are termed the pro-wedge and retro-wedge. The "retro" prefix is consistent with the term "retrocharriage," long used in European literature as the name of the process that creates a retro-wedge.

Wedge growth, in the absence of erosion or isostatic adjustment, is calculated by distributing material accreted to the pro-wedge between the two wedges with the constraint that the pro- and retro-wedges maintain critical geometry and are of equal thickness along the common vertical boundary. Plane strain is assumed and dip sections are calculated independently, therefore, there is no tectonic movement orogen of material in the strike direction. This approximation will be valid provided the average slope in the strike direction at the scale of model wedges is sufficiently small to preclude failure in the strike direction as opposed to the dip direction.

The tectonic mass flux per unit strike length from the converging pro-autochthon into the pro-wedge is,

$$Q_T(x,y,t) = v_T \times d(x,y,t), \quad (2)$$

where v_T is the convergence velocity of the pro-lithosphere, and $d(x,y,t)$ is the thickness of the detached and accreted autochthon at the toe of the pro-wedge. The convergence velocity v_T and the initial thickness of the pro-autochthon $d(x,y,t)$ are specified as input variables (Table 1). $Q_T(x,y,t)$ does, however, vary in space and time as sediment deposition or erosion changes the thickness of the pro-autochthon, $d(x,y,t)$, for example, as shown in Figure 3.

The retro-lithosphere is stationary, therefore, there is no *a priori* tectonic flux into the toe of the retro-wedge. Material is, however, passively accreted to the retro-wedge because its toe advances as the orogen grows.

The same constraints on wedge growth apply when the tectonic model is also flexurally compensated. That is, the accreted mass is distributed iteratively until both wedges are critical for the converged $\beta(x,y,t)$ distribution, they are of equal thickness along the common vertical boundary, they are in the correct isostatic balance for the mass distribution, and mass has been conserved.

Erosion and sedimentation modify the growth of the model tectonic wedges, primarily because they change the surface

Schematic Orogen and Foreland Basins

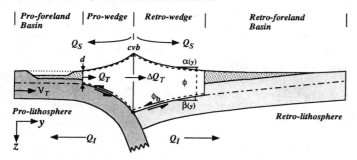

FIG. 3—Schematic section of the doubly-vergent orogen model which opposes two critically tapered Coulomb wedges about a common vertical boundary (cvb). The prefix "pro-" is used to denote elements of the orogen and foreland associated with the converging lithosphere and "retro-" for those elements on the stationary lithosphere. The primary mass flux into the orogen, Q_T, is a result of detachment and accretion of a crustal layer thickness, d, (base shown by long short dashed line) while the lower levels are underthrust and subducted with velocity, v_T. Pro- and retro-wedge geometry is calculated using critical wedge theory, where surface slope, $\alpha(y)$, is a function of the internal coefficient of friction, ϕ, basal coefficient of friction, ϕ_b, and the slope of the basal decollement $\beta(y)$. We assume that ϕ and ϕ_b are constant. In the absence of surface processes ($Q_S = 0$), the pro- and retro-wedge grow in a self-similar fashion (dashed line). With surface processes, denudation or sedimentation alters α and thus the taper of the wedges. Mass flux into the orogen, Q_T, is distributed between pro- and retro-wedge, ΔQ_T, and used to reconstruct the damaged wedges to minimum critical taper. Isostatic adjustment with addition (Q_T) or loss (Q_S) of mass from the orogen creates a horizontal flux of mass in the asthenosphere, Q_I.

TABLE 1.—FORELAND BASIN MODEL PARAMETERS

Size of Planform Model	1050 km × 750 km
Gridsize (*cl*)	15 km
Initial Fractal Surface Relief	Maximum Amplitude 100 m
	Fractal Dimension 2.5
Tectonics	
Coulomb Wedge Coefficients of Friction	Internal $\mu = 0.3249$ ($\phi = 18°$)
	Basal $\mu_b = 0.0524$ ($\phi_b = 3°$)
Initial Potential Detachment Thickness (d)	Ocean 1000 m, Continent 3000 m
Convergence Rate (v_T)	0.033 m/y
Strike Coherence Length (l_s)	75 km
Timestep	250,000 y
Surface Processes	
Hillslope and Marine (diffusive) Transport	
$K_S^{subaerial}$	Bedrock 25 m²/y, Sediment 50 m²/y
K_S^{marine}	Bedrock 250 m²/y, Sediment 500 m²/y
Fluvial (advective) Transport	
River Transport Coefficient (K_f)	0.02
Erosion Length Scale (l_f^{erode})	Bedrock 50 km, Alluvium 25 km
Depositional Length Scale ($l_f^{deposit}$)	25 km ($\sim\sqrt{2}cl$)
Orographic Incident Vapour Flux	
Initial Incident Vapour Flux ($S_W(0)$)	7.875×10^5 m²/y
(*Uniform rainfall distribution equivalence*)	(0.75 m/y)
Extraction Length Scale (l_R)	50 km
Topographic Height Scale (h_R)	1 km
Isostasy	
Lithospheric Flexure	Effective Thickness 30 km
(*Similar to a thin elastic plate*)	Young Modulus 1×10^{11} Pa
	Poisson's Ratio 0.25
Densities	Mantle 3400 kg/m³
	Crust and Sediment 2600 kg/m³

slope, $\alpha(x,y,t)$, and secondarily because $\beta(x,y,t)$ changes with the accompanying isostatic adjustment. If the response of a wedge to a perturbation in α was simply to require redistribution of newly accreted mass in order to reconstruct wedge geometry to the critical taper, then erosion and sedimentation effects would be calculated using the same principles used to grow the wedges kinematically by accretion at their toes. The response is, however, more complex because compressional Coulomb wedges are stable not only when their taper ($\alpha + \beta$) is the minimum critical angle, but also for a range of supercritical tapers where $\alpha + \beta$ is between the minimum and maximum critical angle (Dahlen, 1984). Critical and supercritical regions of the wedge can detach and slide on their base without deformation, but subcritical regions must first deform internally to achieve a critical taper. Consequently, surface processes that reduce the taper angle below critical require the wedge to deform internally, whereas surface processes that create a supercritical taper stabilize that region of the wedge with its supercritical geometry.

The kinematic response of a Coulomb wedge orogen, that has its upper surface roughened or smoothed by surface processes, is therefore potentially quite complex. Details of the doubly-vergent wedge model response to surface processes are explained by Beaumont and Johnson (in prep.). It is sufficient for the purposes of this paper to state that: (i) wedge cross-sections are assumed to deform by plane strain; (ii) the tectonic flux, Q_T, is used to reconstruct subcritical segments in a manner that progresses from the common vertical boundary to the toe of each wedge; and (iii) surplus tectonic flux, after all subcritical segments have regrown to a minimum critical state, is used to grow the orogen in a manner that absorbs supercritical segments as critical segments grow with the minimum critical taper. These kinematic steps agree with the behaviour of the dynamical models mentioned earlier.

It is important that wedge reconstruction correctly preserve or remove surface roughness because the rate of erosion in the coupled orogen-surface processes model is directly related to the roughness, or relief, of the surface. When other conditions are equal a smooth wedge is denuded more slowly than its roughened counterpart.

Plane-strain critical wedge theory (e.g., Dahlen, 1984) provides no insight concerning the mechanics of three dimensional wedges and, in particular, the length scales of coherent deformation in the strike direction. Orogens and sandbox models of orogen mechanics do, however, show coherent deformation in the strike direction at the strike length scales of their respective thrust sheets. Although we have no mechanical understanding of this behaviour, it is necessary for the preservation of surface relief in the present modelling to estimate a strike length scale, l_s, for which we expect the orogen to behave coherently. For scales less than l_s, surface roughness in the strike direction will be preserved because the wedge behaves coherently over l_s during reconstruction. For scales greater than l_s, construction will smooth surface roughness in the strike direction because strike segments of the wedge behave independently at these scales. That is, the wedge will respond mechanically to the long wavelength (longer than l_s) components of surface relief in the strike direction.

In the model, this behaviour is reproduced by filtering strike variations in topography, removing components with wavelengths less than l_s, reconstructing the wedge using the smoothed topography and then restoring the short wavelength components. The choice of l_s is the most difficult aspect of this treatment to constrain. Our choice of $l_s = 75$ km (Table 1) is based on the length scales of thrust sheets in natural orogens. The range of realistic values is approximately 50–200 km. Long values of l_s produce more surface relief in the model orogens and increase the denudation rates, while short values of l_s produce smoother orogens and reduce denudation rates.

Surface Processes Model

The surface processes model (Fig. 4) predicts surface mass flux, Q_S, on geologic time scales by a combination of hillslope, marine and fluvial mechanisms. We summarize the model briefly here because it has already been described by Beaumont and others (1992a) and Kooi and Beaumont (1994) discuss in detail the denudational behaviour of the model.

Hillslope erosion and denudation, that is the processes which remove material from hill and mountain sides and transport it locally to adjacent valleys, is assumed to be a linear diffusion (Culling, 1960, 1965; Flemings and Jordan, 1989) that conserves volume. The local rate of change of

Surface Processes Model

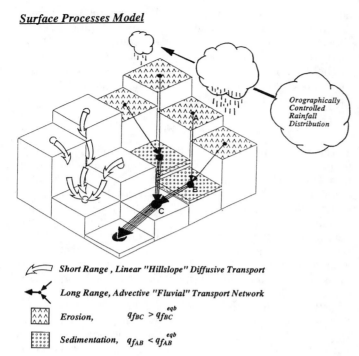

FIG. 4—The surface processes model includes sediment transport, orographic distribution of rainfall and a fluvial discharge network (i.e., a watershed, solid arrows). Sediment transport is the sum of two processes, diffusive (hillslope) transport (unfilled arrows), and advective (fluvial) transport. Fluvial erosion or sedimentation is determined by the sediment load in a river (q_f), and the river's equilibrium carrying capacity (q_f^{eqb}), which is proportional to river power (slope × discharge). Fluvial discharge at any point is the up-slope sum of orographically distributed rainfall collected within a watershed.

height, h is given by:

$$\partial h/\partial t = K_S \nabla^2 h, \qquad (3)$$

where K_s is the diffusivity, which varies with the type of surface material, crust, lithified sediment, or alluvium and with the submarine or subaerial environment (Table 1). Submarine transport is also modelled as linear diffusion (Kenyon and Turcotte, 1985; Syvitski and others, 1988; Jordan and Flemings, 1991), but with an enhanced diffusivity with respect to that for the same material in the subaerial environment. The diffusion of surface material was the only surface transport mechanism employed by Flemings and Jordan (1989, 1990) and Jordan and Flemings (1991). They therefore had to use high diffusivities to predict observed geological mass transport rates. Diffusion plays a smaller, local role in subaerial regions of our models because most material is transported fluvially. It follows that the diffusivities used in the two models cannot be compared directly. Our values are calibrated against small scale observations (Carson and Kirkby, 1972). Both models use submarine diffusion as the sole marine transport mechanism, therefore the diffusivities can be compared. It should, however, be remembered that these are, at best, effective parameters.

Combined suspended and bedload transport is modelled by a network of one dimensional model rivers that drain the current model topography via the steepest slopes (Fig. 4). The local equilibrium sediment carrying capacity, q_f^{eqb}, of a model river is proportional to its power (Begin and others, 1981; Armstrong, 1976; Chase, 1991),

$$q_f^{eqb} = -K_f q_r \partial h/\partial l, \qquad (4)$$

where K_f is the dimensionless transport coefficient, q_r is the local discharge, obtained from upstream collection of runoff, and $\partial h/\partial l$ is the local slope of the river bed. Equation 4 relates the total sediment flux in the model river, when the transport is in equilibrium, to discharge and the local slope. Rivers may not, however, be in equilibrium because the material available for transport may be limited by the ease with which material is abraded, detached and entrained by the model fluvial system. The model rivers may also achieve overcapacity where, for example, slopes decrease without a corresponding increase in discharge. This leads to the situation where sediment is deposited because the sediment transport exceeds the equilibrium carrying capacity.

The approach to equilibrium of the fluvial sediment transport, q_f, is treated in the model as a first order reaction (Eq. 5) that governs sediment entrainment and deposition. This approach is consistent with the concept that a river must do work on the landscape in order to transfer material from the substrate into the transported phase and that this work will increase with decreasing detachability of the substrate. The resistance to detachment is measured by the distance down the water shed, l_f, taken for the detachment reaction to occur, such that

$$\frac{dq_f}{dl} = \frac{1}{l_f}(q_f^{eqb} - q_f) \qquad (5)$$

$q_f^{eqb} - q_f$ is the disequilibrium between the local sediment flux and the equilibrium sediment carrying capacity, at the same location, given by equation 4. The disequilibrium is a measure of the potential that drives the entrainment-deposition reaction such that the disequilibrium is reduced to a factor of $1/e$ over a distance of l_f when q_f^{eqb} is constant. l_f is large for crustal rock in a weathering limited environment where detachment is inefficient. l_f is smaller for lithified sediments that are more easily detached than the crust, and l_f tends to zero for entrainment of disaggregated material such as alluvium. When $l_f = 0$, the reaction no longer limits the model fluvial transport and equation 5 plays no role because $q_f \rightarrow q_f^{eqb}$ as predicted by the transport limited capacity of equation 4. When $q_f > q_f^{eqb}$, for example where slopes decrease, sediment is deposited with an l_f that tends to zero because there is no mechanical restriction on deposition. Deposition will occur rapidly until equilibrium is again achieved. However, in the numerical model, the minimum value of l_f for deposition can be no smaller than the spatial resolution of the model grid (Table 1).

The evolving model records the spatial distribution of exposed crust and sediment created by the model so that fluvial entrainment at a given location first removes the accumulated sediment with an appropriate value of l_f, before attempting to detach and incise the crust with its corresponding l_f.

The overall fluvial sediment transport is determined by numerically integrating the reaction equation (Eq. 5) down the model river network from the drainage divides. At each location the gradient of entrainment or deposition is equal to the difference between the equilibrium carrying capacity (Eq. 4) and the actual amount of sediment that is being transported (q_f from upstream integration of Eq. 5), weighted by the detachability of the substrate. In areas where the substrate is crust the model rivers may be significantly under capacity reflecting the resistance of the crust to detachment, whereas in sedimentary basins, or in areas where hillslope processes supply sufficient detritus, transport will be at, or near, capacity.

Calculation of the model sediment fluxes requires a knowledge of the discharge of water in the model river network. We assume discharge is related to the precipitation on the model surface and do not address the loss of water by evaporation or infiltration.

The precipitation rate is assumed to have a constant mean value, $w_R(y)$, at a given location during each model time-step and is calculated from a simple orographic model. Rain is extracted from an incident water vapour flux as it passes from the model boundary over the current topography from the pro-ward or retro-ward directions normal to the strike of the orogen.

$$w_R(y) = -\left(\frac{dS_w(y)}{dy}\right) = \frac{h(y)}{a_R} S_w(y), \qquad (6)$$

where $h(y)$ is the local topographic height, $a_R = h_R l_R$ is an extraction efficiency parameter, determined by h_R, a topographic scale height, and l_R, a scale length over which $1/e$ of the available vapour would be converted to precipitation were $h(y) = h_R$. $S_w(y)$ is the residual water vapour flux,

$$S_w(y) = S_w(0) - \int_0^y w_R(y)dy, \qquad (7)$$

and $S_w(0)$ is the incident vapour flux at the model boundary.

The precipitation model is a parametric simplification of the physics of orographic rainfall but it correctly reproduces the essential feature that precipitation is high on windward slopes of a mountain belt and lower on the leeward side. With appropriate choices of $S_w(0)$ and a_R (Table 1), it gives more realistic orogen-scale precipitation distributions than, for example, a model in which precipitation is proportional to air mass decompression as measured by local topographic slopes in the direction of the prevailing wind.

We characterize the model climate by the precipitation, which is fundamentally controlled by the value of the initial incident water vapour flux, $S_w(0)$. $S_w(0)$ is given in Table 1 as the equivalent precipitation rate were precipitation distributed uniformly over the model. We use the terms wet and dry to refer to high and low $S_w(0)$ respectively. A "wet" model will have more precipitation on average than a "dry" model. The amount of precipitation at any location on the model is, however, strongly influenced by the model topography. The orographic control of precipitation means that regions of a model may vary between high (>4 m/y) and low (<1 m/y) precipitation depending on the model topography in the upwind direction and $S_w(0)$.

The diffusive (hillslope and marine) and advective (fluvial) model components, when combined, predict the surface processes mass flux, Q_S.

Isostasy Model

The planform response of the model lithosphere to tectonic and sediment loading and unloading during denudation is calculated using the Green function-convolution techniques (Beaumont, 1978) that were later used for a thermally-activated viscoelastic lithosphere (Courtney and Beaumont, 1983; Quinlan and Beaumont, 1984). In this paper, we simplify the flexural calculations and assume a continuous, uniform thickness, elastic lithosphere (Table 1).

In order that models may be compared with regard to mass fluxes of the contributing processes, we interpret isostatic subsidence in terms of the associated mass flux, Q_I, of asthenospheric material. Q_I is the horizontal flux of asthenosphere displaced by vertical motions of the lithosphere (Fig. 3). This is a useful interpretation because it converts vertical movements to a horizontal flux that can be compared with the two other horizontal fluxes, Q_S and Q_T. The model does not consider the part of the pro-lithosphere that is subducted. In reality, this subduction adds to the flow in the asthenosphere and therefore to Q_I, but asthenospheric circulation is not considered in this model.

RELATIONSHIPS AMONG TECTONIC, SURFACE AND ISOSTATIC PROCESSES

Mass Fluxes, Model Geometry and Synthetic Stratigraphy

The geometry of the model orogen and stratigraphy of the foreland basins result from the cumulative interactions among the mass fluxes at each location. We use $Q_T(x,y,t)$, $Q_S(x,y,t)$ and $Q_I(x,y,t)$ to indicate the value of the fluxes at a specific location.

In any column within the isostatically balanced model orogen or foreland basin (Fig. 5), the accumulation or loss of mass is a function of the horizontal gradients of the three mass fluxes across the column. These relationships are the connection between the fluxes and the model geometry and stratigraphy. If we assume a constant sea level, the height of the model surface, $h(x,y,t)$, with respect to sea level, changes as the sum of all three flux gradients. The thickness of the sediment changes with the sediment flux gradient. The position of the base of the sedimentary column changes as the sum of the tectonic and isostatic flux gradients. Correspondingly, the position of the lithosphere beneath the orogen or foreland basin changes with the isostatic flux gradient.

Consequently, the evolution of a model can be entirely described by the fluxes. This however represents a large amount of data without simplifying the analysis of model behaviour. We suggest, but have not proven, that the primary features of synthetic stratigraphy created by the model can be related to the fluxes between large-scale regions of

Flux Competition, Retro-Foreland Basin

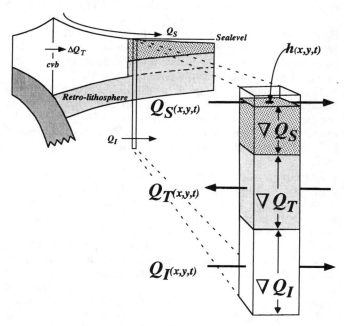

FIG. 5—Model geometry and synthetic foreland basin stratigraphy is a consequence of the interactions among tectonic (Q_T), surface (Q_S) and isostatic (Q_I) mass fluxes. In any column the loss or gain of mass is a function of the horizontal gradients of the three mass fluxes across the column (∇Q_T, ∇Q_S, and ∇Q_I). Surface elevation ($h(x,y,t)$), with respect to sea level, changes as the sum of all three flux gradients ($\nabla Q_T + \nabla Q_S + \nabla Q_I$). Sediment thickness changes with the sediment flux gradient, ∇Q_S, the base of the sedimentary column changes as the sum of the tectonic and isostatic flux gradients ($\nabla Q_T + \nabla Q_I$) and the position of the lithosphere beneath the orogen or foreland basin changes with the isostatic flux gradient, ∇Q_I.

the model. With this in mind, we use Q_T, Q_S and Q_I to mean the fluxes between the retro-wedge and retro-foreland basin or pro-wedge and pro-foreland basin. This is apparently a useful approach to the quantification of model behaviour and synthetic stratigraphy in terms of fluxes, but at present remains conceptual and qualitative. The intent of future work is to analyze the couplings and feedbacks among model processes using both the detailed behaviour of fluxes and a 'box model' of their regional exchange fluxes.

The Graded River Profile

The evolution of the models is dictated by the interaction of tectonic and surface processes in the presence of isostatic compensation, and is therefore strongly influenced by the behaviour of the model rivers. The explanation of denudation or aggradational behaviour is made easier using the concept of the graded river profile. We use the term grade or graded profile to mean an equilibrium topographic profile such that the carrying capacity (Eq. 4) of a model river that runs along the profile is equal to the sediment that is actually transported. Under these circumstances no deposition or entrainment occurs and the disequilibrium, $q_f^{eqb} - q_f$, vanishes. In our models, the entrainment-deposition reaction (Eq. 5) controls the model river's approach to a si-

multaneous equilibrium and graded state in the sense used by Mackin (1948). The concept of a graded river is, of course, an idealization, a state toward which rivers are considered to evolve, but never attain.

One reason that grade is not attained in the models is that vertical movements caused by other model components perturb the tendency of the model rivers to become graded. The resulting rates of erosion or deposition by the model rivers reflect the dynamics of this competition between the rate at which vertical movements create disequilibrium and the rate at which the river can respond. In a model sedimentary basin that is aggrading because vertical movements have reduced slopes along the profile below grade, for example, a major river with large discharge and sediment flux can regrade its profile more rapidly than an equivalent minor river. This is because the disequilibrium (Eq. 5) is larger for the major river because it has the larger discharge. Consequently, it "reacts" with the higher rate of deposition.

The Filled Foreland Basin

In planform, we consider the portion of a foreland basin within a catchment, and delimited by the deformation front and the peripheral bulge, to be *filled* when the peripheral bulge acts as effective base level, and all rivers in the catchment have reached grade. When the rivers are at grade there is no net deposition in the catchment and by implication no space for sediment accumulation ("accommodation," Jervey, 1988). In natural foreland basins, however, rivers never attain grade, but are always regrading in response to changes in the orogen, the climate, vertical motions of the lithosphere or eustatic adjustments to base level. Foreland basins are, therefore, always either underfilled or overfilled. The distinction between underfilled and overfilled is in part determined by whether or not any portion of a foreland basin has been filled to the level where drainage is deflected across the peripheral bulge. When drainage has not breached the bulge, the basin is considered underfilled. Once some portion of the drainage network is captured and deflected across the peripheral bulge, then the distinction between underfilled and overfilled in the affected catchment depends on base level and whether river regrading results in net erosion or net deposition. If the peripheral bulge acts as effective base level for drainage in a catchment, then the portion of a foreland basin within the catchment is considered underfilled when there is net deposition and overfilled when there is net erosion. Once the peripheral bulge no longer acts as effective base level, then river regrading in the foreland basin portion of the catchment responds to the base level of the drainage network beyond the peripheral bulge, and this always results in an overfilled basin. The foreland basin portion of the catchment becomes overfilled either because alluvial facies aggrade across the peripheral bulge, or because incision of the peripheral bulge and subsequent regrading of the catchment results in net denudation of the foreland basin fill. These definitions of filled, underfilled and overfilled encompass Flemings and Jordan's (1989) and DeCelles and Burden's (1992) usage of underfilled and overfilled, and imply that the concept of accommodation (Jervey, 1988) in a foreland basin does not depend simply on sea level, but instead on how tectonism, climate, isostasy and eustasy influence river grade and the elevation of the peripheral bulge.

DESCRIPTION OF MODELS 1 AND 2

There are now a number of published results from process-based foreland basin models, some of which were mentioned in the introduction. All of these models have one thing in common: they represent a cross-section of the foreland basin. Therefore, they do not address the consequences of strike variations in tectonics or surface processes, both of which may be responsible for the commonly observed longitudinal sedimentary filling of foreland basins. The model presented here has some differences in tectonics and surface processes by comparison with other models. The greater range of couplings and feedbacks among the tectonic, surface and isostatic processes also make the model more internally consistent than previous models. It is, however, the ability to calculate the complete planform evolution of the orogen and sedimentary facies filling of the foreland basins that is the most interesting aspect of this model, and this will allow a greater range of geologically important problems to be investigated.

We have therefore chosen to illustrate the planform capabilities of the model by contrasting basins that form in two tectonic settings, both of which involve oceanic subduction leading to closure of an ocean basin and collision between two continents. The planform and sections of the initial conditions for the two models (Fig. 6) show the juxtaposition of two continental margins (surface slope angle ~2.5°), the sense and velocity of convergence, and the subduction axis. Other model parameters are given in Table 1 or explained in the figure caption.

The values of parameters (Table 1) and model design were chosen to reproduce average conditions within a small, rapidly evolving, collisional orogen. The geometrical parameters and the wedge properties were estimated by comparison with the geometry and taper angles of natural orogens, but other properties, for example those of the surface processes model, are not measured at the spatial and temporal scales of the model.

The dependence of effective diffusivities on the scale at which they are measured is discussed by Kooi and Beaumont (1994), who show both observations and theory imply that the values to be used in model experiments are related to the resolution of the model. The values in the present models are consistent with this view and are appropriate for the scale of the model grid (15 km, Table 1). When comparing these values with measurements at smaller scales, they must be reduced by approximately the ratio of the measurement scales. For example, diffusivities (K_s, Table 1) decrease by a factor of 10^4 if they are to be used at a scale of a few metres. The diffusivity values are also reasonable because model sensitivity experiments show these values to predict correctly the average slopes and volumes of sediment transported on geological time scales in natural systems.

The values of K_f and l_f (Table 1) were chosen from model sensitivity experiments to give fluvial denudation rates and sediment fluxes in the rivers in agreement with natural systems. That similar values were also found to be acceptable in modelling studies of escarpment evolution (Kooi and Beaumont, 1994) and for models of denudation of the Southern Alps, New Zealand (Beaumont and others, 1992a), helps establish internal consistency, but does not prove the surface process model to be correct.

The precipitation model is, as discussed above, a simple parametric model, therefore it is difficult to compare its parameter values with observations except to note that the model rainfall predictions are comparable to those observed in small low to moderate latitude orogens (cf. Landsberg, 1981).

Both models have exactly the same parameter values except for the initial inclination angle θ between the continental margins. The convergence velocity (3.3 cm/y[1]) is uniform throughout the 15 My of the model collision. The wind, which is from the retro-side (Fig. 6), has a constant but relatively high vapour flux which is capable of creating a "wet" climate for the model. While oceanic crust is subducted, the value of d corresponds to the offscraping and accretion of ~1000 m of pre-existing oceanic sediment and crust above the detachment (Fig. 6), plus sediment deposited by the model in the trench. The detachment remains at the same level in the crust (Fig. 6) as the convergent continental margin enters the subduction zone, but its elevation relative to sealevel changes as it is flexed downward. d progressively increases to 3000 m and to greater values with sedimentation, and there is a corresponding increase in Q_T. The orogens begin their evolution as submarine accretionary wedges which grow above the subduction axis and become progressively more subaerially exposed as the model continental collision progresses.

In Model 1, both continental margins are parallel to each other and perpendicular to the convergence direction. There is no strike variation in the model geometry, convergence velocity or incident vapour flux. Dip sections of Model 1 results can therefore be compared with the previously published cross-sectional models. The differences will reflect the contrast between models with regard tectonics, surface processes and isostasy. The short wavelength orogen topography in Model 1 is, however, a consequence of the planform fluvial network and is not reproduced by the purely cross-sectional models.

Model 2 is more interesting because the continental margin on the subducting plate is inclined to the subduction zone and the other continental margin at an azimuth θ ~25°. The model predicts significant strike variation in tectonics and, consequently, longitudinal drainage and filling of the retro-foreland basin.

Each model represents part of an orogen and the reflective strike boundary conditions imply that the model is reflectively continued along strike (Fig. 6). Therefore, at a large scale, Model 2 represents the collision of a sawtooth margin with a linear margin. The model has the potential to address the consequences of the collision of promontories and salients on the evolution of the retro-foreland ba-

---

[1] "y"—year, "My"—10^6 years (mega-year or million years)

Initial Conditions

Model 1

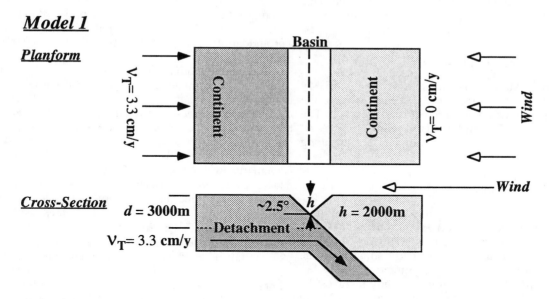

Model 2

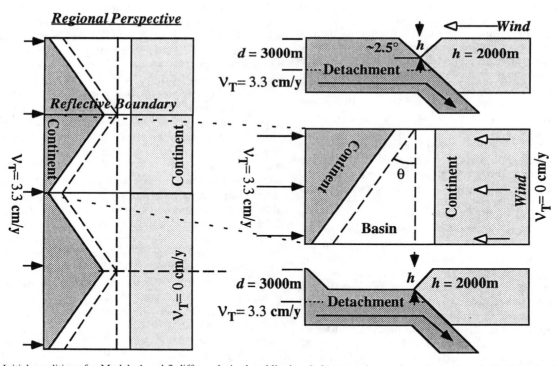

FIG. 6—Initial conditions for Models 1 and 2 differ only in the obliquity, θ, between the continental margins. In Model 1 (top), the continental margins are parallel and their slopes meet to form a marine basin of depth h. With convergence, a surface layer of thickness d is detached from the pro-lithosphere while the lower layer is underthrust and subducted with velocity v_T. In Model 2 (lower right), the continental margins are oblique which respect to each other (θ ~25°), thus the pro- and retro-continental slopes are juxtaposed at one end of the model, and separated by a marine basin of depth h, at the other. With tectonic convergence, orogen growth and pro-lithospheric subduction proceeds as in Model 1. Where continental slopes are juxtaposed, the detached surface layer thickness, d, is ~3000 m, and where oceanic pro-lithosphere is juxtaposed with continental retro-lithosphere, d is initially ~1000 m. Model boundaries perpendicular to the strike of the subduction axis are reflective (lower left). Therefore, when we view the collision of oblique continental margins in Model 2 from a regional perspective, the model represents collision between straight and sawtooth continental margins at the right and left sides of the planfom (lower left), respectively. Only the region between the short dashed lines (lower left) is actually modelled but the ends actually correspond to promontories and recesses in the continent at left. The surfaces of the model continents (grey tones) are initially at sea level except for the addition of positive fractal noise relief with a maximum height of 100 m.

sin. It may, therefore, have application in the analysis of the Paleozoic Appalachian foreland basin (cf. Thomas, 1977), where these features are interpreted to have been important, in addition to other zones of diachronous convergence, like Taiwan.

MODEL 1 RESULTS

The evolution of Model 1 is shown at 5, 10 and 15 My (Figs. 7, 8, 9) in planform and for one section A-A' at the model scale and with the retro-foreland basin enlarged. The planform also shows the rainfall distribution along A-A' and this distribution is representative of the rainfall along strike.

Model 1–5 My

At 5 My (Fig. 7), the 165 km of convergence is sufficient to have created an orogen with a maximum height of ~1.5 km and subaerially exposed width of ~135 km. Flexural subsidence of the foreland basin is ~1 km adjacent to the orogen, and sedimentary filling is asymmetric, partly because the windward side of the orogen already receives the most precipitation (Fig. 7). Both foreland basins still have significant marine regions and drainage from the peripheral bulges into and away from the basins can be clearly seen from the river network. The drainage patterns on the peripheral bulges reflect minor incision and capture by the model rivers during gentle flexural warping of the fractally roughened initial surface. Elsewhere, the initial surface roughness has had little influence on the evolution of the model drainage network as can be seen from the self-organized dip-parallel linear drainage off the rapidly rising orogen. At this stage, incision within the orogen is insufficient for the model rivers to have been captured and for a new network to develop. It should be remembered that the grid size of the model is only 15 km (Table 1) and that model rivers connect the grid centres. Consequently, the grid size limits the resolution of the drainage network and smaller scale details corresponding to those of natural drainage networks are not seen.

The regional section A-A' (Fig. 7) shows the orogen, lithospheric flexure and asymmetric sedimentation in the basins. The pro-foreland basin appears sediment starved, in part due to its leeward setting, but primarily because sediments are tectonically recycled. As the pro-autochthon converges with respect to the orogen, sediments deposited in the pro-foreland basin are also carried toward and accreted to the toe of the model orogen, a process that keeps the pro-foreland basin relatively empty unless Q_S into the basin is larger than the recycled sediment flux. Tectonic recycling of sediment is responsible for the primary difference between stratigraphy of the model pro- and retro- foreland basins. In this example, the asymmetry is extreme, but the recycling is of first order and should therefore also be important in natural basins such as the Molasse, the western Taiwan foreland basin, and the Irian Jaya/Papua-New Guinea foreland basin. A natural pro-foreland basin with significant tectonic recycling may not be as underfilled as suggested by Model 1, because carbonate deposition, chemical precipitation or hemipelagic sedimentation may take place there.

The enlarged section (Fig. 7) shows a prograding sediment wedge in the retro-foreland basin. Progradation occurs because $Q_S > Q_I$ within the basin and the rate of progradation is controlled by the relative magnitudes of Q_S and Q_I.

The chronostratigraphy is shown at 0.5 My intervals and the model sediments are divided into marine, near-shore, coastal, and subaerial facies averaged over 0.25 My intervals. A marine facies designation means that during the entire timestep, the model environment remained marine. A near-shore facies designation means dominantly marine sedimentation, but also indicates that this location has been nonmarine within this 0.25 My interval. Conversely, a coastal facies designation is used where deposits are predominantly subaerial but some marine sedimentation has occurred during the interval (Note that only color figures distinguish near-shore and coast facies. In grey-scale figures, near-shore and coast facies have been combined and called coast facies.). Subaerial facies in the model are described by the power, the slope-discharge product of the river from which the sediments were deposited. River-power provides the best estimate of the energy of a fluvial depositional environment that is available from the model. We do not consider grain size, because there is no simple way to predict the availability of clast sizes for entrainment by a model river at any point. We cannot consider fluvial architecture because the model lacks the spatial and temporal resolution to create bedding structures. We suggest that "higher power" river facies in the model describe the distribution of higher energy fluvial environments that in nature are associated with coarser grain sizes and bed forms commonly associated with upper regime flow (Simons and others, 1965); while "lower power" model river facies describe the distribution lower energy fluvial environments, that in nature are associated with finer grain sizes and bed forms commonly associated with lower regime flow. The weakness of this approach is the assumption that a complete range of clast sizes is available for deposition and that flow regime can be related to discharge without knowledge of sectional stream dimension. It would be incorrect to assume, for example, that "high power" model facies necessarily represent coarse sands and gravels or that "low power" facies are characterized by lower flow regime bed forms. However, for the present model formulation, "river power facies" is the best measure available to associate properties of the model sediment with those of natural systems.

The prograding sedimentary wedge in the retro-foreland basin (Fig. 7) is ~2.1 km thick adjacent to the orogen and 180 km wide. Surface slopes on the alluvial plain vary from $>0.4°$ adjacent to the orogen to $<0.2°$ adjacent to the marine basin. Sediment transport in the marine basin is less efficient than on the fluvially dominated alluvial plain and, as a consequence, surface slopes on the marine prograding wedge (~0.5°) are steeper than slopes of the adjacent coastal plain. The marine basin is ~250 m deep and has maintained this depth for the last 3 My as shown by the uniform thickness of marine facies at the base of the prograding wedge.

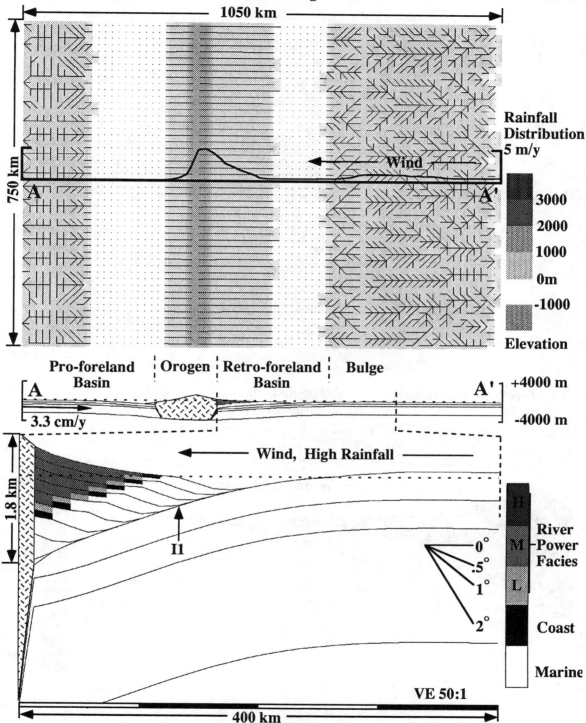

Fig. 7—Model 1 after 5 My. The upper panel shows the planform surface elevation (grey scale at right) and fluvial drainage network (black lines) of the pro-foreland basin, orogen and retro-foreland basin. A representative rainfall distribution is shown along A-A'. Cross-section A-A', middle panel, is a 10:1 vertical exaggeration of the orogen and foreland basins. The model orogen is denoted by hatch marks. Sea level is the dotted line. Convergence is from the left, and pro-lithospheric subduction occurs beneath the peak of the orogen (*not shown*). The lower panel is an enlargement of the retro-foreland basin showing chronostratigraphic horizons at 0.5 My intervals (black lines) and facies distribution (grey scale at right) at model resolution of 0.25 My and 15 km. Model facies are explained in the text. Note that grey-scale figures combine near-shore and coast facies into coast facies. I1 is the initial position of the distal retro-foreland basin shoreline.

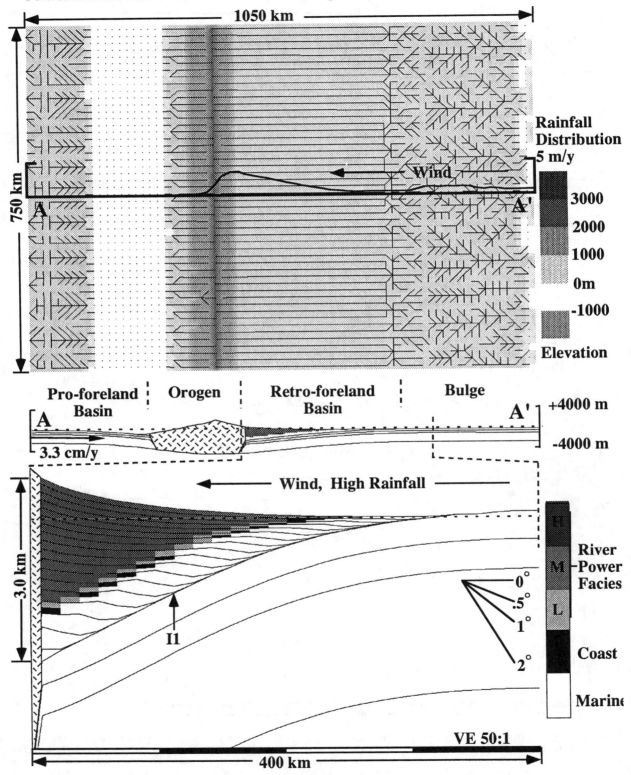

FIG. 8—Model 1 results after 10 My are presented using the format in Figure 7. Note asymmetry in rainfall distribution and internal drainage of underfilled retro-foreland basin (upper panel); asymmetry in pro- and retro-foreland basin filling (middle panel); and tapering out of marine facies, plus spatial and temporal convergence of time lines with basin filling (lower panel).

The erosional surface created by uplift and migration of the peripheral bulge extends across the peripheral bulge from I1, where it is preserved as an unconformity beneath marine strata (Fig. 7). I1 marks the initial position of the distal edge of the marine basin which, when compared with the present position of the distal coast line, shows that the peripheral bulge has migrated ~120-km away from the orogen. The bulge has migrated in response to the increasing width of the orogenic load and sedimentary filling of the retro-foreland basin.

Model 1–10 My

The next 5 My results in a total convergence of 330 km, significant growth of the orogen (~2.2-km maximum height and 195-km subaerial width), evolving asymmetry of rainfall across the orogen, increased loading and flexure, deepening of both basins, and filling of the retro-foreland basin (Fig. 8). The prograding sediment load in the retro-foreland basin and the increased orogen load have caused the crest of the peripheral bulge to migrate another ~60 km away from the orogen. The basin is no longer marine, but has not yet filled to the level that the drainage direction reverses along the incised valleys on the peripheral bulge. No preferred axial drainage develops in the retro-foreland basin because there is little along strike variation in the entire model. This indeterminate drainage pattern would result in the development of a lowland that might be characterized by chain-lakes and marshes.

The width of the retro-foreland basin is now 350 km, yet its elevation at the thrust front has remained at ~500 m throughout the 15 My of basin evolution (compare Figs. 7, 8, 9). By implication, the average river slope decreases as the basin fills, an effect that is also coupled to similar decreases in aggradation rates of fluvial sediments. This evolution is an expression of the rate at which model rivers attain their graded profile in the face of subsidence caused by tectonic loading and downward flexure of the basin. In the early stages of the model evolution (Fig. 7), Q_S into the retro-foreland basin is too small to allow the rivers to approach grade. The rates of aggradation demonstrate that this is true. Later in the evolution, denudation rates in the orogen increase and Q_S becomes sufficient to allow a progressive approach to a graded state.

There are two positive feedbacks in the model that reinforce a decrease in surface slope of the alluvial plain during this later phase. The first is that a decrease in river slope decreases the spatial gradient in the river carrying capacity toward its equilibrium defined by the graded profile. Ultimately, a low enough slope is attained that no downstream over capacity develops, the river attains grade and aggradation correspondingly ceases. The second feedback is that the spatial gradient of orographic precipitation on the windward retro-foreland basin also decreases as the surface slope declines (compare Figs. 7, 8, 9). This effect reduces the spatial gradient of the carrying capacity through its discharge dependence and reinforces the first effect. Such feedbacks may be even more important in natural rivers where carrying capacity has a nonlinear dependence on slope and discharge. The decrease in slope toward a graded profile also causes a decline in sedimentation rate adjacent to the orogen, and a greater proportion of Q_S is transported to more distal parts of the basin which speeds progradation. These relationships have only a subtle influence on the model stratigraphy, but they indicate that models which represent sediment transport in basins as diffusion with a constant diffusivity (e.g., Flemings and Jordan, 1989, 1990; Sinclair and others, 1991) may be inadequate.

In summary, progressive filling of the retro-foreland basin means that the net change in the Q_S across the basin is larger than the corresponding change in Q_I; basin wide sedimentation outpaces flexural subsidence. The orogen grows rapidly in the first 5 My. Surface slopes and spatial gradients in rainfall are large, and the associated high gradients in fluvial carrying capacity deposit a thick, rapidly aggrading, high surface slope, subaerial sedimentary wedge adjacent to the orogen. Between 5 and 10 My the fluvial controls act to make the sedimentation progressively more progradational and less aggradational. Any tendency for surface slopes to decline makes rainfall more uniform across the fluvial plain and reinforces the tendency for progradation and a further decline in surface slopes (Figs. 7, 8, 9). If interpreted as an effectively diffusive process, this evolution means that the diffusivity varies spatially and evolves with time.

These interactions can also be stated in a different way. Basin filling is partly controlled by the evolution of the subaerial accommodation space which has its upper surface defined by the evolving graded river profile. Although we do not dwell on this point, the same interactions are probably important in natural foreland basins where sedimentation will respond to the climate-mediated fluvial control of accommodation space.

Model 1–15 My

By 15 My the total convergence is 500 km, the orogen has grown (~3.0-km maximum height and ~240-km subaerial width), rainfall is more uniform over the retro-foreland but the pro-wedge is drier, the basin depths have increased but at a reduced rate, and the retro-foreland basin has now filled to the degree that drainage crosses the peripheral bulge (Fig. 9). Erosional incision within the orogen has created a river network on the pro-wedge and river capture is beginning on the retro-wedge. These networks are however, partly an artifact because there is clearly a regular pattern in the strike direction which is related to the length scale, l_s, at which the wedges are forced to be coherent.

The alluvial plain retains the same linear drainage normal to the orogen which we interpret as a combination of two effects. First, there is very little difference between any of the rivers as they emerge from the orogen. Second, the depositional characteristics of the rivers tend to build a uniform alluvial plain. Any tendency for irregularities to develop is countered by river avulsion which regrades the plain.

Drainage on the distal alluvial plain is partly internal but elsewhere is controlled by antecedent river channels that cross the peripheral bulge. Progressive deposition in the distal axial depression steadily fills valleys on the basin side of the peripheral bulge and raises base levels of the rivers to

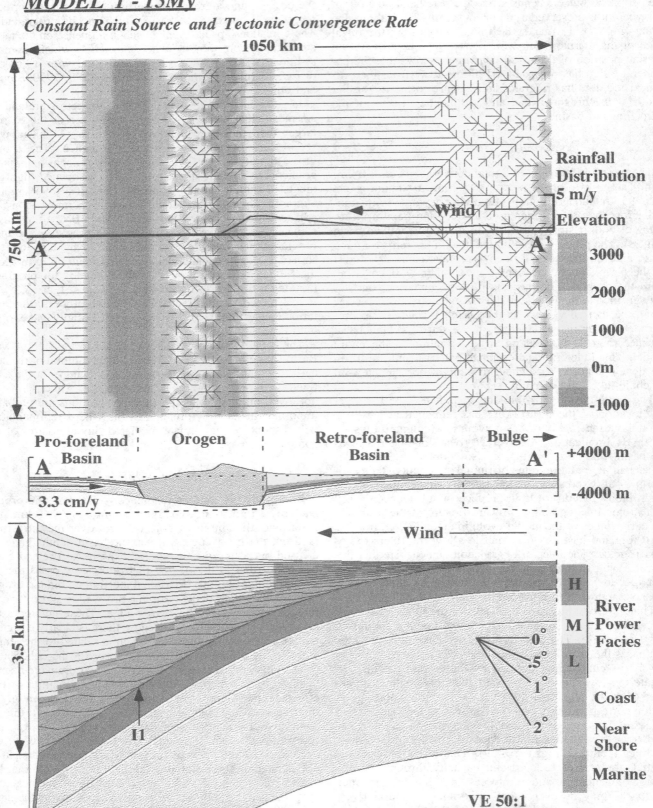

the point that they breach the peripheral bulge along the antecedent valleys. In principle, this process would create characteristic fluvial sedimentary architecture within the valleys that would be diagnostic of a reversal in river flow direction when a foreland basin becomes overfilled. Once the crest of the peripheral bulge is breached, sedimentation will be controlled by the base levels of the rivers that drain the distal flank of the bulge.

The model-scale section (A-A', Fig. 9) illustrates the growth of the orogen and confirms that Q_T into the orogen remains larger than Q_S from the orogen. Tectonic recycling also continues to dominate over sedimentation in the pro-foreland basin.

There is no indication that this particular model will achieve an equilibrium in which Q_T into the orogen will be balanced by Q_S from the orogen. Such a steady state occurs in other models when denudation (Q_S) increases sufficiently rapidly through the increase in orographic precipitation as the model orogen grows. If, however, the model orogen achieves a threshold size at which rainfall becomes progressively more focused on one flank (compare Figs. 7, 8 and 9), then denudation can no longer limit the overall size of the orogen. Under these circumstances, Q_S from the orogen always remains less than Q_T into the orogen, and climate merely mediates the growth of the orogen. Analogous behaviour will occur in natural orogens, and this is presumably what has happened in large orogens like the Andes and the Himalayan-Tibetan system.

The enlarged section of the retro-foreland basin (Fig. 9) shows that as the orogen grows, rivers adjacent to the orogen are most strongly perturbed from a graded state, because this is where sedimentation rates are the highest. The strong influence of orographic precipitation on the alluvial plain increases the efficiency of sediment transport as discussed above. On the distal alluvial plain, the small difference between river slope and the graded river profile is demonstrated by the low rate of aggradation (~ 0.01–0.02 mm/yr), river gradients of ~ 0.0005 ($\sim 0.04°$), and the region of low river power facies.

MODEL 2 RESULTS

The evolution of Model 2 (Figs. 10–12) is shown at the same 5, 10 and 15 My intervals as Model 1. The planform panel is the same as that of previous figures, but the model scale sections are omitted in favour of three magnified sections A-A', B-B', C-C' that illustrate the strike variation of the model. These sections are shown truncated and offset from one another to display the retro-foreland basin stratigraphy, but align with the linear model orogen shown in the planform.

The differences among the sections are that along C-C' continental collision begins at the onset of convergence as in Model 1. However, along B-B' and A-A' there is ~ 180 km and ~ 360 km of convergence, respectively, before collision starts as a consequence of the $\theta \sim 25°$ inclination of the continental margins. Collision is therefore diachronous and spans ~ 10 My along the 750 km strike of the model.

While oceanic lithosphere is being subducted, the pro-wedge and pro-foreland basin are created on flexurally depressed oceanic crust which has a nominal initial depth of 2000 m in this model. There is, therefore, a strong asymmetry in the geometry of the model orogen because there is a large difference in thickness of the continental and oceanic lithosphere above the detachments upon which the oppositely vergent orogenic wedges are constructed (Fig. 10, section A-A').

Model 2–5 My

At 5 My (Fig. 10), the orogen basin section along C-C' has grown to a size similar to that of Model 1. The small difference is a consequence of the strike parallel or longitudinal component of the diffusive marine sediment transport in the foreland basins of Model 2. Along strike, however, the progressively smaller tectonic fluxes have limited orogen growth so that on A-A' the orogen is most like a doubly-vergent accretionary prism with the small subaerial exposure corresponding to the outer-arc high and the retro-foreland basin corresponding to a forearc basin. There is no volcanic arc in the model, but, for typical angles of subduction, this would be located near the retro-peripheral bulge in the lower region of the planform between sections B-B' and C-C'. The model section along C-C' therefore resembles the relationship between the Coast Range and the Great Valley forearc basin in California, for example.

The rainfall ranges from <0.5 m/yr on the retro-peripheral bulge to a maximum of ~ 5 m/yr at the highest point of the orogen on C-C' (see rainfall distribution, Fig. 10). Rainfall is relatively symmetric over the orogen and, therefore, Q_S remains significant on the pro-wedge. There is, however, relatively little sediment accumulation in the pro-foreland basin as a result of the same tectonic recycling of sediment that occurs in Model 1. The sediment accumulation on the pro-autochthon is <400 m (Fig. 10), therefore, Q_T, is not enhanced significantly where $d = \sim 3000$ m. It is only in the vicinity of A-A' that the longitudinal sediment transport into the trench has increased Q_T by ~ 20 percent, because $d = 1000$ m where oceanic lithosphere subducts.

The orogen surface slopes are only $\sim 1.5°$ and correspond closely with the observed average slopes in subaerial orogens. Submarine slopes are similar because in these models the wedge properties are not modified for a marine environment.

FIG. 9—Model 1 after 15 My use the same format as Figures 7 and 8, except in color. The surface elevation color scale is shown to the right of the planform, and the facies legend color scale is to the right of section A-A'. Note drainage networks in the orogen, and partial drainage of the retro-foreland basin across the peripheral bulge (upper panel); surface relief through the watershed on the leeward slope of orogen (middle panel); and aggradation of low river power facies as the rivers build toward grade (lower panel).

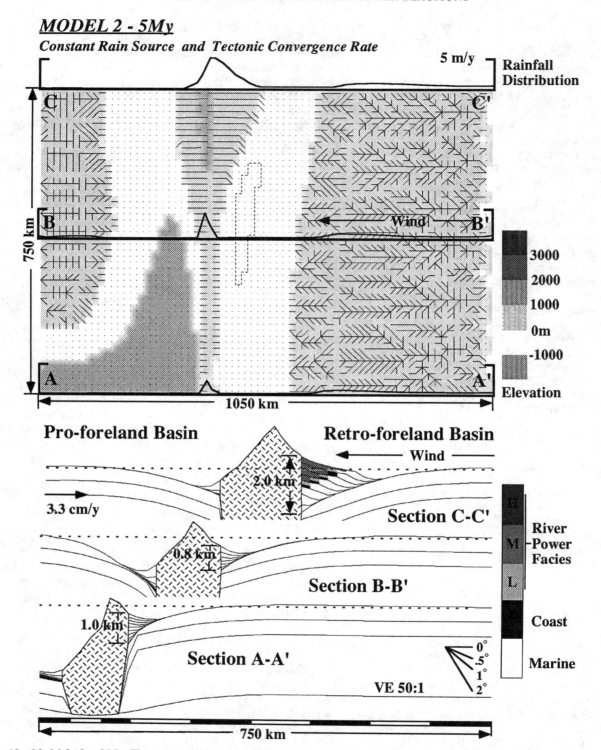

FIG. 10—Model 2 after 5 My. The top panels shows, in planform, the surface elevation (grey scale at right) and fluvial drainage network (black lines) of the orogen and foreland basins. The dotted line delimits the depocentre in the retro-foreland basin. Rainfall distributions are shown along sections A-A′, B-B′ and C-C′. Cross-sections A-A′, B-B′ and C-C′ show orogen topography, plus chronostratigraphy (0.5 My intervals, black lines) and facies distribution (0.25 My intervals, grey scale at right) in the pro- and retro-foreland basins. Sections which align in the model are offset in the figure to display sedimentary filling of basins. Note sediment deposited in the retro-foreland basin during the first 2 My is absent along B-B′ having been entrained into the retro-wedge with the onset of continent/continent collision. Sediments deposited in the first 2 My are present along section A-A′ because the Q_T associated with ocean/continent collision over the last 5 My is sufficiently small that wedge growth has not entrained the sediments. Sediments deposited in the first 2 My along C-C′ have not been entrained into the retro-wedge because Q_S in the first 2 My was larger adjacent to the incipient orogen along C-C′ than Q_S adjacent to the accretionary prism along B-B′, and because increasing rates of erosion with increasing precipitation on the windward retro-wedge slowed the rate of retro-wedge toe advance during the last 3 My.

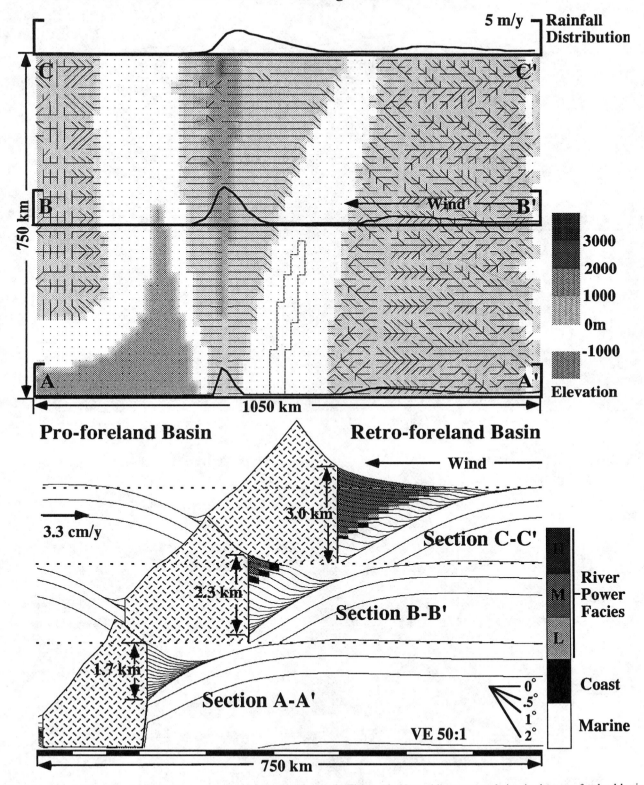

FIG. 11—Model 2 results after 10 My are presented using the format in Figure 10. Note oblique progradation in the retro-foreland basin (upper panel), evolution of rainfall distribution (upper panel), lack of preserved sediment in the pro-foreland basin (lower panel) and increasing thickness of marine facies along strike in the retro-foreland basin (C-C' and B-B', lower panel).

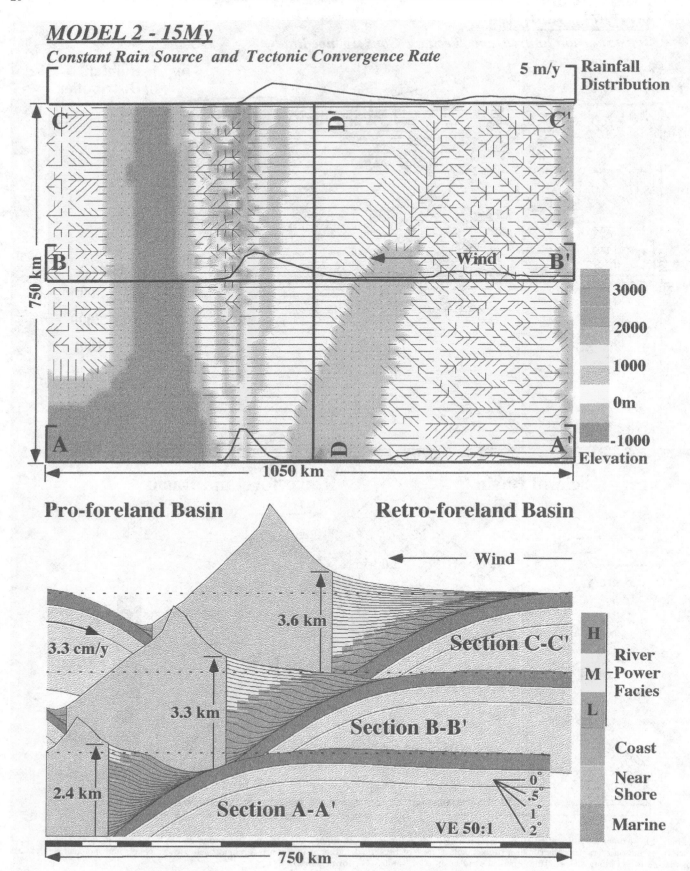

The planforms of Figures 7 and 10 show similar drainage networks which demonstrate that the model predicts dip slopes that are larger than strike slopes. Therefore, except for local deflections on the proximal side of the retro-foreland basin, rivers in Model 2 drain the orogen in the dip direction. The crest of the peripheral bulge is somewhat inclined to the strike of the orogen because lithospheric flexure changes with the strike variation in orogenic and sediment loads. The underfilled basin on A-A' is, therefore, narrower than the nearly-filled basin on C-C'.

The depocentre of the retro-foreland basin lies adjacent to the position where continent/continent collision is just beginning (B-B', Fig. 10). The basin is deepest in this location because the ratio of the isostatic subsidence to sediment flux into the basin is greatest while the orogen is small. Figures 10, 11, and 12 show that the migration of the retro-foreland basin depocentre with correlates onset of continent/continent collision.

Model 2–10 My

As in Model 1, the 330 km of total convergence by 10 My leads to significant growth of the orogen and deepening of the retro-foreland basin. Along C-C', the two models are similar except that longitudinal diffusive transport of sediment leaves a small marine region in Model 2, whereas, Model 1 has a totally nonmarine retro-foreland basin.

Rainfall distributions along sections A-A', B-B', and C-C' characterize both the variation in rainfall along the strike of the orogen, and how rainfall distributions changes across an orogen as it grows. While the orogen is low and narrow, the rainfall distribution is essentially symmetric as in sections A-A' and B-B'. Along B-B', the higher rainfall rates and approximately symmetric rainfall distribution show that rainfall across the retro-foreland has not significantly depleted the incident vapour flux. In contrast to B-B', development of the alluvial plain with growth of the orogen has resulted in significant rainfall over the alluvial plain in section C-C'. Rainfall varies from ~0.75 m/yr on the retro-peripheral bulge to a maximum of ~3.5 m/yr high on the retro-wedge adjacent to the topographic high of the orogen on C-C'. The rainfall is now quite asymmetric where the orogen divide exceeds a height of 1500 m and there is a significant rain shadow which reduces the sediment flux from the orogen into the pro-foreland basin. Sediment recycling therefore does not increase Q_T on C-C', but does increase on A-A' where the longitudinal (strike parallel) marine diffusion has significantly thickened sediments on the subducting plate.

There is a significant strike variation in the model sediment facies distribution in the retro-foreland basin. On C-C' (Fig. 11) the large tectonic flux ($d = 3000$ m) caused the orogen to grow rapidly and to develop sufficient subaerial exposure that Q_S became larger than Q_I. The shoreline prograded and, therefore, a large proportion of the basin sediments are nonmarine. The size of the orogen and retro-foreland basin decreases along strike reflecting the cumulative decrease in Q_T into the orogen due to the obliquity of the collision (Fig. 6). The sediment flux into the basin was, however, disproportionately even smaller than Q_I because basin facies become more marine along strike. On B-B' (Fig. 11), for example, the shoreline has only just begun to prograde, and on C-C' the basin remains marine. Flexural deepening of the basin along strike by larger parts of the orogen is certainly responsible for some of the relative increase in the proportion of marine sediments, but the strike variation in Q_S/Q_I also contributes to the facies distribution.

Model 2–15 My

At 15 My, the 500 km of convergence has created a basin on C-C' (Fig. 12) that is similar to that of Model 1 (Fig. 9) except that instead of depositing sediment in the distal depression of what is now an alluvial retro-foreland basin, the drainage system has become oblique and an axial or trunk river system transports sediment longitudinally along strike. The drainage network of the distal alluvial plain shows that with axial filling of the retro-foreland basin, low gradient rivers are deflected toward the marine basin. The net effect of this drainage pattern is longitudinal fluvial transport sub-parallel to the basin axis. There is now a competition between fluvial sediment transport and deposition in the dip direction, that leads to aggradation and favours drainage that will breach the peripheral bulge, and axial transport, which favours longitudinal progradation and filling of the remaining marine regions of retro-foreland basin.

The trunk river system that drains the axis of the retro-foreland basin in the model has a gradient of ~0.0001, which is approximately that found for modern anastomosed river systems (Smith, 1986; Rust, 1981). Sedimentation rates are low along the basin axis, but sediments with low river power facies continue to aggrade as rivers build toward equilibrium (C-C', Fig. 12) and keep pace with basin subsidence which is driven by tectonic and sedimentary loading. Aggradation of fluvial facies along the basin axis in the model may be analogous to aggrading axial river deposits found, for example, in the modern Magdalena foreland basin (Smith, 1986), the Tertiary Alpine foreland basins (Homewood and others, 1986), or the upper Cretaceous Alberta foreland basin (Leckie, 1989). Overall, progradational filling of the retro-foreland basin, in the model, switches from the dip to the longitudinal direction and limits the rate of aggradation as shown on section D-D' of the fence diagram in Figure 13.

It is not clear in natural systems whether progradational filling of the marine basin must be complete before suffi-

---

Fig. 12—Model 2 results after 15 My are presented using the format in Figures 10 and 11, except in color. The surface elevation color scale is shown to the right of the planform, and the facies legend color scale is to the right of the lower panel. Note partial filling of the retro-foreland basin, development of a trunk river system collecting drainage off the alluvial plain and peripheral bulge (upper panel), and aggradation of low river power facies in the basin axis (C-C', lower panel).

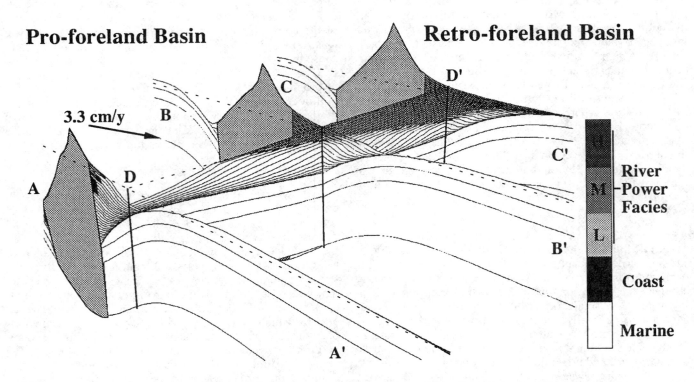

Fig. 13—Fence diagram of Model 2 results after 15 My. Section D-D' shows how longitudinal progradation relates to lateral progradation (Sections A-A', B-B' & C-C') of marine and alluvial facies in the retro-foreland basin. Note that D-D' is strike parallel to the orogen, and does not coincide with the trunk river system.

cient aggradation will occur to allow drainage reversal across the peripheral bulge. However, if the low slope of the trunk river system in the model is maintained, the basin must be filled for ~1000 km in the strike direction before aggradation will raise base levels over a 100-m high peripheral bulge. By implication axial drainage will persist in situations like the Himalayan foreland basins where the base levels of the Indus, Ganges and Brahmaputra river systems are fixed at sea level by sediment transport into the deep ocean. When other conditions are equal, only for rather long orogens, for example, the Andes, the Cordillera of western North America, the Appalachians and possibly the Permian Urals, would aggradation in the distal foreland basin be expected to be sufficient to favour drainage across the peripheral bulge as seen in Model 1.

Rainfall distributions across the model show an increase in asymmetry with the development of the alluvial plain between 10 and 15 My (B-B' and C-C', Figs. 11, 12). There is a strong rainshadow over the leeward side of the orogen and fluvial erosion rates are on the order 0.1 mm/yr. Sediment supply to the pro-foreland basin no longer contributes significantly to Q_T. In contrast, erosion rates on the windward slope of the orogen are 5 to 10-fold larger, and sedimentation across the alluvial plain varies from ~0.2 mm/yr to <0.005 mm/yr. The high precipitation rates on the alluvial plain have increased the carrying capacity of the rivers across the plain, and much of the sediment is, therefore, transported to the marine basin where sedimentation rates are of ~0.4 mm/yr.

SUMMARY AND CONCLUSIONS

The Model 1 and Model 2 results are preliminary illustrations of a model that links orogen tectonics, lithospheric flexure and climate-mediated surface processes in an internally consistent first-order manner. We have confined the presentation to the model results, except to make general reference to natural examples that may be analogous, because it is not clear that the processes are modelled accurately enough, or the model is sufficiently complete, for direct comparison of results with detailed stratigraphy of particular foreland basins. The dilemma is, therefore, between choices of using the model to investigate synthetic stratigraphy for ranges of parameter values that can be estimated for parts of natural systems, as employed in this paper, versus the more difficult problem of overall model validation, which will require studies of complete modern orogen-foreland basin systems.

We believe that the underlying approach to our model development is, however, valid. It is clear that the feedbacks in the system, such as those between model climate and model topography, and between tectonics and sediment recycling, are important to the model results and may be equally important in natural foreland basins. It is also clear that any model of equivalent natural systems on geological time scales must contain at least tectonic, surface processes and isostatic components if it is to have general applicability.

Coupling surface processes to a climate model also appears to be an appropriate way to overcome the difficulties associated with the spatial and temporal scaling of variables that govern mass transport by surface processes. In this model, climate is just rainfall, but it can be generalized to include type, seasonality and variance of precipitation, plus factors that affect weathering rates. A generalized climate model will, however, require more information on paleoclimates at the orogen-foreland basin scale if it is to be useful.

In summary, the results presented in this paper are preliminary and incomplete. Therefore, the following conclusions are of a general nature:

1. Feedbacks among the component models have been shown to be important which implies that process-based models of the type we have described should be linked in a internally consistent manner.
2. Orogens commonly have two foreland basins, one of which may be an oceanic subduction zone. The two basins, which we term pro- and retro-foreland basins, are asymmetric and are predicted by the model to evolve in contrasting ways under tectonic and climatic controls.
3. The asymmetry in the tectonics of oceanic or continental subduction is a primary control on the overall asymmetry of the orogen and basins. It leads to preferential tectonic recycling of sediments in the pro-foreland basin and to a stratigraphic architecture characterized by convergence between the foreland and the orogen. In contrast, the evolution of the retro-foreland basin is more passive and the deformation front advances into the basin sediments only when the orogen achieves a critical state.
4. Climatic asymmetry couples with tectonic asymmetry in a way that depends on prevailing wind directions. Models 1 and 2 emphasize climate effects that reinforce tectonic asymmetry because the sediment flux into the retro-foreland basin dominates. If the prevailing wind direction were reversed, some of the symmetry would be restored because the dominant sediment transport into the pro-foreland basin would offset tectonic recycling of sediments. When analyzing natural foreland basins it is therefore important to establish the tectonic and climatic setting in which the basin formed.
5. The fluvial sediment transport model that we have used is not merely an unnecessary complication of the simple diffusive models. The model creates the networks that are the minimum basis required to address the hierarchy of scales of natural river systems and their sediment carrying capacities.
6. Coupling of fluvial transport and a climate model also has the potential to address the related problem of appropriate scaling of transport coefficients in time and space. It can help avoid *ad hoc* choices of coefficient values, or choices that preclude natural variability.
7. The tectonic model has some of the properties required to represent the competition between processes that roughen an orogen and increase relief and denudation rates, versus processes that smooth topography and reduce sediment yield. A better understanding of tectonic roughening is, however, required if a correct scaling of sediment inputs into the foreland basin is to be achieved.
8. The difference in results between Models 1 and 2 indicates that longitudinal transport will be important in foreland basins when evolution of an orogen is spatially diachronous and/or when drainage base levels are controlled along strike. These conditions are common and most foreland basins are drained by longitudinal trunk river systems. It follows that it is important to establish that longitudinal sediment fluxes are insignificant if dip sections of a natural basin and orogen are to be compared with equivalent cross-sectional models that do not include longitudinal mass fluxes.
9. Process-based models of foreland basin stratigraphy are potentially powerful tools to test conceptual cause and effect relationships as a means to unravel the controls on natural stratigraphic signatures. It is, however, clear that as the "realism" of the models increases so does the complexity of model responses; model responses may also not be unique. It follows that care is needed when attributing characteristics of natural foreland basins to causal mechanisms. Only robust conclusions from internally consistent models should be used for comparison with natural systems.

ACKNOWLEDGMENTS

Funded by NSERC Operating and Lithoprobe Supporting Geoscience grants to Beaumont. Lithoprobe contribution #459. We thank Sean Willett and Glen Stockmal for insights on wedge mechanics, Bob Courtney for generating Green functions used in the isostasy calculations, Juliet Hamilton, Mark Paton and David McCurdy for technical assistance, Felix Frey for his comments on the manuscript, and Ronald Johnson for his perspectives on orogen-foreland basin systems. Reviews by Peter Flemings and Paul Heller caused us to rethink and recast aspects of the paper to make it more accessible to geologists. We thank them and Steve Dorobek for their critiques.

REFERENCES

ARMSTRONG, A. C., 1980, Soils and slopes in a humid environment: Catena, v. 7, p. 327–338.

BEAUMONT, C., 1978, The evolution of sedimentary basins on a viscoelastic lithosphere: theory and examples: Geophysical Journal of the Royal Astronomical Society, v. 55, p. 471–498.

BEAUMONT, C., 1981, Foreland basins: Geophysical Journal of the Royal Astronomical Society, v. 65, p. 291–329.

BEAUMONT, C., FULLSACK, P., AND HAMILTON, J., 1992a, Erosional control of active compressional orogens, *in* McClay, K. R., ed., Thrust Tectonics: London, Chapman and Hall, p. 1–18.

BEAUMONT, C., FULLSACK, P., HAMILTON, J., AND WILLETT, S., 1992b, Preliminary results from a mechanical model of the tectonics of compressive crustal deformation, *in* Ross, G. M., ed., Alberta Basement

Transects, Workshop Report v. 28: Vancouver, Lithoprobe Secretariat, University of British Columbia, p. 22–61.

BEAUMONT, C., QUINLAN, G., AND HAMILTON, J., 1988, Orogeny and stratigraphy: numerical models of the Paleozoic in the eastern interior of North America: Tectonics, v. 7, p. 389–416.

BEGIN, S. B., MEYER, D. F., AND SCHUMM, S. A., 1981, Development of longitudinal profiles of alluvial channels in response to base-level lowering: Earth Surface Processes and Landforms, v. 6, p. 49–68.

CARSON, M. A. AND KIRKBY, M. J., 1972, Hillslope Form and Processes: London, Cambridge University Press, 475 p.

CHAPPLE, W. M., 1978, Mechanics of thin-skinned fold-and-thrust belts: Geological Society of America Bulletin, v. 89, p. 1189–1198.

CHASE, C. G., 1992, Fluvial lansculpting and the fractal dimension of topography: Geomorphology, v. 5, p. 39–57.

COLEMAN, S. M. AND WATSON, K., 1983, Ages estimated from a diffusion equation for scarp degradation: Science, v. 221, p. 263–265.

COURTNEY, R. C. AND BEAUMONT, C., 1983, Thermally activated creep and flexure of the oceanic lithosphere: Nature, v. 305, p. 201–204.

CULLING, W. E. H., 1960, Analytical theory of erosion: Journal of Geology, v. 68, p. 336–344.

CULLING, W. E. H., 1965, Theory of erosion on soil covered slopes: Journal of Geology, v. 73, p. 230–254.

DAHLEN, F. A., 1984, Noncohesive critical Coulomb wedges: an exact solution: Journal of Geophysical Research, v. 89, p. 10125–10133.

DAHLEN, F. A. AND SUPPE, J., 1988, Mechanics, growth, and erosion of mountain belts, in Clark, S. P., ed., Processes in Continental Lithospheric Deformation: Boulder, Geological Society of America Special Publication 218, p. 161–178.

DAHLEN, F. A., SUPPE, J., AND DAVIS, D., 1984, Mechanics of fold-and-thrust belts and accretionary wedges: cohesive Coulomb theory: Journal of Geophysical Research, v. 89, p. 10087–10101.

DAVIS, D., SUPPE, J., AND DAHLEN, F. A., 1983, Mechanics in fold-and-thrust belts and accretionary wedges: Journal of Geophysical Research, v. 88, p. 1153–1172.

DECELLES, P. G. AND BURDEN, E. T., 1992, Non-marine sedimentation in the overfilled part of the Jurassic-Cretaceous Cordilleran foreland basin: Morrison and Cloverly formations, central Wyoming, USA: Basin Research, v. 4, p. 291–313.

DICKINSON, W. R., 1974, Plate tectonics and sedimentation, in Dickinson, W. R., ed., Tectonics and Sedimentation: Tulsa, Society of Economic Paleontologists and Mineralogists Special Publication 22, p. 1–27.

FLEMINGS, P. B. AND JORDAN, T. E., 1989, A synthetic stratigraphic model of foreland basin development: Journal of Geophysical Research, v. 94, p. 3851–3866.

FLEMINGS, P. B. AND JORDAN, T. E., 1990, Stratigraphic modelling of foreland basins: interpreting thrust deformation and lithosphere rheology: Geology, v. 18, p. 430–434.

HANKS, T. C., BUCKNAM, R. C., LAJOIE, K. R., AND WALLACE, R. E., 1984, Modification of wave-cut and faulting-controlled landforms: Journal of Geophysical Research, v. 89, p. 5771–5790.

HOMEWOOD, P., ALLEN, P. A., AND WILLIAMS, G. D., 1986, Dynamics of the Molasse Basin of western Switzerland, in Allen, P. A., and Homewood, P., eds., Foreland Basins: Oxford, International Association of Sedimentologists Special Publication 8, Blackwell, p. 199–218.

JERVEY, M. T., 1988, Quantitative geological modeling of siliciclastic rock sequences and their seismic expression, in Wilgus, C. K., Hastings, B. S., Posamentier, H., Van Wagoner, J., Ross, C. A., and Kendall, C. G., eds., Sea-level Changes: An Integrated Approach: Tulsa, Society of Economic Paleontologists and Mineralogists Special Publication 42, p. 46–69.

JORDAN, T. E., 1981, Thrust loads and foreland basin evolution, Cretaceous Western United States: American Association of Petroleum Geologists Bulletin, v. 65, p. 2506–2520.

JORDAN, T. E. AND FLEMINGS, P. B., 1991, Large-scale stratigraphic architecture, eustatic variation, and unsteady tectonism: a theoretical evaluation: Journal of Geophysical Research, v. 96, p. 6681–6699.

KARNER, G. D. AND WATTS, A. B., 1983, Gravity anomalies and flexure of the lithosphere at mountain ranges: Journal of Geophysical Research, v. 88, p. 10449–10477.

KENYON, P. M. AND TURCOTTE, D. L., 1985, Morphology of delta prograding by bulk sediment transport: Geological Society of America Bulletin, v. 96, p. 1457–1465.

KOOI, H. AND BEAUMONT, C., 1994, Escarpment evolution on higher-elevation rifted margins: Insights derived from a surface processes model that combines diffusion, advection and reaction: Journal of Geophysical Research, v. 99, p. 12191–12209.

LANDSBERG, H. E., 1981, World Survey of Climatology: Amsterdam, Elsevier.

LECKIE, D. A., 1989, Upper Zuni Sequence: middle Cretaceous to lower Tertiary, in Ricketts, B. D., ed., Western Canada Sedimentary Basin, A Case Study: Calgary, Canadian Society of Petroleum Geologists.

MACKIN, J. H., 1948, Concept of the graded river: Geological Society of America Bulletin, v. 59, p. 463–512.

MALAVIEILLE, J., 1984, Modélisation expérimentale des chevauchements imbriqués: application aux chaînes de montagnes: Bulletin de la Société Géologique de France, v. 26, p. 129–138.

QUINLAN, G. M. AND BEAUMONT, C., 1984, Appalachian thrusting, lithospheric flexure and Paleozoic stratigraphy of the eastern interior of North America: Canadian Journal of Earth Science, v. 21, p. 973–996.

PAOLA, C., HELLER, P. L., AND ANGEVINE, C. L., 1992, The large-scale dynamics of grain-size variation in alluvial basins. 1. Theory: Basin Research, v. 4, p. 73–90.

RUST, B. R., 1981, Sedimentation in an arid-zone anastomosing fluvial system: Cooper's Creek, central Australia: Journal of Sedimentary Petrology, v. 51, p. 745–755.

SIMONS, D. B., RICHARDSON, E. V., AND NORDIN, C. F., 1965, Sedimentary structures generated by flow in alluvial channels, in Middleton, G. V., ed., Primary Sedimentary Structures and Their Hydrodynamic Interpretation: Tulsa, Society of Economic Paleontologists and Mineralogists Special Publication 12, p. 34–52.

SINCLAIR, H. D., COAKLEY, B. J., ALLEN, P. A., AND WATTS, A. B., 1991, Simulation of foreland basin stratigraphy using a diffusion-model of mountain belt uplift and erosion—an example from the central Alps, Switzerland: Tectonics, v. 10, p. 599–620.

SMITH, D. G., 1986, Anastomosing river deposits, sedimentation rates and basin subsidence, Magdalena River, Northwestern Columbia, South America: Sedimentary Geology, v. 46, p. 177–196.

STOCKMAL, G. S., 1983, Modeling of large-scale accretionary wedge deformation: Journal of Geophysical Research, v. 88, p. 8271–8287.

STOCKMAL, G. S., BEAUMONT, C., AND BOUTILIER, R., 1986, Geodynamic models of convergent margin tectonics: transition from rifted margin to overthrust belt and consequences for foreland-basin development: American Association of Petroleum Geologists Bulletin, v. 70, p. 181–190.

SYVITSKI, J. P. M., SMITH, J. N., CALABRESE, E. A., AND BOUDREAU, B. P., 1988, Basin sedimentation and growth of prograding deltas: Journal of Geophysical Research, v. 93, p. 6895–6908.

THOMAS, W. A., 1977, Evolution of Appalachian-Ouachita salients and recesses from reentrants and promontories in the continental margin: American Journal of Science, v. 277, p. 1233–1278.

WILLETT, S. D., BEAUMONT, C., AND FULLSACK, P., 1993, Mechanical model for the tectonics of doubly vergent compressional orogens: Geology, v. 21, p. 371–374.

IMPLICATIONS OF OROGENIC WEDGE GROWTH, INTRAPLATE STRESS VARIATIONS, AND EUSTATIC SEA-LEVEL CHANGE FOR FORELAND BASIN STRATIGRAPHY—INFERENCES FROM NUMERICAL MODELING

TIM PEPER, RONALD VAN BALEN, AND SIERD CLOETINGH

Institute of Earth Sciences, Netherlands Research School of Sedimentary Geology, Vrije Universiteit, De Boelelaan 1085, 1081 HV, Amsterdam, The Netherlands

ABSTRACT: In this paper we investigate the relative effects of intraplate stress level fluctuations and eustatic sea-level changes on foreland basin stratigraphy using forward numerical models. The role played by growth of an evolving orogenic wedge is incorporated. The models show that the effect of stress level fluctuations can be as significant as the effect of eustatic sea-level change. Stress level variations and eustasy can be discriminated in models without orogenic wedge growth because of the asymmetric stratigraphic patterns produced by stress. Models with growth of the orogenic wedge, in contrast, show that asymmetric patterns can also be produced by orogenic wedge growth accompanied by an eustatic sea-level drop. Models adopting a stress level relaxation predict patterns compatible to those produced by isostatic rebound. Therefore, it is suggested that backward reconstructions of generic mechanisms for stratigraphic patterns are preferably not based on these patterns only, and require a combination with sediment provenance, petrological, and structural studies.

INTRODUCTION

Foreland basin stratigraphy records tectonic, eustatic, and climatic changes at convergent plate margins. The formation of erosional unconformities and onlapping surfaces is the result of the interplay of temporal variations in the erosion and lateral progradation rates of the orogenic wedge, as well as tectonic and eustatic sea-level changes. Many studies have investigated the role of one of these factors in foreland basin development (e.g., Jordan, 1981; Beaumont, 1981; Schedl and Wiltschko, 1984; Peper and others, 1992). Beaumont and others (1990) discussed the combined effect of climatic change, erosion and sedimentation, flexure, and evolution of the shape of the orogenic wedge. Flemings and Jordan (1989) elaborated the effects of wedge advance, alternating with periods of tectonic quiescence. They proposed that such a mechanism can lead to asymmetries in the stratigraphy, characterized by: (1) the formation of drowning sequences in the proximal areas of the orogenic wedge and shallowing sequences in the area around the peripheral bulge, followed by (2) shallowing in the proximal area and subsidence at the peripheral bulge. Geological studies of the Alpine Molasse basin have suggested that similar asymmetrical patterns may be due to eustatic sea-level changes combined with progradation of an orogenic wedge (e.g., Allen and others, 1991). Kinematic models on intraplate stress level changes suggest that intraplate stresses may also produce asymmetric changes (Cloetingh, 1988; Peper and others, 1992). The relative effects of eustatic sea-level changes and intraplate stresses in foreland basins, however, have barely been investigated. In the present paper, we study these differences by means of a dynamic model. We show that eustatic sea-level changes and in-plane stress level fluctuations can equally affect foreland basin stratigraphy and may be hard to distinguish in the stratigraphic record. We also discuss the effect of the interference of tectonic growth of the orogenic wedge with eustatic sea-level changes and stress level fluctuations on foreland basin stratigraphy.

NUMERICAL MODEL FOR STRATIGRAPHIC PATTERNS

Below, we briefly discuss a dynamic model for the stratigraphic fill of foreland basins, based on flexure of an infinite elastic plate as a consequence of vertical and horizontal loads. The model incorporates time-dependent sea-level and stress level fluctuations, as well as time, dependent changes in the shape of the orogenic wedge, erosion and sedimentation, and compaction. The present model ignores the presence of peripheral basins potentially affecting the foreland basin development (e.g., Quinlan and Beaumont, 1984; Peper and others, 1992). We focus instead on the effect of loading redistribution, eustatic sea-level and stress level fluctuations.

Flexure

The flexural calculations used in our model are based on Bodine (1981) and Kooi (1991), where the flexural response of the plate depends on the flexural rigidity (defined in terms of the effective elastic thickness (EET)), the Young's modulus, and the Poisson ratio. The plate is assumed to be infinite to both its edges. This is consistent with the notion that the vertical motions of a lithospheric plate underlying a foreland basin are frequently constrained by the presence of an underthrusting plate and consequently are mechanically coupled (see also Wortel and Cloetingh, 1985; Cook and others, 1988; ECORS Pyrenees team, 1988; Rait and others, 1991).

Loading

The loads are formed by an evolving orogenic wedge, sediments, sea water, and intraplate stress. These loads are represented by vertical and horizontal forces respectively.

Temporal Changes in the Horizontal and Vertical Loads: Intraplate Stress and Eustatic Sea-Level Fluctuations.—

Studies of intraplate seismicity and large scale flexural folding of continental and oceanic lithosphere provide evidence for intraplate stress regimes, reflecting significant horizontal forces (e.g., Stephenson and Cloetingh, 1991; Govers and others, 1992). Temporal fluctuations in the intraplate stress regimes have been demonstrated by paleo-stress studies in a number of foreland basins, including the Betic Cordilleras (De Ruig, 1990) and the Appalachians

(Evans, 1989). Stress level changes are incorporated in the model by temporally varying, horizontal forces acting on the edge of the plate. Flexural effects induced by sea-level changes are also incorporated. Lithospheric density changes due to in-plane stress variations are ignored (Cathles and Hallam, 1991).

Temporal Changes in the Vertical Load: Growth of the Orogenic Wedge.—

Following studies of accretionary wedges in modern convergent zones (e.g., Davis and others, 1983; Dahlen and others, 1984; Barr and Dahlen, 1989; Willet and others, 1993), we adopt the approximation of a critical wedge shape with time. Contrary to Flemings and Jordan (1989), we assume an orogenic wedge with a slope that varies with time, dependent on the amount of accretion to the wedge (e.g., Koons, 1990; Colletta and others, 1991; Willet and others, 1993). Three stages of wedge dynamics are adopted: (1) Stage of tectonic quiescence during which there is little accretion compared to the erosion rates of the wedge. The wedge diminishes as it is not able to maintain a critical taper. (2) Stage of tectonic activity and approximation of a critical taper during which accretion outpaces erosion. The wedge builds up and achieves a mechanical equilibrium marked by a critical taper. (3) Stage of tectonic activity and maintenance of a critical taper. Continued accretion leads to vertical and horizontal growth of the wedge. After approximation of a maximum thickness, lateral growth of the orogenic wedge dominates (c.f. Koons, 1989, 1990; Colletta and others, 1991; Dahlen and others, 1984; Willet, 1991; Suppe and Connors, 1992). We assume that the orogenic wedge achieves this maximum thickness due to the increase of temperature with depth, causing a loss of ductile strength at the base of the orogenic wedge (Birch, 1966; Goetze and Evans, 1979; Carter and Tsenn, 1987). As a consequence, the slope of the wedge decreases when the orogenic wedge reaches its critical thickness, approximating a horizontal surface (see also Willet, 1991; Suppe and Connors, 1992; Fig. 1). The height of this surface is taken to be equal to the height at the location with both critical thickness and minimum topography (Fig. 1). The growth of the wedge is defined by input of mass volumes with time. These volumes are added to the orogenic wedge, whereby the addition is proportional to the deviation of the eroded wedge from the critical wedge geometry (Fig. 2). With a combination of wedge growth and erosion, this proportional addition implies that vertical wedge growth is initiated in those parts of the orogenic wedge where erosion has been the most effective.

We ignore time-dependent changes in the rheological properties of the orogenic wedge caused by temporal strain-rate variations which might change the critical thickness. These changes can be expected because of the control on the yield strength of the rocks incorporated in the wedge and the rate of incorporation of relatively cool rocks in the wedge. This latter process influences the temperatures and consequently the creep rates of the wedge (see also Barr and Dahlen, 1989). There is little insight, however, into how and to what extent temporal variations of the strain rate control the dynamics of the orogenic wedge, and the

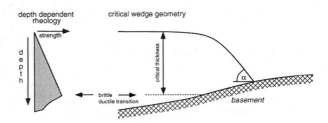

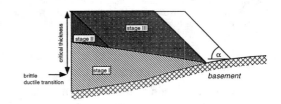

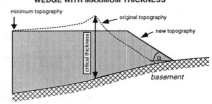

FIG. 1.—(A) Geometric model for the orogenic wedge in relation to the rheological properties of the wedge. The toe of the wedge maintains a maximum angle α which rapidly decreases with approximation of a critical thickness. (B) Cartoon of model adopted to approximate the critical wedge geometry. Stage I: arbitrary topography after a phase of erosion. Stage II: approximation of a wedge with a critical thickness and a critical angle α. Stage III: lateral out building of the wedge after the critical thickness is reached. (C) Cartoon of model adopted for the geometrical modifications of the orogenic wedge when the wedge has reached its critical thickness. Dashed line envelopes an arbitrary initial topography. Continuous line marks the newly approximated geometry of the orogenic wedge.

modeling is not substantially affected by ignoring variations in strain rates (Barr and Dahlen, 1989). The present model does not incorporate changes of the gradient of the orogenic wedge as a consequence of temporal variations of the internal friction, the dip of the lithosphere basement, the cohesion, and the density of the constituent rocks during the evolution of the orogenic wedge (see also Davis and others, 1983, Dahlen and others, 1984; Barr and Dahlen, 1989). The model also ignores the possibility of a rheological and mechanical coupling of the orogenic wedge and the lithosphere across the basal decollement of the orogenic wedge.

Temporal Changes in the Vertical Load: Erosion of the Orogenic Wedge and Sedimentation in the Basin.—

The destruction of the orogenic wedge and the growth and shape of the adjacent sedimentary wedge are calculated

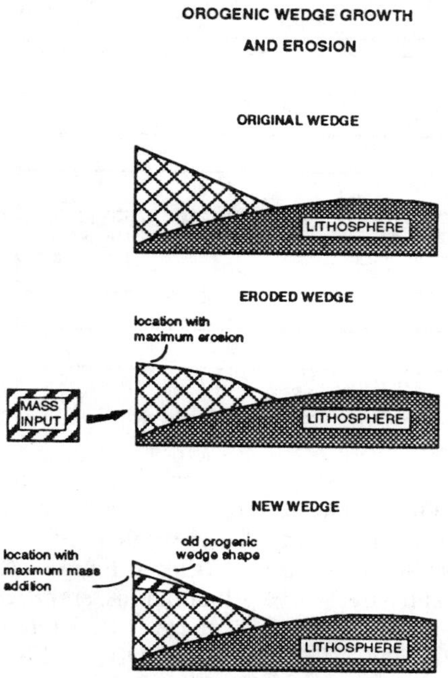

OROGENIC WEDGE GROWTH AND EROSION

FIG. 2.—Cartoon showing modification of geometry of orogenic wedge as a result of mass input to the wedge after a phase of erosion.

by means of a diffusion equation (c.f. Kenyon and Turcotte, 1985; Flemings and Jordan, 1989).

$$\frac{\partial h}{\partial t} = \frac{\partial}{\partial x}\left(K(x)\frac{\partial h}{\partial x}\right) \quad (1)$$

where: h = height of the topography, t = time, K = transportation coefficient, and x = lateral distance. This equation is solved by a one-dimensional, implicit finite difference scheme. The boundary conditions are defined by specified sediment fluxes at both sides of the basin.

The equation can be applied to a basin model for clastic environments incorporating river-dominated, deltaic, and shelf-slope domains (e.g., Flemings and Jordan, 1989), whereby each domain has its own characteristic K value (see also Hanks and others, 1984; Kenyon and Turcotte, 1985; Kenter, 1990; Kaufman and others, 1991). We ignore lateral changes in the transportation constant for basement erosion. Some studies (Summerfield, 1991) provide evidence for lateral changes of K in subaerial environments, but there are too few data from these studies to provide a useful range of K-values for our models (e.g., Leeder, 1991). We also ignore submarine erosion of the lithospheric basement. We ignore temporal changes of K as a result of exhumation of rocks with different rheological properties with time as there is little insight into this process. The model incorporates compaction of the sedimentary units (e.g., Reynolds and others, 1991; Bond and Kominz, 1984). Because decompaction of earlier compacted sedimentary units as a result of vertical uplift in the basin is not significant (Skempton, 1970), we do not allow porosity to increase. Finally, we ignore sediment porosity changes due to in-plane stress level fluctuations.

MODELING PARAMETERS

A 40 km EET (effective elastic thickness) of the lithosphere is adopted. This value corresponds to a strong, relatively old, continental lithosphere (see also Pilkington, 1991; Jordan, 1981; McNutt and others, 1988), implying that the effects of intraplate stress level fluctuations on the stratigraphy are minimized (e.g., Cloetingh and others, 1989; Peper and others, 1992; Peper, 1994). The critical thickness of the orogenic wedge is taken to be 15 km and is estimated from depth-dependent rheological profiles for a wedge which is shortened with a mean strain rate of 10^{-15} s^{-1} (see also Brace and Kohlstedt, 1980; Fig. 3). We assume that the wedge is composed of quartzitic-granitic rocks with a mean density of 2.6 g/cm^3 and a temperature of approximately 600°C at a depth of 40 km (see also Van den Beukel, 1992; Birch, 1966; Brace and Kohlstedt, 1980; Carter and Tsenn, 1987). We adopt a value of 10° for the maximum slope of the orogenic wedge (e.g., Koons, 1989; Willet, 1991). The values of the transportation coefficients are given in Table 1 and are adopted mainly from Hanks and others (1984). Maximum erosion rates at the top of the orogenic wedge obtained with these values are 0.4 mm/yr. The vertical growth rate of the orogenic wedge in the models adopting mass input maximally amounts to 0.75 mm/yr. Both values are within limits of observations (e.g., Stephenson, 1981; Kukal, 1990; Leeder, 1991; Burbank and Beck, 1991). The initial surface porosity and the porosity decay constants of the sediments are taken to be 30% and 3×10^{-4} m^{-1}, respectively. These values correspond to values of observations of porosity in sandstones and shales (e.g., Rieke and Chilingarian, 1974; Bond and Kominz, 1984). We adopt a maximum intraplate stress level of 150 MPa, which is a moderate value, considering the 40 km EET of the plate underlying the foreland basin. This EET corresponds to a more than 60 my old, rheologically stratified, continental lithosphere (e.g., Carter and Tsenn, 1987; Cloetingh and others, 1989). Figure 4 shows that differential horizontal

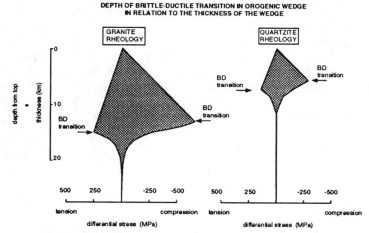

FIG. 3.—Depth-dependent rheological profiles for an orogenic wedge. Different profiles are for wedges composed by granite (left) and quartzite (right). Adopted temperature at 40 km = 600°C, strain rate = 10^{-15} s^{-1}. Abbreviation: BD transition = brittle-ductile transition.

TABLE 1.—MODELING PARAMETERS ADOPTED FOR CALCULATIONS OF SYNTHETIC STRATIGRAPHY AS DISPLAYED IN FIGURES 7, 9.

plate parameters	EET (km) 40	Young's mod. (Pa) 5.10^{10}	Poisson ratio 0.25	
densities (g/cm^3)	orogenic wedge 2.6	mantle 3.3	crust 2.6	sea water 1.0
surface porosities (%)	orogenic wedge 0	sediment 30		
porosity decay constant (1/m)	orogenic wedge 0	sediment 3.10^{-4}		
transport coefficients (m^2/yr)	coastal plain/river environment $2\ 10^4$	shelf 1.10^3	slope $1\ 10^2$	accretionary wedge 50	crustal lithosphere 5.10^{-2}
transitional depth for environments (m)	coastal plain/river to shelf 0	shelf to slope 200		
orogenic wedge parameters	critical thickness (m) 15000	critical angle (°) 10		
time step (yrs)	5000			

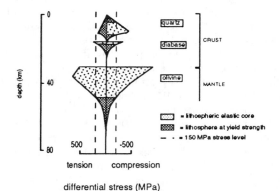

FIG. 4.—Depth-dependent rheological profile for 60 my old, multiple layered continental lithosphere (adopted from Stephenson and Cloetingh, 1991). Vertical line marks 150 MPa stress level. Lightly dotted areas mark elastic core of the lithospheric plate.

stresses of 150 MPa in such a plate do not exceed the yield strength. In addition, in-plane stresses of 150 MPa actually produce larger vertical displacements in models adopting depth-dependent rheological processes than models with an elastic rheology (Cloetingh and others, 1989; Stephenson and Cloetingh, 1991). Maximum sea-level change of 50 m is within the range of most of sea-level changes interpreted by Haq and others (1987); however, maximum sea-level may be as great as 150 m. Modeling parameters are summarized in Table 1 and Figure 5. The starting configuration for all models is shown in Figure 6. It represents a foreland basin formed by an orogenic and a sedimentary load. The initial topographic height is 3750 m at most. The maximum basin depth is 450 m.

MODELING RESULTS

In the following section, we discuss the effect of short-term (4 my) changes in intraplate stress levels and eustatic sea-level fluctuations on foreland basin stratigraphy, with a diminishing and a growing orogenic wedge, respectively (Figs. 7, 8 vs 9, 10). For comparison, reference models ignoring tectonic and eustatic changes other than orogenic wedge modifications are also presented (Figs. 7A, 9A).

A comparison of Figures 7B-E and Figure 7A shows that both sea- and stress level fluctuations significantly affect the model output. However, there is a difference between the models (Figs. 7D, E vs Figs. 7 B, C; Figs. 7C, D vs Figs. 7A, B). For a time interval of $T = 2$ my, the model with an initial eustatic sea-level rise predicts a basin wide aggradation, whereas the model with an initial increase in the compressional stress shows progradation at the basinward side of the peripheral bulge. The opposite is predicted by the models adopting a eustatic sea-level drop and a stress decrease, respectively.

Modeling results for $T = 4$ my (Fig. 7) show the effect of a reversal at $T = 2$ my of the gradient of tectonic stress level changes or eustatic sea-level fluctuations. Stratigraphic patterns that form from 2 to 4 my are opposite to the earlier formed patterns. Furthermore, the models incorporating an initial eustatic sea-level rise or a stress level increase predict a nearly similar shape of the sedimentary wedge directly adjacent to the orogenic wedge, where both patterns formed at $T = 2$ My have been modified by later erosion and sedimentation. Other types of stratigraphic patterns near the orogenic wedge are predicted with adoption of an initial eustatic sea-level or an intraplate stress level drop, as now the stratigraphy is preserved (Figs. 7 B, D). On the whole, the signals of stress and eustatic sea-level fluctuations are clearly recognizable in the modeled stratigraphic record.

Figures 9A-E and 10 display models for the combined effect of in-plane stress or eustatic sea-level fluctuations with growth of the orogenic wedge. We relate the stress level fluctuations to the mass input history and investigate both a transition from an increase in the level of compression to a decrease at the time of mass input stoppage (Fig. 9B), and vice versa (Fig. 9C). A stress increase can occur during periods of mass input because of the shortening which accompanies the mass input. Subsequent stoppage of mass input corresponds to stoppage of shortening and hence a stress drop may occur. However, growth of the orogenic wedge amongst others takes place as a consequence of thrust displacement and may, therefore, induce a significant stress drop (e.g., Peper and others, 1992). The subsequent phase

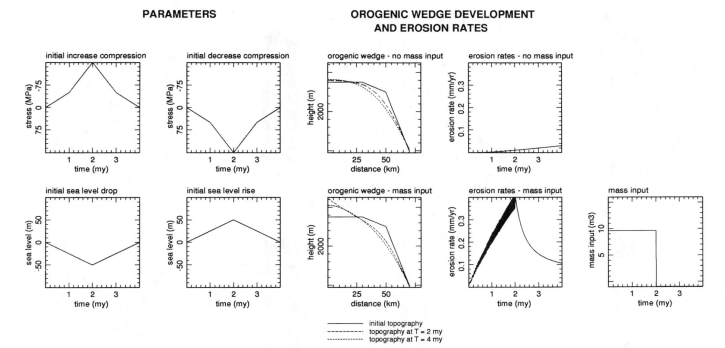

FIG. 5.—Changes of parameters used in the modeling, with time. Diagrams with erosion rates show erosion rate calculated at the top of the orogenic wedge. Diagrams with text "orogenic wedge" reflect the development of the orogenic wedge geometry as a function of time. The diagrams with the erosion rates and the orogenic wedge development are representative for all models discussed in this paper.

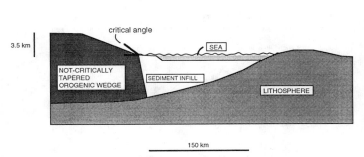

FIG. 6.—Basin configuration adopted at the start of the modeling. Topography reaches a height of at most 3750 m. The basin is filled with sediments to a depth of 450 m.

then is defined by a build up of compression. There is insufficient knowledge of the importance of either processes, and, therefore, we investigate both types of stress level changes. We adopted a simple shape for the temporal stress fluctuation to reduce the number of assumptions to a minimum. Chronostratigraphic charts of the models are displayed in Figure 8. They show the migration of the shelf-slope nick point and on- and off-lap patterns in time. Again the asymmetry in the models with intraplate stress distinguishes that model from the eustasy model.

A comparison of Figures 7 and 9 demonstrates that the effect of tectonic growth of the orogenic wedge on foreland basin stratigraphy significantly interferes with the effect of intraplate stress and eustatic sea-level fluctuations. Figure 7C and Figure 9C show that the uplift as a consequence of decreased compression for $T = 0 - 2$ my is damped by the subsidence as a consequence of tectonic growth of the orogenic wedge, reflected by the lesser progradation of the sedimentary wedge. Similarly, the large-scale aggradation at 2 my in the model with stress increase indicates that the signal produced by compression is also affected (Fig. 7B vs Fig. 9B). The foregoing also applies to models with eustatic sea-level fluctuations. Inspection of Figures 7A vs 9A and Figures 7B, C vs 9B, C demonstrates that the effects of stress level fluctuations show up more clearly in the stratigraphy modeled with mass input than in the stratigraphy modeled without.

During the subsequent period of zero mass input ($T = 2$ to $T = 4$ my), the signals from sea- and stress level rises and subsequent falls are destroyed by erosion while the signals from the models adopting opposite sea- or stress level fluctuations to a large extent remain preserved.

DISCUSSION AND CONCLUSIONS

The models demonstrate that intraplate stress and eustatic sea-level fluctuations produce recognizable stratigraphic patterns in foreland basins. In the models without orogenic wedge growth, the effect of stress level fluctuations can be distinguished from the effect of eustatic sea-level fluctuations because stress level fluctuations produce an asymmetric change in the stratigraphic patterns. In contrast, eustatic sea-level variations produce symmetric changes.

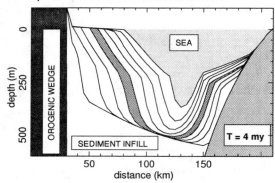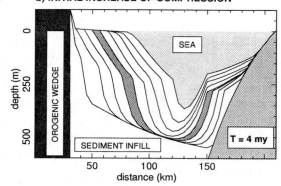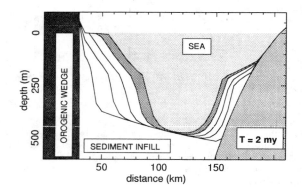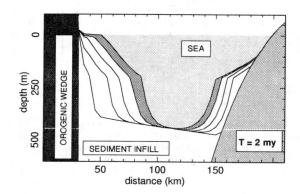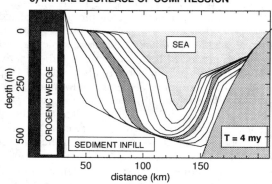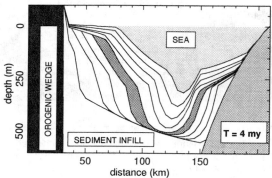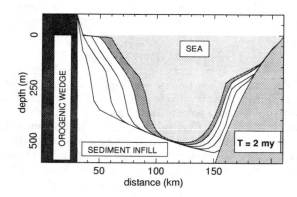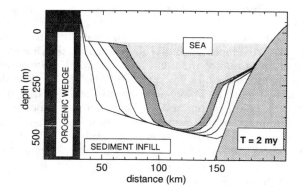

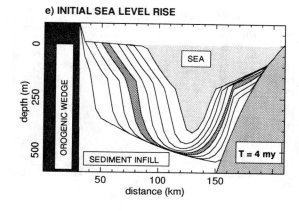

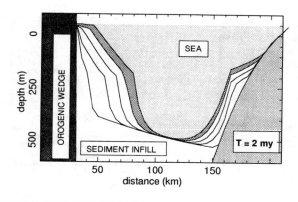

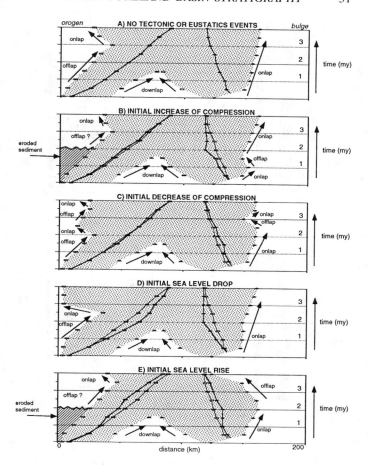

Fig. 7.—Model predictions for the stratigraphy in foreland basins: (A) without tectonic or eustatic events, (B) with initial increase of intraplate compression, (C) with initial decrease of intraplate compression, (D) with initial sea-level rise, and (E) with initial sea-level drop. Time lines are given for each 500,000 yr. Note the large vertical exaggeration. The orogenic wedge has a slope of 10° at most. Abbreviation: T = time.

Fig. 8.—Chronostratigraphic chart of modeled stratigraphy predicted with tectonic quiescence. Chart constructed from stratigraphic patterns predicted for $T = 4$ my (Fig. 6). (A) No additional tectonic or eustatic events, (B) initial increase of compression, (C) initial decrease of compression, (D) initial sea-level drop, and (E) initial sea-level rise. Thick lines reflect migration of shelf/slope nick point, which is compared to migration in reference model without inplane stress variations and sea-level fluctuations (thin lines). Convergence and divergence of lines reflect more aggradational or progradational character of stratigraphic sequence compared to reference sequence.

This symmetry disappears in the models with growth of the orogenic wedge (Fig. 7D vs Fig. 9D; Fig. 8D vs Fig. 10D). Consequently, eustasy as a causal mechanism cannot be discriminated from stress level fluctuations in these models. As these processes are the unknowns when starting backward reconstructions of tectonic or eustatic processes from stratigraphic patterns, we argue that such a reconstruction may not always be possible (see also Peper, 1993; 1994). Therefore, observations of truncations and basinward shelf migration around the flexural bulge simultaneously with aggradation close to the orogenic wedge in for example the Appalachian foreland basin or the French/Swiss molasse basin can be related to several generic processes: (1) visco-elastic relaxation of bending stress in the lithosphere, (2) episodic thrusting, (3) stress level changes, (4) eustatic sea-level changes combined with orogenic wedge growth (Tankard, 1986; Mugnier and Menard, 1986; Allan and others, 1991; Beaumont, 1981; Flemings and Jordan, 1989; see also Fig. 11). The predictions from the stress-decrease models provide an alternative for isostatic rebound induced motions, showing truncation and uplift in the area at the foot of the orogenic wedge and aggradation at the peripheral bulge (Flemings and Jordan, 1989; Fig. 11). For an adequate reconstruction of generic processes it is, therefore, necessary to combine analyses of stratigraphic patterns with studies of sediment provenance, petrology, structural geology, and numerical modeling. Studies of the Western Canada Foreland Basin have demonstrated the potential of this approach, providing evidence for the large control of continental tilting and eustasy on the stratigraphic record (Peper, 1994).

The stratigraphic patterns in foreland basins are equally dependent on sea-level changes or stress level fluctuations and tectonic loading for moderate, vertical growth rates and denudation rates of the orogenic wedge. The models demonstrate that the effects of eustatic sea-level changes and intraplate stress level fluctuations are not necessarily dominated by the effect of growth of the orogenic wedge. Progradational sequences, covered by aggradational sequences

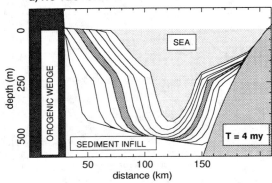
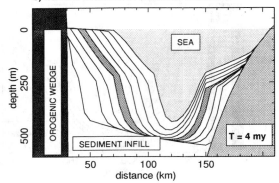
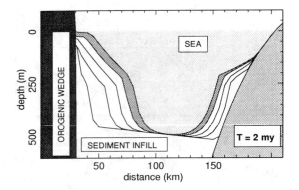
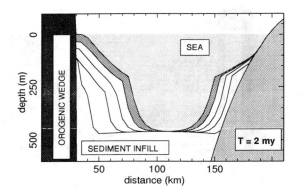
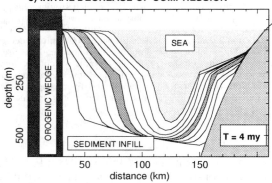
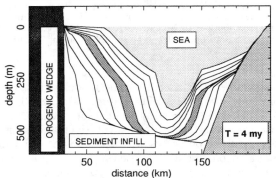
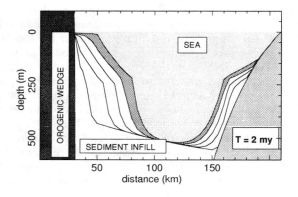
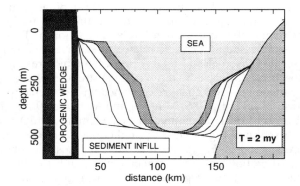

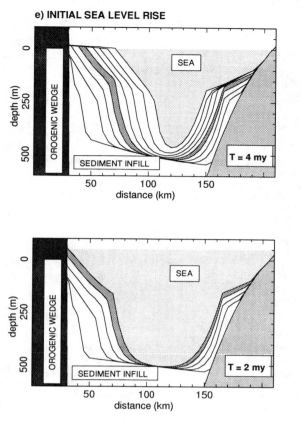

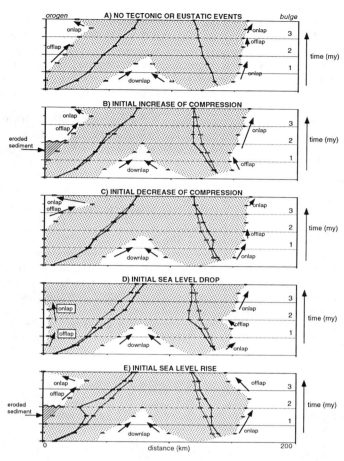

Fig. 9.—Model predictions for foreland basin stratigraphy. These models incorporate mass input into the orogenic wedge. Note absence of shift in onlap onto orogenic wedge due to the relatively large angle of this wedge. Figure conventions as in Figure 7.

Fig. 10.—Chronostratigraphic chart of modeled stratigraphy predicted with growth of orogenic wedge. Figure conventions as in Figure 8.

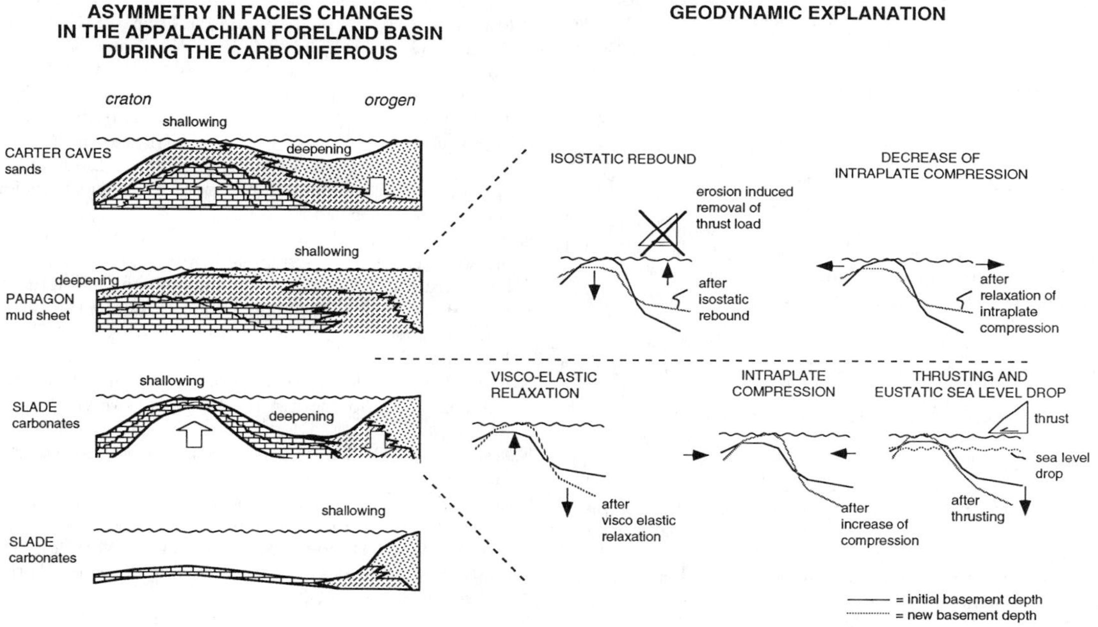

Fig. 11.—Cartoon showing alternative explanations for observations of asymmetric facies changes in the Appalachian foreland basin. Parts of this figure are modified after Tankard (1986).

predicted by models incorporating subsequent sea/stress-level falls and rises, are preserved, indicating that relative subsidence as a consequence of these processes may exceed the uplift related to erosion and isostatic rebound. This again illustrates that stratigraphic patterns cannot always be attributed to specific tectonic conditions. In addition, when the stratigraphic patterns at the foot of the orogenic wedge are not preserved, inferences of their generic mechanism will be even more complicated. Hence, the importance of combining stratigraphic, sedimentologic, and structural studies is emphasized.

ACKNOWLEDGMENTS

TP is funded by the IBS (Integrated Basin Studies) project, part of Joule II research programme funded by the Commission of European Communities (contract No JOU2-CT 92–0110). Chris Paola and an anonymous reviewer are thanked for critical reviews and useful comments. Randell Stephenson is thanked for discussion of the model and for reviewing an earlier version of this paper. Reinie Zoetemeijer, Jan Diederik van Wees and Fred Beekman are thanked for valuable suggestions. IBS contribution No 1. Publication No 2 of the Netherlands Research School of Sedimentary Geology.

REFERENCES

ALLEN, P., CRAMPTON, S., AND SINCLAIR, H. D., 1991, The inception and early evolution of the North Alpine foreland basin, Switzerland: Basin Research, v. 3, p. 3–14.

BARR, T. D. AND DAHLEN, F. A., 1989, Brittle frictional mountain building 2. Thermal structure and heat budget: Journal of Geophysical Research, v. 94, p. 3923–3947.

BEAUMONT, C., 1981, Foreland basins: Geophysical Journal of the Royal Astronomical Society, v. 65, p. 291–329.

BEAUMONT, C., FULSACK, P., WILLET, S., HAMILTON, J., JOHNSON, D., ELLIS, S., AND PATON, M., 1990, Coupling of climate, surface processes and tectonics in orogens and their associated sedimentary basins, in Lithoprobe East: Report of Transect Meeting, October, 1990: St. John's, Memorial University.

BIRCH, F., 1966, Compressibility; elastic constants, in Clark, S. P., Jr., ed., Handbook of Physical Constants: Boulder, Geological Society of America Memoir, 97, p. 97–174.

BODINE, J. H., 1981, The thermal mechanical properties of the oceanic lithosphere: Unpublished Ph.D. Thesis, Columbia University, New York, p.

BOND, G. AND KOMINZ, M. A., 1984, Construction of tectonic subsidence curves for the Early Paleozoic miogeocline, southern Canadian Rocky Mountains: implications for subsidence mechanisms, age of break up, and crustal thinning: Geological Society of America Bulletin, v. 95, p. 155–173.

BRACE, W. F. AND KOHLSTEDT, D. L., 1980, Limits on lithospheric stress imposed by laboratory experiments: Journal of Geophysical Research, v. 85, p. 6248–6252.

BURBANK, D. W. AND BECK, R. A., 1991, Rapid, long-term rates of denudation: Geology, v. 19, p. 1169–1172.

CARTER, N. L. AND TSENN, M. C., 1987, Flow properties of continental lithosphere: Tectonophysics, v. 136, p. 27–63.

CATHLES, L. M. AND HALLAM, A., 1991, Stress-induced changes in plate density, Vail sequences, epeirogeny, and short-lived global sea-level fluctuations: Tectonics, v. 10, p. 659–671.

CLOETINGH, S., 1988, Intraplate stress fluctuations: a new element in basin analyses, in Kleinspehn K. L., and Paola C., eds., New Perspectives in Basin Analyses: New York, Springer, p. 205–230.

CLOETINGH, S., KOOI, H., AND GROENEWOUD, W., 1989, Intraplate stress and sedimentary basin evolution, in Price, R. A., ed., Origin and Evolution of Sedimentary Basins and their Energy and Mineral Resources: Washington, D. C., American Geophysical Union Geophysical Monograph 48, p. 1–16.

COLLETTA, B., LETOUZEY, J., PINEDO, R., BALLARD, J. F., AND BALE, P., 1991, Computerized X-ray tomography analysis of sandbox models: examples of thin-skinned thrust systems: Geology, v. 19, p. 1063–1067.

COOK, F. A., GREEN, A. G., SIMONY, P. S., PRICE, R. A., PARRISH, R. R., MILKEREIT, B., GORDY, P. L., BROWN, R. L., COFLIN, K. C., AND PATENAUDE, C., 1988, Lithoprobe seismic reflection structure of the southeastern Canadian Cordillera: initial results: Tectonics, v. 7, p. 157–180.

DAHLEN, F. A., SUPPE, J., AND DAVIS, D., 1984, Mechanics of fold and thrust belts: cohesive Coulomb theory: Journal of Geophysical Research, v. 89, p. 10,087–10,101.

DAVIS, D., SUPPE, J., AND DAHLEN, F. A., 1983, Mechanics of fold and thrust belts and orogenic wedges: Journal of Geophysical Research, v. 88, p. 1153–1172.

DE RUIG, M. J., 1990, Fold trends and stress deviation in the Alicante fold belt, SE Spain: Tectonophysics, v. 184, p. 393–403.

ECORS Pyrenees team, 1988, The ECORS deep reflection seismic survey across the Pyrenees: Nature, v. 331, p. 508–511.

EVANS, K. F., 1989, Appalachian stress study 3. Regional scale stress variations and their relation to structure and contemporary tectonics: Journal of Geophysical Research, v. 94, p. 17,619–17,645.

FLEMINGS, P. B. AND JORDAN, T. E., 1989, A synthetic stratigraphic model of foreland basin development: Journal of Geophysical Research, v. 94, p. 3851–3866.

GOETZE, C. AND EVANS, B., 1979, Stress and temperature in the bending lithosphere as constrained by experimental rock mechanics: Geophysical Journal of the Royal Astronomical Society, v. 59, p. 463–478.

GOVERS, R., WORTEL, M. J. R., CLOETINGH, S. A. P. L., AND STEIN, C. A., 1992, Stress magnitude estimates from earthquakes in oceanic plate interiors: Journal of Geophysical Research, v. 97, p. 11,749–11,760.

HANKS, T. C., BUCKNAM, R. C., LAJOIE, K. R., AND WALLACE, R. E., 1984, Modification of wave-cut and faulting-controlled landforms: Journal of Geophysical Research, v. 89, p. 5771–5790.

HAQ, B. U., HARDENBOL, J., AND VAIL, P. R., 1987, Chronology of fluctuating sea levels since the Triassic: Science, v. 235, p. 1156–1167.

JORDAN, T. E., 1981, Thrust loads and foreland basin evolution, Cretaceous, western United States: American Association of Petroleum Geologists Bulletin, v. 65, p. 2506–2520.

KAUFMAN, P., GROTZINGER, J. P., AND MCCORMICK, D. S., 1991, Depth-dependent diffusion algorithm for simulation of sedimentation in shallow marine depositional systems, in Franseen, E. K., Watney, W. L., Kendall, G. C. St. C., and Ross, W., eds., Sedimentary Modeling: Computer Simulations and Methods for Improved Parameter Definition: Lawrence, Kansas Geological Survey Bulletin 233, p. 489–508.

KENTER, J. A. M., 1990, Geometry and declivity of submarine slopes: Published Ph.D. Thesis, Vrije Universiteit, Amsterdam, 128 p.

KENYON, P. M. AND TURCOTTE, D. L., 1985, Morphology of a delta prograding by bulk sediment transport: Geological Society of America Bulletin, v. 96, p. 1457–1465.

KOOI, H., 1991, Tectonic modeling of extensional basins—the role of lithospheric flexure, intraplate stress and relative sea-level change: Published Ph.D. Thesis, Vrije Universiteit, Amsterdam, 183 p.

KOONS, P. O., 1989, The topographic evolution of collisional mountain belts: a numerical look at the southern Alps, New Zealand: American Journal of Science, v. 289, p. 1041–1069.

KOONS, P. O., 1990, Two-sided orogen: collision and erosion from the sandbox to the Southern Alps, New Zealand: Geology, v. 18, p. 679–682.

KUKAL, Z., 1990, The rate of geological processes: Earth Science Review, v. 28, p. 8–284.

LEEDER, M. R., 1991, Denudation, vertical crustal movements and sedimentary basin infill: Geologische Rundschau, v. 80, p. 441–458.

MCNUTT, M. K., DIAMENT, M., AND KOGAN, M. G., 1988, Variations of elastic plate thickness at continental thrust belts: Journal of Geophysical Research, v. 93, p. 8825–8838.

MUGNIER, J. L. AND MENARD, G., 1986, Le developpement du basin molassique suisse et l'evolution des Alpes externes: Bulletin des Recherches d' Exploration et de Production Elf-Aquitane, v. 10, p. 167–180.

PEPER, T., 1993, Tectonic control on the sedimentary record in foreland basins—inferences from quantitative subsidence analyses and stratigraphic modeling: Published Ph.D. Thesis, Vrije Universiteit, Amsterdam, 188 p.

PEPER, T., 1994, Tectonic and eustatic control on Late Albian shallowing (Viking and Paddy formations) in the Western Canada Foreland Basin: Geological Society of America Bulletin, v. 106, p. 254–263.

PEPER, T., BEEKMAN, F., AND CLOETINGH, S., 1992, Consequences of thrusting and intraplate stresses for foreland basin development and peripheral areas: Geophysical Journal International, v. 111, p. 104–126.

PILKINGTON, M., 1991, Mapping elastic lithospheric thickness variations in Canada: Tectonophysics, v. 190, p. 283–297.

QUINLAN, G. M. AND BEAUMONT, C., 1984, Appalachian thrusting, lithospheric flexure, and the Paleozoic stratigraphy of the Eastern Interior of North America: Canadian Journal of Earth Sciences, v. 21, p. 973–996.

RAIT, G., CHANIER, F., AND WATERS, D. W., 1991, Landward- and seaward-directed thrusting accompanying the onset of subduction beneath New Zealand: Geology, v. 19, p. 230–233.

REYNOLDS, D. J., STECKLER, M. S., AND COAKLEY, B. J., 1991, The role of the sediment load in sequence stratigraphy: the influence of flexural isostasy and compaction: Journal of Geophysical Research, v. 96, p. 6931–6949.

RIEKE, H. H. AND CHILINGARIAN, G. V., 1974, Compaction of Argillaceous Sediments: Amsterdam, Elsevier, 424 p.

SCHEDL, A. AND WILTSCHKO, D. V., 1984, Sedimentological effects of a moving terrain: Journal of Geology, v. 92, p. 273–287.

SKEMPTON, A. W., 1970, The consolidation of clays by gravitational compaction: Geological Society of London Quarterly Journal, v. 125, p. 373–411.

STEPHENSON, R. A., 1981, Continental topography and gravity: Unpublished Ph.D. Thesis, Dalhousie University, Halifax, 341 p.

STEPHENSON, R. A. AND CLOETINGH, S., 1991, Some examples and mechanical aspects of continental lithosphere folding: Tectonophysics, v. 188, p. 27–37.

SUMMERFIELD, M. A., 1991, Sub-aerial denudation of passive margins: regional elevation versus local relief models: Earth and Planetary Science Letters, v. 102, p. 460–469.

SUPPE, J. AND CONNORS, C., 1992, Critical taper wedge mechanics of fold-and-thrust belts on Venus: Initial results from Magellan: Journal of Geophysical Research, v. 97, p. 13,545–13,562.

TANKARD, A. J., 1986, On the depositional response to thrusting and lithospheric flexure: examples from the Appalachian and Rocky mountain basins, in Allen, P.A., and Homewood, P., eds., Foreland Basins: Oxford, International Association of Sedimentologists Special Publication 18, p. 369–394.

VAN DEN BEUKEL, J., 1992, Some thermomechanical aspects of the subduction of continental lithosphere: Tectonics, v. 11, p. 316–329.

WILLET, S. D., 1991, Dynamic and kinematic growth and change of a Coulomb wedge, in McClay, K. R., ed., Thrust Tectonics: London, Chapman and Hall, p. 19–32.

WILLET, S., BEAUMONT, C., AND FULLSACK, P., 1993, Mechanical model for the tectonics of double vergent compressional orogens: Geology, v. 21, p. 371–374.

WORTEL, M. J. R. AND CLOETINGH, S. A. P. L., 1985, Accretion and lateral variations in tectonic structure along the Peru-Chile Trench: Tectonophysics, v. 112, p. 443–462.

THE PERMIAN BASIN OF WEST TEXAS AND NEW MEXICO: FLEXURAL MODELING AND EVIDENCE FOR LITHOSPHERIC HETEROGENEITY ACROSS THE MARATHON FORELAND

KENN-MING YANG[1] AND STEVEN L. DOROBEK
Department of Geology, Texas A&M University, College Station, TX 77843

ABSTRACT: The Permian Basin of west Texas and southern New Mexico is located in the foreland of the late Paleozoic Marathon-Ouachita orogenic belt. This complex foreland area consists of several sub-basins that are separated by intraforeland uplifts. In an accompanying paper in this volume, regional structural and stratigraphic relationships are used to constrain the tectonic history of the Permian Basin region. A kinematic model for the origin of the Central Basin Platform (CBP), a prominent intraforeland uplift that separates the Midland and Delaware Basins, is also presented. In this paper, we show the results of two-dimensional flexural models for the Permian Basin region.

The Marathon orogenic belt is generally considered to be the dominant topographic load that caused flexural subsidence in the Permian Basin region. Basement shortening and uplift associated with the CBP, however, require that the CBP is an additional topographic load that must be considered when modeling the late Paleozoic evolution of the adjacent basinal areas. The CBP was treated as a distributed tectonic load that produced lithospheric flexure of the adjacent basinal areas. To calculate the effects of this load, loading geometries were determined from the excess basement material measured on structural cross sections. These loading geometries were then used to calculate static profiles of the deflected lithosphere using flexural models for elastic lithosphere. Results from flexural analyses compare well with reconstructed synorogenic geometries of the Midland, Delaware, and Val Verde Basins, indicating that subsidence of these basins most likely was produced by the combined topographic loads of the Marathon orogenic belt, the CBP, and probably structures associated with the Diablo Platform and cryptic loads near the Eastern Shelf on the eastern side of the Midland Basin.

Best-fit model profiles for the Val Verde and Delaware Basins indicate that values for the flexural rigidity of lithosphere in the Permian Basin region are lowest near the southwest corner of the CBP. The Val Verde Basin is narrowest and the Delaware Basin is deepest near this corner. The apparently low rigidities at this corner also coincide with a prominent salient in the Marathon orogenic belt and with the greatest amount of shortening measured along the boundaries of the CBP. These low calculated rigidities reflect locally weaker lithosphere that might be related to inherent lateral strength variations in the Marathon foreland. Alternatively, high bending stresses produced by the combined loads of the southwest corner of the CBP and the prominent salient of the Marathon orogenic belt may also have weakened the lithosphere in this area.

INTRODUCTION

Foreland basins (or foredeeps) form as a flexural response to vertical loading of continental lithosphere by thrust sheets (Beaumont, 1981; Jordan, 1981; Turcotte and Schubert, 1982). Foreland basins typically have asymmetric cross-sectional profiles; the deepest part of a foreland basin is located immediately adjacent to the orogenic wedge that borders one side of the basin. The geometry of foreland basins is dependent on the magnitude of applied loads, both vertical and horizontal, and on the flexural rigidity of the foreland lithosphere. Typically, only vertical loads (i.e., the orogenic wedge and basin-filling sediments) are considered in modeling foreland basin geometries (Quinlan and Beaumont, 1984; Stockmal and others, 1986; Flemings and Jordan, 1989).

In most two-dimensional flexural models of foreland basins, lithospheric heterogeneity in the footwall plate is not considered, and the orogenic wedge along one side of the basin plus basin-filling sediments and water generally are regarded as the only loads responsible for flexure. Synorogenic crustal deformation *within* the foreland, however, may also affect foreland basin geometry and stratigraphic relationships. Some synorogenic intraforeland uplifts may become additional loads that complicate the evolution of foreland basins (Hagen and others, 1985; Flemings and others, 1986; Beck and others, 1988; Shuster and Steidtmann, 1988; McConnell and others, 1990; Yang and Dorobek, 1992).

The Permian Basin of west Texas and southeastern New Mexico is a structurally complex area located in the foreland of the Marathon-Ouachita orogenic belt (Fig. 1). The Permian Basin actually consists of several sub-basins that are separated by complex, fault-bounded uplifts. Despite a great deal of previous study, there is no general consensus regarding the tectonic mechanisms that formed the Permian Basin during late Paleozoic time (cf. Kluth and Coney, 1981; Dewey and Pittman, 1982; Elam, 1984). This lack of a consensus undoubtedly is due to the diverse structural styles, subsidence histories, and geometries of the sub-basins and uplifted areas that comprise the Permian Basin.

In a companion paper, we describe regional structural and stratigraphic relationships that constrain the tectonic history of the Permian Basin region (Yang and Dorobek, this volume). We also present a kinematic model for the origin of the Central Basin Platform (CBP), a prominent intraforeland uplift that separates the Midland and Delaware Basins. In this paper, we show the results of two-dimensional flexural models for the Permian Basin region. The model results reasonably fit reconstructed synorogenic profiles for the Val Verde, Midland, and Delaware Basins. More importantly, our flexural models demonstrate that these sub-basins were affected by the combined tectonic loads of the Marathon orogenic belt, CBP, Diablo Platform, and inferred structures on the eastern side of the Midland Basin. Our model results are, to our knowledge, the first attempt to quantitatively model the late Paleozoic geodynamic history of the Permian Basin region. In addition, this study also shows that late Paleozoic basin geometries from the Permian Basin are affected by lateral variations in lithospheric rigidity across the Marathon-Ouachita foreland. Our

---

[1]Current address: Exploration & Development Research Institute, Chinese Petroleum Corporation, 1 Ta Yuan, Wen Shan, Miaoli, Taiwan 36010.

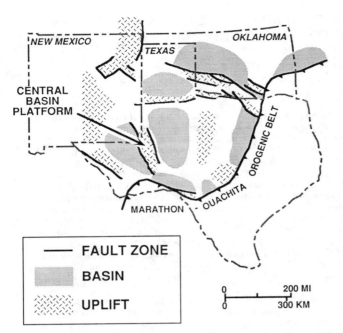

Fig. 1.—Regional base map of the Marathon-Ouachita foreland. Note partitioning of the Marathon-Ouachita foreland into a series of late Paleozoic sub-basins that are separated by high relief, fault-bounded uplifts. Modified from McConnell (1989).

Distal parts of the Marathon-Ouachita foreland area also experienced significant deformation at the same time that shortening occurred in the orogenic belt to the south and east. In the Permian Basin, the greatest amount of deformation across the distal foreland occurred during late Pennsylvanian to early Permian time (Yang, 1993; Yang and Dorobek, this volume). The antecedent Tobosa Basin was tectonically differentiated at this time into the crustal uplifts and sub-basins that now characterize the Permian Basin region (Fig. 2). One of these uplifts, the Central Basin Platform (CBP), is a prominent, fault-bounded paleotopographic high that separates the Midland and Delaware Basins and had its greatest uplift during late Pennsylvanian to early Permian time. It has been suggested that the CBP is located over a failed rift in the center of the antecedent Tobosa Basin. Inferred faults that were produced during late Precambrian-early Cambrian extension were reactivated during the Marathon-Ouachita orogeny and resulted in uplift of the CBP (Wuellner and others, 1986).

PREVIOUS GEODYNAMIC MODELS AND RATIONALE FOR FLEXURAL MODELING OF THE PERMIAN BASIN

Previous explanations for late Paleozoic subsidence in the Midland and Delaware Basins are based on the characteristic models demonstrate that some foreland basins are extremely complex and may be influenced by the combined effects of several topographic loads (cf. Jordan and others, 1988) and lateral variations in the strength of the loaded plate (cf. Waschbusch and Royden, 1992).

OVERVIEW OF THE TECTONIC HISTORY OF THE PERMIAN BASIN

A brief overview of the tectonic history of the Permian Basin is given here. A more detailed description of structural and stratigraphic relationships that were used to constrain the timing of tectonic events can be found in Yang (1993) and Yang and Dorobek (this volume).

During early to middle Paleozoic time, the present Permian Basin region was occupied by a relatively shallow, semi-circular basin called the Tobosa Basin. The Tobosa Basin probably formed initially during an extensional event in late Precambrian-early Cambrian time (Walper, 1977; Shurbet and Cebull, 1980; Thomas, 1983; Wuellner and others, 1986). Gradual subsidence occurred throughout the Tobosa Basin from early Cambrian to middle Mississippian time (Horak, 1985; Yang and Dorobek, this volume).

In late Mississippian time, initial collision occurred between Laurussia and Gondwanaland, and the Marathon-Ouachita orogenic belt first started to form. Mississippian-Pennsylvanian strata of the Tesnus, Haymond, and Gaptank Formations record early stages in the evolution of the Marathon foredeep in west Texas (McBride, 1978, 1989; Wuellner and others, 1986; Laroche and Higgins, 1990). These foredeep facies were caught up in or buried by thrust sheets that were emplaced during later stages of the Marathon-Ouachita orogeny (Reed and Strickler, 1990).

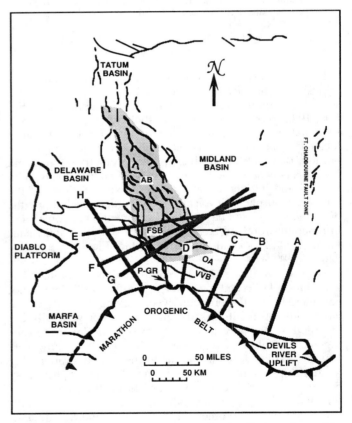

Fig. 2.—Generalized tectonic map of Permian Basin region and location of cross sections shown in Figures 3, 5, 6, 7, and 9. Modified from Ewing (1990), GEOMAP (1983), and Reed and Strickler (1990). The various sub-basins and uplifted areas that comprise the Permian Basin region are labeled. The shaded area represents the Central Basin Platform; AB = Andector Block; FSB = Fort Stockton Block; P-GR = Puckett-Grey Ranch Fault Zone.

istics of the boundary faults between the basins and adjacent uplifts and on the form of poorly constrained subsidence curves (Flawn and others, 1961; Walper, 1977; Dewey and Pitman, 1982; Hills, 1970, 1984; Elam, 1984; Horak, 1985; Denison, 1989). Earlier studies have suggested that these basins formed either by "vertical movement" along preexisting high angle faults (Walper, 1977; Hills, 1984, 1985; Wuellner and others, 1986) or by synorogenic extension in the foreland area of the Marathon orogenic belt (Dewey and Pitman, 1982; Elam, 1984). No previous models, however, have attempted to explain the coeval, late Paleozoic subsidence in all of the sub-basins that comprise the Permian Basin.

More recently, the Permian Basin region or parts of it have been regarded as a foreland basin that was produced by tectonic loading from the Marathon orogenic belt (Wuellner and others, 1986; Ross, 1986; Hanson and others, 1991). Our flexural models for the Val Verde Basin and southern parts of the Delaware Basin show that these limited areas are within the flexural wavelength produced by the final position of the Marathon orogenic belt (see below and Yang and Dorobek, 1992). Some of the deepest parts of the Midland and Delaware Basins, however, are far north of the observed flexural profile produced by the Marathons and thus could not be caused by flexure from that orogenic wedge. These model results and observed basin geometries force us to consider other possible explanations for subsidence throughout much of the Midland and Delaware Basins.

Interpretation of the CBP as a basement-cored uplift with reverse- and thrust-faulted margins (Galley, 1958; Ewing, 1984, 1991; Yang and Dorobek, 1991, 1992; Yang and Dorobek, this volume) implies that subsidence in the Midland and Delaware Basins may be due to flexure beneath the combined loads of the Marathon orogenic belt, the CBP, and other uplifts that border the basins (cf. Ewing, 1991). Areas of known late Paleozoic crustal shortening include the Marathon orogenic belt, segments of the boundary fault zones that border the CBP (Ewing, 1984, 1991; Yang and Dorobek, 1991, 1992, Yang and Dorobek, this volume; Shumaker, 1992), east-verging, steeply dipping faults along the eastern border of the Diablo Platform (Ewing, 1991), and possibly along the Fort Chadbourne fault zone on the eastern side of the Midland Basin; any of these areas of shortening may have produced topographic uplifts that caused flexure across the Permian Basin region. Other areas of crustal shortening can be documented across the Permian Basin, but they are very local in extent or do not have significant amounts of shortening associated with them. Thus, we assume that they did not contribute significantly to the tectonic loads that produced flexure across the Permian Basin. We have focused our flexural models instead on the effects of the Marathon orogenic belt and CBP, the two topographic highs in the Permian Basin region that are best constrained structurally and kinematically.

FLEXURAL MODELS FOR THE PERMIAN BASIN REGION

As described above, we view the entire Permian Basin as a "composite foreland basin" (cf. Jordan and others, 1988) that consists of sub-basins produced by several different topographic loads. This hypothesis can be tested by using well-established flexural models for elastic plates. Some parts of the Permian Basin can be modeled as elastic plates that are flexed beneath distributed topographic loads (Turcotte and Shubert, 1982; Speed and Sleep, 1982). In this section, we describe the methods used and basic results from our flexural models.

Values for Parameters Used in Flexural Models

The values for parameters used in the flexural models are: mantle density = 3.3 gm/cc; basement density = 2.75 gm/cc; sediment density = 2.2 gm/cc. The density of basement rocks was derived from model crustal profiles that were generated from regional gravity data (Keller and others, 1980). Sediment density was estimated from average values for decompacted synorogenic strata using empirical porosity-depth curves shown in Sclater and Christie (1980). Values from 10^{21} to 10^{23} N-m were used in the flexural models for the rigidity of foreland lithosphere. This range of rigidities produced model profiles that best approximated the observed profiles in all of the sub-basins that comprise the Permian Basin.

Method for Determining the Distributed Topographic Load of the CBP

Flexural models show that subsidence in many sedimentary basins is the result of loading by uplifted basement blocks that have been thrust onto the adjacent basin floor (Jordan, 1981; Hagen and others, 1985; Flemings and others, 1986; Beck and others, 1988; Shuster and Steidtmann, 1988). Deflection of the basin floor beneath the uplifted, relatively high density basement rocks and development of an asymmetric basin next to the uplifted basement reflect regional isostatic compensation.

Steeply dipping reverse faults in the boundary fault zone of the Fort Stockton Block indicate that uplift of this block was produced by basement-involved shortening (Yang and Dorobek, 1992, this volume). Basement-involved shortening, in turn, created a topographic load that caused flexure of the adjacent lithosphere. Several simplifying assumptions must be made to reconstruct the synorogenic cross-sectional geometry of the CBP so that flexural analyses can be done. We first assume that lower Paleozoic strata originally were horizontal and were uplifted and tilted during late Paleozoic basement shortening. We also assume that no significant deformation affected the area after the Early Permian and that the lithosphere beneath the Permian Basin behaved as a thin, unbroken elastic plate.

On structural cross sections across the southern part of the Permian Basin (Fig. 3), the Fort Stockton Block of the CBP appears as a wedge-shaped mass. The eastern end of the wedge extends into the central part of the Midland Basin. The upper surface of the wedge gradually rises westward to the crest of the Fort Stockton Block and then drops sharply down into the Delaware Basin across a narrow boundary fault zone. Lower Paleozoic strata on our reconstructed structural profiles gradually rise away from the structural boundaries on both sides of the wedge (Fig. 3).

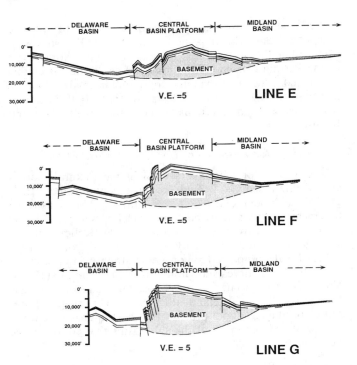

FIG. 3.—Reconstructed structural cross-sections across the Fort Stockton Block and adjacent Midland and Delaware Basins (Lines E, F, and G). Lower Paleozoic strata are shown by unlabeled, subparallel lines. Shaded area indicates the model load geometry of the Fort Stockton Block that was used in the flexural analyses described in the text. The geometry of the bottom of the load was obtained by the iterative method illustrated in Figure 4 and as discussed in the text.

We assume the western limit of the topographic load coincides with the westernmost boundary fault along the western side of the CBP. Uplifted, lower Paleozoic strata on the CBP dip eastward away from the western boundary fault zone, and form the eastern backlimb of the structure. This structural profile is similar to Laramide-type foreland uplifts (Stearns, 1978; Smithson and others, 1979; Lowell, 1983; Brown, 1984; Stone, 1984; Erslev, 1986). Geometric constraints demand that the east-dipping backlimb joins the original, horizontal baseline (Erslev, 1986). We take the eastern limit of the model load along this cross section as the intersection point between the tilted backlimb and the original untilted baseline (Fig. 3).

The resulting wedge-shaped mass beneath the uplifted lower Paleozoic strata was used as the topographic load for calculating the deflection in the flanking basins. The topographic load was treated as a distributed load and divided into 10 to 11 units, each 9.45 km wide. The deflections produced by each load unit were calculated and summed to form the final cumulative deflection profile. The solution for the flexural profile produced by this distributed load is determined by the method of Hetenyi (1946) and the simple algorithm for the solution is from Jordan (1981), that is:

$$W_k = \sum_{j=1}^{N} w_{kj}$$

W_k = total deflection at point k, and w_{kj}: deflection from a load unit j

N = flexural rigidity of plate

$w_{kj} = \rho_j h_j G_{k-j}$

ρ_j = density of load

h_j = thickness of load unit j

$G_{k-j} = \{\exp[-(|k - j| - 1/2)\Delta x/\alpha] \cos[(|k - j| - 1/2)\Delta x/\alpha] - \exp[-(|k - j| + 1/2)\Delta x/\alpha] \cos[(|k - j| + 1/2)\Delta x/\alpha]\}/2\rho_a$

$\alpha = [4 N/(\rho_a g)]^{1/4}$

ρ_a = density of asthenosphere

g = gravity

Δx = width of each load unit

We also assume that the floor of the antecedent Tobosa Basin was a horizontal surface prior to uplift of the CBP. This is a reasonable assumption because no major variations in lithofacies or dramatic thickness changes have been documented in middle Paleozoic strata from the central part of the Tobosa Basin. The originally horizontal floor on both sides of the CBP must have been deflected downward when the CBP was uplifted. However, the amount of the downward deflection produced by the CBP is grossly underestimated if the lower part of the topographic load between the deflected baseline and the original horizontal baseline is ignored. Thus, this part of the load must be estimated indirectly.

We estimated the geometry of the deflected baseline directly beneath the CBP by first assuming that the flexural profile reflects isostatic equilibrium. Several iterations were made to obtain the load geometry shown in Figure 3. We first drew a horizontal line between the two edges of the load geometry in Figure 3 and used the load above this line for the first iteration of the flexural model (Fig. 4). The original, horizontal baseline beneath this initial load geometry then was deflected in the first iteration. The model load above this deflected baseline was larger than the initial load and was used for the second iteration. We also assumed that sediments completely filled the adjacent flexural basins that were produced by the "revised" topographic load. The enlarged topographic load plus the sediments in the basins then were used together to calculate a third deflection profile. This procedure was repeated for approximately fifteen iterations until a final, steady state profile was attained (Fig. 4).

Flexural Models for the Val Verde Basin: Lithospheric Flexure Produced by the Marathon Orogenic Belt

Stratigraphic relationships within thrust sheets of the Marathon orogenic belt indicate that an ancestral late Mississippian to middle Pennsylvanian foredeep was located in the region now occupied by the Marathon orogenic belt (Yang and Dorobek, this volume). It is impossible to model the

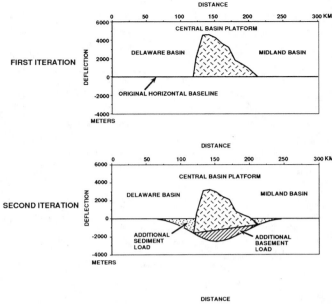

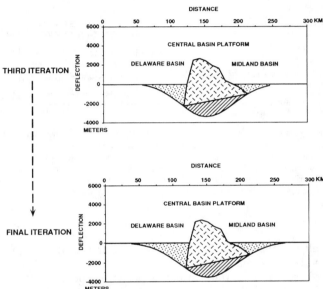

Fig. 4.—Sequential iterations of our flexural model showing how the final load geometry of the CBP was obtained and how this distributed load affected flexure in the adjacent basins.

flexural profile of this ancestral Marathon foredeep because of problems with palinspastic reconstruction. Flexural subsidence across this ancestral foredeep, however, slowed down considerably during middle Pennsylvanian time when shallow-water carbonate facies of the Strawn Limestone were deposited (Yang, 1993; Yang and Dorobek, this volume). Thus, the Strawn Limestone probably formed during a period of relative tectonic quiescence during the protracted Marathon-Ouachita orogeny.

Flexural subsidence across the entire Val Verde Basin increased again after middle Pennsylvanian time. Upper Pennsylvanian shale and carbonate facies onlap the Strawn Limestone. Upper Pennsylvanian through Wolfcampian strata in the Val Verde Basin also thicken dramatically toward the south. These stratigraphic relationships indicate that a second phase of tectonic activity produced flexure across the Val Verde Basin.

Decompacted upper Pennsylvanian to Wolfcampian strata on four cross sections through the Val Verde Basin were used to represent flexural profiles produced by the final episode of loading by the Marathon orogenic belt (Fig. 5). These cross sections are nearly perpendicular to the axis of the Val Verde Basin and the front of the Marathon orogenic belt. The cross sections also are located from east to west along the axis of the Val Verde Basin to test for any variation in the properties of the lithosphere across the basin. Line A, which is in the eastern part of the basin, extends from the Eastern Shelf, across the eastern side of the Midland Basin, and to the Devils River Uplift, which is regarded as part of the Marathon orogenic belt. Lines B and C extend from the southern part of the Midland Basin, across the Ozona Arch, and to the Marathon thrust front. Line D, in the western part of the Val Verde Basin, extends from the southern flank of the Fort Stockton Uplift to the proximal part of the foredeep.

The Marathon orogenic belt was treated as a semi-infinite load, and the lithosphere beneath the Val Verde Basin was assumed to be a thin, unbroken elastic plate. If the width of the orogenic belt is larger than 1α (i.e., where the flexural parameter has dimensions of length), then treating the orogenic belt as a semi-infinite sheet probably is appropriate (Speed and Sleep, 1982). In our case, for flexural rigidities of 10^{21}, 10^{22}, and 10^{23} N-m, α is equal to 25, 44, and 78 km, respectively. The Marathon orogenic belt extends for more than 200 km to the south of its frontal thrust (Handschy and others, 1987), hence we chose to treat it as a semi-infinite sheet. The governing equation for calculating lithospheric flexure based on these boundary conditions is from Speed and Sleep (1982):

$w(x) = (p/2) \cos(x/\alpha) \exp(-x/\alpha)$, where,

$w(x)$ = vertical deflection

p = mass of the topographic load per unit area

$k = (\rho_a - \rho_f)$

$\alpha = (4 N/k g)^{1/4}$

N = flexural rigidity of plate

ρ_a = density of asthenosphere

ρ_f = density of sediments

g = gravity

The calculated lithospheric deflection is determined by the rigidity of the loaded plate and the elevation of the topographic load; these can be obtained once the position of the leading edge of the Marathon orogenic belt is determined. The position of the frontal thrust of the Marathon orogenic belt is shown in several previous studies (Moore and others, 1981; Reed and Strickler, 1990) and on several tectonic maps (GEOMAP, 1983; Ewing, 1990, 1991).

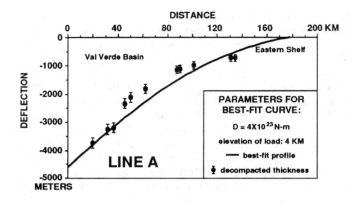
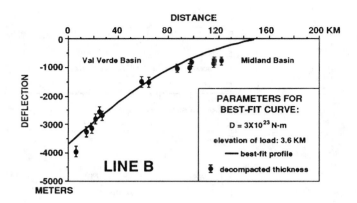
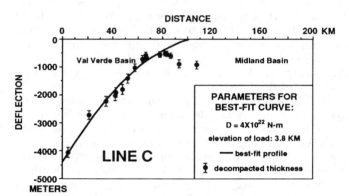
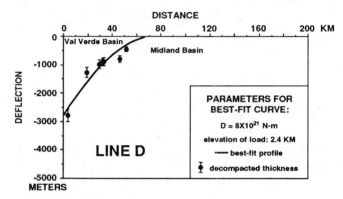

Fig. 5.—Model flexural profiles versus observed basin profiles for four cross sections across the Val Verde Basin. Lines A through D are distributed from east to west across the Val Verde Basin. Decompacted synorogenic strata were used to reconstruct the synorogenic basin profiles. Vertical error bars at each control point along the observed profiles reflect the maximum and minimum decompaction constants applied to each point. Note the reasonable fit between the model and observed profiles except for northern parts of the basin where the observed basin depth is greater than that predicted by the flexural models. Also note that progressively lower rigidities from east to west are needed to produce a reasonable fit between the model and observed basin profiles. See text for more complete discussion of the model results.

Our model flexural profiles match observed Val Verde Basin geometries very well, except on Lines B and C where the observed basin geometry in the distal part of the Val Verde Basin is deeper than the model profile (Figs. 5B, C). The model profiles end at the southern terminus of the theoretical forebulge, whereas the observed profiles show that the topography of the forebulge has been eliminated. It is important to note that Lines B and C actually extend into southern parts of the Midland Basin. Differences between the model results and the observed Val Verde Basin profiles suggest that an additional tectonic load(s) affected the geometry in distal parts of the Val Verde Basin and eliminated the forebulge created by the Marathon orogenic belt.

Comparison of the observed profiles from the Val Verde Basin (Fig. 5) shows that the synorogenic basin geometry becomes narrower from east to west along the axis of the basin. The model flexural profiles (Fig. 5) indicate the rigidity of lithosphere beneath the Val Verde Basin is from 10^{21} to 10^{23} N-m and progressively decreases toward the western end of the basin. The western end of the basin is bordered to the north by the Fort Stockton Uplift. The estimated late Paleozoic elevation of the Marathon orogenic belt is about 4 km, a value that compares well with many modern mountain ranges.

It also is important to note that the forebulge of the Val Verde Basin along Lines C and D seems to have stayed in more or less the same place over approximately 30–35 my, from pre-Desmoinesian to late Wolfcampian time (Fig. 6). This is indicated by the coincidence between the area of maximum erosion of pre-Desmoinesian strata and the region of maximum thinning of upper Pennsylvanian through upper Wolfcampian strata (Fig. 6). These stratigraphic relationships indicate that the forebulge north of the Val Verde Basin locally remained fixed over one area in the Marathon foreland lithosphere. As the Marathon orogenic belt advanced northward from middle Pennsylvanian to late Wolfcampian time, the forebulge apparently did not migrate.

Flexural Models for the Midland and Delaware Basins: Lithospheric Flexure Produced by the CBP and Other Loads

In our companion paper in this volume, we show that the CBP is an intraforeland uplift that is bounded by complex

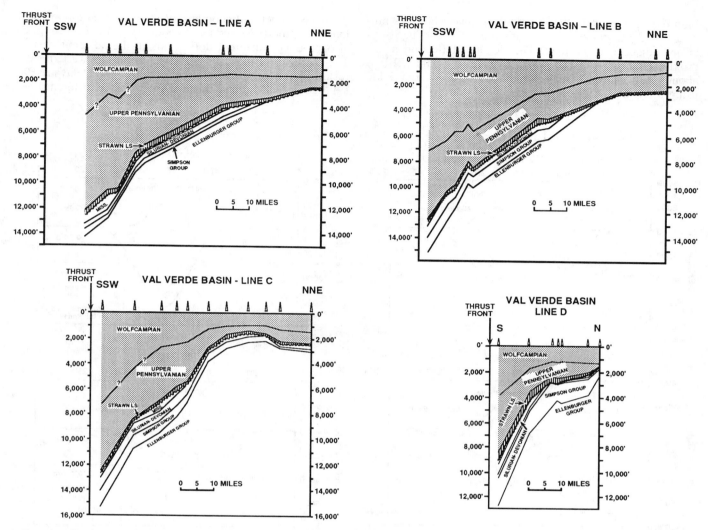

Fig. 6.—Stratigraphic cross sections across the Val Verde Basin; small triangles at top of cross sections indicate well control points. See Figure 2 for locations of cross sections. Note the regional unconformity that separates lower Paleozoic strata and upper Pennsylvanian to Lower Permian strata on all the cross sections. The Strawn Limestone above this unconformity surface has relatively uniform thickness across each cross section. Upper Pennsylvanian to Lower Permian strata, however, thicken dramatically toward the Marathon thrust front. Also note that maximum erosion of pre-Strawn strata is at the same approximate location where synorogenic strata are thinnest.

fault zones. Large segments of the CBP's boundary faults are steeply dipping, reverse and thrust faults, indicating that crustal shortening was associated with uplift of the CBP. The greatest amount of shortening along the CBP's boundary fault zones also coincides with the largest structural relief, greatest basin asymmetry, and thickest accumulation of synorogenic strata in the adjacent basins. This is especially true for local parts of the Delaware Basin. These relationships imply that uplift of the CBP caused flexure in the adjacent Midland and Delaware Basins during late Paleozoic time. In addition, late Paleozoic, clockwise rotation of the Fort Stockton and Andector Blocks, the blocks that comprise the CBP, caused unequal amounts of shortening and uplift along the boundaries of the CBP (Yang and Dorobek, this volume). Thus, the loading effects of the CBP were not equally distributed and flexure due to these loads resulted in variable geometries in the adjacent basins. Where

several topographic loads are in close proximity (e.g., the southern end of the CBP and the northern salient of the Marathon orogenic belt; Fig. 2), their flexural profiles interfered with one another and created additional variability in basin geometries.

In this section, we show the results of two-dimensional flexural models applied on three cross sections across the southern part of the Delaware Basin, CBP, and Midland Basin. These model results might explain the late Paleozoic geometries of the Midland and Delaware Basins. The models also indicate that some variation in lithospheric rigidity existed across the Delaware Basin and affected the geometry of that basin during late Paleozoic time.

Model Flexural Profiles for the Midland and Delaware Basins.—

Different lithospheric rigidities across the Permian Basin were used to generate the model flexural profiles shown in

Figure 7. The models show that using lower rigidities produces a larger deflection beneath the CBP and narrower basins adjacent to the load. The flexural deflection in the proximal part of the Delaware Basin next to the CBP is about 2 km using a rigidity of 10^{23} N-m and about 3 km using a rigidity of 10^{21} N-m.

The asymmetrical geometry of the estimated topographic load (i.e., the cross-sectional geometry of the CBP; Fig. 3) also produces different deflections in the adjacent basins. The geometry of the CBP is skewed toward its western side, resulting in a larger flexural deflection in the Delaware Basin than in the Midland Basin. However, the difference be-

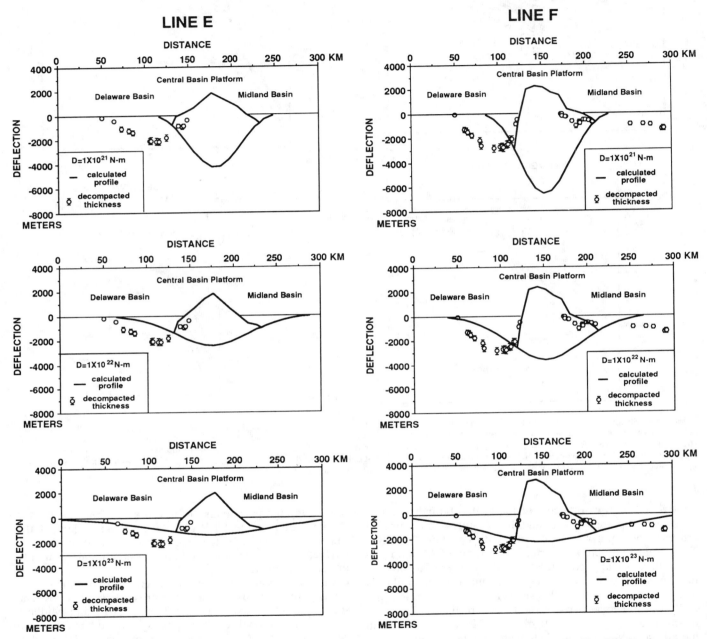

FIG. 7.—Model flexural profiles and observed basin profiles along Line E (Fig. 7A), Line F (Fig. 7B), and Line G (Fig. 7C). Load geometry of the Fort Stockton Block is shown in Figure 3. Decompacted synorogenic strata were used to reconstruct the synorogenic basin profiles; error bars on control points along the observed profiles reflect maximum and minimum decompaction constants applied to each point. Different profiles and load geometries were generated by using the flexural rigidities indicated in each profile. Note on Line E (Fig. 7A) that the best estimated rigidities for the model flexural profiles are between 10^{22} N-m and 10^{23} N-m and that the load from the CBP is insufficient to create subsidence in the proximal part of the Delaware Basin and in the boundary fault zone. Note on Line F (Fig. 7B) that the best estimated rigidity for the model flexural profile is 10^{22} N-m and that the load from the CBP is insufficient to create subsidence in the distal part of the Delaware and the Midland Basins. Note on Line G (Fig. 7C) that the best estimated rigidity for the model flexural profile is 10^{21} N-m and that the load from the CBP is insufficient to create subsidence in the distal part of the Delaware Basin. See text for additional discussion of the different profiles.

LINE G

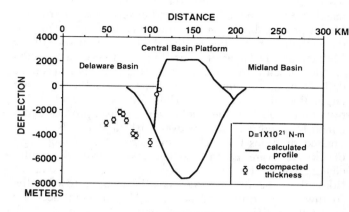

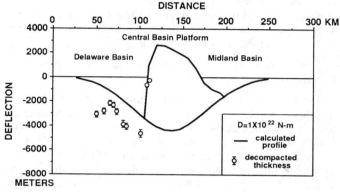

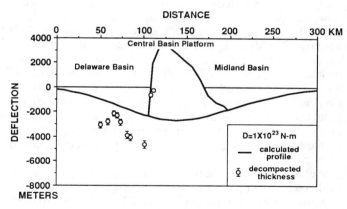

Fig. 7.—Continued.

tween the deflection in the Delaware and Midland Basins diminishes if increasingly higher rigidity values are used to model the flexural profile (Fig. 7). The deflection in the deepest part of both basins is almost equal if the rigidity of lithosphere beneath the basins is larger than 10^{23} N-m (Fig. 7).

DISCUSSION

Our flexural models provide a reasonable explanation for some aspects of the geometries of the sub-basins that comprise the Permian Basin region. The fit between our model results and observed basin profiles is affected by the appropriate selection of parameters used in the flexural calculations, the chosen boundary conditions, and limitations in the data set used to construct the basin profiles.

Potential Sources of Error in the Two-dimensional Flexural Models

There are several possible sources of error in our model results. Changing the parameters and boundary conditions for the two-dimensional flexural models causes the largest differences in the proximal parts of sub-basins in the Permian Basin. These differences would decrease away from the load in any of the models. This is important when comparing discrepancies between the calculated and observed basin geometries.

The first assumption that we made is that the lithosphere beneath the Permian Basin area behaved as an unbroken elastic plate. However, as suggested by previous studies, preexisting zones of weakness in the Marathon foreland may have been reactivated during late Paleozoic deformation to form the CBP and the adjacent basins (Walper, 1977; Hills, 1984; Wuellner and others, 1986; Keller and others, 1989). If these zones of weakness actually are faults that extend through the entire lithosphere, it would be better to model the lithosphere beneath the Permian Basin area as a broken plate. The geometry of the flexural deflection caused by the CBP, however, would be quite different from our model results if the lithosphere is treated as a broken plate (cf. Turcotte and Shubert, 1982). Unfortunately, we do not have deep reflection seismic data that might constrain the location of lithospheric-scale faults across the Permian Basin which might affect flexure over this region. It is also important to note that there are a few high-angle basement faults with vertical displacement in the basinal areas (Hills, 1970; Walper, 1977; GEOMAP, 1983; Gardiner, 1990; Ewing, 1991). There may be some component of shortening associated with these faults that may produce local topographic loads, and ignoring any possible shortening associated with these faults may lead to an underestimation of the flexure along the profile.

Another source of error is the loading geometry of the CBP used in flexural models of the Midland and Delaware Basins. Lower Paleozoic strata that were eroded from the top of the CBP when it was uplifted during the Marathon-Ouachita orogeny were not removed on the restored structural profiles (Fig. 3). The difference between the deflection in the proximal Delaware Basin using the uneroded structural profile versus the actual profile is about 10% of the total deflection (Fig. 8), a relatively negligible amount. Given the uncertainty in the estimation of the basement load, the potential error introduced by using eroded versus reconstructed stratigraphy on top of the CBP seems even less critical.

The estimated basement density used in flexural models for the Midland and Delaware Basins may also be a source of error. Several wells penetrate basement rocks on the Fort Stockton Block; basement lithology varies from granitic to mafic rocks with densities between 2.78 to 3.08 gm/cc (Keller and others, 1989). The value used in our models is

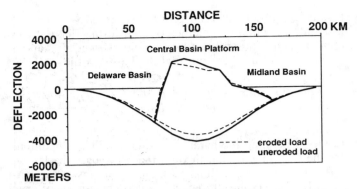

Fig. 8.—Comparison between the model profiles using eroded and uneroded lower Paleozoic strata on the Central Basin Platform. Note that the difference in the model basin profiles is less than 10%, which we consider to be negligible. See text for discussion of the profile.

derived from gravity models for the Delaware Basin and CBP (Keller and others, 1980) which suggest basement density beneath the CBP is on average about 2.75 gm/cc. The gravity models also suggest a mass of mafic rocks with a density of 3.0 gm/cc is located beneath the CBP. Therefore, a basement density of 2.75 gm/cc may be conservative. If a higher basement density for the CBP is used, a deeper deflection would be produced in the adjacent basins.

The observed basin profiles were constructed by using decompacted thicknesses of upper Pennsylvanian through Wolfcampian strata. Early cementation of these basinal strata might have prevented the maximum amount of compaction possible. Applying maximum decompaction values to these strata would overestimate the original stratal thicknesses. We have attempted to avoid these problems by using decompaction constants that are half-way between those that assume total early cementation and those that assume no early cementation. Error bars on the decompacted sediment thicknesses shown in Figures 5 and 7 indicate that the selection of a decompaction constant is not very significant for generating a better fit to the model profiles.

Comparison of the Theoretical and Observed Basin Profiles

The model basin profiles can be compared to the observed profiles to evaluate the effects of various loads on the adjacent basins (Figs. 5, 7). The model profiles accurately predict several characteristics of the observed basin profiles: (1) the asymmetrical Delaware Basin deepens toward the western side of the CBP whereas the more symmetrical Midland Basin is located east of the CBP; (2) there is a larger vertical deflection in the proximal part of the Delaware Basin than in the Midland Basin; and (3) the Val Verde Basin progressively narrows from east to west along strike.

Estimation of Flexural Rigidity Across the Permian Basin.—

Appropriate values for the rigidity of lithosphere beneath the Permian Basin can also be obtained by comparing the theoretical and observed profiles (Figs. 5, 7). Basin width and maximum deflection in the proximal part of a flexural basin are two important criteria for determining the best-fit match between model profiles and observed basin profiles. Model profiles based on rigidities less than 10^{21} N-m produce much narrower basins than the observed widths of the Midland and Delaware Basins (Fig. 7). In contrast, model profiles based on rigidities greater than 10^{23} N-m produce deflections in the central part of the Midland Basin that are larger than those observed. Thus, the best overall fit between the model profiles and observed basin profiles is obtained by using rigidities between 10^{21} and 10^{23} N-m. Estimated rigidities for each profile across the southern part of the Delaware Basin also decrease toward the south, changing from 10^{23} N-m to 10^{21} N-m, indicating that the lithosphere becomes progressively weaker toward the southwestern terminus of the CBP. This apparent decrease in flexural rigidity toward the southwest corner of the CBP is also supported by our model results for flexure in the Val Verde Basin, which suggest that rigidity decreases from east to west along the axis of the basin (Fig. 5).

Composite Tectonic Loads and Flexure Across the Permian Basin Region.—

Some differences between the theoretical and observed profiles can not be explained by our choices for parameters used in the calculations. The greatest differences are found mostly in distal parts of the sub-basins where the observed deflection is larger than that predicted by the models. For example, a better fit to the vertical deflection in *distal* parts of the basins is produced along Lines E, F, and G by using a rigidity greater than 10^{23} N-m. Using higher rigidities, however, produces an unrealistically wide Delaware Basin along these cross sections. In addition, the model profiles for both the Midland and Delaware Basins become more similar if larger rigidities are used. This clearly does not match the observed profiles where the Delaware Basin is more asymmetric and deeper overall than the Midland Basin (Fig. 7). The greatest differences between observed and model profiles across the Val Verde Basin also are located in the distal part of the basin where the forebulge created by the Marathon orogenic belt apparently has been removed (Fig. 5).

We attribute the observed profiles in some parts of the Midland, Val Verde, and Delaware Basins to flexural interactions between additional tectonic loads from surrounding areas. Subsidence produced by these additional loads depends on how far flexure due to these loads can reach into the basin. North-south stratigraphic cross sections down the axis of the Delaware Basin show that upper Pennsylvanian through Wolfcampian strata thicken slightly toward the Marathon orogenic belt, except in the area immediately adjacent to the thrust front where the thickness of synorogenic strata increases abruptly (Fig. 9; Yang and Dorobek, this volume). This thickening of strata is very different from the observed synorogenic profiles for the Val Verde Basin (Fig. 6) where upper Pennsylvanian through Wolfcampian strata thicken dramatically toward the Marathon orogenic belt. The calculated flexural rigidities of the lithosphere beneath the Delaware and Val Verde Basins are similar so the width of the Val Verde Basin can be used to infer the flex-

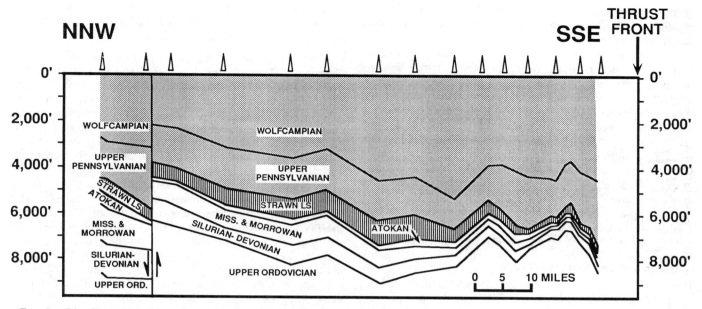

Fig. 9.—Line H, stratigraphic cross section from the southern part of the Delaware Basin. Cross section extends perpendicular to the frontal thrust of the Marathon orogenic belt into the southern part of the Delaware Basin. Note that synorogenic strata do not thicken dramatically toward the Marathon orogenic belt.

ural effects in the Delaware Basin that were caused by the Marathon orogenic belt. The distance between the frontal thrust of the Marathon orogenic belt and the area immediately west of the Fort Stockton Block is similar to the width of the Val Verde Basin. Therefore, the additional deflection at the *northern end* of the Delaware Basin profile (Fig. 9) must be due to the additional load of the CBP, and essentially none of the deflection is produced by the Marathon orogenic belt.

Structures associated with the Diablo Platform west of the Delaware Basin may provide additional loads that affected Pennsylvanian to Early Permian subsidence in the western part of the Delaware Basin. The boundary between the Diablo Platform and the Delaware Basin is characterized by high angle faults and a monoclinal structure called the Huapache monocline (GEOMAP, 1983; McKnight, 1983; Ewing, 1991). Many of these structures formed during Pennsylvanian to late Wolfcampian time (McKnight, 1983), coeval with synorogenic subsidence in the Delaware Basin. Crustal shortening near the boundary of the Diablo Platform and the resultant topographic load might have caused flexure in the adjacent Delaware Basin. In summary, we suggest that the geometry of the southern Delaware Basin is a composite of the flexural effects caused by the Marathon orogenic belt, southern end of the CBP, and possibly some cryptic structures along the western side of the basin.

The Fort Stockton Block is not the only part of the CBP that acted as a load and caused subsidence in the Delaware Basin. Subsidence in central parts of the Midland and Delaware Basins is probably also influenced by loading at the reverse-faulted SW and NE corners of the Andector Block. Flexure caused by uplifted corners of the Andector Block probably interfered with flexure produced by the Fort Stockton Block. We did not, as discussed below, attempt three-dimensional flexural analyses to test this hypothesis.

Differences between the observed and model profiles from eastern parts of the Midland Basin are not easily attributed to additional topographic loads because there are no obvious features that indicate late Paleozoic crustal shortening occurred along the eastern side of the Midland Basin. These differences also can not be explained by loading from the Marathon orogenic belt because this part of the Midland Basin is far north of the flexural profile produced by the Marathon orogenic belt. Shortening along the Fort Chadbourne fault zone may contribute loads that affected subsidence in eastern parts of the Midland Basin. The Fort Chadbourne fault zone, however, is generally considered to be a zone of strike-slip deformation (Ewing, 1991), and it is difficult to identify much shortening along this fault zone. Minor amounts of crustal shortening might also be associated with an E-W trending, high angle fault along the southern margin of the Midland Basin that has been interpreted as a regional strike-slip fault (Gardiner, 1990). Shortening along this fault could form an additional load and produce subsidence north of it. However, the distance between this fault and much of the rest of the Midland Basin is too great to explain significant amounts of the subsidence in those areas. In summary, the cause and tectonic implications of subsidence in the eastern Midland Basin remain enigmatic and need to be studied further.

Variation in Lithospheric Rigidity across the Permian Basin Region

Our model results not only demonstrate that late Paleozoic subsidence in the sub-basins of the Permian Basin is at least partly flexural in origin, but they also suggest that rigidity of lithosphere was highly variable across the Marathon foreland.

The rigidity of oceanic lithosphere is mainly controlled by its age or thermal state (Watts and others, 1982). Factors that affect the rigidity of continental lithosphere, however, are still controversial (Karner and others, 1983; Willett and others, 1985), primarily owing to the complex mechanical properties of continental lithosphere (Goetze and Evans, 1979; Kirby, 1983). McNutt and others (1988) have shown that the rigidity of continental lithosphere is directly related to the curvature of the orogenic belt that acts as a topographic load. Highly curved orogenic belts in map view generate highly bent lithosphere on vertical profiles. Highly bent lithosphere, in turn, induces high bending stresses which further weaken the flexed plate.

Our flexural analyses suggest that the lithosphere near the southwestern corner of the CBP has the lowest rigidity in the Permian Basin (Figs. 5, 7). Here, the Fort Stockton Block is characterized by the greatest amount of vertical uplift along any margin of the CBP and by the greatest amount of crustal shortening across the platform. This area of apparently weak lithosphere also is adjacent to the sharply curved salient of the Marathon fold-and-thrust belt. In map view, the angle between the two limbs of the salient is about 90° (Fig. 2). Coincidence of the sharp bend in the Marathon orogenic belt and maximum crustal deformation in the foreland area suggests a mechanical and causal relationship. The arcuate geometry of the foredeep basin around the prominent salient of the Marathon orogenic belt may have been produced by very high bending stresses that contributed to the lower flexural rigidity in this area.

Lastly, it is important to note that the forebulge north of the Val Verde Basin along Lines C and D seems to have stayed in more or less the same place from pre-Desmoinesian to late Wolfcampian time (~30–35 my). This is indicated by the coincidence between the area of maximum erosion of pre-Desmoinesian strata and the region of maximum thinning of upper Pennsylvanian through upper Wolfcampian strata (Fig. 6). These stratigraphic relationships indicate that as the Marathon orogenic belt advanced northward from middle Pennsylvanian to late Wolfcampian time, the forebulge apparently did not migrate with it. Recent models of forebulge behavior by Waschbusch and Royden (1992) suggest that the position of a forebulge may stay fixed as an orogenic wedge advances toward it if there are preexisting lateral strength variations in the flexed lithosphere. We can not rule this out as a possibility for the behavior of the forebulge north of the Val Verde Basin. During late Pennsylvanian to Wolfcampian time, however, additional loads formed to the north of this forebulge and apparently caused yoking of different flexural profiles. In effect, the forebulge produced by the Marathon orogenic belt was removed by the flexural profile caused by topographic loads to the north. Thus, the apparently fixed position of the forebulge may be entirely fortuitous and is unrelated to inherent lateral strength variations across the Marathon foreland.

Rationale for Using a Two-dimensional Flexural Model versus a Three-dimensional Flexural Model for the Permian Basin Region

As discussed above, it is likely that the geometries of the Delaware, Midland, and Val Verde Basins reflect flexure due to several distributed loads along various sides of each basin. Quinlan and Beaumont (1984) and Beaumont and others (1988) have shown, however, that variations in the subsidence and transverse flexural profiles of some foreland basins can be modeled by varying the magnitude of distributed loads along the strike of the orogenic belt. Similar variations in foreland basin profiles can be obtained, however, if the rigidity of the foreland plate varies along strike. We chose to do two-dimensional flexural analyses with variable plate rigidities along strike to model flexure across the Permian Basin because of the simple observation that the thickness of synorogenic strata in the proximal part of the Val Verde Basin is nearly constant along the axis of that basin (Fig. 5, Lines A-C). This constant stratal thickness suggests that the distributed load in the Marathon orogenic belt was rather uniform, and that the synorogenic, transverse basin profiles were due instead to lateral variations in plate rigidity. Thus, we opted for this simpler approach in our flexural models.

CONCLUSIONS

Flexural models provide an important test for our hypothesis that the Permian Basin is a composite foreland area caused by the combined effects of several different topographic loads. Results from our flexural models also suggest that the rigidity of the lithosphere across the Marathon foreland may have been variable. The main conclusions from this study include:

1. Reconstructed structural profiles that were used for flexural models show that the CBP is a wedge-shaped, uplifted feature in cross section. Its eastern edge extends into the central part of the Midland Basin, and its western side is locally characterized by subparallel, west-verging, basement-involved thrust and reverse faults. The wedge-shaped, uplifted block was treated as a distributed topographic load that caused deflection in the adjacent basins. Model profiles are similar to reconstructed synorogenic profiles of these basins which show a deeper, asymmetrical Delaware Basin and a shallower, more symmetrical Midland Basin.
2. Comparison between the model and observed profiles suggests that deflection caused by the load of the CBP could have produced most of the synorogenic subsidence in the adjacent Midland and Delaware Basins. The model results also show that there would be no significant difference in the geometries of the Delaware and Midland Basins if the flexural rigidity of the lithosphere beneath the entire Permian Basin region was larger than 10^{23} N-m. Thus, the best estimate of lithospheric rigidity across the Delaware and Midland Basins is between 10^{21} and 10^{23} N-m. Estimated rigidities obtained from flexural profiles for the Val Verde Basin are also within this range.

3. Generally, there is a greater deflection in the distal parts of the observed profiles than can be produced by the model loads. In the Delaware Basin, the additional deflection might be produced by structures associated with the Diablo Platform that provided an additional topographic load on the western side of the Delaware Basin. The synorogenic Delaware Basin profile probably was produced largely by loading from the CBP and Diablo Platform area, with the only exception to this being the area immediately adjacent to the Marathon orogenic belt. The additional deflection in distal parts of the observed Midland Basin profiles can not be attributed to any tectonic loads along the eastern side of the basin as there are no well-documented thrust faults there. The exact causes and tectonic implications of the additional subsidence are still unknown.

4. The width of the Val Verde Basin decreases from east to west. In its western parts, the Val Verde Basin is bordered to the north by the most uplifted corner of the CBP and to the south by the prominent salient in the Marathon orogenic belt. The best-fit model profiles for the western part of the Val Verde Basin and for the southern part of the Delaware Basin are obtained by using the lowest values for lithospheric rigidity in the entire Permian Basin region. Weak lithosphere in this part of the Permian Basin might be related to inherent lateral strength variations in the foreland lithosphere prior to the Marathon-Ouachita orogeny. Alternatively, high bending stresses may have been generated in the Marathon foreland when the prominent salient in the Marathon orogenic belt was thrust onto the southern part of Laurasia during late Paleozoic time. These high bending stresses may have further weakened the footwall plate and caused the apparently lower rigidity in this area.

ACKNOWLEDGMENTS

Chevron U. S. A. donated the seismic profiles used in this study. We would like to thank Ron Genter at Chevron for helping us acquire the data and for giving us permission to publish it. This study was supported by the American Chemical Society, Petroleum Research Fund (Grant #23269-AC2 to S. L. Dorobek), the National Science Foundation (Grant #EAR-9117935 to S. L. Dorobek), and the Center for Energy and Mineral Resources at Texas A&M University. Additional support was provided by the American Association of Petroleum Geologists Grant-in-Aid to K.-M. Yang. We also thank Garry Quinlan for providing constructive comments which improved the manuscript.

REFERENCES

BEAUMONT, C., 1981, Foreland basins: Geophysical Journal of the Royal Astronomical Society, v. 65, p. 291–329.

BEAUMONT, C., QUINLAN, G., AND HAMILTON, J., 1988, Orogeny and stratigraphy: numerical models of the Paleozoic in the eastern interior of North America: Tectonics, v. 7, p. 389–416.

BECK, R. B., VONDRA, C. F., FILKINS, J. E., AND OLANDER, J. D., 1988, Syntectonic sedimentation and Laramide basement thrusting, Cordilleran foreland; timing of deformation, in Schmidt, C. J., and Perry, W. J., Jr., eds., Interaction of the Rocky Mountain Foreland and the Cordilleran Thrust Belt: Boulder, Geological Society of America Memoir 171, p. 465–487.

BROWN, W. G., 1984, Basement Involved Tectonics: Foreland Areas: Tulsa, American Association of Petroleum Geologists Continuing Education Course Note Series 26, 92 p.

DENISON, R. E., 1989, Foreland structure adjacent to the Ouachita foldbelt, in Hatcher, R. D., Jr., Thomas, W. A., and Viele, G. W., eds., The Appalachian-Ouachita Orogen in the United States: Boulder, Geological Society of America, The Geology of North America, v. F-2, p. 681–688.

DEWEY, J. F. AND PITMAN, W. C., III, 1982, Late Palaeozoic basins of the southern U. S. continental interior (abs.), in The Evolution of Sedimentary Basins: Philosophical Transactions, Royal Society of London, series A, v. 305, p. 145–148.

ELAM, J. G., 1984, Structural systems in the Permian Basin: West Texas Geological Society Bulletin, v. 24, p. 7–10.

ERSLEV, E. A., 1986, Basement balancing of Rocky Mountain foreland uplifts: Geology, v. 14, p. 258–262.

EWING, T. E., 1984, Late Paleozoic structural evolution of the Permian Basin (abs.): American Association of Petroleum Geologists Bulletin, v. 68, p. 474–475.

EWING, T. E., 1990, The tectonic map of Texas: Austin, Bureau of Economic Geology, The University of Texas at Austin.

EWING, T. E., 1991, The tectonic framework of Texas: Text to accompany "The Tectonic Map of Texas": Austin, Bureau of Economic Geology, The University of Texas at Austin, 36 p.

FLAWN, P. T., GOLDSTEIN, A., JR., KING, P. B., AND WEAVER, C. E., 1961, The Ouachita System: Austin, Bureau of Economic Geology, University of Texas, Publication No. 6120, 401 p.

FLEMINGS, P. B., JORDAN, T. E., AND REYNOLDS, S. A., 1986, Flexural analysis of two broken foreland basins: The Late Cenozoic Bermejo Basin and the Early Cenozoic Green River Basin (abs.): American Association of Petroleum Geologists Bulletin, v. 70, p. 591.

FLEMINGS, P. B. AND JORDAN, T. E., 1989, A synthetic stratigraphic model of foreland basin development: Journal of Geophysical Research, v. 94, p. 3851–3866.

GALLEY, J. E., 1958, Oil and geology in the Permian Basin of Texas and New Mexico, in Weeks, L. G., ed., Habitat of Oil—A Symposium: Tulsa, American Association of Petroleum Geologists, p. 395–446.

GARDINER, W. B., 1990, Fault fabric and structural subprovinces of the Central Basin Platform: A model for strike-slip movement, in Flis, J. E., and Price, R. C., eds., Permian Basin Oil and Gas Fields: Innovative Ideas in Exploration and Development: Midland, West Texas Geological Society, Publication No. 90–87, p. 15–27.

GEOMAP, 1983, Pre-Pennsylvanian Subcrop Map of the Permian Basin of West Texas and Southeast New Mexico: Plano, GEOMAP EXECUTIVE REFERENCE MAP 502.

GOETZE, C. AND EVANS, B., 1979, Stress and temperature in the bending lithosphere as constrained by experimental rock mechanics: Geophysical Journal of the Royal Astronomical Society, v. 59, p. 463–478.

HAGEN, E. S., SHUSTER, M. W., AND FURLONG, K. P., 1985, Tectonic loading and subsidence of intermontane basins: Wyoming foreland province: Geology, v. 13, p. 585–588.

HANDSCHY, J. W., KELLER, G. R., AND SMITH, K. J., 1987, Ouachita System in northern Mexico: Tectonics, v. 6, p. 323–330.

HANSON, B. M., POWERS, B. K., GARRETT, C. M., JR., MCGOOKEY, D. E., MCGLASSON, E. H., HORAK, R. L., MAZZULLO, S. J., REID, A. M., CALHOUN, G. G., CLENDENING, J., AND CLAXTON, B., 1991, The Permian basin, in Gluskoter, H. J., Rice, D. D., and Taylor, R. B., eds., Economic Geology, U. S.: Boulder, Geological Society of America, The Geology of North America, v. P-2, p. 339–356.

HETENYI, M., 1946, Beams on Elastic Foundation: Ann Arbor, The University of Michigan Press, 255 p.

HILLS, J. M., 1970, Paleozoic structural directions in southern Permian Basin, west Texas and southeastern New Mexico: American Association of Petroleum Geologists Bulletin, v. 54, p. 1809–1827.

HILLS, J. M., 1984, Sedimentation, tectonism, and hydrocarbon generation in Delaware Basin, west Texas and southeastern New Mexico: American Association of Petroleum Geologists Bulletin, v. 68, p. 250–267.

HILLS, J. M., 1985, Structural evolution of the Permian Basin of west Texas and New Mexico, in Dickerson, P. W., and Muehlberger, W. R., eds., Structure and Tectonics of Trans-Pecos Texas: Midland, West Texas Geological Society Publication 85-81, p. 89–99.

HORAK, R. L., 1985, Trans-Pecos tectonism and its effect on the Permian Basin, *in* Dickerson, P. W., and Muehlberger, W. R., eds., Structure and Tectonics of Trans-Pecos Texas: Midland, West Texas Geological Society Publication 85-81, p. 81-87.

JORDAN, T. E., 1981, Thrust loads and foreland basin development, Cretaceous, western United States: American Association of Petroleum Geologists Bulletin, v. 65, p. 2506-2520.

JORDAN, T. E., FLEMINGS, P. B., AND BEER, J. A., 1988, Dating thrust-fault activity by use of foreland-basin strata, *in* Kleinspehn, K. L., and Paola, C., eds., New Perspectives in Basin Analysis: New York, Springer-Verlag, p. 307-330.

KARNER, G. D., STECKLER, M. S., AND THORNE, J. A., 1983, Long-term thermo-mechanical properties of the lithosphere: Nature, v. 304, p. 250-252.

KELLER, G. R., HILLS, J. M., AND DJEDDI, R., 1980, A regional geological and geophysical study of the Delaware Basin, New Mexico and west Texas: Socorro, New Mexico Geological Society Guidebook, 31st Field Conference, Trans-Pecos Region, p. 105-111.

KELLER, G. R., HILLS, J. M., BAKER, M. R., AND WALLIN, E. T., 1989, Geophysical and geochronological constraints on the extent and age of mafic intrusions in the basement of west Texas and eastern New Mexico: Geology, v. 17, p. 1049-1052.

KIRBY, S. H., 1983, Rheology of the lithosphere: Reviews of Geophysics and Space Physics, v. 21, p. 1458-1487.

KLUTH, C. F. AND CONEY, P. J., 1981, Plate tectonics of the Ancestral Rocky Mountains: Geology, v. 9, p. 10-15.

LAROCHE, T. M. AND HIGGINS, L., eds., 1990, Marathon Thrust Belt: Structure, Stratigraphy, and Hydrocarbon Potential: Midland, West Texas Geological Society and Permian Basin Section-SEPM Field Seminar, 148 p.

LOWELL, J. D., 1983, Foreland Deformation, *in* Lowell, J. D., ed., Rocky Mountain Foreland Basins and Uplifts: Denver, Rocky Mountain Association of Geologists, p. 243-256.

MCBRIDE, E. F., 1978, Tesnus and Haymond Formations – siliciclastic flysch, *in* Mazzullo, S. J., ed., Tectonics and Paleozoic Facies of the Marathon Geosyncline, West Texas: Midland, Permian Basin Section, Society of Economic Paleontologists and Mineralogists, Publication 78-17, p. 131-148.

MCBRIDE, E. F., 1989, Stratigraphy and sedimentary history of pre-Permian Paleozoic rocks of the Marathon uplift, *in* Hatcher, R. D., Jr., Thomas, W. A., and Viele, G. W., eds., The Appalachian-Ouachita Orogen in the United States: Denver, Geological Society of America, The Geology of North America, v. F-2, p. 603-620.

MCCONNELL, D. A., 1989, Determination of offset across the northern margin of the Wichita uplift, southwest Oklahoma: Geological Society of America Bulletin, v. 101, p. 1317-1332.

MCCONNELL, D. A., GOYDAS, M., SMITH, G. N., AND CHITWOOD J. P., 1990, Morphology of the Frontal fault zone, southwest Oklahoma: Implications for deformation and deposition in the Wichita uplift and Anadarko basin: Geology, v. 18, p. 634-637.

MCKNIGHT, C. L., 1983, Structural evolution of the Guadalupe Mountains, southern-central New Mexico and west Texas (abs.): American Association of Petroleum Geologists Bulletin, v. 67, p. 511.

MCNUTT, M. K., DIAMENT, M., AND KOGAN, M. G., 1988, Variations of elastic plate thickness at continental thrust belt: Journal of Geophysical Research, v. 93, p. 8825-8838.

MOORE, G. E., MENDENHALL, G. V., AND SAULTZ, W. L., 1981, Northern extent of Marathon thrust Elsinore area, Pecos County, Texas, *in* Jons, R., ed., Marathon-Marfa Region of West Texas Symposium and Guidebook: Midland, Permian Basin Section, Society of Economic Paleontologists and Mineralogists, 81-20, p. 131-133.

QUINLAN, G. M. AND BEAUMONT, C., 1984, Appalachian thrusting, lithospheric flexure and the Paleozoic stratigraphy of the eastern interior of North America: Canadian Journal of Earth Sciences, v. 21, p. 973-996.

REED, T. A. AND STRICKLER, D. L., 1990, Structural geology and petroleum exploration of the Marathon thrust belt, west Texas, *in* Laroche, T. M., and Higgins, L., eds., Marathon Thrust Belt: Structure, Stratigraphy, and Hydrocarbon Potential: Midland, West Texas Geological Society and Permian Basin Section-Society of Economic Paleontologists and Mineralogists, 1990, p. 39-64.

ROSS, C. A., 1986, Paleozoic evolution of southern margin of Permian basin: Geological Society of America Bulletin, v. 97, p. 536-554.

SCLATER, J. G. AND CHRISTIE, P. A. F., 1980, Continental stretching: an explanation of the post-mid-Cretaceous subsidence of the central North Sea basin: Journal of Geophysical Research, v. 85, p. 3711-3739.

SHUMAKER, R. C., 1992, Paleozoic structure of the Central Basin uplift and adjacent Delaware Basin, west Texas: American Association of Petroleum Geologists Bulletin, v. 76, p. 1804-1824.

SHURBET, D. H. AND CEBULL, S. E., 1980, Tobosa-Delaware basin as an aulacogen: Texas Journal of Science, v. 32, p. 17-21.

SHUSTER, M. W. AND STEIDTMANN, J. R., 1988, Tectonic and sedimentary evolution of the northern Green River basin, western Wyoming, *in* Schmidt, C. J., and Perry, W. J., Jr., eds., Interaction of the Rocky Mountain Foreland and the Cordilleran Thrust Belt: Boulder, Geological Society of America Memoir 171, p. 515-530.

SMITHSON, S. B., BREWER, J. A., KAUFMAN, S., OLIVER, J. E., AND HURICH, C. A., 1979, Structure of the Laramide Wind River uplift, Wyoming, from, COCORP deep reflection data and from gravity data: Journal of Geophysical Research, v. 84, p. 5955-5971.

SPEED, R. C. AND SLEEP, N. H., 1982, Antler orogeny and foreland basin: A model: Geological Society of America Bulletin, v. 93, p. 815-828.

STEARNS, D. W., 1978, Faulting and forced folding in the Rocky Mountain foreland, *in* Matthews, V., III, ed., Laramide Folding Associated with Basement Block Faulting in the Western United States: Denver, Geological Society of America Memoir 151, p. 1-37.

STOCKMAL, G. S., BEAUMONT, C., AND BOUTILIER, R., 1986, Geodynamic models of convergent tectonics: the transition from rifted margin to overthrust belt and consequences for foreland-basin development: American Association of Petroleum Geologists Bulletin, v. 70, p. 181-190.

STONE, D. S., 1984, The Rattlesnake Mountain, Wyoming debate: A review and critique of models: Mountain Geologist, v. 21, p. 37-46.

THOMAS, W. A., 1983, Continental margins, orogenic belts, and intracratonic structures: Geology, v. 11, p. 270-272.

TURCOTTE, D. L. AND SCHUBERT, G., 1982, Geodynamics—Applications of Continuum Physics to Geological Problems: New York, John Wiley and Sons, 450 p.

WALPER, J. L., 1977, Paleozoic tectonics of the southern margin of North America: Gulf Coast Association of Geological Societies Transactions, v. 27, p. 230-241.

WASCHBUSCH, P. J. AND ROYDEN, L. H., 1992, Episodicity in foredeep basins: Geology, v. 20, p. 915-918.

WATTS, A. B., KARNER, G. D., AND STECKLER, M. S., 1982, Lithospheric flexure and the evolution of sedimentary basins: Philosophical Transactions, Royal Society of London, series A, v. 305, p. 249-281.

WILLETT, S. D., CHAPMAN, D. S., AND NEUGEBAUER, H. J., 1985, A thermomechanical model of continental lithosphere: Nature, v. 314, p. 520-523.

WUELLNER, D. E., LEHTONEN, L. R., AND JAMES, W. C., 1986, Sedimentary-tectonic development of the Marathon and Val Verde basins, West Texas, U. S. A.: a Permo-Carboniferous migrating foredeep, *in* Allen, P., and Homewood, P., eds., Foreland Basins: Oxford, International Association of Sedimentologists Special Publication 8, p. 15-39.

YANG, K.-M., 1993, Late Paleozoic Synorogenic Stratigraphy, Tectonic Evolution, and Flexural Modeling of the Permian Basin, West Texas and New Mexico: Unpublished Ph. D. Dissertation, Texas A&M University, College Station, 142 p.

YANG, K.-M. AND DOROBEK, S. L., 1991, The tectonic mechanism for uplift and rotation of crustal blocks in the Central Basin Platform, Permian Basin, Texas and New Mexico (abs.): American Association of Petroleum Geologists Bulletin, v. 75, p. 698.

YANG, K.-M. AND DOROBEK, S. L., 1992, Mechanisms for late Paleozoic synorogenic subsidence of the Midland and Delaware Basins, Permian Basin, Texas and New Mexico, *in* Mruk, D., and Curran, B., eds., Permian Basin Exploration and Production Strategies–Applications of Sequence Stratigraphic and Reservoir Characterization Concepts: Midland, West Texas Geological Society, 1992 Fall Symposium, p. 45-60.

PART II
PROVENANCE STUDIES IN FORELAND BASINS

PROVENANCE OF MUDSTONES FROM TWO ORDOVICIAN FORELAND BASINS IN THE APPALACHIANS

C. BRANNON ANDERSEN
Department of Geology, Heroy Geology Laboratory, Syracuse University, Syracuse, NY 13244

ABSTRACT: Mudstones from the Taconic and Blountian foreland basins were analyzed for whole-rock chemical composition and clay mineral composition. These foreland basins formed during the middle and late Ordovician when exotic terranes collided with Laurentia. The purpose was to determine if the mudstones record first-order trends in provenance that are related to tectonic history.

In both basins, the ratio of chlorite to illite and the ratio of the concentration of three "mafic" elements (Ti, Cr, Ni) to Nb (a "felsic" element) increases with time. However, mudstones from the Taconic foreland basin have a higher proportion of Ti, Cr, and Ni than those from the Blountian foreland basin. Results from the Blountian foreland basin showed the greater amount of scatter.

The outboard terranes that collided with Laurentia were the most important sources of siliciclastic sediment because carbonate platforms fringed the continental margin of each basin. Thus, compositional trends in each basin reflect an increase in the proportion of sediment eroded from mafic source rocks within the colliding terrane. In presently accepted tectonic models for the Taconic foreland basin, the colliding terrane is an arc system. The provenance signature in the mudstones suggests the sediment source shifts from a non-magmatic outer arc to the inner volcanic arc during the collision. The comparatively lower concentration of mafic elements in the Blountian foreland basin mudstones may indicate that the colliding terrane was composite or that its angle of convergence was more oblique.

INTRODUCTION

The tectonics and stratigraphy of peripheral-margin foreland basins, which form during attempted subduction of continental crust associated with arc-continent collision, are relatively well understood (Dickinson, 1977; Jacobi, 1981; Quinlan and Beaumont, 1984; Covey, 1986; Tankard, 1986). During collision, the outboard terrane should be a major source of sediment to the foreland basin. However, few studies have examined the relationship between sediment provenance indicators and tectonic evolution of the outboard terrane. Such relationships may be particularly important in Paleozoic foreland basins where the outboard terrane may be obscured due to later tectonic events.

Provenance studies of foreland basins have focused mainly on sandstones (e.g., McBride, 1962; Rowley and Kidd, 1981; Mack, 1985), but recent work by Schwab (1986, 1991) indicates that the framework grain composition of sandstones does not constrain tectonic models very well. In part, this may be due to diagenetic processes that remove or alter framework grains (e.g., McBride, 1985; Grantham and Velbel, 1988; Milliken and others, 1989; Johnson and Stallard, 1989; Girty, 1991; Ramseyer and others, 1992). Mudstones therefore provide at least three major advantages over sandstones for provenance studies: (1) mudstones are the predominant rock type of foreland basin sequences, (2) mudstones are less subject to changes in bulk composition during diagenesis (Shaw, 1956), and (3) mudstones better reflect the *average* composition of regional source areas (Heller and Frost, 1988).

The present study focuses on provenance of mudstones from the Ordovician Blountian and Taconic foreland basins in the eastern United States. In both basins, stratigraphic changes in clay mineral composition and chemical composition point to an increase through time in the proportion of sediment derived from a mafic source. Thus, the composition of mudstones apparently reflects the tectonic evolution of the source terrane. The addition of compositional data helps refine the tectonostratigraphic model of Ettensohn (1991).

TECTONIC SETTING

The temporal and spatial distribution of Ordovician foreland basins in the Appalachians of the eastern United States indicates that the "Taconic" orogeny was diachronous (Rodgers, 1971). Drake and others (1989) subdivided the orogeny into the Blountian and Taconic phases (Fig. 1). The Blountian foreland basin sequence (Fig. 2A) formed during the early Middle Ordovician (*Nemagraptus gracilus* graptolite zone) in response to collision of an exotic terrane with the Virginia promontory (Kay, 1942; Rodgers, 1953). The Taconic foreland basin sequence formed somewhat later, in the Middle Ordovician (*Corynoides americanus* through *Climacograptus pygmaeus* zones), in response to collision of a magmatic arc with the New York promontory (Riva, 1974; Stanley and Ratcliffe, 1985; Ettensohn, 1991) (Fig. 2B). Carbonate platforms bounded the western edge of both basins; this requires that all siliciclastic sediments must have been derived from outboard terranes (Fig. 3). Subsequent tectonic deformation during the Acadian and Alleghenian orogenies resulted in the present, poor spatial definition of the terranes (Williams and Hatcher, 1983; Zen, 1983; Hatcher, 1987; Drake and others, 1989).

Ettensohn (1991) has developed a tectonostratigraphic model for the Blountian and Taconic foreland basins. This four-part process-response model describes the sedimentary responses to tectonic events. In the initial tectonic phase, a peripheral bulge forms as an outboard terrane collides with the continent. As the bulge migrates towards the interior of the craton, erosion creates a widespread unconformity within the carbonate shelf sequence. During the second tectonic phase, emplacement of thrust sheets onto the lithosphere causes deformational loading and consequent deepening of the basin. The sedimentary record of this deepening is deposition of transgressive carbonates, drowning of the shelf, and burial by siliciclastic muds of the outer-slope facies. During the third tectonic phase, the accreted terrane erodes in response to "loading-type" lithospheric relaxation, and the foreland basin fills with flysch deposits of the basin-axis facies. In the fourth (and final) tectonic phase, unloading-type lithospheric relaxation causes uplift of the eroded

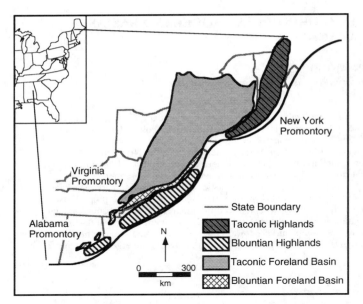

FIG. 1.—Map showing location of Ordovician foreland basins and associated highlands in the southern and central Appalachians. Modified from Ettensohn (1991).

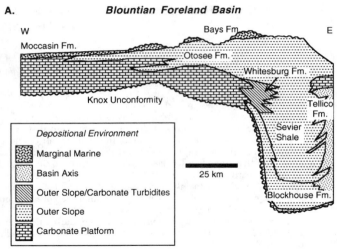

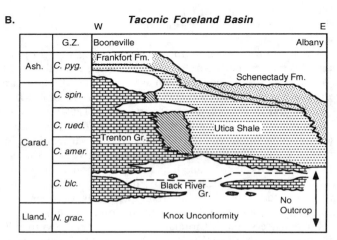

FIG. 2.—Generalized facies of Blountian and Taconic foreland basins. (A) Blountian foreland basin stratigraphy modified from Shanmugam and Walker (1978). (B) Taconic foreland basin stratigraphy modified from Bradley (1989). G.Z. = Graptolite zone.

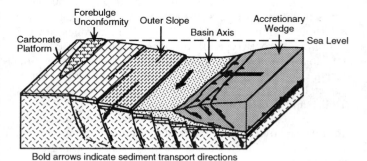

FIG. 3.—Schematic block diagram of a peripheral-margin foreland basin showing the major facies (modified from Bradley, 1989). Dominant directions of sediment transport are indicated by bold arrows (the basin axis transport direction is for the Taconic foreland basin and based on graptolite orientations from Ruedemann, 1897). Transport directions along the basin axis imply the possibility of a more distal source of sediment. The carbonate platform precludes detrital input from the continent.

terrane. The sedimentary response is deposition of subaerial molasse in marginal-marine and terrestiral environments.

Thus, the evolution of the colliding terrane produces a foreland basin with three distinct siliciclastic facies. The question is whether or not mudstone provenance provides additional information about the tectonic evolution of the outboard terrane as it collides with the continent.

METHODS

Sample Locations

The mudstone samples examined in this study represent nearly complete stratigraphic traverses of both the Blountian and Taconic foreland basins (Fig. 4). Although some samples were collected from the marginal-marine and terrestrial redbed facies, the focus is on sediments from the outer-slope and basin-axis facies. Samples from redbed facies have limited value because subaerial diagenesis alters mudstone composition (Thompson, 1970).

Mudstone samples were collected sequentially, though not neccessarily at equally spaced intervals, through each outcrop. At each location, one to two kilograms of fresh sample were collected after the sampling site had been cleared of weathered bedrock. In the laboratory, samples were scrubbed to remove soil and rinsed with distilled water. The mudstone chips then were hand-picked, and any obviously weathered material was discarded. Samples then were powdered for analysis of chemical composition.

Mineral Composition

Preparation of mudstones for clay mineral analysis generally followed the procedures of Moore and Reynolds (1989) and Drever (1973). However, due to the presence of dolomite in many samples, 1N HNO_3 was substituted for pH

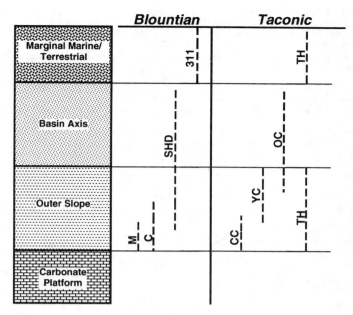

FIG. 4.—Stratigraphic position of sampled outcrops within the foreland basin facies. Blountian foreland basin: M = Blockhouse Formation, Moshiem, TN; C = Rich Valley Formation, Chillhowie, VA; SHD = Blockhouse and Tellico Formations, South Holston Dam, TN; and 311 = Bays Formation, near Blacksburg, VA. Taconic foreland basin: CC = Utica Shale, Canajoharie Creek, NY; YC = Utica Shale, Yatesville Creek, NY; and OC = Utica Shale and Frankfort Formation, Ohisa Creek, NY. TH is a unique location (Thorn Hill, TN) where Taconic foreland basin sediments (Martinsburg Formation and Juniata Formation) onlap directly over sediments of the Blountian foreland basin sequence redbeds (Moccasin Formation).

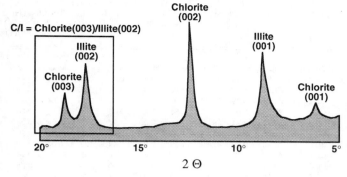

FIG. 5.—Representative X-ray diffraction pattern of a mudstone from Taconic foreland basin outer-slope facies. Chlorite/illite ratios were calculated using the chlorite(003)/illite(002) peaks, as discussed in the text. Samples that are from the outer-slope facies and have a chlorite/illite ratio of zero do not show a chlorite(002) peak. Samples that are from the marginal-marine redbed facies and have a chlorite/illite ratio of zero typically have a very small 7Å peak. This peak disappears after treatment with HCl, indicating that trace chlorite is present.

= 5.0 Na-acetate-buffered acetic acid to dissolve carbonates. Experiments showed that this treatment had no effect on the clay diffraction patterns, a finding consistent with previous studies of clay mineral separation from carbonates (Ostrom, 1961; Cook, 1992).

X-ray diffraction scans of clay mounts were run at 0.25° 2-theta, using Cu-K-alpha radiation. Whole rock powders were also analyzed by x-ray diffraction scans run at 1.0° 2-theta. Petrographic analysis of selected samples provided additional mineral composition data and information about textural and paragenetic relationships.

Diffractograms of air-dried and ethylene-glycol-saturated samples were compared to determine if smectite was present. To determine whether kaolinite was present, selected samples were boiled in 1N HCl to remove chlorite (Martin-Vivaldi and Gallego, 1961). Any remaining 7 Å peak was considered to be kaolinite, although Mg-chlorite might resist dissolution if present (Moore and Reynolds, 1989).

Chlorite(003)/illite(002) peak-area ratios were calculated for more than 120 foreland-basin mudstones using the height-width method (Moore and Reynolds, 1989; Fig. 5). These peaks have the dual advantage of being close together yet avoiding low-angle reflections (Moore and Reynolds, 1989). Errors in peak-area ratios are about ±5%.

Chemical Composition

Chemical composition was determined by X-ray fluorescence at McGill University and X-ray Assay Laboratories in Canada, using an internal standard of Austin Glen Shale (Table 1). Detection limits were 10 ppm for trace elements analyzed at X-ray Assay Laboratories, except for Ni, which was analyzed separately and had a detection limit of 5 ppm. Detection limits for trace elements analyzed at McGill University were 10 ppm for Cr and Ni and 5 ppm for other trace elements. Detection limit for major oxides was 0.01% at both laboratories. The precision of the analyses varies with the element, based on multiple analyses of the Austin Glenn Shale. The standard deviation for major oxides is less than 0.5%. The standard deviation for trace elements, on average, is less that 15% of the concentration of the element. Analyses included all major oxides as well as Y, Nb, Cr, Ni, Sr, Rb, Zr, and Ba. Some samples were analyzed for V, Pb, Th, and U.

Of these elements, Ni, Cr, Ti and Nb showed the most significant trends. Ni, Cr, and Ti are transition metals that commonly substitute for iron and magnesium in mafic igneous rocks and are therefore suitable to represent "mafic component" elements (Krauskopf, 1979). Although iron and magnesium are logical choices to represent the "mafic component," their concentrations tend to be anomalously high in the Taconic foreland basin due to the abundance of ferroan dolomite. Nb, an incompatible element, is most abundant in felsic igneous rocks and can be used to represent "felsic component" elements (Krauskopf, 1979). One key assumption is that these elements remain immobile during diagenesis. A number of studies have shown that mudstones are nearly isochemical during diagenesis with the exception of volatiles and potassium (e.g., Shaw, 1956; Hower and others, 1976; Jennings and Thompson, 1986; Argast and Donnelly, 1987; Awwiller, 1993). Ague (1991) showed that Ti is relatively immobile during regional metamorphism. A comparison of Ti with Cr, Ni, and Nb in the results of Whitfield and Turner (1979) shows that Cr, Ni, and Nb have a very low seawater-crustal rock partition coefficient similar to that of Ti. Thus, it is assumed that diagenesis has not significantly changed bulk chemical composition and

TABLE 1—RAW GEOCHEMICAL DATA FOR BLOUNTIAN AND TACONIC FORELAND BASIN MUDSTONES. SAMPLES ARE ORDERED BY STRATIGRAPHIC POSITION; THE FIRST SAMPLE IS LOWEST IN THE SEQUENCE, AND THE LAST SAMPLE IS HIGHEST IN THE SEQUENCE. SAMPLES MARKED WITH AN ASTERISK AND IN ITALICS WERE TREATED WITH HYDROCHLORIC ACID TO REMOVE CARBONATES PRIOR TO ANALYSIS

Taconic Foreland Basin

Sample #	SiO2	Al2O3	CaO	MgO	Na2O	K2O	Fe2O3	MnO	TiO2	P2O5	Cr	Ni	Ba	Rb	Sr	Nb	Zr	Y	LOI	Total
CC-1	12.59	3.22	42.48	1.38	0.10	0.66	2.70	0.05	0.15	0.10	<10	<10	114	32	1067	6	71	7	35.91	99.66
*CC-1**	*51.70*	*15.40*	*1.54*	*1.65*	*0.42*	*3.46*	*9.95*	*0.01*	*0.70*	*0.05*	*85*	*63*	*173*	*170*	*74*	*35*	*160*	*10*	*15.50*	*100.50*
CC-2	24.38	5.79	32.47	1.97	0.38	1.31	2.18	0.04	0.33	0.20	<10	13	211	57	1206	9	116	15	30.79	99.87
CC-3	24.38	3.98	34.89	1.55	0.27	0.86	2.13	0.04	0.24	0.14	<10	14	169	35	1395	6	93	14	31.15	99.68
CC-4	28.91	7.21	29.17	1.80	0.35	1.67	2.32	0.04	0.38	0.14	32	22	183	77	1011	10	119	17	27.73	99.73
CC-5	16.85	3.68	39.64	1.72	0.24	0.77	1.34	0.03	0.20	0.15	<10	<10	146	36	1600	<5	83	13	34.98	99.63
CC-6	7.40	2.32	47.89	1.39	0.04	0.44	1.69	0.04	0.10	0.13	<10	<10	71	24	1257	<5	57	6	38.06	99.51
*CC-6**	*53.90*	*14.60*	*1.40*	*1.57*	*0.49*	*2.76*	*8.63*	*0.02*	*0.74*	*0.04*	*76*	*87*	*196*	*137*	*79*	*30*	*199*	*10*	*16.10*	*100.30*
CC-7	19.57	5.12	36.75	2.08	0.23	1.10	1.82	0.03	0.25	0.14	<10	14	171	24	1374	7	93	13	32.63	99.76
CC-8	19.90	5.52	36.50	1.94	0.23	1.21	1.59	0.03	0.26	0.12	<10	<10	121	58	1242	6	91	12	32.58	99.90
*CC-8**	*62.60*	*14.40*	*0.72*	*1.61*	*1.06*	*3.11*	*3.14*	*0.02*	*0.87*	*0.06*	*94*	*50*	*262*	*166*	*99*	*25*	*272*	*10*	*12.70*	*100.30*
CC-9	20.88	3.61	37.37	1.75	0.30	0.81	1.31	0.03	0.19	0.24	<10	16	82	34	1405	<5	87	15	33.23	99.72
CC-10	13.46	4.22	41.39	1.60	0.12	0.90	1.27	0.05	0.18	0.20	<10	14	140	47	1489	<5	70	12	36.23	99.67
CC-11	17.28	5.20	37.92	1.63	0.26	0.93	1.77	0.04	0.23	0.37	<10	<10	<10	--	--	--	--	--	33.75	99.54
YC-1	32.65	7.55	25.93	1.93	0.39	1.58	3.16	0.04	0.37	0.17	11	23	194	--	--	--	--	--	25.75	99.58
YC-2	36.89	9.05	22.77	1.41	0.48	1.92	3.21	0.04	0.46	0.18	<10	47	266	--	--	--	--	--	23.18	99.72
*YC-2**	*61.40*	*15.70*	*0.97*	*1.17*	*1.15*	*3.29*	*2.96*	*0.03*	*0.84*	*0.07*	*80*	*71*	*566*	*157*	*94*	*32*	*185*	*10*	*11.70*	*99.40*
YC-3	14.28	3.81	30.13	8.00	0.27	0.69	7.42	0.11	0.19	0.15	<10	<10	108	--	--	--	--	--	35.24	100.40
YC-4	36.93	9.05	19.90	3.04	0.40	1.97	4.11	0.05	0.45	0.19	16	28	284	--	--	--	--	--	23.57	99.74
YC-5	13.02	3.69	28.55	10.34	0.24	0.66	6.74	0.09	0.18	0.13	<10	11	68	--	--	--	--	--	36.53	100.21
YC-6	31.55	6.31	29.49	0.97	0.37	1.21	1.82	0.03	0.36	0.16	<10	26	126	--	--	--	--	--	27.26	99.70
YC-7	60.83	12.34	4.44	1.73	0.58	2.88	6.70	0.05	0.59	0.13	67	71	405	110	177	13	120	25	9.99	100.32
YC-8	58.47	12.58	4.83	2.00	0.64	2.92	6.73	0.05	0.61	0.16	97	109	474	196	111	13	127	31	11.22	100.28
YC-9	57.18	13.17	4.86	1.90	0.68	3.04	7.11	0.06	0.64	0.14	65	81	402	117	188	14	135	30	10.93	99.77
YC-10	56.75	13.32	5.03	2.05	0.68	3.10	6.97	0.06	0.66	0.15	56	73	454	118	197	14	135	30	11.39	100.24
YC-12	55.15	12.93	6.20	2.24	0.69	3.03	7.11	0.07	0.66	0.15	64	57	435	116	213	13	133	30	11.46	99.76
YC-13	59.60	15.28	2.09	2.17	0.77	3.73	7.85	0.07	0.83	0.16	75	44	495	144	127	16	130	26	7.76	100.40
YC-14	16.11	6.01	22.53	10.33	0.12	1.17	10.40	0.30	0.32	0.13	30	<10	973	52	330	7	104	28	33.22	100.83
YC-16	58.73	16.68	2.38	2.90	0.92	3.71	7.81	0.08	1.01	0.17	95	32	390	--	--	--	--	--	6.53	101.01
YC-17	19.22	6.62	20.55	9.65	0.33	1.29	9.95	0.28	0.38	0.14	30	<10	1054	--	--	--	--	--	31.24	99.76
OC-3	57.30	16.50	2.02	2.99	0.86	3.76	7.33	0.11	0.91	0.15	70	42	386	161	121	11	150	17	7.15	99.20
OC-7	57.50	18.50	0.74	2.84	0.78	4.33	6.94	0.07	1.07	0.15	90	44	423	177	113	19	176	37	6.00	99.00
OC-8	63.20	9.48	4.24	2.75	1.32	1.82	6.41	0.15	0.72	0.17	31	17	11900	82	640	12	198	37	7.54	99.20
OC-12	57.70	11.50	5.26	3.31	1.15	2.36	6.64	0.20	0.82	0.17	44	23	243	90	122	32	267	30	9.15	98.40
OC-15	58.90	16.90	1.38	2.73	1.02	3.77	6.75	0.10	1.05	0.18	78	39	343	163	90	26	163	35	5.80	98.70
OC-16	59.50	16.00	1.73	2.80	1.11	3.49	6.95	0.10	1.02	0.17	66	44	359	146	100	35	218	40	6.20	99.20
OC-17	58.40	9.41	5.31	3.15	0.92	1.70	8.68	0.21	0.74	0.23	36	18	200	78	172	13	241	44	9.55	98.40

Blountian Foreland Basin

Sample #	SiO2	Al2O3	CaO	MgO	Na2O	K2O	Fe2O3	MnO	TiO2	P2O5	Cr	Ni	Ba	Rb	Sr	Nb	Zr	Y	LOI	Total
C-1	14.00	3.38	43.90	1.12	0.11	0.52	1.27	0.02	0.11	0.09	<10	9	<10	26	1310	<10	14	<10	35.40	100.20
C-2	25.00	6.24	33.80	1.34	0.17	1.21	2.26	0.06	0.20	0.17	<10	15	<10	41	520	15	31	12	27.90	98.40
C-5	40.40	14.40	17.40	1.45	0.31	2.61	4.34	0.04	0.50	0.14	22	29	206	98	314	26	76	32	18.70	100.40
M-1	37.90	12.70	20.00	1.54	0.23	2.55	3.98	0.03	0.53	0.17	18	33	192	94	311	13	94	23	20.60	100.30
M-2	48.00	14.20	12.80	1.44	0.21	2.84	4.25	0.03	0.55	0.18	43	43	358	109	253	<10	77	29	15.40	100.00
SHD-3	59.50	14.80	5.07	1.49	2.09	2.03	6.40	0.08	1.03	0.16	43	29	357	88	129	19	343	51	7.10	99.90
SHD-5	50.50	18.40	5.59	2.49	0.99	3.35	7.93	0.09	0.95	0.19	61	41	578	149	122	16	181	27	8.50	99.10
SHD-6	48.50	16.60	8.82	1.85	0.85	2.96	7.76	0.07	0.89	0.22	52	37	480	125	158	39	172	34	11.30	99.90
SHD-9	51.50	16.80	6.65	2.46	1.22	2.77	8.25	0.12	0.97	0.21	57	44	491	118	156	31	237	53	8.90	100.00
SHD-10	45.20	15.20	11.40	2.49	0.88	2.82	7.01	0.09	0.80	0.17	41	39	424	122	179	15	168	33	12.50	98.70
SHD-14	40.00	9.33	19.40	3.03	1.25	1.25	5.78	0.14	0.56	0.12	10	27	146	58	242	13	214	17	18.70	99.50
SHD-15	51.10	18.20	5.45	1.94	0.70	3.46	7.51	0.05	0.96	0.19	60	43	608	131	138	17	193	47	9.60	99.30
SHD-16	54.60	4.62	19.90	0.88	1.21	0.46	2.28	0.05	0.31	0.08	<10	10	<10	24	217	<10	244	<10	15.90	100.40
SHD-21	51.70	12.60	12.50	1.78	0.62	2.93	4.13	0.05	0.47	0.17	12	21	347	106	166	43	124	20	12.90	99.90
SHD-22	50.50	15.70	8.67	2.23	1.12	2.98	6.75	0.06	0.88	0.17	48	34	433	114	164	13	205	34	10.90	99.60
SHD-23	47.00	5.04	22.80	1.44	1.10	0.50	2.91	0.08	0.33	0.08	<10	13	<10	21	289	<10	181	23	18.70	100.00
SHD-24	54.70	15.50	6.60	1.96	1.42	2.71	6.87	0.06	0.92	0.17	88	42	420	115	146	18	295	39	8.60	99.60
SHD-30	53.00	21.90	0.40	1.79	0.39	5.35	7.51	0.05	1.14	0.17	84	43	865	193	58	22	199	48	7.15	99.00

that Ti, Cr, Ni, and Nb were relatively immobile during diagenesis.

The proportion of sediment derived from a mafic source, the mafic/felsic ratio, was calculated as:

$$\sqrt[3]{Cr*Ni*Ti/Nb} \qquad (1)$$

The presence of abundant calcite and dolomite in samples from the lower, outer-slope facies of the Taconic foreland basin complicated the analysis of results. Abundant carbonates can dilute trace element concentrations to below detection limits. In such cases, the detection limit of 10 ppm was arbitrarily used for calculating the mafic/felsic ratio. Four samples with abundant calcite were leached with 0.5 N HCl to remove carbonates and were re-analyzed to confirm that ratio calculations using detection limits were reasonable. Besides the problem of abundant carbonates, previous studies have shown composition may vary with grain size (Argast and Donnelly, 1987). For this reason, Zr, which is concentrated in the sand fraction as zircon and therefore sensitive to grain size, was not used as a measure of the proportion of "felsic" sediment. Because mudstones are a mixture of silt and clay sized particles, Cr, Ni, Nb, and Ti have the potential to be concentrated in one or the other size fraction. The observed values would then represent a hydraulic (rather than a provenance) signature. Cr, Ni, Ti, and Nb concentrations were plotted against the Si/Al ratio to see whether the concentrations of these elements are independent of grain-size distribution. The Si/Al ratio increases with increasing average grain size (Argast and Donnelly, 1987), so a positive correlation would suggest the concentration of an element in the heavy mineral fraction of the silt size. Trace-element composition calculated using Cr, Ni, Ti, and Nb is unrelated to grain size in either foreland basin (Figure 6).

RESULTS

Whole-Rock Mineral Composition

Petrographic and X-ray diffraction results show that whole-rock mineral composition of mudstones from the Taconic and Blountian foreland basins consists mostly of silt-sized quartz, feldspar, carbonates, chlorite, and muscovite in a clay matrix or as laminae. Biotite is present in small amounts. Plagioclase (albite) is the dominant feldspar. Potassium feldspar is rare in both mudstones and sandstones. The significance of low K-feldspar content is equivocal, as this may be a diagenetic effect (McBride, 1985; Miliken, 1992). Calcite is ubiquitous and abundant in the lower outer-slope facies. Dolomite is abundant in the Taconic foreland basin but rare in the Blountian foreland basin. There appears to be no stratigraphic trend in whole-rock mineral composition apart from an upward decrease in carbonate content.

Petrographic analysis and field observations in the Taconic foreland basin indicate that silt layers increase in thickness up-section from laminations in the outer-slope facies to centimeter-scale beds of siltstone turbidites in the basin-axis facies. This upward change in texture is similar to that found in the Blountian foreland basin (Shanmugam and Walker, 1978).

Clay Mineral Composition

The clay mineral assemblage of the Taconic and Blountian foreland basin mudstones is diagenetic. Illite in these mudstones contains interlayered smectite which suggests that illite formed diagenetically from smectite. Chlorite can also form diagenetically from smectite if enough iron and magnesium are present (Ahn and Peacor, 1985). For Cambrian and Ordovician shelf mudstones from Norway, Björlykke (1974) interpreted the increase in the chlorite/illite ratio as recording an influx of volcanically-derived sediment. In the Taconic and Blountian foreland basins, the chlorite/illite ratio assumes greater importance because ferroan dolomite in the mudstones dramatically increases the abundances of Fe, Mg, and Mn in some samples. Fe and Mg therefore cannot be used to calculate a mafic/felsic ratio charateristic of detritus. Chlorite/illite ratio trends from the Taconic and Blountian foreland basins are similar to one another, although the Blountian shows a greater degree of variation. The Taconic foreland basin has a relatively simple burial history and is not thrust faulted; the chorite/illite ratio clearly reflects compositional change in the mud. The Blountian foreland basin has a more complex burial history and is telescoped by thrust faults. This complex burial history may be the cause of the greater chlorite/illite ratio variation in the Blountian foreland basin.

Taconic Foreland Basin.—

In New York, the transition from the outer-slope facies to the basin-axis facies is gradational. Mudstones in the outer-slope facies contain silt laminae. In the basin-axis facies, however, these laminae have thickened to become individual beds of siltstone characteristic of a flysch deposit. The gradational nature of this facies transition is similar to the facies transition in the Blountian foreland basin (Shanmugam and Walker, 1978). A change in clay mineral composition occurs within this transition. The chlorite/illite ratio of mudstones from the lower, outer-slope facies is zero (Fig. 7). The absence of a 7 Å peak in samples from the lower outer-slope facies indicates that chlorite was not present even in trace amounts. The chlorite/illite ratio begins to increase steadily in the upper, outer-slope facies to a maximum of 0.5 in the basin-axis facies (Fig. 7).

In Tennessee, the southern Taconic foreland basin sequence onlaps the western side of the Blountian foreland

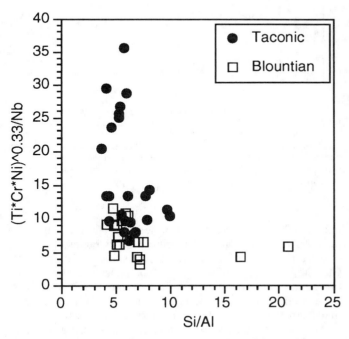

FIG. 6.—Calculated ratio of Ti, Cr, and Ni to Nb (proxy for composition) versus ratio of Si to Al (proxy for grain size). The ratio of Ti, Cr, and Ni to Nb reflects relative contribution from mafic (Ti, Cr, and Ni) and felsic (Nb) sediment sources. Trace element composition appears unrelated to grain size.

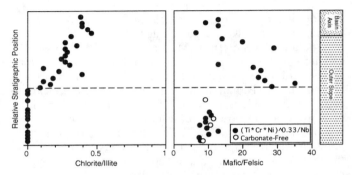

FIG. 7.—Clay mineral and chemical compositions for the Taconic foreland basin. This figure represents a composite sequence. Stratigraphic columns indicate facies type (symbols as in figure 2). Clay mineral composition from the Taconic foreland basin !References!Referencesshows an increase in the chlorite/illite ratio beginning within the outer-slope facies, as shown by the dashed line. Taconic foreland-basin mudstones show a distinct change in chemical composition at about the same time the chlorite/illite ratio begins to increase.

basin sequence (Fig. 1). The Taconic sequence here contains a conformable sequence of shelf and outer-slope mudstones overlain by marginal-marine and terrestrial redbeds, indicating the maximum water depth was much shallower at this location. The basin-axis facies is missing. The Taconic sequence conformably overlies marginal-marine redbeds of the Blountian foreland basin. The chlorite/illite ratios for these rocks are interesting for two reasons. First, the outer-slope to shelf shales of the Taconic foreland basin sequence have a constant chlorite/illite ratio of about 0.4 which abruptly changes at both contacts with redbeds (Fig. 8). Second, the marginal-marine and terrestrial redbeds at the top of both sequences, with few exceptions, have a chlorite/illite ratio of zero and have a clay mineral assemblage of illite+minor chlorite (the (001) and (002) peaks confirming the presence of chlorite). This abrupt change in the chlorite/illite ratio is attributable to leaching of chlorite during subaerial diagenesis of marginal-marine and terrestrial molasse, a process which also can modify the chemical composition of the sediments (e.g., Thompson, 1970).

Analysis of illite peaks for samples from the northern Taconic foreland basin indicates R1 to R3 ordering with less than 20% smectite interlayers. Illite peaks for the southern Taconic foreland basin indicate R3 ordering with less than 10% smectite interlayers. Kaolinite is not present.

Blountian Foreland Basin.—

In the Blountian foreland basin, the facies transition from outer slope to basin axis is also gradational (Shanmugam and Walker, 1978). Much of the outer-slope facies is composed of distal turbidites (Shanmugam and Walker, 1978). Chlorite/illite ratios from the eastern Blountian foreland basin show a trend similar to that from the Taconic foreland basin but exhibit greater second-order variation (Fig. 9). The chlorite/illite ratio in the outer-slope facies just above the limestone is zero, increasing upward and reaching a maximum of 1.2 in the basin-axis facies. Although the chlorite/illite ratio increases steadily in the outer-slope facies as observed in the Taconic foreland basin, the data for the basin-

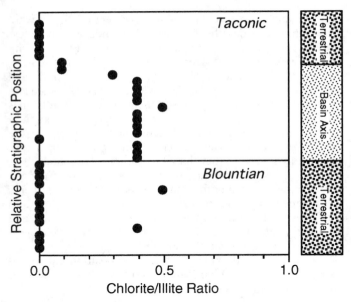

FIG. 8.—Clay mineral composition at Thorn Hill, TN. At this outcrop, sediments from the Taconic foreland basin overlie sediments from the Blountian foreland basin. Samples from redbeds at Thorn Hill having chlorite/illite ratios of zero probably have been modified by subaerial diagenesis in the marginal marine environment.

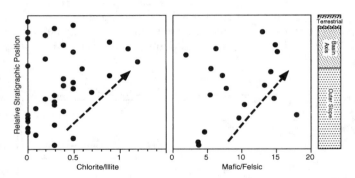

FIG. 9.—Clay mineral and chemical compositions for the Blountian foreland basin. The Blountian foreland basin clay mineral composition shows a general upward increase in the chlorite/illite ratio, as shown by the dashed arrow, but there is more scatter than for the Taconic foreland basin. Chemically, Blountian foreland basin mudstones show a general shift to a higher mafic proportion (dashed arrow), but as with the clay mineral composition, there is a great deal of scatter.

axis facies show a great deal of scatter. The ratio drops to near zero in the marginal-marine redbed facies, probably due to leaching of chlorite during subaerial diagenesis. Chlorite/illite ratios of the outer-slope-facies mudstones from the western side of the Blountian foreland basin are somewhat higher (0.0 to 0.3), but this ratio is still lower than that from the upper outer-slope and basin-axis facies.

Analysis of the illite peak indicates R3 ordering with less than 10% smectite interlayers for all samples. Samples from the marginal-marine redbed facies with a chlorite/illite ratio of zero typically contain trace amounts of chlorite, as indicated by presence of a low intensity 7 Å peak. Disappearance of the 7 Å peak after treatment with HCl indicates that kaolinite is not present.

Chemical Composition

Trends in the mafic/felsic ratio are interpreted as reflecting changes in source-area chemistry. Two other processes that may affect the chemical composition of mudstones are the intensity of chemical weathering (climate) and diagenesis. Studies of weathering profiles suggest that the composition of the parent rock exerts a greater control over bulk chemical composition of detrital sediment than does chemical weathering (e.g., Nesbitt and Young, 1989). As previously discussed in the methods section, burial diagenesis typically does not affect the bulk composition of mudstones until high-grade metamorphism. On the other hand, subaerial diagenesis of mudstones in the marginal-marine environment may greatly change their bulk chemical composition (Thompson, 1970), reducing the value of these rocks for provenance determination. Thus, although weathering and burial diagenesis are recongized as processes that may affect the chemical composition of mudstones, the primary control over the bulk chemical composition of mudstone is the composition of the source rock.

In the Taconic foreland basin, the large proportion of carbonate in the mudstones from the outer-slope facies diluted the concentrations of trace metals. However, leaching the samples to remove carbonate did not appreciably affect the mafic/felsic ratio (Fig. 7). Examination of the mafic/felsic ratio trend shows that the mudstone composition changes abruptly within the upper outer-slope facies at about the same time the chlorite/illite ratio begins to increase (Fig. 7). The mafic/felsic ratio then decreases during the transition from outer-slope to basin-axis facies. This decrease in the mafic/felsic ratio is difficult to reconcile with the continued increase of the chlorite/illite ratio (thus presumably iron and magnesium), which suggests an increase in the proportion of sediment derived from a source with a mafic signature.

In the Blountian foreland basin, there is a general increase in the mafic/felsic ratio which occurs within the outer-slope facies (Fig. 9). However, like the chlorite/illite ratio, the mafic/felsic ratio shows a great deal of variability. The scatter in the chemical data begins within the outer-slope facies, which is somewhat before scatter began for the chlorite/illite ratio. Overall, like the chlorite/illite ratio, chemical composition is more variable in the Blountian foreland basin than in the Taconic foreland basin.

When the Ni/Nb and Cr/Nb ratios are plotted against the Ti/Nb ratio, the data from each basin forms a mixing line between "mafic" and "felsic" sources. A comparison of data from the two basins shows the Taconic foreland basin mudstones have a higher proportion of sediment derived from a mafic source than the Blountian foreland basin mudstones (Fig. 10).

DISCUSSION

The stratigraphic variations of clay mineral composition and chemical composition of mudstones from the Taconic and Blountian foreland basins apparently provide records

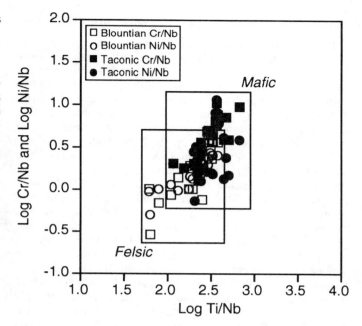

FIG. 10.—Chemical compositions from both basins form a mixing line between a felsic source and a mafic source. Taconic foreland basin mudstones have a higher mafic component than do Blountian foreland basin mudstones.

of tectonic events in their evolving source areas. Recent studies of mudstones have focused on the relationship between chemistry and tectonic setting (e.g., Bhatia, 1985; Roser and Korsch, 1986, 1988), but few have incorporated petrographic, mineralogic, chemical, and isotopic data from mudstones to obtain provenance information related to tectonics (McLennan and others, 1990); this is especially true for foreland basins. More importantly, few studies have examined the significance of stratigraphic variation of mudstone composition in individual basins. Björlykke (1974) is one of the few who integrated clay mineralogy with chemical composition in the context of a stratigraphic framework, albeit across a much longer time span (Cambrian through Ordovician). The tectonostratigraphic model of Ettensohn (1991) provides an excellent process-response framework for interpretation of the mudstone composition from the two basins examined in this study.

Significance of Mudstone Composition

The stratigraphic increase in the chlorite/illite and mafic/felsic ratios indicates a change in the composition of the sediments derived from the colliding terranes. The mixing line formed by the combined chemical data set suggests that the mudstones are composed of a mixture of sediments derived from felsic and mafic sources. Because the only significant sources for the siliciclastic sediment were the colliding outboard terranes, the change in mudstone composition in each basin appears to reflect the tectonic evolution of its source area. The changes in mudstone composition are interpreted as a response to the tectonic events proposed by Ettensohn (1991).

Interpretation of the Taconic Foreland Basin Sequence

Interpretations of the regional geology and geochronology of the northeastern United States indicate the Bronson Hill Anticlinorium in Massachusetts and Vermont represents the magmatic arc which collided with Laurentia to form the Taconic foreland basin (Stanley and Ratcliffe, 1985). All the regional elements of the Taconic Orogeny are present: the Ammonoosuc magmatic arc (Bronson Hill Anticlinorium), the forearc basin (Moretown Formation), and the non-volcanic outer arc (Taconic thrust sheets). The arc system was active tectonically at least until the end of the Ordovician (Tucker and Robinson, 1990). The tectonic reconstructions of Stanley and Ratcliffe (1985) constrain interpretation of mudstone provenance. The outer arc was composed of the Late Proterozoic and Cambrian rift and passive-margin sequence rocks. The Laurentian continent was probably the ultimate source for the passive-margin sediments. Sediments, eroded from both the outer arc and magmatic arc, filled the forearc basin. The magmatic arc rocks are predominantly mafic and have the chemical composition of arc basalts (Schumacher, 1988). The felsic composition of the lower outer-slope facies mudstones is consistant with an outer-arc source. During collision, the interarc basin either closed or filled progressively, increasing the amount of sediment contributed by the magmatic arc to the foreland basin (Andersen and others, 1992). This is reflected by the increase in the chlorite/illite and mafic/felsic ratios of the mudstones.

Interpretation of the Blountian Foreland Basin Sequence

The Blountian foreland basin, which itself constitutes the main evidence for the Blountian orogeny, presents more problems because the colliding terrane is unknown (Drake and others, 1989). Interpretation of provenance change is therefore not as straightforward as for the Taconic foreland basin. The increase of both the chlorite/illite and mafic/felsic ratios indicates an increase in the proportion of sediment eroded from a mafic source. The data, however, show significantly more scatter than in the Taconic foreland basin and a much smaller mafic component. The lower mafic/felsic ratio, relative to the Taconic foreland basin, corresponds to a lower Cr and Ni content and a lower percentage of volcanic rock fragments in sandstones from the basin-axis facies (Hiscott, 1984; Mack, 1985). However, the similarity in trends of mudstone composition between the Taconic and Blountian foreland basins indicates that the overall tectonic evolution of the Blountian source terrane was roughly parallel to that of the Taconic source terrane.

Implications for the Foreland Basin Tectonostratigraphic Model

The previous interpretations have important implications for the tectonostratigraphic model proposed by Ettensohn (1991) for Ordovician foreland basins in the Appalachians. This process-response model links foreland basin stratigraphy to tectonic events that are associated with collision of an outboard terrane with the continent. Thus, the model provides a framework for understanding the tectonic history of the foreland basin.

Mudstone provenance data add to the model by revealing the tectonic history of the outboard terrane that collided with the continent (Fig. 11). The source of sediment apparently shifted from an outer, non-volcanic arc to a magmatic arc. Outer-arc rocks were ultimately derived from sediments eroded from Laurentia (Fig. 11A). The low mafic/felsic and chlorite/illite ratios of the earliest outer-slope mudstones reflect the inherited continental signature. During early deformational loading, outer-slope mudstones continued to have a continental signature because mafic sediments eroded from the volcanic arc were trapped in the forearc basin (Fig. 11B). Importantly, mudstone provenance began to change within the outer-slope facies during the process of active deformational loading. Toward the end of this tectonic phase, the forearc basin probably closed due to shortening within the accretionary wedge (Chapple, 1978), causing the change in mudstone provenance. After closure of the forearc basin, active tectonism ceased, and erosion of the colliding terrane during loading-type relaxation filled the basin with flysch (Fig. 11C). As a result, the compositional change in foreland-basin mudstones *preceded* the facies change from outer-slope to basin axis.

Although the tectonostratigraphic model explains stratigraphic similarities of the Taconic and Blountian basins (Fig. 11), the more mafic composition of the Taconic foreland basin mudstones (Fig. 9) may reflect differences between the collisional histories of the two orogens. First, the terrane that collided with Laurentia to form the Blountian foreland basin may have been a composite terrane, unlike the Bronson Hill magmatic arc. The variable sources of a composite terrane, which are typically older sedimentary and metamorphic rocks, may also account for the greater scatter of the data. The high proportion of metasedimentary clasts in conglomerates and metamorphic rock fragments in basin-axis sandstones seems to support this scenario (Mack, 1985; Drake and others, 1989). Second, as proposed by Hiscott (1984), the difference in chemical composition could reflect a difference in collisional style; a more oblique angle of collision would result in less obduction of oceanic crust and, consequently, reduced mafic source area.

Further development of the tectonostratigraphic model will require studies of foreland basins that differ in age and tectonic style from the Taconic and Blountian foreland basins. Additional techniques, such as analysis of neodymium isotopic composition, might be used to test interpretations of mudstone provenance derived from mineral and chemical composition (e.g., Nelson and DePalao, 1988). Still another approach might use uranium-lead isotope systematics of quartz grains to constrain provenance age (Scott McLennan, pers. commun., 1993).

CONCLUSIONS

The chlorite/illite ratio and mafic/felsic ratio of mudstones can be used to track changes in foreland basin provenance. The mafic/felsic ratio indicates that mudstone composition reflects a mixture of sediments eroded from the outer arc and magmatic arc of the colliding arc system. Sediment eroded from the outer arc has a low chlorite/illite ratio and a felsic composition whereas sediment eroded from

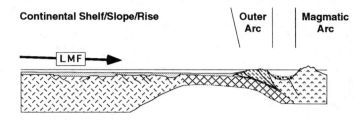
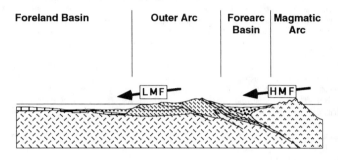
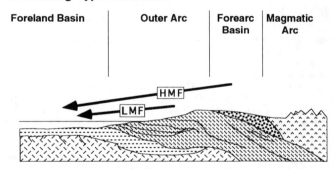

FIG. 11.—Schematic diagram of tectonostratigraphic model for foreland basins, including changes in mudstone composition. Cross sections are modified from Stanley and Ratcliffe (1985). LMF equals low mafic/felsic ratio and HMF equals high mafic/felsic ratio. (A) Prior to the collision, sediment having a low mafic/felsic ratio (continental provenance) is eroded from the continent and deposited in the shelf/slope system. This sediment source is shut off with the development of a carbonate platform. The sediment derived from the continent is accreted to form the outer arc, along with sediment derived from the magmatic arc. (B) After initial collision and deformational loading, sediment derived from the outer arc is contributed to the foreland basin. This sediment inherits its low mafic/felsic ratio (continental signature) from the outer arc. Sediment eroded from the magmatic arc has a high mafic/felsic ratio (arc signature) and is trapped in forearc basins. (C) During loading-type relaxation, the forearc basin closes and/or fills. Sediment reaching the foreland basin is now a mixture of sediment with a continental signature (eroded from the outer arc) and sediment with an arc signature (eroded from the magmatic arc/forearc basin). This mixture results in a change in the chemical and clay mineral composition in the foreland basin.

the magmatic arc has a higher chlorite/illite ratio and a higher mafic/felsic ratio. Stratigraphic variation of the provenance signature reflects the tectonic evolution of the colliding outboard terrane. This change in provenance as a response to tectonics integrates well with the tectonostratigraphic model for Ordovician Appalachian foreland basins proposed by Ettensohn (1991).

Comparison of the Taconic and Blountian foreland basins indicates that both basins received increasing amounts of mafic magmatic-arc-derived sediment as arc-continent collision progressed. However, the Taconic foreland basin apparently received the greater influx of arc-derived sediment. The accreting terranes apparently differed with respect either to style of collision or terrane composition.

ACKNOWLEDGMENTS

This project was funded by grants from Sigma Xi, the Geological Society of America, Exxon Production Research, and the Syracuse University Department of Geology. This paper is part of dissertation research completed at Syracuse University. Discussions with Ken Walker, Fred Read, Gene Rader, and Avery Drake greatly improved my understanding of foreland basin stratigraphy in Ordovician Appalachian basins. Any mistakes, however, are my own. Discussions with Frank Florence improved my understanding of the structure and petrology of the Bronson Hill anticlinorium. I would especially like to thank John Delano, SUNY-Albany for providing me with many chemical analyses of the Utica shale and for sharing his enthusiasm for the crazy task of understanding mudstones. I am indebted to my advisor, Bryce Hand, Frank Florence, and two anonomous reviewers all who made suggestions that greatly improved this paper.

REFERENCES

AGUE, J. J., 1991, Evidence for major mass transfer and volume strain during regional metamorphism of pelites: Geology, v. 19, p. 855–858.

AHN, J. H. AND PEACOR, D. R., 1985, Transmission electron microscope study of diagenetic chlorite in Gulf Coast argillaceous sediments: Clays and Clay Minerals, v. 33, p. 228–236.

ANDERSEN, C. B., HAND, B. M., AND DELANO, J. W., 1992, Marine mudstones of the Taconic foreland basin: a record of collision zone evolution (abs.): Geological Society of America Abstracts with Programs, v. 24., p. A319.

ARGAST, S. AND DONNELLY, T. W., 1987, The chemical discrimination of clastic sedimentary components: Journal of Sedimentary Petrology, v. 57, p. 813–823.

AWWILLER, D. N., 1993, Illite/smectite formation and potassium mass transfer during burial diagenesis of mudrocks: A study from the Texas Gulf Coast Paleocene-Eocene: Journal of Sedimentary Petrology, v. 63, p. 501–512.

BHATIA, M. R., 1985, Composition and classification of Paleozoic flysch mudrocks of eastern Australia: Implications in provenance and tectonic setting interpretation: Sedimentary Geology, v. 41, p. 249–268.

BJÖRLYKKE, K., 1974, Geochemical and mineralogical influence of Ordovician island arcs on epicontinental clastic sedimentation: a study of lower Paleozoic sedimentation in the Oslo Region: Sedimentology, v. 21, p. 251–272.

BRADLEY, D. C., 1989, Taconic plate kinematics as revealed by foredeep stratigraphy, Appalachian Orogen: Tectonics, v. 8, p. 1037–1049.

CHAPPLE, W. M., 1978, Mechanics of thin-skinned fold-and-thrust belts: Geological Society of America Bulletin, v. 89, p. 1189–1198.

COOK, R. J., 1992, A comparison of methods for the extraction of smectites from calcareous rocks by acid dissolution techniques: Clay Minerals, v. 27, p. 73–80.

Covey, M., 1986, The evolution of foreland basins to steady state: evidence from the western Tiawan foreland basin, in Allen, P. A., and Homewood, P., eds., Foreland Basins: Oxford, International Association of Sedimentologists Special Publication 8, p. 77–90.

Dickinson, W. R., 1977, Tectono-stratigraphic evolution of subduction-controlled sedimentary assemblages, in Talwani, M., and Pittman, W. C., III, eds., Island Arcs, Deep Sea Trenches, and Back Arc Basins: Washington, D. C., American Geophysical Union Maurice Ewing Series, v. 1, p. 33–40.

Drake, A. A., Jr., Sinha, A. K., Laird, J., and Guy, R. E., 1989, The Taconic orogen, in Hatcher, R. D., Jr., Thomas, W. A., and Viele, G. W., eds., The Appalachian-Ouachita Orogen in the United States: Boulder, Geological Society of America, The Geology of North America, v. F-2, p. 101–177.

Drever, J. I., 1973, The preparation of oriented clay mineral specimens for x-ray diffraction analysis by a filter-membrane peel technique: American Mineralogist, v. 58, p. 553–554.

Ettensohn, F. R., 1991, Flexural interpretation of relationships between Ordovician tectonism and stratigraphic sequences, central and southern Appalachians, U.S.A., in Barnes, C. R., and Williams, S. H., eds., Advances in Ordovician Geology: Ottawa, Geological Survey of Canada, Paper 90–9, p. 213–224.

Girty, G. H., 1991, A note on the composition of plutoniclastic sand produced in different climatic belts: Journal of Sedimentary Petrology, v. 61, p. 428–433.

Grantham, J. H. and Velbel, M. A., 1988, The influence of climate and topography on rock-fragment abundance in modern fluvial sands of the southern Blue Ridge Mountains, North Carolina: Journal of Sedimentary Petrology, v. 58, p. 219–227.

Hatcher, R. D., Jr., 1987, Tectonics of the southern and central Appalachian internides: Annual Review of Earth and Planetary Sciences, v. 15, p. 337–362.

Heller, P. L. and Frost, C. D., 1988, Isotopic provenance of clastic deposits: application of geochemistry to sedimentary provenance studies, in Kleinspehn, K. L., and Paola, C., eds., New Perspectives in Basin Analysis: New York, Springer-Verlag, p. 27–42.

Hiscott, R. N., 1984, Ophiolitic source rocks for Taconic age flysch: trace element evidence: Geological Society of America Bulletin, v. 95, p. 1261–1267.

Hower, J., Eslinger, E. V., Hower, M. E., and Perry, E. A., 1976, Mechanism of burial metamorphism of argillaceous sediments: Geological Society of America Bulletin, v. 87, p. 725–737.

Jacobi, R. D., 1981, Peripheral bulge—a causal mechanism for the Lower/Middle Ordovician disconformity along the western margin of the northern Appalachians: Earth and Planetary Science Letters, v. 56, p. 245–251.

Jennings, S. and Thompson, G. R., 1986, Diagenesis of Plio-Pleistocene sediments of the Colorado River Delta, southern California: Journal of Sedimentary Petrology, v. 56, p. 89–98.

Johnson, M. J. and Stallard, R. F., 1989, Physiographic controls on the composition of sediments derived from volcanic and sedimentary terrains on Barro Colorado Island, Panama: Journal of Sedimentary Petrology, v. 59, p. 768–781.

Kay, G. M., 1942, Development of the northern Alleghany synclinorium and adjoining region: Geological Society of America Bulletin, v. 95, p. 1601–1658.

Krauskopf, K. B., 1979, Introduction to Geochemistry (2nd ed.): New York, McGraw-Hill, 617 p.

Mack, G. H., 1985, Provenance of the Middle Ordovician Blount clastic wedge, Georgia and Tennessee: Geology, v. 13, p. 299–302.

Martin-Vivaldi, J. L. and Gallego, M., 1961, Some problems in the identification of clay minerals by x-ray diffraction. 1. Chlorite-kaolinite mixtures: Clay Minerals Bulletin, v. 4, p. 288–292.

McBride, E. F., 1962, Flysch and associated beds of the Martinsburg Formation (Ordovician), central Appalachians: Journal of Sedimentary Petrology, v. 32, p. 48–91.

McBride, E. F., 1985, Diagenetic processes that affect provenance determinations in sandstone, in Zuffa, G. G., ed., Provenance of Arenites: Dordrecht, D. Reidel Publishing Company, p. 95–113.

McLennan, S. M., Taylor, S. R., McCulloch, M. T., and Maynard, J. B., 1990, Geochemical and Nd-Sr isotopic composition of deep-sea turbidites: Crustal evolution and plate tectonic associations: Geochimica et Cosmochimica Acta, v. 54, p. 2015–2050.

Milliken, K. L., 1992, Chemical behavior of detrital feldspars in mudrocks versus sandstones, Frio Formation (Oligocene), south Texas: Journal of Sedimentary Petrology, v. 62, p. 790–801.

Milliken, K. L., McBride, E. F., and Land, L. S., 1989, Numerical assessment of dissolution versus replacement in the subsurface destruction of detrital feldspars, Oligocene Frio Formation, south Texas: Journal of Sedimentary Petrology, v. 59, p. 740–757.

Moore, D. M. and Reynolds, R. C., Jr., 1989, X-ray Diffraction and the Identification and Analysis of Clay Minerals: Oxford, Oxford University Press, 332 p.

Nelson, B. K. and DePaolo, D. J., 1988, Comparison of isotopic and petrographic provenance indicators in sediments from Tertiary continental basins of New Mexico: Journal of Sedimentary Petrology, v. 58, p. 348–359.

Nesbitt, H. W. and Young, G. M., 1989, Formation and diagenesis of weathering profiles: Journal of Geology, v. 97, p. 129–147.

Ostrom, M. E., 1961, Separation of clay minerals from carbonate rocks by using acid: Journal of Sedimentary Petrology, v. 31, p. 123–129.

Quinlan, G. M. and Beaumont, C., 1984, Appalachian thrusting, lithospheric flexure, and Paleozoic stratigraphy of the eastern interior of North America: Canadian Journal of Earth Science, v. 21, p. 973–996.

Ramseyer, K., Boles, J. R., and Lichtner, P. C., 1992, Mechanism of plagioclase albitization: Journal of Sedimentary Petrology, v. 62, p. 349–357.

Riva, J., 1974, A revision of some Ordovician graptolites of eastern North America: Paleontology, v. 17, p. 1–40.

Rodgers, J., 1953, Geologic map of east Tennessee with explanatory text: Tennessee Division Geology Bulletin 58, Part II, 168 p.

Rodgers, J., 1971, The Taconic Orogeny: Geological Society of America Bulletin, v. 82, p. 1141–1178.

Roser, B. P. and Korsch, R. J., 1986, Determination of tectonic setting of sandstone-mudstone suites using SiO_2 content and K_2O/Na_2O ratio: Journal of Geology, v. 94, p. 635–650.

Roser, B. P. and Korsch, R. J., 1988, Provenance signatures of sandstone-mudstone suites determined using discriminant function analysis of major-element data: Chemical Geology, v. 67, p. 119–139.

Rowley, D. B. and Kidd, W. S. F., 1981, Stratigraphic relationships and detrital composition of the medial Ordovician flysch of western New England: implications for the tectonic evolution of the Taconic Orogeny: Journal of Geology, v. 89, p. 199–218.

Ruedemann, R., 1897, Evidence of current action in the Ordovician of New York: American Geologist, v. 19, p. 367–391.

Schumacher, J. C., 1988, Stratigraphy and geochemistry of the Ammonoosuc volcanics, central Massachusetts and southwestern New Hampshire: American Journal of Science, v. 288, p. 619–663.

Schwab, F. L., 1986, Sedimentary "signatures" of foreland basin assemblages: real or counterfeit?, in Allen, P.A., and Homewood, P., eds., Foreland Basins: Oxford, International Association of Sedimentologists Special Publication 8, p. 91–102.

Schwab, F. L., 1991, Detrital modes of late Precambrian—early Paleozoic sandstones across Newfoundland: Do they constrain Appalachian tectonic models?: Geological Society of America Bulletin, v. 103, p. 1317–1323.

Shanmugan, G. and Walker, K. R., 1978, Tectonic significance of distal turbidites in the Middle Ordovician Blockhouse and lower Sevier Formations in east Tennessee: American Journal of Science, v. 278, p. 551–578.

Shaw, D. M., 1956, Geochemistry of pelitic rocks. III. Major elements and general geochemistry: Geological Society of America Bulletin, v. 67, p. 919–934.

Stanley, R. S. and Ratcliffe, N. M., 1985, Tectonic synthesis of the Taconian orogeny in western New England: Geological Society of America Bulletin, v. 96, p. 1227–1250.

Tankard, A. J., 1986, On the depositional response to thrusting and lithospheric flexure: examples from the Appalachian and Rocky Mountain Basins, in Allen, P. A., and Homewood, P., eds., Foreland Basins: Oxford, International Association of Sedimentologists Special Publication 8, p. 369–392.

Thompson, A. M., 1970, Geochemistry of color genesis in red-bed sequence, Juniata and Bald Eagle Formations, Pennsylvania: Journal of Sedimentary Petrology, v. 40, p. 599–615.

Tucker, R. D. and Robinson, P., 1990, Age and setting of the Bronson Hill magmatic arc: a re-evaluation based on U-Pb zircon ages in southern New England: Geological Society of America Bulletin, v. 102, p. 1404–1419.

WHITFIELD, M. AND TURNER, D. R., 1979, Water-rock partition coefficients and the composition of seawater and river water: Nature, v. 278, p. 132–137.

WILLIAMS, H. AND HATCHER, R. D., JR., 1983, Appalachian suspect terranes, in Hatcher, R.D., JR., Williams, H., and Zietz, I., eds., Contributions to the Tectonics and Geophysics of Mountain Chains: Boulder, Geological Society of America Memoir 158, p. 33–54.

ZEN, E-AN, 1983, Exotic terranes in the New England Appalachians—limits, candidates and ages: A speculative essay, in Hatcher, R.D., JR., Williams, H., and Zietz, I., eds., Contributions to the Tectonics and Geophysics of Mountain Chains: Boulder, Geological Society of America Memoir 158, p. 55–81.

PROVENANCE OF THE UPPER CRETACEOUS NANAIMO GROUP, BRITISH COLUMBIA: EVIDENCE FROM U-PB ANALYSES OF DETRITAL ZIRCONS

PETER S. MUSTARD
Geological Survey of Canada, 100 West Pender Street, Vancouver, British Columbia, V6B 1R8, Canada
AND
RANDALL R. PARRISH AND VICKI McNICOLL
Geological Survey of Canada, 601 Booth Street, Ottawa, Ontario, K1A 0E8, Canada

ABSTRACT: The Nanaimo Group of southwest British Columbia overlies the Wrangellia terrain and the western Coast Belt and is in fault contact with the northwestern margin of the Cascades to the southeast. Generally interpreted as deposits of a Late Cretaceous forearc basin, a foreland basin model is preferred for the Nanaimo Group, in large part due to the recent recognition of major, westerly-directed thrust systems in the Coast Belt to the east and northwestern Cascades, coupled with new age constraints which indicate that thrusting in part overlaps with Nanaimo Group sedimentation. As a test of the foreland basin model, U-Pb ages of twenty-two detrital zircons from three formations of the Nanaimo Group provide new evidence about changing source areas with time for Nanaimo Group deposition. Zircons from the lower Campanian Extension and Protection Formations indicate that Coast Belt and San Juan thrust systems were the dominant source areas, and Wrangellia was not a major source of detritus. Submarine fan sandstone of the uppermost Gabriola Formation (Maastrichtian age) contains abundant detrital titanite, epidote, and zircon with ages as follows: Precambrian zircons, zircons of late Mesozoic age with Precambrian inheritance, concordant 87 Ma zircons, and a predominant 72–73 Ma population of zircons. These results indicate derivation from varied sources which may include the eastern Coast Belt (87 Ma), Paleozoic or late Precambrian sedimentary rocks (recycled Precambrian zircons), and a combination of the Idaho Batholith and Omineca Belt plutons (72–73 Ma grains and grains with inheritance). The Idaho Batholith as a possible source is consistent with the position of the basin prior to latest Cretaceous-(?)-Eocene transcurrent dextral faulting. Thus, the source areas for upper Nanaimo Group sediments were considerably more widespread than previously believed, suggesting that major fluvial drainage systems were active in the western Cordillera during the late Maastrichtian.

INTRODUCTION

We present the first U-Pb analyses of detrital zircons from the Upper Cretaceous Nanaimo Group, part of the Georgia Basin of southwestern British Columbia (Fig. 1). The Nanaimo Group overlies or is adjacent to three basement entities with generally distinctive stratigraphic components and intrusive ages: Wrangellia terrain on Vancouver Island; the Coast Belt on the mainland of British Columbia; and the San Juan thrust system, the western part of the north Cascade Mountains in northwest Washington State (Fig. 1 and Monger, 1990). The isotopic data from this study demonstrate that thrust systems in the Coast Belt and probably the northwest Cascades became the dominant source areas early in Nanaimo Group sedimentation and provide the first evidence that during late stages of sedimentation, source areas several hundred kilometers southeast of the present Coast Belt were also providing sediment.

STRATIGRAPHIC SETTING

The Nanaimo Group, together with Tertiary sediments, forms the fill of the Georgia Basin which is preserved in a northwest-trending, neotectonic structural and topographic depression encompassing Georgia Strait, eastern Vancouver Island, the Fraser River lowlands of British Columbia and northwest Washington State (Fig. 1). The present Georgia Basin is an erosional remnant, and its configuration is largely the result of post-depositional deformation. For much of its depositional history, the basin appears to have extended considerably farther to the west and perhaps to the east. Sedimentary rocks of the Georgia Basin comprise two main packages: the Upper Cretaceous Nanaimo Group, exposed mainly on the east side of Vancouver Island and the Gulf Islands of Georgia Strait, and Paleocene to Miocene rocks exposed in the Vancouver area and northwest Washington.

The Nanaimo Group comprises more than 4 km of sedimentary rocks of Turonian to Maastrichtian age (ages from abundant macro- and microfossil control in most units; reviewed in Haggart, 1991a, b). A comprehensive review of Nanaimo Group stratigraphy, sedimentology, and basin evolution is provided in Mustard (in press). Formations were first proposed by Clapp (1912) with generally accepted modifications made by Muller and Jeletzky (1970) and by Ward (1978). The stratigraphic nomenclature of Ward (1978) is used in this paper with the group subdivided into eleven formations comprising conglomerate, sandstone, and mudstone, with significant coal deposits in the lower formations (Fig. 2). In most places, all but the basal Nanaimo Group consists of marine sandstone-conglomerate successions separated by thick packages of mudstone and fine-grained sandstone. Both successions occur as sedimentary sequences hundreds of meters to more than a kilometer thick in places. Internal lateral and vertical thickness and facies variations are common with both coarsening and fining-upward trends evident on several scales. Formation boundaries are conformable and generally gradational with common lateral intertonguing, but dramatic facies changes have caused local sharp and erosive formation boundaries which some workers have interpreted as unconformities (e.g., Muller and Jeletzky, 1970) although no major biostratigraphic gaps are apparent.

Sedimentologic conclusions prior to the 1980's suggested fluvial and deltaic models for most Nanaimo Group sedimentation, and interpreted the extensive laminated and thin-bedded mudstone-sandstone facies as prodelta shelf deposits and the sandstone-conglomerate facies as fluvial or proximal deltaic deposits (e.g., Muller and Jeletzky, 1970). Researchers in the 1980's, aided by the explosion of literature on deep marine environments and increasing amounts of paleontologic data from specific formations, reinter-

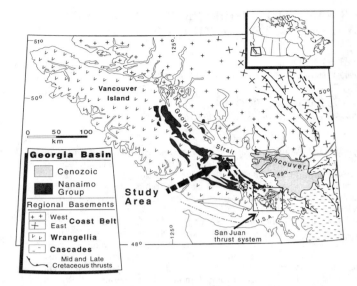

FIG. 1.—Regional setting of the Georgia Basin with the Upper Cretaceous Nanaimo Group shown in dark grey (modified from Monger, 1990). Regional basements to the Nanaimo Group are illustrated, and the San Juan thrust system, part of the larger northwest Cascades system, is outlined. Major thrust faults are simplified from McGroder (1991) and Brandon and others (1988) for the north Cascades, and from Journeay and Friedman (1993) for the southwest B. C. Coast Belt.

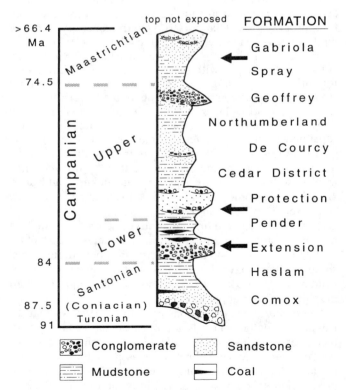

FIG. 2.—Generalized Nanaimo Group stratigraphy using the stratigraphic nomenclature of Muller and Jeletzky (1970) as modified by Ward (1978). Arrows denote relative stratigraphic positions of zircon samples used in this study.

preted most of the Nanaimo Group outside of the coal-bearing areas as submarine fan deposits (e.g., Ward and Stanley, 1982; Pacht, 1984; England and Hiscott, 1992; Mustard, in press). Exceptions include the basal Comox Formation, a coarse-conglomerate and sandstone unit which is generally accepted to be mostly alluvial or coastal marine, and possible coastal or shallow marine fan-delta deposits of the Extension Formation in the southeast part of the basin.

TECTONIC SETTING AND NANAIMO GROUP PROVENANCE

Most researchers have suggested the Nanaimo Group was deposited in a forearc basin (Muller and Jeletzky, 1970; Ward and Stanley, 1982; England, 1990) based on the setting of the basin east of the zone of underthrusting of the Farallon/Kula oceanic plate beneath the North American plate and west of the generally Cretaceous to Tertiary magmatic rocks of the Coast Mountains, part of the morphogeological entity defined as the Coast Belt (Gabrielse and others, 1991). A strike-slip basin model was suggested by Pacht (1984), but faults which he interpreted as synsedimentary and strike-slip have since been shown to be early Tertiary thrusts (England and Calon, 1991).

A foreland basin setting was first proposed by Brandon and others (1988) on the basis of mid-to-Late Cretaceous thrusting which they recognized immediately south of the Nanaimo Group (San Juan thrust system). Problems with this model at the time included poor control on the age of the thrusting, which Brandon and others (1988) could demonstrate was younger than about 100 Ma but could not show overlapped with the time of Nanaimo Group sedimentation. In addition, there was little evidence for coeval thrusting in the Coast Belt, which several studies had suggested was a major source for much of the Nanaimo Group detritus (e.g., Pacht, 1984).

New support for this foreland basin model comes from several recent studies, most importantly the recognition of Late Cretaceous thrust systems in the southern Coast Belt and northwest Cascades, which can be shown to have been active both during initial Nanaimo Group basin formation and major periods of sedimentation. A complex history of late Cretaceous shortening has been demonstrated for a series of northwest-trending, contractional (and in late stages strike-slip) fault systems preserved in the central and eastern Coast Belt. The structural history of these systems is summarized in Journeay and others (1992) and in Journeay and Friedman (1993) which document the well-constrained age controls from related U-Pb zircon geochronologic studies of plutons either crosscutting or crosscut by the thrust system (see also Friedman and Armstrong, in press). The major period of southwest-directed thrusting is evident during the Cenomanian and Turonian (about 96 to 91 Ma) in the central Coast Belt (Lillooet River Fault System of Journeay and Friedman, 1993) and Cenomanian to Campanian (about 100 to 84 Ma) in the northwest Cascades, including the San Juan thrust system, as documented by McGroder (1991), Brandon and others (1988), and Miller and Paterson (1992). This major period of westerly thrusting overlaps with the initial basin subsidence and early basin fill stages of the Nanaimo Group (Turonian to Santonian, reviewed in Mustard, in press). Both northeast and southwest vergent thrusting and associated (but probably late stage) dextral strike-slip faulting can be demonstrated in the eastern Coast Belt for latest Cretaceous time, with southwest directed thrusting also present from about 86 to 68 Ma (Journeay and others, 1992), corresponding to the main period of Nanaimo Group sedimentation.

Recent studies also indirectly link the timing of initial Nanaimo Group sedimentation to synthrust sedimentation in the San Juan thrust system. Garver (1988) recognized a clastic unit of probable Cenomanian-Turonian age (his Obstruction Formation) in the San Juan Islands, which both contains clasts derived from San Juan terrain sources and has been deformed into northwest-verging folds. Garver suggests this clastic unit represents synthrust detritus derived from active thrust sheets and deformed by late stages of thrusting. Significantly, Turonian-age strata have recently been confirmed to be present at the base of the Nanaimo Group immediately west of the San Juan Islands (Haggart, 1991a). If the interpretation of Garver is correct, the proximity of synthrust detritus in the San Juan thrust system to coeval strata at the base of the Nanaimo Group strongly supports a thrust-system control on initial Nanaimo Group sedimentation.

The evidence cited above and a re-examination of the Nanaimo Group basin fill led Mustard and Monger (1991) to resurrect a foreland basin model for the Nanaimo Group. The thick Wrangellia composite terrain of Paleozoic and Jurassic volcanic arcs, oceanic sedimentary piles and major intrusive bodies, and western elements of the Coast Belt provided a semi-rigid foreland basement loaded and deformed in front of the westward-propagating thrust stacks of Coast Belt and Cascade composite terrains to form the initial Nanaimo Basin. Continued uplift, at least partly due to thrusting in the source regions, periodically rejuvenated sediment supplies to the Nanaimo basin throughout its history. Recognition of direct links between specific thrusting and sedimentation events is hampered by two post-depositional effects. First, several kilometers of Late Tertiary uplift of the western Coast Belt (Parrish, 1983; Monger. 1991b) have created a gap between the best preserved thrust systems, near Harrison Lake, and the remaining foreland basin fill with erosion of the original eastern part of the Nanaimo basin, resulting in preservation of only deeper basin marine strata for much of the upper Nanaimo Group. Second, a compressive event of probable mid-Eocene age has deformed the Nanaimo Group into a fold-and-thrust belt which includes a thrust contact between the southern Nanaimo Group and the San Juan terrains (England and Calon, 1991), obscuring the original relationship between the Late Cretaceous thrust system in the San Juan islands and the Nanaimo Group strata.

An indirect test of the foreland basin hypothesis is to examine the provenance of the basin fill. A simple pattern expected for Nanaimo Group sources in the foreland basin model includes: (1) early loading and forebulge migration away (west) from the thrust systems with early thrust-related uplift on the east side of the basin providing a provenance signature of the western Coast Belt and Wrangellia terrain (and perhaps minor forebulge erosion, in this case providing a Wrangellia terrain signature), and (2) continued forebulge migration and basin deepening, submerging the foreland area with the main influx of sediment from the uplifted and eroding thrust stacks, in this case the San Juan terrains and the Coast Belt. An evolution in the importance of the Coast Belt versus the San Juan thrust systems as source areas is also predicted by the current knowledge of timing of thrusting in these areas. If the evidence of Garver (1988) is correct, the San Juan thrust system was most active immediately prior to and during the earliest stages of Nanaimo basin fill. Thrusting in the Coast Belt (and perhaps the northwest Cascades of mainland Washington) was also occurring during later (Campanian to Maastrichtian) deposition of the Nanaimo Group. This suggests that detritus from the San Juan thrust stacks might be most prevalent in lower Nanaimo Group strata and less common in upper strata where a dominance by Coast Belt sources should be apparent. The thrust events in the latest Cretaceous also appear restricted to the eastern Coast Belt and a provenance signature dominated by eastern Coast Belt sources (generally mid to late Cretaceous plutons) might also be expected for upper Nanaimo Group strata. In contrast, western Coast Belt sources mixed with detritus from San Juan terrains might be expected in lower units of the Nanaimo Group.

Previous Nanaimo Group provenance studies have differed somewhat on the postulated source areas. Muller and Jeletzky (1970) concluded that the Wrangellia terrain was the major source of Nanaimo Group sediment for most of the history of the basin with only minor contribution from other sources. More recent studies, based on sandstone and conglomerate compositions and paleocurrent trends, suggest that the provenance of the Nanaimo Group detritus changes with time (Ward and Stanley, 1982; Pacht, 1984). Detritus derived from pre-Late Jurassic Wrangellian rocks are the predominant component only in the basal Comox Formation. The major source are as for all but the Comox Formation appear to have been the Late Cretaceous thrust system of the San Juan Islands to the south and the Coast Belt to the east although in the Nanaimo area a Wrangellia source generally is postulated for the coal-bearing Lower Campanian stratigraphy, including the outcrops of Extension and Protection Formations sampled for this study (Pacht, 1984; England, 1990). Pacht (1984) and England and Hiscott (1992) suggested that the Coast Belt had become the major source region by the time of deposition of the upper Nanaimo Group strata. Thus, recent provenance studies have provided some confirmation for the expected patterns of changing source areas with time predicted by the foreland basin model. The three most likely source areas also have sufficiently distinct stratigraphic and isotopic components to suggest detrital zircons could discriminate individual source areas for the Georgia Basin at various time intervals, a suggestion which prompted this study.

MAJOR SOURCE AREAS

Monger (1991a) reviewed the distribution and age of rocks from the Coast Belt and Wrangellia. Pre-Tertiary Wrangellia comprises stratified and intrusive rocks of predominantly Paleozoic to Jurassic age, cut by intrusions of mainly

Early to Middle Jurassic ages and Devonian or older events (circa 200–175 Ma and 360 Ma or older; Armstrong, 1988; Parrish and McNicoll, 1992). In contrast, the pre-Tertiary Coast Belt is mainly composed of Cretaceous plutonic, volcanic, and sedimentary rocks with relatively minor amounts of Middle to Late Jurassic plutonic and volcanic strata (Armstrong, 1988; Friedman and Armstrong, 1990, in press; Gabrielse and others, 1991). The Coast Belt can be subdivided into several major entities based on dominant ages of granitic rocks and structural styles (tracts of Monger, 1991a). The simplest is a division into western and eastern Coast Belts that was first recognized by Crickmay (1930) and is shown on Figure 1. The western Coast Belt is composed mainly of Late Jurassic to Early Cretaceous granitic rocks with local greenschist grade septa of Triassic through Lower Cretaceous rocks, deformed into discrete shear zones. The eastern Coast Belt comprises penetratively deformed strata metamorphosed up to high amphibolite facies and cut by Late Cretaceous through Early Tertiary granitic rocks. The San Juan thrust system comprises a diverse group of rocks ranging from early Paleozoic to middle Cretaceous in age but commonly of sufficiently dissimilar type or age to allow discrimination from Wrangellia and Coast Belt sources (Pacht, 1984; Brandon and others, 1988).

TABLE 1.—SAMPLE LOCATIONS AND DESCRIPTIONS

Extension Formation, Sample MVM-90–101
LOCATION: Roadcut exposure on southern edge of Nanaimo townsite (Tenth St, 50 m west of Douglas St.) NTS Map 92G/4, Nanaimo Sheet, 1:50,000; UTM zone 10u, 431550E, 5442950N.
DESCRIPTION: Basal pebble conglomerate of a ten meter thick succession of normally graded pebble conglomerates. Chert and volcanic clasts abundant in a coarse-grained lithic wacke matrix. Part of upper Extension Formation (Millstream member of Bickford and Kenyon, 1988).

Protection Formation, Sample MVM-90–100
LOCATION: West-facing 30m high cliff located about 4 km southeast of Nanaimo, about 50 m east of Hwy 1 and 100–150 m north of Cedar Road. NTS Map 92G/4, Nanaimo Sheet, 1:50,000; UTM zone 10u, 433350E, 5441620N.
DESCRIPTION: A chert-rich, small pebble conglomerate with a coarse-grained, lithic arenite matrix. Part of an overlapping succession of pebble conglomerates and coarse sandstones which forms part of the basal Protection Formation (Cassidy Member of Bickford and Kenyon, 1988).

Gabriola Formation, Sample MVM-90–61
LOCATION: Gabriola Island, about 50 m south of Orlebar Point and about 3 meters above the well-exposed base of the formation. NTS Map 92G/4, Nanaimo Sheet, 1:50,000; UTM zone 10u, 440500E, 5450000N.
DESCRIPTION: From base of a massive coarse-grained thick bed of arkosic arenite in a succession of thick-bedded sandstone turbidites and minor interbedded mudstone. Three meters above contact with Spray Formation mudstone.

SAMPLED FORMATIONS

Three samples were collected, two from the type area of the Nanaimo Group at Nanaimo and one from the type area of the Gabriola Formation on Gabriola Island, east of Nanaimo (all defined by Clapp, 1912). The basal sandstone and conglomerate of the Nanaimo Group (Comox Formation) unconformably overlie Wrangellia terrain strata and in many places are obviously composed of angular volcanic and plutonic clasts from the subjacent or nearby sources. For this reason the lowest sandstone sampled was from the Extension Formation. The other samples are from the slightly younger Protection Formation and the youngest unit of the Nanaimo Group, the Gabriola Formation. Table 1 provides location data and sample descriptions.

Extension Formation

The lower Campanian Extension Formation was originally named by Clapp (1912) for conglomerates well-exposed in the area south of Nanaimo and further defined by Muller and Jeletzky (1970) and Ward (1978) to include overlying sandstone in the Nanaimo area. The formation ranges to 480 m thick (Pacht, 1980) in the southern Gulf Islands and is composed mostly of conglomerate and sandstone. In the Nanaimo area, the formation is 100 to 200 m thick and includes significant coal deposits. The depositional environment in the Nanaimo area has been interpreted as a nonmarine, fluvial-floodplain with coastal swamps, and workers have generally considered the source area to be local Wrangellian basement (Muller and Jeletzky, 1970; Bickford, 1989; England, 1990). Pacht (1980) considered the major source for Extension Formation sediments to be imbricated terrains of the San Juan thrust system now exposed southeast of the Nanaimo Group although he postulated a Wrangellia paleohigh in the Nanaimo area which locally contributed detritus and should be the source area for the sample used in this study.

Protection Formation

The lower to middle Campanian Protection Formation was named by Clapp (1912) for a distinctive and persistent unit characterized in the Nanaimo area by greyish-white, quartz-rich arkosic sandstone but also including minor siltstone, coal, and conglomerate. About 200–250 m thick at Nanaimo (Muller and Jeletzky, 1970; Bickford and Kenyon, 1988), the formation is generally thicker in the southern Gulf Islands with a >500-m thick section present on Pender Island (Pacht, 1980). Muller and Jeletzky (1970) proposed a deltaic depositional environment for the Protection Formation with all sediment derived from erosion of Wrangellia terrain. Pacht (1980) interpreted the formation as a product of fluvial deltaic and shallow-marine deposition in the Nanaimo area but suggested that in most of the basin a submarine fan model was more appropriate. England (1990) proposed depositional conditions similar to those of Pacht. Sandstones of the Protection Formation are quartz-rich with abundant potassium feldspar and plagioclase. Pacht (1980) suggested most of the Protection Formation was derived from intrusive rocks of the Coast Belt, except in the Nanaimo area where he speculated intrusive bodies of Wrangellia terrain were the main source. England (1990) also shows Wrangellia as the main source area for the Protection Formation of the Nanaimo area, which includes the area sampled for this study.

Gabriola Formation

The Gabriola Formation (Clapp, 1912) is the top stratigraphic unit of the Nanaimo Group and is preserved only on the outer (eastern) Gulf Islands from Hornby Island in the north to Mayne Island in the south. In most places unfossiliferous, it is considered Maastrichtian, based on a few poorly preserved macrofossils and the lower Maastrichtian age of the conformably underlying Spray Formation (McGugan, 1979; Haggart, 1989). Characterized by massive, thick-bedded, coarse- to fine-grained sandstone (except on Hornby Island where a conglomerate-rich facies is preserved), the Gabriola Formation dips to the northeast into Georgia Strait with only the lower 300 to 800 m of the formation exposed. Muller and Jeletzky (1970) considered the Gabriola Formation to be a deltaic complex with source areas to the west (Wrangellia) and south (San Juan Islands). England and Hiscott (1992) interpreted this formation as part of a submarine fan facies deposited on the outer shelf or slope with sources to the east and southeast (Coast Belt and San Juan thrust system).

Paleocurrents

Regional paleocurrent trends for the formations are shown in Figure 3. In general they support the interpretations of Pacht (1984) for the Extension and Protection Formations, showing strong trends coming from the Coast Belt to the east and from the San Juan thrust system to the southeast. No strong trend is apparent in the Nanaimo area, probably reflecting a mixing of data from non-marine fluvial systems and shallow-marine currents. Paleocurrents from the Gabriola Formation show a radial pattern towards the west but skewed somewhat towards the northwest, supporting the interpretation of submarine fans sourced from the Coast Belt to the east with a basin axis trending northwest.

RESULTS AND SUGGESTED PROVENANCE

Methods of Analysis

Heavy minerals from each of the three samples were concentrated by standard methods used at the Geological Survey of Canada (Parrish and others, 1987). Each of the samples contained zircons although they were much more abundant in the Gabriola Formation sample. Other accessory minerals (titanite, epidote, apatite) were present in all samples and also most common in the Gabriola Formation. An initial finding evident from these data was that the Gabriola Formation had a dominant plutonic source in contrast to the Extension and Protection Formations which contained many clasts of chert. In addition, examination of thin sections cut from several samples of each formation showed that sandstone clasts are generally subangular to subround, feldspars are generally fresh and large, and evidence for sedimentary recycling of grains (e.g., cement overgrowths) is extremely rare, generally absent. This evidence, along with the general coarse-grain size, feldspathic compositions, conglomerate-rich nature of the successions, and high heavy mineral content of the samples all strongly suggest that most sandstone clasts are likely to be first cycle derivatives. For each sample the recovered zircons were divided into fractions based on size, color, and morphology. The grains for isotopic analysis were chosen with the view that only the largest grains likely contain sufficient radiogenic Pb (>25 picograms) for an acceptable analysis, thus introducing a bias to the zircon sample. The largest zircon grains usually were euhedral and well-faceted, indicating a mainly plutonic source. All zircons selected for analysis were

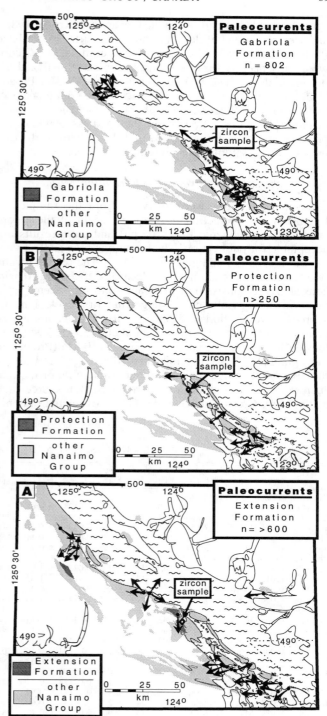

FIG. 3.—Regional paleocurrent trends for the Extension (A), Protection (B), and Gabriola Formations (C). Each arrow represents a vector mean of paleocurrent measurements (10 to 50 per site) located at the arrow initiation point (where arrows are clustered some have been slightly moved for legibility). Paleoflow indicators include sandstone cross-stratification, channel orientations, and conglomerate clast imbrication. Sources for Extension and Protection Formation measurements are about 50% from Pacht (1980, 1984), about 30% from Mustard (unpublished data) and the remainder from unpublished student theses (Hanson, 1976; Ruddiman, 1980; Fahlstrom, 1982) or Cathyl-Bickford (pers. commun. 1992). All Gabriola Formation measurements are by Mustard (unpublished data). All measurements have been corrected for bedding tilt and folding effects.

strongly abraded as part of sample preparation (Krogh, 1982), and we are quite confident that potential surface-correlated Pb loss in the grains has been all but eliminated. Zircons were subsequently dissolved and isotopically analyzed using methods outlined in Parrish and others (1987). Pb and U blanks were approximately 5–15 pg and 1 pg, respectively. Some grains contained abundant inclusions, typical of plutonic zircons. The isotopic data and associated errors of ratios and ages are presented in Table 2. Concordia diagrams for the samples are presented in Figure 4.

Extension and Protection Formation Results

Seventeen zircons were analyzed from the samples of Extension and Protection Formations. Eight of these contained little radiogenic Pb (<10 pg) and gave results unacceptable for publication. The remaining nine analyses are shown in Figure 4 and given in Table 2. The results for the two formations are similar and discussed together. All zircons analyzed consisted of well-faceted to sub-faceted crystals with no obvious core, suggesting first-cycle igneous sources. Two fractions indicate an Early Cretaceous source area, one intersecting concordia at about 120–125 Ma (Extension Fm., grain B) and one at about 110 Ma (Protection Fm., grain D). These Barremian and Albian ages are younger than potential source rocks in either Wrangellia or the San Juan terrains and support a western Coast Belt source such as the Early Cretaceous Gambier Group or related plutonic rocks (Lynch, 1991). Three zircon analyses (grains B, C, and E of Protection Fm.) have relatively large errors due to low U contents of <100 ppm. These grains indicate crystallization ages of about 145–155 Ma. Of the eight unacceptable analyses (not shown on Fig. 4 or Table 2), six had nominal ages consistent with this 145–155 Ma range. Four additional concordant grains from the Extension Formation (grains H, J, C, and F) have $^{206}Pb/^{238}U$ ages of 158.5 ± 0.3 Ma, 167 ± 3 Ma, 221 ± 1 Ma, and 320 ± 1 Ma. The two younger of these four grains (158 and 167 Ma) appear too young for a Wrangellia source, and like the main 145–155 Ma group, were probably derived from the western Coast Belt. Many igneous bodies in the western Coast Belt fall within an age range of about 145 to 170 Ma (e.g., Cloudburst Pluton: 147± 0.5; Parrish and Monger, 1992; Thornborough Pluton: 159 ± 5 Ma; Malibu Diorite: 162 ± 5 Ma; Mt. Jasper Pluton: 167 ± 4 Ma, the latter three ages from Friedman and Armstrong, in press). In the San Juan thrust system, Brandon and others (1988) also reported Late Jurassic zircons from the Fidalgo Igneous Complex of the Decatur terrain (youngest U/Pb zircon age is 160 ± 4 Ma) plus abundant undated tuffs and volcanic flows interbedded with cherts containing Late Jurassic radiolaria. The 220 and 320 Ma grains may have been derived from the San Juan terrains, which contain igneous rocks of both Late Triassic (Haro Formation and Deadman Bay volcanics) and late Mississippian- early Pennsylvanian (East Sound Group) age (Brandon and others, 1988), or from other parts of the

TABLE 2.—U-PB ANALYTICAL DATA FOR DETRITAL ZIRCONS, NANAIMO GROUP

sample,** analysis	wt.## (mg)	U. (ppm)	Pb,+ (ppm)	$^{206}Pb*/^{204}Pb$	$Pb_c,#$ (pg)	$^{208}Pb+/^{206}Pb$	$^{206}Pb++/^{238}U$	$^{206}Pb++/^{238}U$ (Ma)	$^{207}Pb++/^{235}U$	$^{207}Pb++/^{235}U$ (Ma)	$^{207}Pb++/^{206}Pb$	corr. coef.	$^{207}Pb***/^{206}Pb$ (Ma)
Extension Formation, MVM-90–101													
B,c,sf,e	0.014	75.17	1.474	100	15	0.13	0.01941 ± 0.68	123.9 ± 1.7	0.1315 ± 4.09	125.5 ± 9.7	0.04914 ± 3.62	0.72	154.3 ± 162
C,c,f,e	0.009	206.4	7.068	217	20	0.09	0.03486 ± 0.27	220.9 ± 1.2	0.2508 ± 2.20	227.3 ± 9.0	0.05219 ± 2.03	0.69	293.9 ± 90
F,c,f	0.012	144.5	7.557	701	8	0.14	0.05086 ± 0.18	319.8 ± 1.1	0.3694 ± 0.48	319.2 ± 2.6	0.05268 ± 0.41	0.57	314.9 ± 18
H,c,f,e	0.060	163.3	4.040	626	25	0.11	0.02489 ± 0.11	158.5 ± 0.3	0.1687 ± 0.39	158.3 ± 1.2	0.04914 ± 0.34	0.61	154.7 ± 16
J,c,f	0.012	216.8	5.711	89	57	0.12	0.02621 ± 0.75	166.8 ± 2.5	0.1651 ± 6.40	155.1 ± 18	0.04568 ± 5.96	0.63	−19.0 ± 317
Protection Formation, MVM-90–100													
B,c,f	0.006	53.50	1.283	43	19	0.13	0.02367 ± 1.80	150.8 ± 5.4	0.1659 ± 10.9	155.8 ± 32	0.05082 ± 9.76	0.70	232.8 ± 526
C,c,sf	0.010	59.18	1.409	92	12	0.18	0.02367 ± 1.42	150.8 ± 4.2	0.1705 ± 4.57	159.9 ± 13.5	0.05226 ± 3.81	0.65	296.6 ± 184
D,c,f	0.005	542.0	8.920	229	14	0.06	0.01727 ± 0.34	110.4 ± 0.8	0.1170 ± 1.40	112.4 ± 3.0	0.04914 ± 1.23	0.58	154.4 ± 59
E,c,sf	0.006	75.82	1.811	57	17	0.10	0.02420 ± 1.68	154.1 ± 5.1	0.1664 ± 13.7	156.3 ± 40	0.04987 ± 12.6	0.71	188.8 ± 723
Gabriola Formation, MVM-90–61													
A,p,r,e	0.004	103.4	34.83	1045	7	0.17	0.3067 ± 0.15	1724.4 ± 4.7	4.465 ± 0.19	1724.5 ± 3.2	0.1056 ± 0.14	0.72	1724.5 ± 5
B,p,r,e	0.004	386.4	127.8	1780	17	0.11	0.3156 ± 0.10	1768.2 ± 3.0	4.727 ± 0.12	1772.1 ± 2.1	0.1086 ± 0.07	0.81	1776.7 ± 3
C	0.051	286.0	3.307	493	21	0.14	0.01132 ± 0.24	72.6 ± 0.3	0.7215 ± 0.62	70.7 ± 0.8	0.04622 ± 0.49	0.68	9.0 ± 24
D,p,r,e	0.003	166.2	89.93	850	18	0.16	0.4744 ± 0.16	2502.8 ± 6.6	10.810 ± 0.18	2506.8 ± 3.3	0.1653 ± 0.11	0.77	2510.1 ± 4
E,c,f	0.012	220.3	3.062	195	12	0.14	0.01355 ± 0.27	86.8 ± 0.5	0.09000 ± 1.69	87.5 ± 2.8	0.04816 ± 1.51	0.71	107.3 ± 73
F,c,f	0.020	206.1	5.281	677	10	0.05	0.02701 ± 0.17	171.8 ± 0.6	0.1851 ± 0.55	172.5 ± 1.7	0.04971 ± 0.48	0.53	181.3 ± 22
G,c,f	0.015	209.1	2.439	206	12	0.15	0.01133 ± 0.37	72.7 ± 0.5	0.07437 ± 1.67	72.8 ± 2.3	0.04759 ± 1.45	0.64	78.9 ± 71
H,c,f	0.018	163.9	1.882	165	14	0.14	0.01131 ± 0.69	72.5 ± 1.0	0.06882 ± 4.08	67.6 ± 5.3	0.04414 ± 3.85	0.4	−102.9 ± 201
I,c,f,el	0.014	165.9	2.080	86	24	0.22	0.01149 ± 0.60	73.6 ± 0.90	0.07587 ± 3.69	74.3 ± 5.3	0.04789 ± 3.31	0.68	93.8 ± 165
J,c,f,el	0.011	517.2	21.30	927	16	0.06	0.04169 ± 0.17	263.3 ± 0.9	0.5143 ± 0.35	421.4 ± 2.4	0.08948 ± 0.29	0.54	1414.3 ± 11
K,c,f,el	0.010	124.7	1.753	62	20	0.40	0.01130 ± 1.01	72.5 ± 1.5	0.07460 ± 8.17	73.1 ± 11.5	0.04786 ± 7.41	0.78	92.2 ± 395
L,c,f,el	0.024	858.7	41.98	790	77	0.13	0.04660 ± 0.10	293.6 ± 0.60	0.5885 ± 0.19	469.9 ± 1.4	0.09159 ± 0.14	0.68	1458.9 ± 5
M	0.023	233.7	2.711	131	34	0.15	0.01128 ± 0.52	72.3 ± 0.7	0.06743 ± 4.72	66.3 ± 6.1	0.04335 ± 4.43	0.60	−147.4 ± 236

**All grains were originally >100 microns in their longest dimension and very clear with no cracks, inclusions or apparent cores; c = colorless, p = light pink, r = rounded, f = facetted, sf = subfacetted, e = equant, el = elongate.
##Weighing error = 0.001 mg.
+Radiogenic Pb.
*Measured ratio, corrected for spike, and Pb fractionation of 0.09% ± 0.03%/AMU
#Total common Pb in analysis corrected for fractionation and spike.
++Corrected for blank Pb and U, and common Pb (Stacey-Kramers model Pb composition equivalent to the interpreted age of the individual zircons); errors are 1 standard error of the mean in percent for ratios and 2 standard errors of the mean in percent when expressed in Ma.
***Corrected for blank and common Pb, errors are 2 standard errors of the mean in Ma.

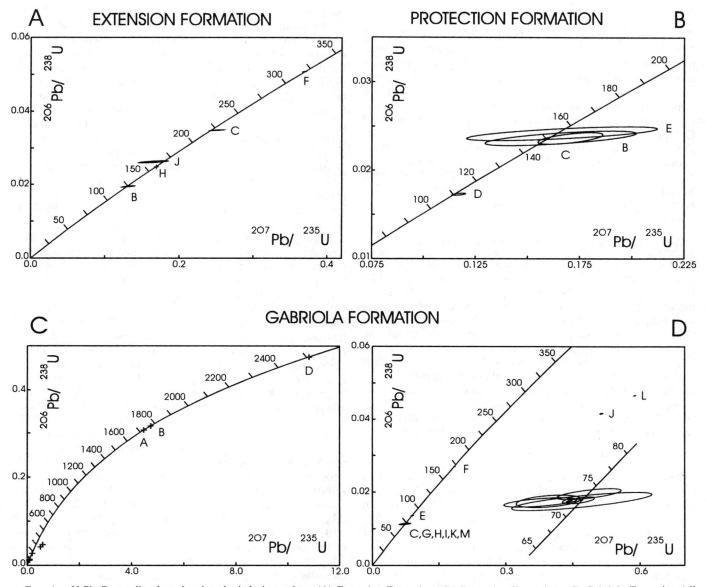

Fig. 4.—U-Pb Concordia plots showing detrital zircon data. (A) Extension Formation; (B) Protection Formation; (C) Gabriola Formation (all data); (D) Gabriola Formation, expanded scale plot of post Pre-Cambrian data.

northwest Cascades (e.g., Marblemount Intrusives with 220 Ma U-Pb age, Mattinson, 1972). Paleozoic igneous rocks in Wrangellia are Devonian or older (e.g., Parrish and McNicoll, 1992), precluding this terrain as a source for the 320 Ma grain. The 220 Ma grain could have come from intrusive rocks associated with the Carnian-Norian age Karmutsen Formation of Wrangellia (Isachsen and others, 1985; Parrish and McNicoll, 1992), but neither the Karmutsen lavas nor related intrusives, which are of mafic composition, contain significant zircon. An eastern to southeastern source, therefore, appears more likely.

Gabriola Formation Results

Zircon, titanite, epidote, and apatite were abundant heavy minerals in the sample of Gabriola Formation, indicating a dominantly plutonic source. Immediately evident from inspection of the zircons was the presence of smaller, rounded pink to reddish grains forming less than 1% of the total zircons. The physical similarity to detrital zircons in the Windermere Supergroup (e.g., Ross and Parrish, 1991) was readily apparent and a Precambrian age suspected. Several of these grains were selected for analysis. The vast majority of grains were clear, sub- to euhedral, and resembled typical igneous zircons from Mesozoic plutonic rocks. In total, thirteen grains were successfully analyzed, and the results are shown in Figures 4C and 4D. The data are concordant with the exceptions of grains J and L, and grains A, B, and D which are Precambrian and show very minor discordance.

Precambrian Grains.—

The Precambrian grains are both Proterozoic (1725 ± 5, 1777 ± 3) and Archean (2510 ± 4 Ma) as given by their nearly concordant ^{207}Pb/^{206}Pb ages. They are distinctly rounded in habit and pink to reddish brown in color, characteristics typical of the detrital zircons of the Windermere Supergroup (Ross and Parrish, 1991) and other lower Paleozoic formations of the eastern Cordillera (Smith and Gehrels, 1991). The bimodal early 1.7–1.8 Ga and Archean zircon ages are also a marked characteristic of Windermere detrital zircons (Ross and Parrish, 1991). We suggest a source of these zircons in the eastern Cordillera, either in British Columbia or adjacent parts of Montana, Idaho, or Washington, and suggest these are recycled sedimentary zircons.

Mesozoic Grains with Precambrian Inheritance.—

Grains J and L are grossly discordant but were very clear and magmatic in their habit with no visible evidence of older cores. Nevertheless, the analyses are typical of Mesozoic granitic zircons with Precambrian inheritance. Plutonic rocks with Precambrian inheritance are virtually absent in the Coast Belt, are present but uncommon in parts of the core zone of the Cascade belt, but are abundant to ubiquitous in the eastern Cordillera of British Columbia (Parrish, 1992a), Idaho, Washington, and Montana. It is not possible to specify the magmatic age of these grains, and they could range in age from 75 to 170 Ma. No local source in the Coast Belt, Wrangellia, or the North Cascades is at all likely, and we suggest transport from considerably farther east.

Concordant Mesozoic Grains.—

The remaining data from the Gabriola Formation contain two grains with concordant analyses of circa 87 ± 1 Ma and 172 ± 1 Ma and a predominant group of 6 grains with concordant ages of 72–73 Ma (Figure 4D). The 72–73 Ma grains appear to make up a very significant proportion of the entire zircon population, possibly up to half the total. The 87 Ma grain could be derived from the eastern Coast Belt where plutons of this age are known (87 ± 1 Ma Hurley Pluton, Friedman and Armstrong, 1990; 87.3 ± 0.3 Ma orthogneiss east of Mount Waddington, Parrish, 1992b; 84 ± 1 Ma Scuzzy Pluton, Parrish and Monger, 1992). The 172 Ma grain could be sourced from the western Coast Belt, easternmost Wrangellia, or from the Omineca Belt farther east. For example, Friedman and Armstrong (in press) report ages of 176 +4/-6 Ma from a sill in the eastern part of the western Coast Belt. The main group of 72–73 Ma grains form much of the zircon population, and it would seem necessary to find a major plutonic source to explain this data.

Armstrong (1988) reports a narrow magmatic belt in the eastern Coast Belt for the period of 85 to 65 Ma, which provides a possible source for the abundant 73 Ma age zircons. However, specific volcanic or intrusive bodies of 72–73 Ma age have yet to be identified despite a recent comprehensive dating program for the southern Coast Belt (Friedman and Armstrong, 1990, in press). This lack of specific candidates in the eastern Coast Belt casts some doubt on it as a major source region. However, several plutons in the Skagit Gneiss Complex of the North Cascades do contain zircons which crystallized at about 70–75 Ma (Haugerud and others, 1991). These are an unlikely source for the Gabriola Formation zircons. Studies of K-Ar cooling ages from the Skagit Gneiss (summarized in Tabor and others, 1989) and from areas to the north (Coleman and Parrish, 1991) clearly show that exhumation of these complexes to depths where the plutons would contribute sediment did not occur until Eocene time. Although cover sequences to the gneiss complex cannot be ruled out as a possible source, the coarse, euhedral nature of the zircons and their abundance suggest a plutonic source. The occurrence of these zircons with others which indicate Omineca Belt sources also suggests a source area in the eastern Cordillera is possible.

Kleinspehn (1985) suggested from studies of the Early Cretaceous Methow-Tyaughton basins that considerable dextral fault movement had occurred on the Fraser-Straight Creek fault systems of southern B. C. and northern Washington. More recently, Parrish and Coleman (1990) and Coleman and Parrish (1991) provided evidence that the granitic rocks of the southernmost Coast Belt, which would have been overlain by the Nanaimo Group, originally were positioned considerably to the southeast of their present position prior to Eocene dextral faulting on both the late Eocene Fraser River-Straight Creek and latest Cretaceous(?) to middle Eocene Yalakom-Hozameen-Ross Lake Fault systems. The amount of dextral displacement on these faults has been estimated by several authors. Monger and Journeay (1992) review studies of the Straight Creek-Fraser River fault system and suggest offset of about 130 to 140 km is most reasonable (other estimates vary from 80 to about 190 km). Riddell and others (1993) provide evidence for about 115 km of dextral offset on the Yalakom Fault (other estimates vary from about 80 to 175 km, reviewed in Riddell and others, 1993). Figure 5 uses a cumulative restoration of about 250 km on these faults, providing a conservative estimate of the Late Cretaceous position of the Nanaimo Group (and adjacent southern Coast Belt and North Cascades), showing them as considerably closer to the Late Cretaceous Idaho Batholith than at present. We speculate that the Idaho Batholith might be a possible source for the 72–73 Ma zircons. Bickford and others (1981) suggest that zircon and monazite U-Pb ages from the northeastern part of this complex of late Cretaceous plutons include zircons with interpreted crystallization ages of about 73 Ma, although a complex pattern of inheritance and thermal resetting is present. This source is especially attractive in light of recent studies of sandstone provenance for several Paleogene sandstone units of the Pacific Northwest; Heller and others (1992b) concluded that the Idaho Batholith was exposed and the major source of detritus for these sandstone units. A combination of nearby Coast Belt and Idaho Batholith-Omineca Belt farther east seems the most reasonable source terrain for the detritus in the Gabriola Formation.

DISCUSSION

Previous studies have suggested that in the Nanaimo area the Extension and Protection Formations comprise non-ma-

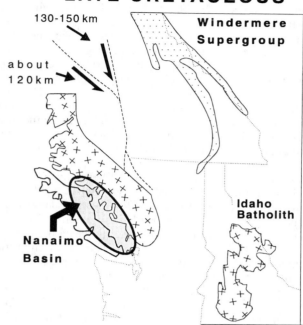

Fig. 5.—Schematic diagrams showing position of Nanaimo Basin and Coast Belt at present (left) and approximate position during latest Cretaceous (right) with about 250 of dextral strike-slip movement restored on Fraser-Straight Creek and Yalakom-Ross Lake fault systems. This minimum estimate of offset does not include offset on the several other early Tertiary strike-slip faults in northwest Washington and southern B. C. or any estimate of the amount of early Tertiary extension.

rine and shallow marine facies and that nearby Wrangellia strata exposed to the north or west were the sole or main source of the sediments. Sandstone clast compositions apparently supported this conclusion, with Pacht (1984) interpreting the volcanic-clast-rich lithic arenite of the Nanaimo area as reflecting derivation from the late Triassic Karmutsen Formation and plagioclase-rich arkose coming from erosion of the Middle Jurassic Island Intrusives (both units forming the sub-Nanaimo Group basement in the Nanaimo area). However, of the nine zircons analyzed from the Extension and Protection Formations, only one Late Triassic zircon compatible with a local (Karmutsen Formation) source was present (although a San Juan terrain source is more likely for this grain as well). The rest of the zircons indicate Late Jurassic to Early Cretaceous western Coast Belt or possible mixed Coast Belt and San Juan thrust system sources. Adding support to the hypothesis of the Coast Belt as the major source area are studies of lower Campanian Nanaimo Group sedimentary outliers within or on the eastern side of Georgia Strait, which have shown them to be coastal non-marine or shallow marine facies of a south or west-facing shoreline (Rouse and others, 1975; Mustard and Rouse, 1991; Mustard, unpublished data). Potential source rocks for these Extension and Protection Formations correlatives include typical Coast Belt intrusives and volcanic-rich Early Cretaceous units such as the Gambier Group (Lynch, 1991). Thus there is both good evidence the Coast Belt was supplying sediment to the Nanaimo Group basin at this time and the source rocks include volcanic types compatible with the sandstone compositions present in the Nanaimo area samples. Reconciling this evidence of an emergent Coast Belt and a west or south facing coastline with the good evidence for terrestrial to shallow marine deposition in the Nanaimo area suggests that a complex but generally west-facing coastline existed which extended west in the Nanaimo area over a wide area of coastal sedimentation. This model suggests the basin was open to the west at this time, a view supported by recent detailed analysis of the internal features of the Nanaimo Group in Mustard (in press) but slightly different from the published paleogeographic reconstructions of Pacht (1984) and in sharp contrast to the facies maps of England (1990), who shows an east-facing basin with Wrangellia emergent, a paleogeography also suggested by Muller and Jeletzky (1970).

Detrital zircons from the Gabriola Formation suggest several interesting interpretations. The strong Coast Belt to North Cascade signature of the 87 Ma and possibly the 72–73 Ma zircons is expected in light of the arkosic nature of the sandstones, previous studies suggesting upper Nanaimo Group rocks were primarily derived from the Coast Belt and the regional paleocurrent trends. However, the presence of the bimodal (Early Proterozoic-Archean) pink zircon population so characteristic of Windermere Supergroup detrital zircons and zircons showing complex inheritance patterns typical of Omineca Belt plutons supports a more easterly source for both these and the younger (72–73 Ma) zircons, the latter probably derived from the Idaho Batholith as outlined earlier.

The presence of Precambrian detrital zircons and of discordant magmatic zircons with a Precambrian inheritance signifies that the source areas for upper Nanaimo Group sedimentary rocks were considerably more widespread than previously suggested. The postulated origin of these zircons from recycled Windermere Supergroup sedimentary rocks and Omineca Belt (or Idaho Batholith) plutons requires a major fluvial system comparable to the main river systems of the present Cordillera (such as the Fraser and Columbia River systems). The present position of the Windermere Supergroup with respect to the Gabriola Formation requires several hundred kilometers of transport. Restoration of dextral transcurrent displacement and major amounts of Eocene extension between the eastern Cordillera and the Georgia Basin would reduce this distance somewhat. Such a major fluvial system was also proposed for several Paleogene units of the Pacific Northwest by Heller and others (1992b) who also concluded that the Idaho Batholith was a major source for detritus of these sandstone units. The very similar interpretations, based on different types of isotopic studies and for different age sedimentary successions, suggest a long-lived, westward draining, fluvial system initiated in the latest Cretaceous and continuing to supply sediment to the western margin of the Pacific Northwest and southwest British Columbia at least into Eocene time.

The abundance of 72–73 Ma zircons from the Gabriola Formation, sampled within a few meters of the formation base, is also significant in that it confirms a maximum age of the Gabriola Formation as Maastrichtian. Given that the source must have been a major plutonic body, possibly located some distance to the east, uplift and erosion of cover rocks to expose the pluton and incorporation of plutonic material into major river systems feeding the Nanaimo Group submarine fan systems would reasonably suggest at least a few million years residence time (the time detritus is stored in the basin before it is brought to the ultimate site of deposition) between pluton crystallization at 72–73 Ma and final sandstone deposition although recent studies show that very short residence times (e.g., <5 my) are possible for similar systems (e.g., Heller and others, 1992a). Even if the residence time were only a few million years, the Gabriola Formation at this sample site can reasonably be inferred as no older than upper Maastrichtian (about 70–65 Ma), a significant conclusion given that Paleocene sedimentary rocks overlie the Nanaimo Group in the southern Gulf Islands and northern San Juan Islands (Mustard and Rouse, 1992). Possibly the Cretaceous and Tertiary parts of the Georgia Basin are not separated by a significant time gap although an unconformity is present in some places (e.g., Mustard and Rouse, 1991).

The evidence and discussion presented above illustrate the diversity of information contained in detrital zircon age populations. This study has shed significant additional light on the complex evolution of Nanaimo Group provenance. The significance of this type of provenance study to the tectonic setting of the basin is also demonstrated. The Nanaimo Group was deposited in a forearc setting only in the broadest sense of this definition such that there was a subduction zone present some unknown distance to the west of the basin, and a magmatic arc was present to the east of the basin. We have argued here (and in more detail in Mustard, in press) that the Nanaimo Group setting is more properly termed a foreland basin, reflecting the primary control on basin initiation and sedimentation of broadly west-directing thrust stacking of Coast Belt and northwestern Cascade rocks in the middle and Late Cretaceous. We would not suggest that the detrital zircon study alone conclusively indicates a foreland basin setting. Rather it provides an important component of the broader basin analysis which depends on a thorough understanding of both the detailed nature of the basin fill and the regional structural evolution. The recent detailed age and structural studies of the thrust events in the Coast Belt and Cascades provided the main regional evidence needed to allow a foreland basin model to be seriously considered (most importantly, Brandon and others, 1988; Journeay and others, 1992; Journeay and Freidman, 1993; and McGroder, 1991). A modern analysis of the stratigraphic fill and sedimentologic evolution of the Nanaimo Group also provided support for an internal foreland basin style of sedimentation (Mustard, in press). The detailed zircon study presented here provides strong evidence to support the expected provenance evolution pattern of such a foreland basin model (outlined in the tectonic setting section of this paper) and has served to demonstrate the more complex evolution of the basin during the final stage of Nanaimo Group sedimentation when major fluvial systems began to provide sediment from well east of the eroded thrust system.

ACKNOWLEDGEMENTS

Field research was supported as part of the Geological Survey of Canada's Georgia Basin Project under the direction of Jim Monger, who also provided valuable advice and discussion of the geology of the southwest Canadian Cordillera and a comprehensive review of an early version of this manuscript. We also thank the staff of the Geochronology Laboratory at the G. S. C. for assistance in generating the U-Pb data, Andy Shoebridge for field assistance, and Paul Heller and Bill McCell for constructive reviews of the manuscript. Geological Survey of Canada Publication Number 10893.

REFERENCES

ARMSTRONG, R. L., 1988, Mesozoic and early Cenozoic magmatic evolution of the Canadian Cordillera: Boulder, Geological Society of America Special Paper 218, p. 55–91.

BICKFORD, C. G. C., 1989, Geology, mining conditions and resource potential of the Wellington coal bed, Georgia Basin, in Geological Fieldwork, 1988: Victoria, British Columbia Ministry of Mines and Petroleum Resources Paper 1989–1, p. 553–558.

BICKFORD, C. G. C. AND KENYON, C., 1988, Coalfield Geology of Eastern Vancouver Island, in Geological Fieldwork, 1987: Victoria, British Columbia Ministry of Mines and Petroleum Resources Paper 1988–1, p. 441–450.

BICKFORD, M. E., CHASE, R. B., NELSON, B. K., SHUSTER, R. D., AND ARRUDA, E. C., 1981, U-Pb studies of zircon cores and overgrowths, and monazite: implications for age and petrogenesis of the northeastern Idaho Batholith: Journal of Geology, v. 89, p. 433–457.

BRANDON, M. T., COWAN, D. S., AND VANCE, J. A., 1988, The Late Cretaceous San Juan thrust system, San Juan Islands, Washington: Boulder, Geological Society of America Special Paper 221. 81 p.

CLAPP, C. H., 1912, Geology of Nanaimo sheet, Nanaimo Coal Field, Vancouver Island, British Columbia: Ottawa, Geological Survey of Canada Summary Report 1911, p. 91–105.

COLEMAN, M. E. AND PARRISH, R. R., 1991, Eocene dextral strike-slip and extensional faulting in the Bridge River Terrain, southwest British Columbia: Tectonics, v. 10, p. 1222–1238.

CRICKMAY, C. H., 1930, The structural connection between the Coast Range of British Columbia and the Cascade Range of Washington: Geological Magazine, v. 67, p. 482–491.

ENGLAND, T. D. J., 1990, Late Cretaceous to Paleogene evolution of the Georgia Basin, southwestern British Columbia: Unpublished Ph.D. Thesis, Memorial University, St John's, 481 p.

ENGLAND, T. D. J. AND CALON, T. J., 1991, The Cowichan fold and thrust system, Vancouver Island, southwestern British Columbia: Geological Society of America Bulletin, v. 103, p. 336–362.

ENGLAND, T. D. J. AND HISCOTT, R. N., 1992, Lithostratigraphy and deep-water setting of the upper Nanaimo Group (Upper Cretaceous), outer Gulf Islands of southwestern British Columbia: Canadian Journal of Earth Sciences, v. 29, p. 574–595.

FAHLSTROM, B. E., 1982, Stratigraphy and depositional history of the Cretaceous Nanaimo Group of the Chemainus area, British Columbia: Unpublished M.Sc. Thesis, Oregon State University, Corvallis, 115 p.

FRIEDMAN, R. M. AND ARMSTRONG, R. L., 1990, U-Pb dating, southern Coast Belt, British Columbia, in Project Lithoprobe, Southern Canadian Cordilleran Workshop Volume: Calgary, University of Calgary, p. 146–155.

FRIEDMAN, R. M. AND ARMSTRONG, R. L., in press, Jurassic and Cretaceous geochronology of the southern Coast Belt, British Columbia, 49°–51°N, in Miller, D., ed., Jurassic Magmatism of the North American Cordillera: Boulder, Geological Society of America Special Paper.

GABRIELSE, H., MONGER, J. W. H., WHEELER, J. O., AND YORATH, C. J., 1991, Part A. Morphogeological belts, tectonic assemblages, and terranes; in Gabrielse, H., and Yorath, C. J., eds., Geology of the Cordilleran Orogen in Canada: Ottawa, Geological Survey of Canada, Geology of Canada 4, p. 15–28.

GARVER, J., 1988, Stratigraphy, depositional setting, and tectonic significance of the clastic cover to the Fidalgo Ophiolite, San Juan Islands, Washington: Canadian Journal of Earth Sciences, v. 25, p. 417–432.

HAGGART, J. W., 1989, New and revised ammonites from the Upper Cretaceous Nanaimo Group of British Columbia and Washington State: Geological Survey of Canada Bulletin, v. 396, p. 181–221.

HAGGART, J. W., 1991a, A new assessment of the age of the basal Nanaimo Group, Gulf Islands, British Columbia: Ottawa, Geological Survey of Canada Paper 91–1E, p. 77–82.

HAGGART, J. W., 1991b, Biostratigraphy of the Upper Cretaceous Nanaimo Group, Gulf Islands, British Columbia, in Smith, P. L., ed., A Field Guide to the Paleontology of Southwestern Canada: Vancouver, Canadian Paleontology Conference I, Geological Association of Canada, Paleontology Division, p. 222–257.

HANSON, W. B., 1976, Stratigraphy and sedimentology of the Cretaceous Nanaimo Group, Saltspring Island, British Columbia: Unpublished Ph.D. Thesis, Oregon State University, Corvallis, 339 p.

HAUGERUD, R. A., VAN DER HEYDEN, P., TABOR, R. W., STACEY, J. S., AND ZARTMAN, R. E., 1991, Late Cretaceous and early Tertiary plutonism and deformation in the Skagit Gneiss Complex, North Cascade Range, Washington and British Columbia: Geological Society of America Bulletin, v. 103, p. 1297–1307.

HELLER, P. L., RENNE, P. R., AND O'NEIL, J. R., 1992a, River mixing rate, residence time, and subsidence rates from isotopic indicators: Eocene sandstones of the Pacific Northwest: Geology, v. 20, p. 1095–1098.

HELLER, P. L., TABOR, R. E., O'NEAL, J. R., PEVEAR, D. R. SHAFIQULLAH, M., AND WINSLOW, N., 1992b, Isotopic provenance of Paleogene sandstones from the accretionary core of the Olympic Mountains, Washington: Geological Society of America Bulletin, v. 104, p. 140–153.

ISACHSEN, C., ARMSTRONG, R. L., AND PARRISH, R. R., 1985, U-Pb, Rb-Sr, and K-Ar geochronometry of Vancouver Island igneous rocks (abs.): Sidney, Geological Association of Canada Victoria Section Symposium, Abstracts, p. 21–22.

JOURNEAY, J. M. AND FRIEDMAN, R. M., 1993, The Lillooet River Fault System: evidence of late Cretaceous shortening in the Coast Belt of SW British Columbia: Tectonics, v. 12, p. 756–775.

JOURNEAY, J. M., SANDERS, C., VAN-KONIJNENBURG, J. H., AND JAASMA, M., 1992, Fault systems of the Eastern Coast Belt, southwest British Columbia: Ottawa, Geological Survey of Canada Paper 92–1A, p. 225–235.

KLEINSPEHN, K. L., 1985, Cretaceous sedimentation and tectonics, Tyaughton-Methow Basin, southwest British Columbia: Canadian Journal of Earth Sciences, v. 22, p. 154–174.

KROGH, T. E., 1982, Improved accuracy of U-Pb ages by the creation of more concordant systems using an air abrasion technique: Geochimica et Cosmochimica Acta, v. 46, p. 637–649.

LYNCH, J. V. G., 1991, Georgia Basin Project, stratigraphy and structure of Gambier Group rocks in the Howe Sound-Mamquam River area, southwest Coast Belt, British Columbia: Ottawa, Geological Survey of Canada Paper 91–1A, p. 49–57.

MATTINSON, J. M., 1972, Ages of zircons from the Northern Cascade Mountains, Washington: Geological Society of America Bulletin, v. 83, p. 3769–3784.

MCGRODER, M. F., 1991, Reconciliation of two-sided thrusting, burial metamorphism, and diachronous uplift in the Cascades of Washington and British Columbia: Geological Society of America Bulletin, v. 103, p. 189–209.

MCGUGAN, A., 1979, Biostratigraphy and paleoecology of Upper Cretaceous (Campanian and Maastrichtian) foraminifera from the Upper Lambert, Northumberland, and Spray Formations, Gulf Islands, British Columbia, Canada: Canadian Journal of Earth Sciences, v. 16, p. 2263–2274.

MILLER, R. B. AND PATERSON, S. R., 1992. Tectonic implications of syn- and post-emplacement deformation of the Mount Stuart batholith for mid-Cretaceous orogenesis in the North Cascades: Canadian Journal of Earth Sciences, v. 29, p. 479–485.

MONGER, J. W. H., 1990, Georgia Basin: Regional setting and adjacent Coast Mountains geology, British Columbia: Geological Survey of Canada Paper 90–1F, p. 95–107.

MONGER, J. W. H., 1991a, Georgia Basin Project: structural evolution of parts of the southern Insular and southwestern Coast belts, British Columbia: Geological Survey of Canada Paper 90–1A, p. 219–228.

MONGER, J. W. H., 1991b, Late Mesozoic to Recent evolution of the Georgia Strait-Puget Sound region, British Columbia and Washington: Washington Geology, v. 19, p. 3–7.

MONGER, J. W. H. AND JOURNEAY, J. M., 1992, Guide to the geology and tectonic evolution of the southern Coast Belt: Bowen Island, Field Guide to 1992 Penrose Conference on Tectonic Evolution of the Coast Mountains Orogen, 97 p.

MULLER, J. E. AND JELETZKY, J. A., 1970, Geology of the upper Cretaceous Nanaimo Group, Vancouver Island and Gulf Islands, British Columbia: Ottawa, Geological Survey of Canada Paper 69–25, 77 p.

MUSTARD, P. S., in press, The Upper Cretaceous Nanaimo Group, Georgia Basin, British Columbia, in Monger, J. W. H., ed., Geology and Geohazards of the Vancouver Region, Southwestern British Columbia: Ottawa, Geological Survey of Canada Bulletin 481.

MUSTARD, P. S. AND MONGER, J. W. H., 1991, Upper-Cretaceous Tertiary Georgia Basin, British Columbia: Forearc or Foreland (abs.): Abstracts with Program, Geological Association of Canada, Annual Meeting, v. 16, p. A88.

MUSTARD, P. S. AND ROUSE, G. E., 1991, Sedimentary outliers of the eastern Georgia Basin margin, British Columbia: Ottawa, Geological Survey of Canada Paper 91–1A, p. 229–240.

MUSTARD, P. S. AND ROUSE, G. E., 1992, Tertiary Georgia Basin, British Columbia and Washington State: a view from the other side: Ottawa, Geological Survey of Canada Paper 92–1A, p. 13–23.

PACHT, J. A., 1980, Sedimentology and petrology of the Late Cretaceous Nanaimo Group in the Nanaimo Basin, Washington and British Columbia: Implications for Late Cretaceous tectonics: Unpublished Ph.D. Thesis, Ohio State University, Columbus, 368 p.

PACHT, J. A., 1984, Petrologic evolution and paleogeography of the Late Cretaceous Nanaimo Basin, Washington and British Columbia: implications for Cretaceous tectonics: Geological Society of America Bulletin, v. 95, p. 766–778.

PARRISH, R. R., 1983, Cenozoic thermal evolution and tectonics of the Coast Mountains of British Columbia; 1. Fission track dating, apparent uplift rates, and patterns of uplift: Tectonics, v. 2, p. 941–994.

PARRISH, R. R., 1992a, U-Pb ages for Jurassic-Eocene plutonic rocks in the vicinity of Valhalla Complex, southeast British Columbia: Ottawa, Geological Survey of Canada Paper 91–2, p. 115–134.

PARRISH, R. R., 1992b, U-Pb ages for Cretaceous plutons in the eastern Coast Belt, southern British Columbia: Geological Survey of Canada Paper 91-2, p. 109–113.

PARRISH, R. R. AND COLEMAN, M. E., 1990, A model for Eocene extension and strike-slip faulting for the Canadian Cordillera and Pacific Northwest (abs.): Abstracts with Program, Geological Association of Canada, v. 15, p. 101.

PARRISH, R. R. AND MCNICOLL, V. J., 1992, U-Pb age determinations from the southern Vancouver Island area, British Columbia: Ottawa, Geological Survey of Canada, Paper 91-2, p. 79–86.

PARRISH, R. R. AND MONGER, J. W. H., 1992, New U-Pb dates from southwestern British Columbia: Ottawa, Geological Survey of Canada, Paper 91-2, p. 87–108.

PARRISH, R. R., RODDICK, J. C., LOVERIDGE, W. D., AND SULLIVAN, R. W., 1987, Uranium-Lead analytical techniques at the geochronology laboratory, Geological Survey of Canada: Ottawa, Geological Survey of Canada Paper 87-2, p. 3–7.

RIDDELL, J., SCHIARIZZA, P., GABA, R. G., CAIRA, N., AND FINDLAY, A., 1993, Geology and mineral occurrences of the Mount Tatlow map area (92O/5, 6, and 12), in Grant, B., and Newell, J. M., eds., Geological Fieldwork 1992: Victoria, British Columbia Ministry of Energy, Mines and Petroleum Resources Paper 1993-1, p. 37–52.

ROSS, G. M. AND PARRISH, R. R., 1991, Detrital zircon geochronology of metasedimentary rocks in the southern Omineca Belt, Canadian Cordillera: Canadian Journal of Earth Sciences, v. 28, p. 1254–1270.

ROUSE, G. E., MATHEWS, W. H., AND BLUNDEN, R. H., 1975, The Lions Gate Member: a new late Cretaceous sedimentary subdivision in the Vancouver area of British Columbia: Canadian Journal of Earth Sciences, v. 12, p. 464–471.

RUDDIMAN, W., 1980, The geology and stratigraphy of the lower Nanaimo Group, Nanaimo, British Columbia: Unpublished M.Sc. Thesis, Oregon State University, Corvallis, 111 p.

SMITH, M. T. AND GEHRELS, G. E., 1991, Detrital zircon geochronology of upper Proterozoic to lower Paleozoic continental margin strata of the Kootenay Arc: implications for the early Paleozoic tectonic development of the eastern Canadian Cordillera: Canadian Journal of Earth Sciences, v. 28, p. 1271–1284.

TABOR, R. W., HAUGERUD, R. A., BROWN, E. H., BABCOCK, R. S., AND MILLER, R. B., 1989, Accreted Terrains of the North Cascades Range, Washington; IGC Field Trip Guidebook T307: Washington, D. C., 28th International Geological Congress, American Geophysical Union, 62 p.

WARD, P. D., 1978, Revisions to the stratigraphy and biochronology of the Upper Cretaceous Nanaimo Group, British Columbia and Washington: Canadian Journal of Earth Sciences, v. 15, p. 405–423.

WARD, P. D. AND STANLEY, K. O., 1982, The Haslam Formation: a Late Santonian-Early Campanian forearc basin deposit in the Insular Belt of southwestern British Columbia and adjacent Washington: Journal of Sedimentary Petrology, v. 52, p. 975–990.

PROVENANCE OF THE DEVONIAN CLASTIC WEDGE OF ARCTIC CANADA: EVIDENCE PROVIDED BY DETRITAL ZIRCON AGES

VICKI J. McNICOLL
Geological Survey of Canada, 601 Booth St., Ottawa, Ontario, K1A 0E8, Canada
AND
J. CHRIS HARRISON, HANS P. TRETTIN, AND RAY THORSTEINSSON
Geological Survey of Canada, 3303–33rd St. NW, Calgary, Alberta, T2L 2A7, Canada

ABSTRACT: U-Pb geochronology data are presented for single detrital zircon grains from Eifelian, Givetian, and Frasnian sandstones of the Bird Fiord Formation of the central Arctic Islands, Hecla Bay, and Fram Formations of south central Ellesmere Island and the Okse Bay Formation of northern Ellesmere Island. The ranges of $^{207}Pb/^{206}Pb$ crystallization ages of the detrital zircons extracted from these sandstones are as follows: eight zircons between 2.62 and 3.0 Ga, four grains between 2.25 and 2.47 Ga, eleven zircons between 1.57 and 2.02 Ga, seven zircons between 1.04 and 1.20 Ga, and one grain of 0.43 Ga.

Potential source terrains include the Precambrian shield areas of Canada and Greenland and the mid-Paleozoic Caledonian-Franklinian orogen of Scandinavia, East and North Greenland, and the Canadian Arctic Islands. The East Greenland Caledonian orogen and its unroofed foreland molasse basin are the most probable primary source for the dated detrital zircons of the Devonian clastic wedge. The geology and geographic location of this region also satisfy other aspects of provenance including paleocurrent measurements, mineralogical and petrographic considerations, and the timing and magnitude of provenance area uplift.

INTRODUCTION

Existing hypotheses regarding provenance of the Middle and Upper Devonian foreland clastic wedge of Canada's Arctic Islands include derivation predominantly from northern tectonic lands, including Pearya terrane, versus transport from the Caledonian mountain belt and the Greenland shield to the east. In order to place additional constraints on these hypotheses, thirty-one U-Pb detrital zircon ages were obtained from four representative sandstones of the central and eastern portions of the foreland wedge. The interpretation of the results is augmented by published and some unpublished paleocurrent data, accessory and heavy fraction mineralogy, and a summary of likely source terrains in the circum-Arctic region. The correlation of relative and absolute geological time scales is based on Cowie and Bassett (1989) and Harland and others (1989) for the Paleozoic and Plumb (1991) for the Precambrian.

REGIONAL GEOLOGY

The Middle and Upper Devonian sediment record in the Canadian Arctic Islands includes one of the largest foreland clastic wedges of its age in North America. It is approximately ten times the size of the age-equivalent "Catskill" wedge of the Appalachian foreland (Embry, 1991). At present the clastic wedge underlies an area of approximately 240,000 km² and ranges in thickness from 3900 m on southwestern Ellesmere Island to over 4200 m on western Melville Island (Fig. 1). Up to 9 or 10 km of section are inferred from thermal maturity data for the combined eroded and preserved portions of the clastic wedge in the Prince Patrick and Banks Island areas (Embry, 1988). Coeval and related deposits are assigned to the Imperial Formation in the northern Yukon (Gordey, 1991).

The widespread appearance of voluminous Devonian siliciclastic detritus in the Arctic Islands began at about 383 Ma within the *costatus* conodont zone of the early Eifelian (Embry, 1991). Foreland sedimentation continued for approximately 25 my and ceased after the mid-Famennian with the advent of the Ellesmerian Orogeny, a widespread, thin-skinned, foreland folding and related deformation within the Franklinian Mobile Belt (Thorsteinsson and Tozer, 1970).

The Arctic Devonian clastic wedge comprises the youngest rocks in a structurally conformable succession ranging from Lower Cambrian to Upper Devonian (Famennian) time. The top of this succession is an angular unconformity marking the passage of the Ellesmerian orogeny. The northwestern erosional limit of the clastic wedge is present on central Ellesmere Island, western Grinnell Peninsula on Devon Island, and on northwestern Melville Island above deep-seated Ellesmerian folds and thrusts that expose mostly sub-Middle Devonian strata typical of the hinterland region of the Franklinian belt.

On the cratonward side, the clastic wedge overlies a lowest Eifelian carbonate platform. Progressively older craton cover is exposed at the present erosion surface further to the south and southeast. The oldest deposits of the Arctic platform include Cambrian and Lower Ordovician platform cover preserved unconformably above Meso-Neoproterozoic sedimentary strata and an Archean to Paleoproterozoic volcanic-plutonic basement complex of eastern Devon, Somerset, Victoria, and northern Baffin islands (Fig. 1). The pre-erosional extent and thickness of the Devonian clastic wedge on the platform is uncertain.

Accumulation of Carboniferous (upper Viséan through Moscovian) and younger, rift-related deposits of the Sverdrup Basin place an upper age limit on the mid-Paleozoic folding and hinterland uplift of the Devonian clastic wedge. Rocks of these ages are widely exposed on the hinterland side of the Devonian clastic wedge on Ellesmere and northern Melville islands and on Grinnell Peninsula.

Two outliers of the Devonian clastic wedge also occur within the hinterland Franklinian orogen near Yelverton Pass and de Vries Glacier on northern Ellesmere Island (Fig. 1). The basal contact of these outliers is not exposed, but these clastic sediments are clearly less deformed than surrounding Silurian and older rocks. Nevertheless, uncertainty exists as to whether these outliers represent deposition in a post-orogenic intermontane setting or a succession conformable with underlying Paleozoic rocks and co-extensive

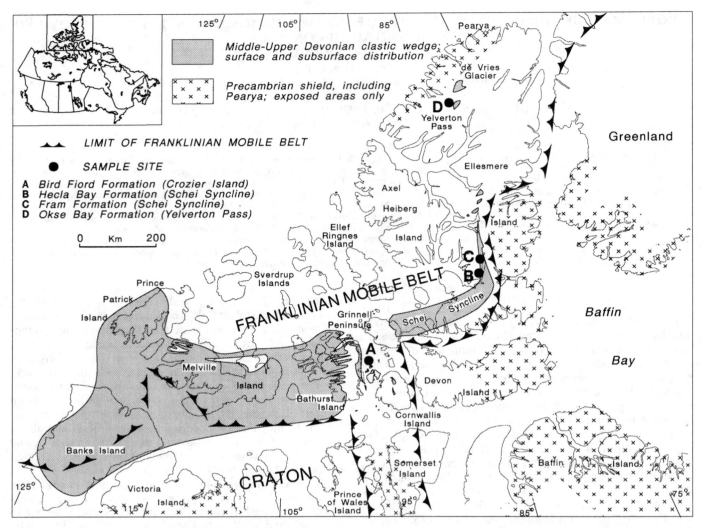

FIG. 1.—Location of the Middle and Upper Devonian clastic wedge and other selected geological features in Arctic Canada.

with the rest of the foreland clastic wedge. The significance of this point lies in the extent to which the intervening foldbelt may have been a potentially exposed source area during deposition of the clastic wedge and its related outliers.

Stratigraphy of the Clastic Wedge

Deposition of the clastic wedge took place in the Eifelian through Frasnian or approximately 385 to 370 Ma. The stratigraphy of the clastic wedge is divided into three sequences, each bounded above and below by regionally significant unconformities and their correlative conformable surfaces (Figs. 2, 3). From base to top these include: the Hecla Bay sequence embracing coeval and laterally linked depositional facies of the Blackley, Cape De Bray, Weatherall, Bird Fiord, Strathcona Fiord, and Hecla Bay Formations; the Beverley Inlet sequence including the Fram, Hell Gate, Nordstrand Point, and Beverley Inlet Formations; and the Parry Islands sequence including the Parry Islands Formation that consists of three regionally-correlated members. In addition to these strata, Embry (1988) also suggests the presence of now-eroded, late Famennian to Tournaisian strata in the Arctic Islands which may have been 2.5 to 3.0 km thick on Bathurst, Melville, and Prince Patrick Islands and 5.0 to 6.0 km thick on Banks Island.

PROVENANCE OF THE DEVONIAN CLASTIC WEDGE

The provenance of the Devonian clastic wedge presents a geological paradox. The geological setting is typical of many foreland basins. A tectonic load lying to the northwest over the site of a Lower Cambrian-middle Devonian marine basin is considered to have caused flexural bending of the lithosphere that localized and preserved an orogen-derived clastic wedge in a southwest-trending, foredeep moat lying on the foreland (southeast) side of the tectonic load (Embry, 1988 and references therein).

The conventional view dictates that the principal source of sediment in the clastic wedge would also be derived from the same tectonic load lying outboard to the northwest. This

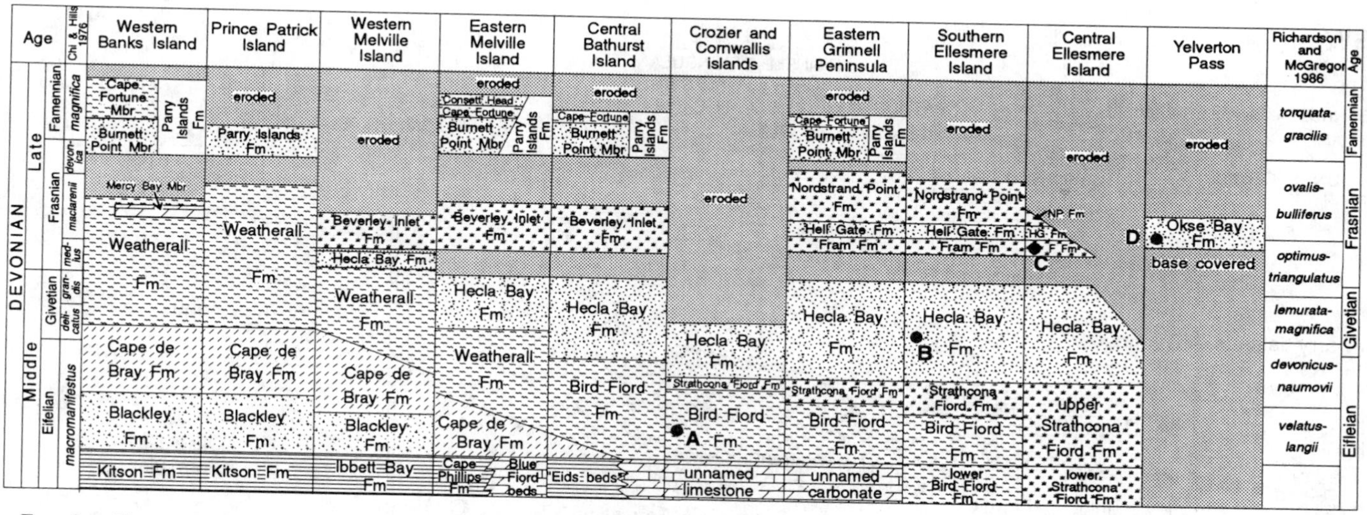

FIG. 2.—Time correlation chart for Middle and Upper Devonian strata of the Arctic Islands redrawn from Embry (1988) and Trettin (1991c). Stratigraphic positions of samples A to D are shown. The legend for the facies symbols is in Figure 3.

idea was first put forward by Thorsteinsson (1960) and expanded by Tozer and Thorsteinsson (1964). Regional depositional facies patterns and petrographic data, however, tend to favor the model first suggested and described in detail by Embry and Klovan (1974, 1976) and more or less supported by Dineley (1975), Trettin (1978), Rice (1987), and Embry (1988, 1991). This model indicates that principal source areas during Middle Devonian (Eifelian and Givetian) time lay on the cratonward side of the foreland basin, and sediment was transported a short distance from the Ellesmere-Greenland shield and/or a long distance and across the entire width of the Greenland shield from the Caledonian mountain belt and a pre-existing molasse basin in the Caledonian foreland. The model also proposes mixed sediment sources for Frasnian (Late Devonian) units with chert, which is abundant in the uppermost Hecla Bay and Beverley Inlet Formations of the western Arctic Islands, derived from the Franklinian orogen to the north and quartz and micas derived from the shield and/or Caledonian mountain belt to the east and northeast. Northerly source regions are indicated by an abundance of chert in the clastic wedge in Famennian (Late Devonian) time and pre-existing Tournaisian (Early Mississippian) of the western Arctic Islands, only.

In order to place additional constraints on hypotheses regarding provenance, detrital zircons were analyzed from four representative samples of the central and eastern portions of the Devonian clastic wedge. The samples examined were from the Bird Fiord, Hecla Bay, Fram, and Okse Bay Formations (labelled A-D, respectively, on Fig. 1).

Paleocurrent Data

There is a general southwesterly direction of sediment transport and accumulation in the clastic wedge. Paleocurrent indicators and previous provenance studies are summarized below and in Figure 4 for zircon sample site areas of the Bird Fiord, Hecla Bay, Fram, and Okse Bay formations.

In the Bird Fiord Formation on Grinnell Peninsula, de Freitas and Mayr (1992) have reported south-directed paleocurrent indicators from the lowermost and uppermost members and northeast-directed indicators from the middle part (Fig. 4). More distal equivalents of the Bird Fiord Formation are assigned to the Cape De Bray and Weatherall Formations on Melville Island. Seismic clinoforms in the Cape De Bray Formation define a southwest direction of delta progradation throughout Eifelian time (Harrison, 1991; in press). Correlative, more proximal, deposits of the Strathcona Fiord Formation include southwest-and northeast-directed paleocurrent indicators from the Schei Syncline of southern Ellesmere Island (Embry and Klovan, 1976; Rice, 1987) and west-directed paleocurrents on central Ellesmere Island (Trettin, 1978; Roblesky, 1979). Trettin (1978), Rice (1987), and Rice (in press) suggest that the Ellesmere-Greenland shield or Caledonian mountain belt are sources for quartz, feldspar, micaceous minerals, and metamorphic rock fragments.

Paleocurrents from the Hecla Bay Formation are west- and southwest-directed in the Schei Syncline on Ellesmere Island (Fig. 4; Embry and Klovan, 1976; Roblesky, 1979; Rice, 1987). However, transport to the northeast, northwest and south is also indicated at some localities (Rice, 1987). Paleocurrent indicators measured in the Hecla Bay Formation on Melville Island are more consistently southwest-directed (Embry and Klovan, 1976). The work of Embry and Klovan (1976), Rice (1987), Embry (1988, 1991), and Mayr and others (in press) suggest a source for quartz, feldspar, and micaceous minerals in the Ellesmere-Greenland shield, Caledonian mountain belt, and a pre-existing Caledonian foreland basin.

Paleocurrent indicators from the Fram Formation of the Schei Syncline are variably toward the west and southwest although other directions have also been documented (Fig. 4; Embry and Klovan, 1976; Roblesky, 1979; Rice, 1987). Embry and Klovan (1976) suggested that the bulk of the quartz and other detrital minerals of the Fram Formation of

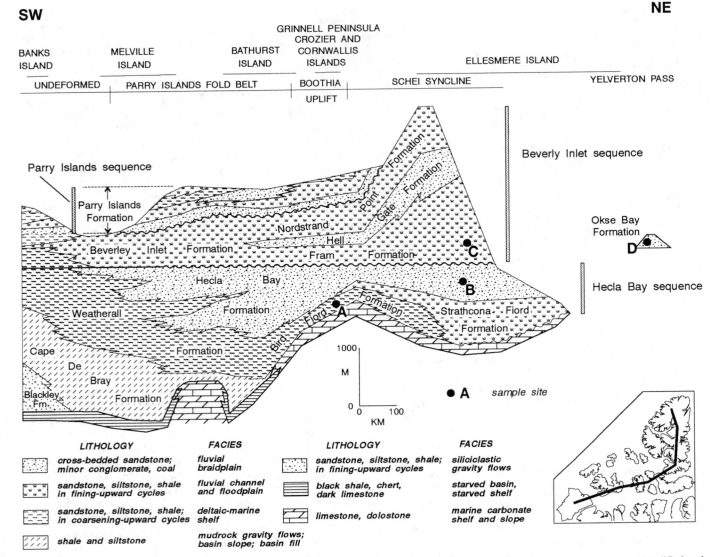

FIG. 3.—Stratigraphic cross-section of the Hecla Bay, Beverley Inlet, and Parry islands sequences of the Devonian clastic wedge. Modified and redrafted from Embry (1988).

the southern Schei Syncline were derived from the Ellesmere-Greenland shield and Caledonian mountain belt, a view also supported by the detailed work in this area by Rice (1987).

Paleocurrent data for the Okse Bay Formation outliers near Yelverton Pass and de Vries Glacier indicate flow from the southeast and east (Maurel, 1989; Fig. 4). The nearest possible source region in this direction is the Grantland Uplift located in the Hazen Plateau and Grantland Mountains area on Ellesmere Island (Trettin, 1971). Potential source formations include the Lower Cambrian Grantland Formation for detrital quartz and the Hazen Formation for detrital chert; however, the existence and extent of uplift in this area during the deposition of the Okse Bay Formation is uncertain.

SAMPLE DESCRIPTIONS

Detailed petrographic descriptions, listed heavy and accessory fraction mineralogy, and sample location data for each of the four samples are provided in Appendix A.

Sample A: Bird Fiord Formation

Sample A is a very fine-grained, green weathering, calcareous, fossiliferous sandstone collected from the Bird Fiord Formation on Crozier Island located 38 km northwest of Cornwallis Island (Fig. 1, Appendix A). The Bird Fiord Formation, first defined by McLaren (1963), is the oldest unit of the clastic wedge in this part of the Arctic Islands. The formation includes about 150 to 200 m of dark grey

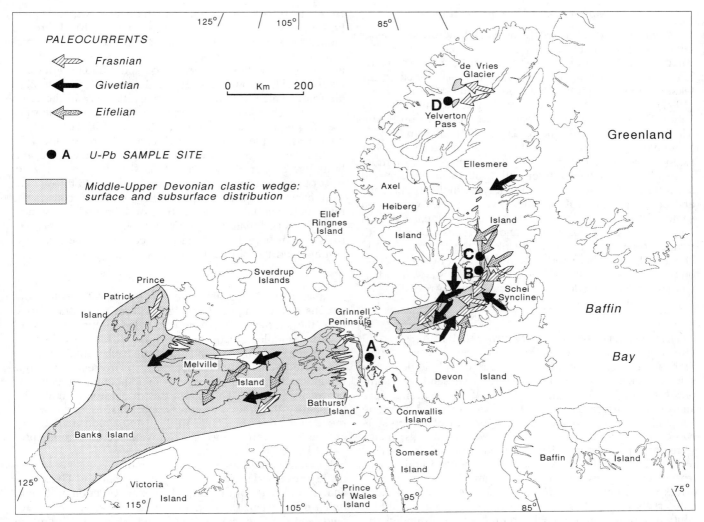

Fig. 4.—Paleocurrent data for the Devonian clastic wedge of the Arctic Islands compiled from Embry and Klovan (1976), Harrison (1991; in press), Maurel (1989), Rice (1987), Roblesky (1979), Trettin (1978), and some unpublished data contributed by de Freitas (pers. commun., 1992).

shale, dark grey-brown siltstone, and fossiliferous micaceous calcareous sandstone. It has been described by Embry and Klovan (1976) and Goodbody (1989) who interpret shallow siliciclastic shelf and marine deltaic depositional environments.

Sample B: Hecla Bay Formation

Sample B, a pale brown, weakly cemented quartz sandstone, was collected from the medial part of the Hecla Bay Formation on the east-dipping upright limb of the Schei Syncline on southern Ellesmere Island (Fig. 1, Appendix A). The Hecla Bay Formation was first proposed by Tozer and Thorsteinsson (1964) and has also been described by Embry and Klovan (1976) and Roblesky (1979). The formation, which is about 638 m in thickness near the sample site, is dominated by white- to pale yellow weathering, moderately resistant, quartz sandstone. The Hecla Bay Formation of the eastern and central Arctic Islands is considered by Embry and Klovan (1976) and Embry (1988) to have been deposited in a fluvial braidplain setting.

Sample C: Fram Formation

Sample C is pale pink weathering, calcite-cemented, fine-grained, quartz sandstone. It was collected from an exposure of the Fram Formation on the east-dipping limb of the Schei Syncline on southern Ellesmere Island (Fig. 1, Appendix A). The Fram Formation was first proposed by Embry and Klovan (1976) and has also been described by Roblesky (1979). The formation, locally 1480 m thick, includes an assemblage of red and green shales, green siltstones, and brownish-yellow and red sandstones, all arranged in fining-upward cycles each 1 to 10 m in thickness. A meandering fluvial channel and overbank floodplain depositional setting have been proposed by Embry and Klovan (1976) and Embry (1988).

Sample D: Okse Bay Formation

Sample D was collected from weakly-deformed, upper Givetian-Frasnian sandstone and conglomerate outliers located in the Yelverton Pass area of northern Ellesmere Island (Fig. 1, Appendix A). These strata, assigned to the Okse Bay Formation by Mayr and Trettin (1990) and Mayr (1991), are overlain, above a low-angle unconformity, by Upper Carboniferous basal deposits of the Sverdrup Basin. The base is concealed, and rocks of latest Silurian to mid-Devonian age are not exposed or preserved in the broader region. The Okse Bay Formation includes 600 m of strata in the Yelverton Pass outlier and 1360 m in the exposed section near de Vries Glacier. Rock types include sandstone and conglomerate with, in the lower part, lesser maroon shale, siltstone, and minor caliche. Depositional environments include various fluvial and overbank settings with braided river deposits dominant in the upper part. Sample D was collected about 150 m below the top of the formation on the southeastern side of Yelverton Pass. It is a well sorted, quartz-cemented, medium grained sandstone (Appendix A).

U-Pb ANALYTICAL TECHNIQUES AND RESULTS

The sandstone samples were processed using standard crushing and mineral separation techniques. Seven or eight, high quality, single grains of various morphologies were selected for analysis from each of the four sandstone samples for a total of 31 grains. Only single zircon grains were analyzed to ensure that each analysis represented a unique age and not a mixture of several ages. All of the crystals chosen for analysis were very clear with no cracks, inclusions, or apparent cores. Several populations of zircon were recognized including pink to colorless rounded grains with frosted and slightly pitted surfaces, and light pink to colorless (and rarely light brown), euhedral, equant to elongate crystals with sharp to somewhat worn facets. Table 1 contains a description of each zircon fraction. Rounded zircons outnumbered facetted ones in all four of the samples. Representative SEM photographs of various morphologies are shown in Figure 5.

U-Pb analytical methods utilized in this study are those outlined in Parrish and others (1987). Techniques included strong air abrasion (Krogh, 1982) for all zircon fractions analyzed, mineral dissolution in microcapsules (Parrish, 1987), a mixed ^{205}Pb-^{233}U-^{235}U isotopic tracer (Parrish and Krogh, 1987), multicollector mass spectrometry (Roddick and others, 1987), and estimation of errors by numerical error propagation (Roddick, 1987). Blank U and Pb levels were about 0–2 and 8–15 pg, respectively. Analytical results are presented in Table 1, where errors on the ages are quoted at the 2 sigma level and displayed in the concordia plots (Figs. 6–8).

Most of the grains produced analyses with low degrees of discordance, and the ^{207}Pb/^{206}Pb ages and their errors are interpreted to represent crystallization ages. For more discordant analyses (those which have undergone more Pb loss), the ^{207}Pb/^{206}Pb ages are interpreted to be minimum crystallization ages.

Collectively, the zircons are all Archean through Mesoproterozoic in age with the exception of a single grain (A2) from the Bird Fiord Formation, the age of which is Late Ordovician to Early Silurian (Table 1, Fig. 6). There are eight grains of Archean age (2.62 to 3.00 Ga) with representatives from all four samples (Fig. 8). Six of these grains are in a narrow range between 2.62 and 2.73 Ga. Four colorless, variably rounded to facetted zircons of 2.24 to 2.48 Ga (early Paleoproterozoic) occur exclusively in the Fram Formation (Fig. 8). Eleven grains from all four of the samples fall within the age range of 1.56 to 2.05 Ga (Fig. 7), with most of the grains evenly scattered through the 1.75 to 2.05 Ga range (inset of Fig. 7). The youngest Precambrian age cluster of 1.04 to 1.2 Ga (late Mesoproterozoic; inset of Fig. 6) contains seven grains with one to three grains from each of the four samples. There are no zircons in the range between 0.4 Ga and the time of deposition of the host sandstones (Eifelian through Frasnian or approximately 385 to 370 Ma). In general, there is no correlation between zircon age and the physical attributes of the grains.

DISCUSSION

Apart from the unique Paleozoic grain A2 and the four zircons in the range 2.24–2.48 Ga from the Fram Formation, the similarities between the dated zircons in each of the four samples greatly outweigh the differences. For this reason, the seven to eight grains of each sample are not considered separately. Rather they are treated as a single group of 31 grains derived from a potentially common provenance region. The aim of this discussion is to attempt to identify the source region which satisfies both the full mix and relative abundance of the dated zircons in the present study as well as the other provenance constraints provided by the host formations.

Possible source areas to consider in this discussion include Precambrian shield areas of Canada and Greenland and various parts of the lower Paleozoic Caledonian-Franklinian mobile belt of Scandinavia, East and North Greenland, and the Canadian Arctic Islands (Fig. 9). Eliminated areas include the Baltic Shield and the Precambrian cratonic fragment and lower Paleozoic platform cover of the Chukchi-Seward massif (Gnibidenko, 1969; Obut, 1977; Till, 1983). These areas are considered to lie on the opposite side of a major Devonian topographic barrier and mountainous drainage divide provided by the intervening Caledonian-Franklinian orogen.

Portions of the Franklinian orogen far to the west and northwest, now located in Alaska and the Russian Far East, are also not considered to be likely source areas for the clastic wedge. Lower Paleozoic and Precambrian rocks of the Alaska-Chukotka region are dominated by rocks which are probably compositionally too immature to account for the high maturity of most sandstones in the Devonian clastic wedge. In addition, paleocurrent indicators do not support a source to the west and northwest. Exceptions may exist in the far western part of the clastic wedge, on Prince Patrick Island for example, where Frasnian sandstones may have been derived from the restored north Alaska region. Detrital zircons from this part of the clastic wedge, however, have not been analyzed.

Viable source areas are discussed in more detail below, and the main conclusions include the following. Shield and

TABLE 1.—U-PB ANALYTICAL DATA FOR DETRITAL ZIRCONS FROM THE DEVONIAN CLASTIC WEDGE

Fraction[a]	Wt.[b] mg	U ppm	Pb[c] ppm	$\frac{^{206}Pb}{^{204}Pb}$[d]	Pb[e] pg	^{208}Pb[f] %	Radiogenic ratios (±1σ, %)[g]			Ages (Ma, ±2σ)[h]		
							$\frac{^{206}Pb}{^{238}U}$	$\frac{^{207}Pb}{^{235}U}$	$\frac{^{207}Pb}{^{206}Pb}$	$\frac{^{206}Pb}{^{238}U}$	$\frac{^{207}Pb}{^{235}U}$	$\frac{^{207}Pb}{^{206}Pb}$
A. Bird Fiord Formation, TC-719B												
A1, c,r,pi,e	0.012	158.4	52.50	2306	16	8.171	0.3188 ± 0.13	4.834 ± 0.15	0.1100 ± 0.06	1784 ± 4	1791 ± 3	1799 ± 2
A2, c,r,el	0.009	399.4	28.52	352	47	13.13	0.06830 ± 0.16	0.5228 ± 0.48	0.05551 ± 0.40	426 ± 1	427 ± 3	433 ± 18
A3, c,r,el	0.008	97.70	19.50	452	21	14.52	0.1842 ± 0.23	1.932 ± 0.42	0.07610 ± 0.33	1090 ± 5	1092 ± 6	1098 ± 13
A4, c,sf	0.010	132.7	47.40	1349	20	8.334	0.3398 ± 0.16	5.624 ± 0.18	0.1201 ± 0.08	1886 ± 5	1920 ± 3	1957 ± 3
A5, p,r,e	0.004	212.6	41.58	501	19	11.68	0.1864 ± 0.16	1.981 ± 0.31	0.07708 ± 0.24	1102 ± 3	1109 ± 4	1123 ± 10
A6, p,r,e	0.005	98.58	39.12	958	10	12.12	0.3604 ± 0.25	6.189 ± 0.25	0.1245 ± 0.16	1984 ± 8	2003 ± 4	2022 ± 6
A7, c,r,el	0.006	62.92	36.38	1044	11	15.59	0.4818 ± 0.38	11.74 ± 0.39	0.1767 ± 0.06	2535 ± 16	2584 ± 7	2622 ± 2
B. Hecla Bay Formation, TC-Hecla Bay												
B1, p,r,e	0.090	102.4	36.02	7889	23	12.91	0.3209 ± 0.09	4.855 ± 0.10	0.1097 ± 0.03	1794 ± 3	1795 ± 2	1795 ± 1
B2, p,r,pi,e	0.014	92.22	57.40	593	71	16.82	0.5093 ± 0.11	12.69 ± 0.16	0.1806 ± 0.11	2654 ± 5	2657 ± 3	2659 ± 4
B3, c,r,e	0.019	68.79	47.18	598	73	20.69	0.5311 ± 0.10	13.84 ± 0.16	0.1890 ± 0.10	2746 ± 4	2739 ± 3	2734 ± 3
B4, c,r,e	0.013	205.2	118.6	1357	63	9.258	0.5154 ± 0.09	12.97 ± 0.11	0.1825 ± 0.05	2680 ± 4	2678 ± 2	2676 ± 2
B5, c,r,el	0.008	248.1	49.83	3198	7	19.92	0.1739 ± 0.09	1.775 ± 0.11	0.07404 ± 0.06	1034 ± 2	1037 ± 2	1043 ± 3
B6, c,f,e	0.005	173.0	41.79	387	34	8.772	0.2335 ± 0.14	3.125 ± 0.30	0.09707 ± 0.24	1352 ± 3	1439 ± 4	1568 ± 9
B7, c,f,el	0.009	74.10	30.87	72	290	14.40	0.3686 ± 1.48	6.290 ± 2.04	0.1238 ± 1.67	2023 ± 51	2017 ± 36	2011 ± 60
B8, b,f,el	0.006	583.5	221.0	3905	18	17.44	0.3258 ± 0.08	5.157 ± 0.10	0.1148 ± 0.04	1818 ± 3	1846 ± 2	1877 ± 1
C. Fram Formation, TC-1105A												
C1, p,r,pi,el	0.004	287.9	63.66	341	51	11.66	0.2102 ± 1.62	2.312 ± 1.90	0.07975 ± 1.33	1230 ± 36	1217 ± 27	1198 ± 55
C2, p,r,e	0.013	263.02	114.1	886	82	25.82	0.3355 ± 0.11	5.228 ± 0.17	0.1130 ± 0.12	1865 ± 3	1857 ± 3	1848 ± 4
C3, c,r,e	0.015	65.36	31.73	995	26	14.36	0.4178 ± 0.13	8.986 ± 0.14	0.1560 ± 0.08	2250 ± 5	2337 ± 3	2413 ± 3
C4, c,r,pi,el	0.006	37.94	12.37	82	68	10.12	0.3092 ± 0.70	4.338 ± 1.99	0.1017 ± 1.60	1736 ± 21	1701 ± 33	1656 ± 60
C5, c,sf,el	0.010	67.58	32.07	900	20	8.744	0.4362 ± 0.19	9.259 ± 0.21	0.1539 ± 0.15	2334 ± 7	2364 ± 4	2390 ± 5
C6, c,f,el	0.004	42.71	22.78	224	21	15.41	0.4512 ± 0.56	10.06 ± 0.58	0.1617 ± 0.38	2400 ± 22	2440 ± 11	2474 ± 13
C7, c,sf,e	0.008	33.11	20.00	291	28	16.30	0.4899 ± 0.74	13.42 ± 0.78	0.1986 ± 0.23	2570 ± 31	2709 ± 15	2815 ± 8
C8, c,sf,e	0.004	158.6	72.91	227	75	11.56	0.4138 ± 0.26	8.103 ± 0.55	0.1420 ± 0.45	2232 ± 10	2243 ± 10	2252 ± 15
D. Okse Bay Group, TM-132D												
D1, p,r,e	0.010	299.2	191.1	5664	15	21.10	0.4933 ± 0.09	12.68 ± 0.10	0.1864 ± 0.03	2585 ± 4	2656 ± 2	2711 ± 1
D2, p,r,e	0.006	1101	202.1	10310	7	2.260	0.1937 ± 0.08	2.082 ± 0.10	0.07796 ± 0.03	1141 ± 2	1143 ± 1	1146 ± 1
D3, c,r,pi,e	0.006	363.0	77.73	2223	12	14.16	0.1980 ± 0.10	2.156 ± 0.13	0.07898 ± 0.08	1164 ± 2	1167 ± 2	1172 ± 3
D4, c,sf,el	0.004	306.0	167.7	3926	10	10.31	0.4828 ± 0.09	12.19 ± 0.10	0.1832 ± 0.03	2540 ± 4	2619 ± 2	2682 ± 1
D5, c,f,e	0.003	54.04	37.71	284	22	11.58	0.5861 ± 0.53	18.02 ± 0.50	0.2230 ± 0.18	2973 ± 25	2991 ± 10	3003 ± 6
D6, p,sf,e	0.002	401.8	132.1	649	18	9.931	0.3108 ± 0.14	4.599 ± 0.21	0.1073 ± 0.16	1745 ± 4	1749 ± 4	1755 ± 6
D7, c,sf,e	0.003	159.0	57.13	349	27	10.84	0.3339 ± 0.27	5.310 ± 0.39	0.1154 ± 0.33	1857 ± 9	1871 ± 7	1886 ± 12
D8, p,f,el	0.011	612.7	119.5	2303	36	7.127	0.1955 ± 0.09	2.097 ± 0.12	0.07777 ± 0.07	1151 ± 2	1148 ± 2	1141 ± 3

[a] All fractions have been abraded following the method of Krogh (1982); c = colourless, p = pink, b = light brown, r = rounded, pi = pitted, f = facetted, sf = subfacetted (somewhat worn facetts), el = elongate, e = equant
[b] Error on weight =±0.001 mg
[c] Radiogenic Pb
[d] Measured ratio corrected for spike and Pb fractionation of 0.09±0.03%/AMU
[e] Total common Pb on analysis corrected for fractionation and spike
[f] Radiogenic Pb
[g] Corrected for blank Pb and U and common Pb (Stacey-Kramers (<2.5Ga) and Cummings-Richards (>2.5Ga) model Pb composition equivalent to the $^{207}Pb/^{206}Pb$ age)
[h] Corrected for blank and common Pb

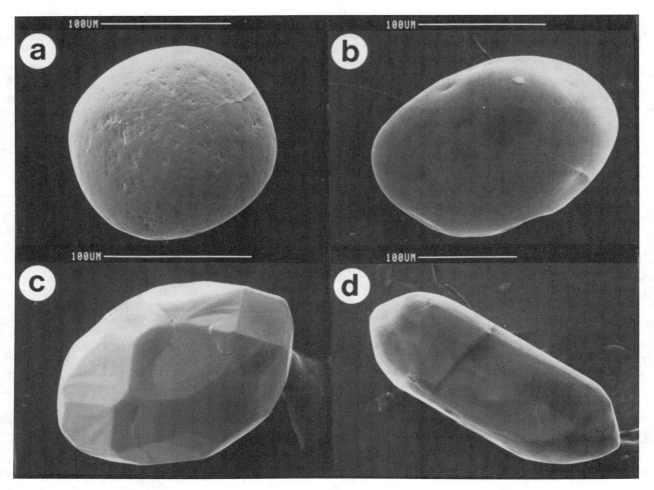

Fig. 5.—Scanning electron microscope photographs of representative detrital zircon grains from the Devonian clastic wedge prior to strong air abrasion (Krogh, 1982). (A) Colorless, round zircon with pitted surface; (B) pink, rounded zircon grain; (C) well-facetted, colourless, equant grain; and (D) elongate, colorless zircon with somewhat rounded facets.

platform cover areas of Canada and West Greenland are unlikely to have been dominant source areas mainly due to the timing of uplift. Potential provenance regions within the mid-Paleozoic orogen of Ellesmere Island, North Greenland, and Svalbard are also considered less viable due to the rather mediocre match of available radiometric ages, as well as inadequately satisfying other provenance constraints. The Scandinavian Caledonides provide an excellent radiometric match; however, this area was probably situated on the east-facing and opposite side of a Devonian drainage divide and consequently would have provided erosion products to catchment areas in central and eastern Europe. The region which best fits the spread of detrital zircon ages in addition to other mineralogical, petrographic, paleocurrent, and timing constraints is the East Greenland Caldonides and its pre-exisiting Late Silurian-Early Devonian molasse basin.

Canadian Shield

The extensive southern shield area of mainland Canada, known as Laurentia, lies between 700 and 3500 km from the present site of the clastic wedge. There is a very favorable match between the ages of detrital zircons analyzed in this study and the ages of igneous and metamorphic rocks of the Archean Slave, Rae, Hearne, and Superior provinces (2.5–3.0 Ga) (Hoffman, 1989 and references therein); subsurface basement of northern and central Alberta (2.0–2.4 Ga) (Villeneuve and others, 1993); Paleoproterozoic orogens including the Thelon-Taltson, Foxe-Rinkian, and Trans-Hudson (1.7–2.0 Ga); and the Mesoproterozoic Grenville province (1.0–1.3 Ga) (Hoffman, 1989 and references therein). Direct transport is unlikely, however, as paleocurrent data indicate a predominant northeast to southwest direction of sediment transport. More significantly, Precambrian rocks of the shield were peneplained before extensive marine overlap by the Cambrian and Ordovician epicontinental seas (Sloss, 1988). The last main phase of erosion of the shield was during the Neoproterozoic and Early Cambrian, and craton-derived siliciclastic deposits of these ages are widely preserved in all flanking miogeoclines of North America (Young, 1979; Bally, 1987). Younger uplifts involving these shield rocks, including the

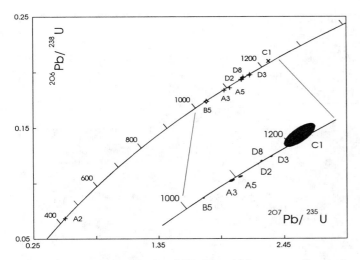

FIG. 6.—U-Pb concordia plot of Silurian and Mesoproterozoic, single grain zircon analyses. Points correspond to analyses presented in Table 1. Error ellipses reflect the 2 sigma uncertainty. Inset in lower right shows detail for Mesoproterozoic (ie., Grenville-aged) ages.

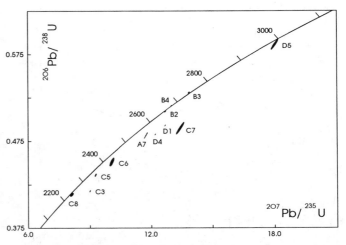

FIG. 8.—U-Pb concordia plot of Paleoproterozoic and Archean single grain zircon analyses. Points correspond to analyses presented in Table 1. Error ellipses reflect the 2 sigma uncertainty.

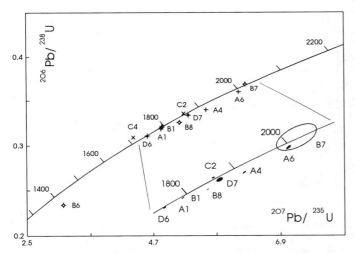

FIG. 7.—U-Pb concordia plot of early Meso- and late Paleoproterozoic, single grain zircon analyses. Points correspond to analyses presented in Table 1. Error ellipses reflect the 2 sigma uncertainty. Inset in lower right shows detail of analyses.

Minto Inlier, Brock Inlier, and Boothia Uplift (Fig. 9), exist but are inadequate provenance areas considering their small size, location, and incorrect or uncertain timing of uplift (Trettin, 1991a; Trettin and others, 1991).

Ellesmere-Greenland Shield

Potential provenance areas located on the shield of southeast Ellesmere Island and Greenland range from a few tens of kilometres to about 2000 km in transport distances from the clastic wedge. Unlike the southern shield areas, a Greenland shield provenance is compatible with the bulk of available paleocurrent indicators. However, reliable matching of the detrital zircon ages to ages in the shield is hindered by the scarcity of reliable data from this large area and by the enormous gap in geological knowledge posed by the presence of the Greenland ice cap. The northeasterly-trending, Precambrian metamorphic-plutonic "Inglefield," Committee, Foxe-Rinkian, and Nagssugtoqidian belts of Baffin and southeast Ellesmere islands continue in West Greenland, and all contain a scattering of Paleoproterozoic and Late Archean U-Pb zircon and Rb-Sr ages (Hoffman, 1989 and references therein). Specific matches can be made between the detrital grain A4, for example (1957 ± 3 Ma), and the age of a 1960 ± 5 Ma orthopyroxene tonalite on southeast Ellesmere Island (Frisch and Hunt, 1988). Similarly, Late Archean grains A7 (2622 ± 2 Ma) and D1 (2711 ± 1 Ma) are comparable to U-Pb zircon ages of circa 2620 and 2710 Ma from basement gneisses at Melville Bugt and Upernavik in Northwest Greenland (Kalsbeek, 1986).

There is no known source on the Greenland shield for 40% of the detrital zircons in this study including the 8 Mesoproterozoic and 4 early Paleoproterozoic grains and the unique Late Ordovician-Silurian grain. In addition, common accessory minerals in the sandstones, including detrital chlorite, muscovite and staurolite, suggest a significant greenschist and lower amphibolite grade metamorphic bedrock component in the source area. Vast areas of the eastern Devon, southeastern Ellesmere and Greenland shield are underlain by granulite facies metamorphic rocks. The closest, lower amphibolite facies rocks on the shield occur on northern Baffin Island and in the Foxe-Rinkian belt of central Baffin Island and central West Greenland (Escher and Pulvertaft, 1976; Jackson and Morgan, 1978; Dawes, 1988; Frisch, 1988).

Although it may have supplied some of the detritus to the clastic wedge, the Greenland shield, including multicyclic sources such as the Meso-Neoproterozoic Thule Group, was probably not a dominant source area due to the timing of erosional unroofing. Erosional remnants of Cambro-Ordovician platform limestones exposed in northwest

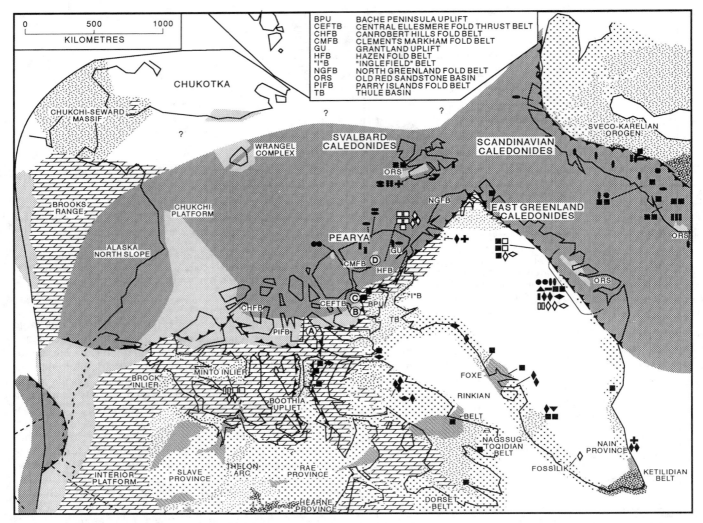

Fig. 9.—Generalized pre-Carboniferous geologic map of the circum-Arctic region, illustrating potential provenance areas and selected U-Pb zircon ages. This Late Devonian plate reconstruction is based loosely on the maps of Ziegler (1988) for the North Atlantic region and Embry (1988) for the position of north Alaska and Chukotka with respect to the Canadian Arctic Islands. The stretching of Ellesmere and Axel Heiberg islands takes into account the palinspastic restoration of shortening associated with the mid-Tertiary Eurekan Orogeny (Okulitch and Trettin, 1991). Geological boundaries are taken primarily from Okulitch and others (1989) and Hoffman (1989). Sources for the U-Pb ages include: Rainbird and others (1992) for Minto Arch; Frisch and Sandeman (1991) for Boothia Uplift; Pidgeon and Howie (1975) and Jackson and others (1990) for Baffin Island; Frisch (1988) and Frisch and Hunt (1988) for southeast Ellesmere and Devon islands; Kalsbeek and others (1984), Escher and others (1986), Kalsbeek (1986), Hansen and others (1987b), Kalsbeek and others (1988), and Taylor and Kalsbeek (1990) for the Greenland shield; Ohta and others (1989), Ohta (1992) and references therein for Svalbard; Sinha and Frisch (1976), Trettin and others (1987; 1992) and unpublished data of V. J. McNicoll for Pearya, Clements Markham Fold Belt, and Grantland Uplift on northern Ellesmere Island; Roberts and Gee (1985) and references therein, Tucker and others (1990a) and references therein, Tucker and others (1990b), Daly and others (1991), and Schouenborg and others (1991) for the Scandinavian Caledonides; and Hansen and others (1978), Henriksen (1985) and references therein, Jepsen and Kalsbeek (1985), Peucat and others (1985), Hansen and Friderichsen (1987), Hansen and others (1987a), Hansen and others (1987c), and Tucker and others (1993) for the East Greenland Caledonides.

Greenland (Higgins and others, 1991) and at Fossilik in southwest Greenland (Stouge and Peel, 1979) suggest a final, major erosion and peneplanation cycle prior to the Devonian. Subsequent epeirogenic activity in the lower Paleozoic is unlikely to have been significant enough to account for the large volume of siliciclastic sediment preserved in the Devonian foreland basin.

Hazen, Central Ellesmere, and Canrobert Hills Fold Belts

The Hazen, Central Ellesmere, and Canrobert Hills fold belts represent the immediate Franklinian hinterland of the Devonian clastic wedge to the north and northwest and are located 100–500 km from the sample sites (Fig. 9). Miogeocline and basin-facies strata dominate the geology of the

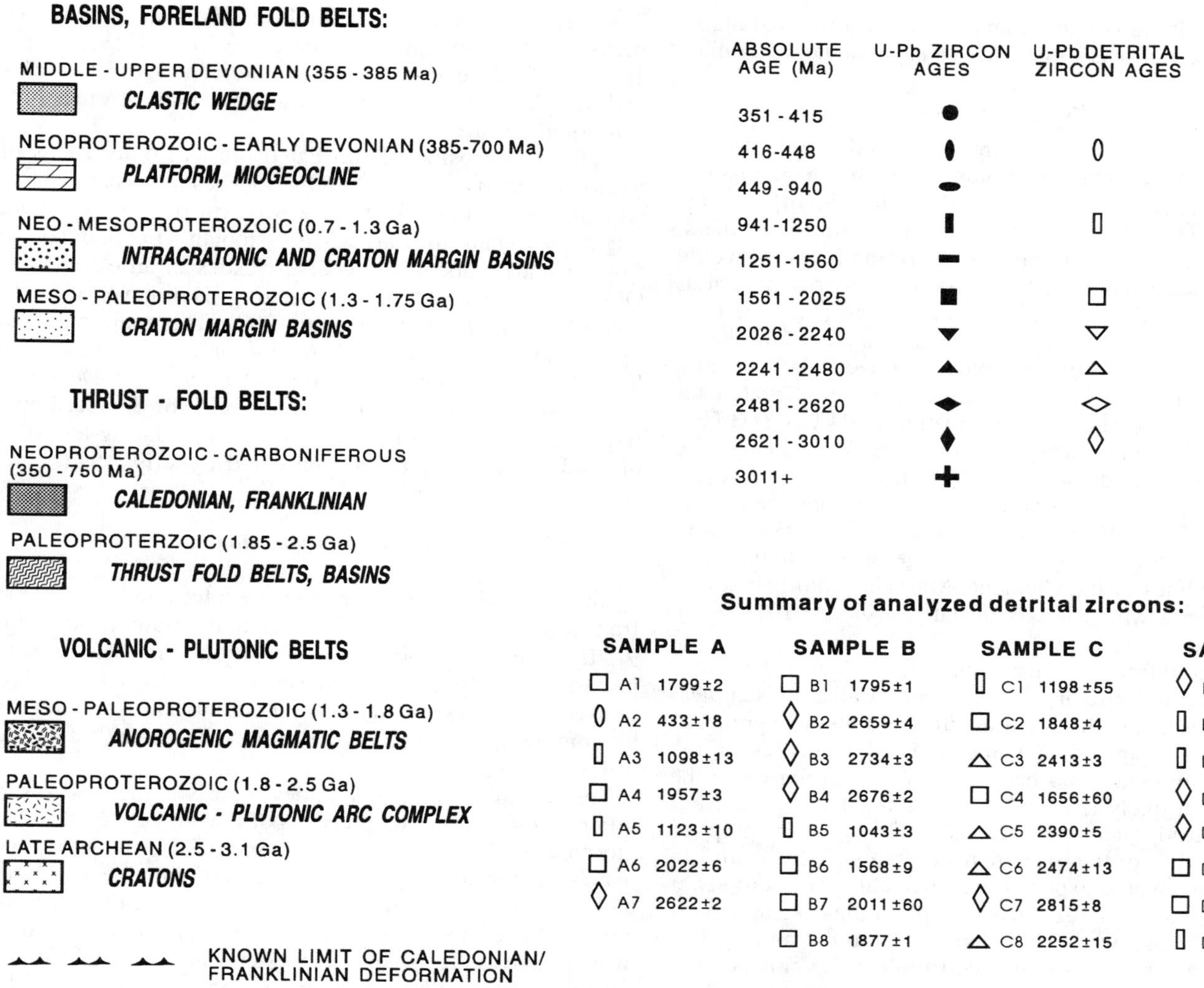

BASINS, FORELAND FOLD BELTS:

MIDDLE - UPPER DEVONIAN (355 - 385 Ma)
CLASTIC WEDGE

NEOPROTEROZOIC - EARLY DEVONIAN (385-700 Ma)
PLATFORM, MIOGEOCLINE

NEO - MESOPROTEROZOIC (0.7 - 1.3 Ga)
INTRACRATONIC AND CRATON MARGIN BASINS

MESO - PALEOPROTEROZOIC (1.3 - 1.75 Ga)
CRATON MARGIN BASINS

THRUST - FOLD BELTS:

NEOPROTEROZOIC - CARBONIFEROUS (350 - 750 Ma)
CALEDONIAN, FRANKLINIAN

PALEOPROTEROZOIC (1.85 - 2.5 Ga)
THRUST FOLD BELTS, BASINS

VOLCANIC - PLUTONIC BELTS

MESO - PALEOPROTEROZOIC (1.3 - 1.8 Ga)
ANOROGENIC MAGMATIC BELTS

PALEOPROTEROZOIC (1.8 - 2.5 Ga)
VOLCANIC - PLUTONIC ARC COMPLEX

LATE ARCHEAN (2.5 - 3.1 Ga)
CRATONS

KNOWN LIMIT OF CALEDONIAN/FRANKLINIAN DEFORMATION

ABSOLUTE AGE (Ma)	U-Pb ZIRCON AGES	U-Pb DETRITAL ZIRCON AGES
351 - 415	●	
416 - 448	◐	○
449 - 940	⬬	
941 - 1250	▮	☐
1251 - 1560	▬	
1561 - 2025	■	☐
2026 - 2240	▼	▽
2241 - 2480	▲	△
2481 - 2620	◆	◇
2621 - 3010	♦	◊
3011+	+	

Summary of analyzed detrital zircons:

SAMPLE A	SAMPLE B	SAMPLE C	SAMPLE D
A1 1799±2	B1 1795±1	C1 1198±55	D1 2711±1
A2 433±18	B2 2659±4	C2 1848±4	D2 1146±1
A3 1098±13	B3 2734±3	C3 2413±3	D3 1172±3
A4 1957±3	B4 2676±2	C4 1656±60	D4 2682±1
A5 1123±10	B5 1043±3	C5 2390±5	D5 3003±6
A6 2022±6	B6 1568±9	C6 2474±13	D6 1755±6
A7 2622±2	B7 2011±60	C7 2815±8	D7 1886±12
	B8 1877±1	C8 2252±15	D8 1141±3

FIG. 9.—Continued.

region and include Lower Cambrian sandstones of the Grantland Formation, Cambrian-Silurian carbonates, strata of the Ibbett Bay and Hazen Formations, and turbidite deposits such as the Blackley, Imina, and Danish River Formations (Trettin and others, 1991). The basin-facies deposits and turbidites are generally too fine grained to provide a secondary source for the coarser grade, Devonian clastic wedge sandstones. Magmatic and high grade metamorphic rocks are absent.

The Grantland Formation, exposed in the Grantland Uplift, contains detrital zircons of both Archean (2.69–2.75 Ga) and Paleoproterozoic (1.75–1.85 Ga) ages (McNicoll, unpub. data). Some of the paleocurrent indicators, notably those in the Yelverton Pass area, suggest a source in this direction; however, the exact timing of exposure and erosion of this region is uncertain. The region of unroofed Grantland Formation is small and would have been volumetrically insufficient to account for the total sediment supplied to the clastic wedge; however, it may have provided some detritus to the Okse Bay Formation in the outliers on northern Ellesmere Island.

North Greenland Fold Belt

Sediment from the North Greenland fold belt area would have been transported southwest for 500 to 1500 km to reach sample sites in the Devonian clastic wedge (Fig. 9). Transport direction is consistent with the paleocurrent data. There are no magmatic or high grade metamorphic rocks in this belt with ages comparable to the detrital zircon analyses of this study. Important potential sources of second- and multicycle zircon include Lower Cambrian sandstones and siliciclastic turbidites of the Skagen and Polkorridoren groups (2–3 km thick) and Silurian turbidites of the Peary Land Group (3.5 km+ thick) (Higgins and others, 1991); however, the ages of contained detrital zircon are unknown. Metamorphic grade of rocks in North Greenland is related to mid-Paleozoic loading and ranges from unmetamorphosed in the south through a wide belt of sub-greenschist and greenschist grade rocks in the bulk of the fold belt to low amphibolite facies in the north (Higgins and others, 1981; Surlyk, 1991). Pelitic and psammo-pelitic rocks of these grades could account for the phyllosilicates and staurolite in the clastic wedge samples but not unreworked higher grade minerals such as garnet, sillimanite, kyanite, and rutile.

Pearya Terrane

Pearya terrane comprises Grenville-aged or middle Mesoproterozoic basement complex with prominent assemblages of Ordovician volcanics and a scattering of younger, intrusive rocks (Trettin, 1987, 1991b; Trettin and others, 1987, 1992). Sediment transported from this area would have travelled between 50 and 1000 km to reach various sample sites in the Devonian clastic wedge (Fig. 9). Data from the basement complex of Pearya include an age of 1.04 ± 0.025 Ga from the Deuchars Glacier belt (Trettin and others, 1987) which is a particularly close match for grain B5 from the Hecla Bay Formation (1.04 ± 0.003 Ga). Dated volcanic and intrusive granitoid rocks of Pearya are Early to Middle Ordovician and Devonian in age; however, all of these ages are either older or younger than the unique grain A2 (433 ± 15 Ma) from the Bird Fiord Formation. A closer match is provided by the age of the Mount Rawlinson volcanics of the Clements Markham fold belt southeast of Pearya which has been dated at 455 +10/−5 Ma (Trettin and others, 1987).

Pearya could have been a source for some of the Grenville-age detrital zircons which occur in all four samples, and it is reasonable that Silurian zircons similar to grain A2 could have originated in some undated volcanics of the Clements Markham fold belt. However, these areas to the north are not likely first cycle sources for the Archean, Paleo- and Mesoproterozoic zircons which together comprise more than 70% of the dated detrital grains of this study. In addition, one would expect a large percentage of both lower Paleozoic and Grenville-aged zircons in these sediments if the clastic wedge was mainly derived from the north. Most of the paleocurrent data collected from the Devonian clastic wedge do not point to a sediment source in this region. Many exposed rock units in the Pearya and Clements Markham belts are compositionally immature and do not readily provide a source for high maturity sands typical of the Devonian clastic wedge.

Svalbard

Svalbard in its pre-Atlantic restored position lies 700 to 1800 km northeast of the sample sites in the Devonian clastic wedge (Fig. 9). A complex lower Paleozoic and Precambrian history for the region is indicated by a wide scatter of U-Pb zircon and Rb-Sr ages (Ohta and others, 1989; Ohta, 1992). Data from Svalbard is comparable to ages of the Ordovician-Silurian detrital grain A2 and the 1.04–1.20 Ga zircons from the clastic wedge. The youngest ages in Svalbard are from late- to post-Caledonian granites which range from about 414 to 440 Ma. Grenville-aged deformation (950–1270 Ma) has been well established by radiometric ages (U-Pb and Rb-Sr) of synorogenic igneous activity and by the presence of unconformities within the circa 20 km thick Hecla Hoek succession.

Neoproterozoic and Cambrian anorogenic, possible rift-related, igneous rocks are 500–550 and 600–660 Ma; detrital zircons of these ages were not analyzed in this study. There is also a scarcity of Archean and Paleoproterozoic rocks on Svalbard although there is some evidence for a 1670–1750 Ma event, and upper intercept ages of ca. 2.1 and 3.2 Ga (U-Pb zircon) suggest reworking of older Precambrian basement.

Uplift and erosion of the Paleozoic and older rocks of Svalbard occurred at about 400–430 Ma and was associated with the accumulation of circa 8–9 km of post-orogenic, Early Devonian to Early Mississippian (Lochkovian to Tournaisian) redbeds and coarse clastics in an extensional Old Red Sandstone-type basin setting (Dallmeyer and others, 1990; Manby and Lyberis, 1992). This main phase of uplift and erosion on Svalbard appears to predate, by at least 15 my, the onset of sedimentation in the Devonian clastic wedge of the Arctic Islands; much of Svalbard may have been in a state of thermal subsidence during the Middle and Upper Devonian (Manby and Lyberis, 1992).

Scandinavian Caledonides

The Scandinavian Caledonides are traceable for 1500 km from southern Norway and Sweden to the Barents Sea (Fig. 9). In its pre-Atlantic restored position, this belt lies approximately 1800 to 2500 km east and northeast of the clastic wedge and is separated from the Canadian Arctic region by contiguous portions of the same orogenic belt exposed in East Greenland.

The Scandinavian Caledonides are a volumetrically significant source region with rocks of a wide variety of metamorphic grades. The ages of these rocks provide a good match for the ages of the analyzed detrital zircons. Geological mapping, together with Rb-Sr and U-Pb geochronology, have shown that the Precambrian geological provinces of the Baltic shield also extend into the Caledonian nappes. From south to north the following belts occur: the allochthonous equivalents of the Sveco-Norwegian ("Grenvillian") belt with ages of 0.9 to 1.2 Ga, the allochthonous Trans-Scandinavian magmatic belt (1.55–1.75 Ga), the transported Sveco-Karelian Orogen (1.75–2.1 Ga), and a reactivated Archean craton (2.7–3.1 Ga) in the far north (Gorbatschev, 1985; Schouenborg and others, 1991). Widely scattered Ordovician to Lower Devonian volcanics and small intrusives are variably between 386 and 490 Ma in age, with most between 411 and 490 Ma (Roberts and Gee, 1985; Stephens and others, 1985).

Uplift and erosion associated with the extensional collapse and gravitational spreading of the Scandinavian Caledonian orogen and foreland occurred between about 400 and 365 Ma (Anderson and others, 1991, 1992; Dallmeyer and others, 1992). This timing is consistent with the timing of deposition of the Devonian clastic wedge in the Arctic Islands. There is a significant uncertainty, however, about the extent to which this distant portion of the entire Caledonide belt actually drained to the west. It is more probable that the Scandinavian Caledonides lay along and east of the ancestral Caledonian mountain divide and that most of its erosional products were shed far to the east into northeastern Europe and Russia (Dineley, 1975).

East Greenland Caledonides

The Caledonides in East Greenland are exposed in a north-trending belt 1350 km long between Scoresby Sund and North Greenland (Fig. 9). Sediment eroded from this area would have been transported 900 to 2200 km to reach sample sites in the Devonian clastic wedge. Rocks within the orogen include a variety of amphibolite grade, Archean and Paleoproterozoic gneiss complexes occurring in thrust sheets 5–6 km thick; a 15 km+ thick, impure quartzite-rich, siliciclastic succession of Mesoproterozoic age (Eleanore Bay Group) variably affected by greenschist to upper amphibolite facies, Caledonian metamorphism; Neoproterozoic tillites and related sediments (500–800 m); and remnants of a 3 km thick Cambrian to Middle Ordovician platform sequence (Higgins and others, 1981; Henriksen 1985; Caby and Bertrand-Sarfati, 1988). The Eleanore Bay Group, which locally grades to migmatized sillimanite gneiss and occurs with granitoid rocks and older Precambrian paragneisses such as the Krummedal Sequence (Higgins, 1988), has been interpreted as a foreland molasse succession derived from the coeval unroofing of the Grenville Orogen of southeastern Canada (Caby and Bertrand-Sarfati, 1988). Post-orogenic, syn-rift Frasnian-Givetian redbeds are preserved in a single, 7 km thick, Old Red Sandstone-type basin in central East Greenland (Larsen and Bengaard, 1991).

Age determinations from the gneiss complexes of central East Greenland include Rb-Sr and some U-Pb zircon ages in the range 2.30–2.96 Ga and 1.70–1.98 Ga (Steiger and others, 1979; Rex and Gledhill, 1981; Higgins and others, 1981; Hansen and others, 1987a). Grenville-aged Rb-Sr whole-rock and U-Pb ages in the range 1.0–1.2 Ga are reported for some granites and Krummedal Sequence rocks (Steiger and others, 1979; Rex and Gledhill, 1981; Higgins, 1988). U-Pb zircon ages from northeast Greenland include ages in the 1.70–2.0 Ga range with some lower intercept reset ages of circa 400 Ma (Tucker and others, 1993; Jepsen and Kalsbeek, 1985). Detrital zircons from the same area include Archean ages of circa 2.5–3.0 Ga (Tucker and others, 1993).

Caledonian plutonic rocks represent only a small part of the exposed geology of East Greenland. Rb-Sr whole-rock isochron and U-Pb zircon ages range from 560–377 Ma (Rex and Gledhill, 1981; Henriksen, 1985). Some migmatites and granitic gneisses also contain U-Pb lower intercept ages of circa 400–440 Ma (Steiger and others, 1979; Peucat and others, 1985; Higgins, 1988; Tucker and others, 1993). The main Caledonian intrusive event occurred at circa 420 Ma (Hansen and others, 1987c).

Two phases of uplift are indicated for the East Greenland Caledonides by various phyllosilicate cooling ages. The earlier uplift phase, identified in shallowly buried rocks, is between 440 and 405 Ma (Larsen and Bengaard, 1991) and is approximately coeval with the foreland accumulation in North Greenland of easterly-derived Silurian Peary Land Group turbidites presumably derived from the Caledonides mountain belt (Higgins and others, 1991). The second uplift phase, featuring some faults with up to 10 km of vertical throw (Larsen and Bengaard, 1991), is dated by cooling ages obtained from more deeply buried portions of the orogen. These ages range from 365 to 405 Ma with a mean age close to 385 Ma (Larsen and Bengaard, 1991). This second uplift of East Greenland, analogous to the eduction phase of the Scandinavian Caledonides described by Anderson and others (1991), is closely linked to the extensional collapse of a tectonically overthickened orogen and the related accumulation of Givetian-Frasnian, Old Red Sandstone basin deposits (Larsen and Bengaard, 1991). The 385 and 365 Ma ages also closely match the time of first appearance in the early Eifelian and last appearance in the mid-Famennian, respectively, of the preserved Devonian clastic wedge of Arctic Canada.

Precambrian and lower Paleozoic rocks of the East Greenland Caledonides offer all the necessary sources for the dated detrital zircons of the clastic wedge reported in this study. The region satisfies many of the other criteria for a suitable provenance region. The sediment transport direction from the east and northeast is consistent with the bulk of the paleocurrent indicators compiled for the Devonian clastic wedge. The distance from source to site of deposition is consistent with the uniform grain size and overall maturity of the clastic wedge in the Arctic Islands. There is a wide range of metamorphic grades (to upper amphibolite facies) represented in the East Greenland belt similar to the wide range implied by the accessory and heavy minerals in the clastic wedge samples. Uplift was locally on the order of 11 km in central East Greenland, and the off-stripped section includes not only basement gneisses but also the 15+ km thick Eleanore Bay Group which is dominated by coarse, quartz-rich siliciclastic strata. The volume of eroded rock sections is matched only by that preserved in the Devonian clastic wedge of the Arctic Islands. The mix of impure quartzites and gneisses in the East Greenland Caledonides could also provide a potentially suitable mix of first- and second-cycle quartz, potash feldspar, zircon, and other minerals to the clastic wedge. Finally the correct timing of uplift and erosion of East Greenland is provided by a remarkable match of younger phyllosilicate cooling ages in East Greenland and the first appearance of foreland sands in the Eifelian in the Arctic Islands.

Caledonian Foreland in East Greenland

The Caledonian foreland in East Greenland, comprising Archean and Paleoproterozoic gneisses, Mesoproterozoic sandstones and basalts, and Neoproterozoic to Silurian sedimentary cover, is exposed in northeasternmost Greenland and in nunataks to the south between the orogen and the inland icecap (Henriksen, 1985; Sønderholm and Jepsen, 1991; Higgins and others, 1991). Potential second-cycle sources in this region include the pre-existing sediment load derived by erosion of the East Greenland and Scandinavian Caledonide mountain belt in the Silurian and Early Devonian and deposited in a foreland basin setting west of the Caledonian thrust front (Surlyk, 1991). Pre-erosional thicknesses of circa 2 to 10 km or more are inferred from grades of regional metamorphism in North Greenland. Similarly, conodont alteration indices from Lower Ordovician carbonates of the East Greenland foreland point to thick, pre-existing cover in this area (Smith, 1991). Less certain is the extent to which thermal maturity and metamorphic grade have been induced by thrust sheet loading. The zircon signature of these eroded strata will always remain unknown but, logically, must have matched that of the ultimate source terrane in the Caledonian mountain belt described above (Hurst and others, 1983).

CONCLUSIONS

Thirty-one detrital zircons have been extracted and dated by the U-Pb technique from four representative sandstone samples collected from the eastern half of the Devonian clastic wedge of Canada's Arctic Islands. Thirty of the grains are Precambrian in age and occur in four age groupings: 1.04–1.20 Ga (7 grains from all four samples), 1.56–2.05 (11 grains from all four samples), 2.24–2.48 Ga (4 grains from one sample only), and 2.62–3.00 (8 grains from all four samples). A single zircon from the Bird Fiord Formation was dated at 433 ± 15 Ma (Late Ordovician—Silurian).

Metamorphic and metasedimentary rocks of the East Greenland Caledonides provide the best match as the ultimate source of dated detrital zircons in this study. This region and its unroofed foreland also satisfy the other criteria including relative distance of sediment transport, paleocurrent directions, minor and heavy fraction mineralogy, and the timing and magnitude of uplift.

ACKNOWLEDGMENTS

The staff of the GSC geochronology laboratory are thanked for their assistance in generating the U-Pb data. R. R. Parrish provided valuable advice during the course of the analytical work. The authors would also like to thank A. F. Embry, R. Rainbird, and T. Lawton for constructive reviews of the manuscript, and T. de Freitas for a healthy exchange of ideas generated prior to and during manuscript preparation and for providing some unpublished paleocurrent data from Prince Patrick and Melville Islands. Additional thanks are extended to Peter Neelands for assistance with the drafting of text figures and to Carmen Lee for compiling and word processing the references. Samples were collected in conjunction with regional geological mapping efforts supported by the Polar Continental Shelf Project. However, this research would not have been possible without the financial support of the Geological Survey of Canada.

REFERENCES

ANDERSON, M. W., BARKER, A. J., BENNETT, D. G., AND DALLMEYER, R. D., 1992, A tectonic model for Scandian terrane accretion in the northern Scandinavian Caledonides: Journal of the Geological Society (London), v. 149, p. 727–741.

ANDERSON, T. B., JAMTVEIT, B., DEWEY, J. F., AND SWENSSON, E., 1991, Subduction and eduction of continental crust: major mechanisms during continent-continent collision and orogenic extensional collapse, a model based on the south Norwegian Caledonides: Terra Nova, v. 3, p. 303–310.

BALLY, A. W., 1987, Phanerozoic Basin Evolution in North America: Episodes, v. 10, p. 248–253.

CABY, R. AND BERTRAND-SARFATI, J., 1988, The Eleonore Bay Group (central East Greenland), in Winchester, J. A., ed., Later Proterozoic Stratigraphy of the Northern Atlantic Regions: London, Chapman and Hall, p. 212–236.

COWIE, J. W. AND BASSETT, M. G., 1989, 1989 Global stratigraphic chart with geochronometric and magnetostratigraphic calibration: Episodes, v. 12, p. 79–83 (with supplement).

DALLMEYER, R. D., JOHANSSON, L., AND MOLLER, C., 1992, Chronology of Caledonian high-pressure granulite-facies metamorphism, uplift, and deformation within northern parts of the Western Gneiss Region, Norway: Geological Society of America Bulletin, v. 104, p. 444–445.

DALLMEYER, R. D., OHTA, Y., AND PEUCAT, J. J., 1990, Tectonothermal evolution of contrasting metamorphic complexes in northwest Spitsbergen (Biskayerhalvoya): Evidence from Ar/Ar and Rb-Sr mineral ages: Geological Society of American Bulletin, v. 102, p. 653–663.

DALY, J. S., AITCHESON, S. J., CLIFF, R. A., GAYER, R. A., AND RICE, A. H. N., 1991, Geochronological evidence from discordant plutons for a late Proterozoic orogen in the Caledonides of Finnmark, northern Norway: Journal of the Geological Society (London), v. 148, p. 29–40.

DAWES, P. R., 1988, Etah meta-igneous complex and the Wulff structure: Proterozoic magmatism and deformation in Inglefield Land, North-west Greenland: Gronlands Geologiske Undersogelse Rapport Nr. 139, 24 p.

DE FREITAS, T. AND MAYR, U., 1992, The middle Paleozoic sequence of northern Devon Island, Northwest Territories: Ottawa, Geological Survey of Canada Paper 92–1B, p. 53–63.

DINELEY, D. L., 1975, North Atlantic Old Red Sandstone-Some implications for Devonian Paleogeography, in Yorath, C. J., Parker, E. R., and Glass, D. J., eds., Canada's Continental Margins and Offshore Petroleum Exploration: Calgary, Canadian Society of Petroleum Geologists Memoir 4, p. 773–790.

EMBRY, A. F., 1988, Middle-Upper Devonian Sedimentation in the Canadian Arctic Islands and the Ellesmerian Orogeny, in McMillan, N. J., Embry, A. F., and Glass, D., eds., Proceedings of the Second International Symposium on the Devonian: Calgary, Canadian Society of Petroleum Geologists Memoir 14, p. 15–28.

EMBRY, A. F., 1991, Middle-Upper Devonian Clastic Wedge of the Arctic Islands, in Trettin, H. P., ed., Geology of the Innuitian Orogen and Arctic Platform of Canada and Greenland: Ottawa, Geological Survey of Canada, The Geology of North America, v. E, p. 261–280.

EMBRY, A. F. AND KLOVAN, J. E., 1974, The Devonian clastic wedge of the Canadian Arctic Archipelago (abs.): Geological Society of America, Abstracts with Programs, v. 6, p. 721–722.

EMBRY, A. F. AND KLOVAN, J. E., 1976, The Middle-Upper Devonian clastic wedge of the Franklinian Geosyncline: Bulletin of Canadian Petroleum Geology, v. 24, p. 485–639.

ESCHER, J. C., KALSBEEK, F., LARSEN, O., NIELSEN, T. F. D., AND TAYLOR, P. N., 1986, Reconnaissance dating of Archean rocks from Southeast Greenland: Gronlands Geologiske Undersogelse Rapport Nr. 130, p. 90–95.

ESCHER, A. AND PULVERTAFT, T. C. R., 1976, Rinkian mobile belt of West Greenland, in Escher, A., and Watt, W. S., eds., Geology of Greenland: Gronlands Geologiske Undersogelse, p. 104–119.

FRISCH, T., 1988, Reconnaissance Geology of the Precambrian Shield of Ellesmere, Devon and Coburg Islands, Canadian Arctic Archipelago: Ottawa, Geological Survey of Canada Memoir 409, 102 p.

FRISCH, T. AND HUNT, P. A., 1988, U-Pb zircon and monazite ages from the Precambrian Shield of Ellesmere and Devon Islands, Arctic Archipelago: Ottawa, Geological Survey of Canada Paper 88-2, p. 117–125.

FRISCH, T. AND SANDEMAN, A. I., 1991, Reconnaissance geology of the Precambrian Shield of the Boothia Uplift, northwestern Somerset Island and eastern Prince of Wales Island, District of Franklin: Ottawa, Geological Survey of Canada Paper 91–1C, p. 173–178.

GNIBIDENKO, G. S., 1969, Metamorficheskie kompleksky v strukturakh severo-zapadnogo sektora Tikhookeanskogo poyasa: Akademiya Nauk SSSR, Sibirskoe Otdelenie, Izdatel'stva Nauka, 128 p.

GOODBODY, Q. H., 1989, Stratigraphy of the Lower to Middle Devonian Bird Fiord Formation, Canadian Arctic Archipelago: Bulletin of Canadian Petroleum Geology, v. 37, p. 48–82.

GORBATSCHEV, R., 1985, Precambrian basement of the Scandinavian Caledonides, in Gee, D. G., and Sturt, B. A., eds., The Caledonide Orogen- Scandinavia and Related Areas, Part 1: Chichester, John Wiley and Sons, p. 197–212.

GORDEY, S. P., 1991, Upper Devonian to Middle Jurassic Assemblages, in Gabrielse, H., and Yorath, C. J., eds., Geology of the Cordilleran

Orogen in Canada: Ottawa, Geological Survey of Canada, The Geology of North America, v. G-2, p. 219–328.

HANSEN, B. T. AND FRIDERICHSEN, J. D., 1987, Isotopic dating in Liverpool Land, East Greenland: Gronlands Geologiske Undersogelse Rapport Nr. 134, p. 25–37.

HANSEN, B. T., HIGGINS, A. K., AND BÄR, M., 1978, Rb-Sr and U-Pb age patterns in polymetamorphic sediments from the southern part of the East Greenland Caledonides: Bulletin of the Geological Society of Denmark, v. 27, p. 55–62.

HANSEN, B. T., HIGGINS, A. K., AND BORCHARDT, B., 1987a, Archaean U-Pb zircon ages from the Scoresby Sund region, East Greenland: Gronlands Geologiske Undersogelse Rapport Nr. 134, p. 19–24.

HANSEN, B. T., KALSBEEK, F., AND HOLM, P. M., 1987b, Archaean age and Proterozoic metamorphic overprinting of the crystalline basement at Victoria Fjord, North Greenland: Gronlands Geologiske Undersogelse Rapport Nr. 133, p. 159–168.

HANSEN, B. T., STEIGER, R. H., HENRIKSON, N., AND BORCHARDT, B., 1987c, U-Pb and Rb-Sr age determinations on Caledonian plutonic rocks in the central part of the Scoresby Sund region, East Greenland: Gronlands Geologiske Undersogelse Rapport Nr. 134, p. 5–18.

HARLAND, W. B., ARMSTRONG, R. L., COX, A. V., CRAIG, L. E., SMITH, A. G., AND SMITH, D. G., 1989, A Geologic Time Scale 1989: Cambridge, Cambridge University Press, 131 p.

HARRISON, J. C., 1991, Melville Island's salt-based fold belt (Arctic Canada): Unpublished Ph.D. Thesis, Rice University, Houston, 783 p.

HARRISON, J. C., in press, Melville Island's salt-based fold belt (Arctic Canada): Ottawa, Geological Survey of Canada, Bulletin 472.

HENRIKSEN, N., 1985, The Caledonides of central East Greenland 70°–76°N, in Gee, D. G., and Sturt, B. A., eds., The Caledonide Orogen-Scandinavia and Related Areas: Chichester, John Wiley and Sons, p. 1095–1110.

HIGGINS, A. K., 1988, The Krummedal supracrustal sequence in East Greenland, in Winchester, J.A., ed., Later Proterozoic Stratigraphy of the Northern Atlantic Regions: London, Chapman and Hall, p. 86–96.

HIGGINS, A. K., FRIDERICHSEN, J. D., AND SOPER, N. J., 1981, The North Greenland fold belt between Central Johannes V. Jensen Land and Eastern Nansen Land: Gronlands Geologiske Undersogelse Rapport Nr. 106, p. 35–45.

HIGGINS, A. K., INESON, J. R., PEEL, J. S., SURLYK, F., AND SONDERHOLM, M., 1991, Lower Paleozoic Franklinian Basin of North Greenland: Gronlands Geologiske Undersogelse Bulletin 160, p. 71–139.

HOFFMAN, P. F., 1989, Precambrian geology and tectonic history of North America, in Bally, A. W., and Palmer, A. R., eds., The Geology of North America: An Overview: Boulder, Geological Society of America, v. A, p. 447–512.

HURST, J. M., MCKERROW, W. S., SOPER, N. J., AND SURLYK, F., 1983, The relationship between Caledonian nappe tectonics and Silurian turbidite deposition in North Greenland: Journal of the Geological Society (London), v. 140, p. 123–131.

JACKSON, G. D. AND MORGAN, W. C., 1978, Precambrian Metamorphism on Baffin and Bylot Islands, in Fraser, J. A., and Heywood, W. W., eds., Metamorphism in the Canadian Shield: Ottawa, Geological Survey of Canada Paper 78-10, p. 249–267.

JACKSON, G. D., HUNT, P. A., LOVERIDGE, W. D., AND PARRISH, R. R., 1990, Reconnaissance geochronology of Baffin Island, N.W.T.: Ottawa, Geological Survey of Canada Paper 89-2, p. 123–148.

JEPSEN, H. F. AND KALSBEEK, F., 1985, Evidence for non-existence of a Carolinidian fold belt in eastern North Greenland, in Gee, D. G., and Sturt, B. A., eds., The Caledonide Orogen-Scandinavia and Related Areas: Chichester, John Wiley and Sons, p. 1071–1076.

KALSBEEK, F., 1986, The tectonic framework of the Precambrian shield of Greenland. A review of new isotopic evidence: Gronlands Geologiske Undersogelse Rapport Nr. 128, p. 55–64.

KALSBEEK, F., TAYLOR, P. N., AND HENRIKSEN, N., 1984, Age of rocks, structures, and metamorphism in the Nagssugtoqidian mobile belt, West Greenland—field and Pb-isotopic evidence: Canadian Journal of Earth Sciences, v. 21, p. 1126–1131.

KALSBEEK, F., TAYLOR, P. N., AND PIDGEON, R. T., 1988, Unreworked Archaean basement and Proterozoic supracrustal rocks from northeastern Disko Bugt, West Greenland: implications for the nature of Proterozoic mobile belts in Greenland: Canadian Journal of Science, v. 25, p. 773–782.

KROGH, T. E., 1982, Improved accuracy of U-Pb ages by the creation of more concordant systems using an air abrasion technique: Geochimica et Cosmochimica Acta, v. 46, p. 637–649.

LARSEN, P. AND BENGAARD, H., 1991, Devonian basin initiation in East Greenland: a result of sinistral wrench faulting and Caledonian extensional collapse: Journal of the Geological Society (London), v. 148, p. 355–368.

MANBY, G. AND LYBERIS, N., 1992, Tectonic evolution of the Devonian Basin of northern Svalbard: Norsk Geologisk Tidsskrift, v. 72, p. 7–19.

MAUREL, L. E., 1989, Geometry and evolution of the Tanquary Structural High and its effects on the paleogeography of the Sverdrup Basin, northern Ellesmere Island, Canadian Arctic: Ottawa, Geological Survey of Canada Paper 89-1G, p. 177–189.

MAYR, U., 1991, Reconnaissance and interpretation of Upper Devonian to Permian stratigraphy of northeastern Ellesmere Island, Canadian Arctic Archipelago: Ottawa, Geological Survey of Canada Paper 91-8, 117 p.

MAYR, U., PACKARD, J. J., GOODBODY, Q., RICE, R. J., GOODARZI, F., AND STEWART, K. R., in press, Phanerozoic geology of southern Ellesmere and North Kent islands, Canadian Arctic Archipelago (Craig Harbour, Baad Fiord and Cardigan Strait map areas, NTS 49A, 49B, and 59A): Ottawa, Geological Survey of Canada Bulletin 470.

MAYR, U. AND TRETTIN, H. P., 1990, Revised geological map, Tanquary Fiord, District of Franklin (NTS 340D), scale 1:125,000: Ottawa, Geological Survey of Canada, Open File 2135.

MCLAREN, P. J., 1963, Devonian stratigraphy, in Fortier, Y. O., Blackadar, R. G., Glenister, B. F., Greiner, H. R., McLaren, D. J., McMillan, N. J., Norris, A. W., Roots, E. F., Souther, J. G., Thorsteinsson, R., and Tozer, E. T., Geology of the North-central Part of the Arctic Archipelago, N.W.T. (Operation Franklin): Ottawa, Geological Survey of Canada Memoir 320, p. 57–65.

OBUT, A. M., 1977, Stratigrafia i fauna Ordovika i Silura, Chukotskogo Poluoostrova: Moskva, Trudy Instituta Geologii i Geofizikii, v. 351, 222 p.

OHTA, Y., 1992, Recent understanding of the Svalbard basement in the light of new radiometric age determinations: Norsk Geologisk Tidsskrift, v. 72, p. 1–5.

OHTA, Y., DALLMEYER, R. D., AND PEUCAT, J. J., 1989, Caledonian terranes in Svalbard: Boulder, Geological Society of America Special Paper 230, p. 1–15.

OKULITCH, A. V., LOPATIN, B. G., AND JACKSON, H. R., 1989, Circumpolar geological map of the Arctic, scale 1:6,000: Ottawa, Geological Survey of Canada, Map 1765A.

OKULITCH, A. V. AND TRETTIN, H. P., 1991, Late Cretaceous-Early Tertiary Deformation, Arctic Islands, in Trettin, H.P., ed., Geology of the Innuitian Orogen and Arctic Platform of Canada and Greenland: Ottawa, Geological Survey of Canada, The Geology of North America, v. E, p. 469–489.

PARRISH, R. R., 1987, An improved micro-capsule for zircon dissolution in U-Pb geochronology: Chemical Geology (Isotope Geoscience Section), v. 66, p. 99–102.

PARRISH, R. R. AND KROGH, T. E., 1987, Synthesis and purification of 205Pb for U-Pb geochronology: Chemical Geology (Isotope Geoscience Section), v. 66, p. 103–110.

PARRISH, R. R., RODDICK, J. C., LOVERIDGE, W. D., AND SULLIVAN, R. W., 1987, Uranium-lead analytical techniques at the geochronology laboratory, Geological Survey of Canada: Ottawa, Geological Survey of Canada Paper 88-2, p. 3–7.

PEUCAT, J. J., TISSERANT, D., CABY, R., AND CLAUER, N., 1985, Resistance of zircons to U-Pb resetting in a prograde metamorphic sequence of Caledonian age in East Greenland: Canadian Journal of Earth Sciences, v. 22, p. 330–338.

PIDGEON, R. T. AND HOWIE, R. A., 1975, U-Pb age of zircon from a charnokitic granulite from Pangnirtung on the east coast of Baffin Island: Canadian Journal of Earth Sciences, v. 12, p. 1046–1047.

PLUMB, K. A., 1991, New Precambrian time scale: Episodes, v. 14, p. 139–140.

RAINBIRD, R. H., HEAMAN, L. M., AND YOUNG, G. M., 1992, Sampling Laurentia: detrital zircon geochronolgy offers evidence for an extensive Neoproterozoic river system originating from the Grenville orogen: Geology, v. 20, p. 351–354.

Rex, D. C. and Gledhill, A. R., 1981, Isotopic studies in the East Greenland Caledonides (72–74N)- Precambrian and Caledonian ages: Copenhagen, Gronlands Geologiske Undersogelse Rapport Nr. 104, p. 47–72.

Rice, R. J., 1987, The Sedimentology and Petrology of the Okse Bay Group (Middle and Upper Devonian) on S.W. Ellesmere Island and North Kent Island in the Canadian Arctic Archipelago: Unpublished Ph.D. Thesis, McMaster University, Hamilton, 769 p.

Roberts, D. and Gee, D. G., 1985, An introduction to the structure of the Scandinavian Caledonides, in Gee, D. G., and Sturt, B. A., eds., The Caledonide Orogen- Scandinavia and Related Areas: Chichester, John Wiley and Sons, p. 55–68.

Roblesky, R. F., 1979, Upper Silurian (Late Ludlovian) to Upper Devonian (Frasnian?) Stratigraphy and Depositional History of the Fiord Region, Ellesmere Island, Canadian Arctic: Unpublished M.S. Thesis, University of Calgary, Calgary, 230 p.

Roddick, J. C., 1987, Generalized numerical error analysis with applications to geochronolgy and thermodynamics: Geochimica et Cosmochimica Acta, v. 51, p. 2129–2135.

Roddick, J. C., Loveridge, W. D., and Parrish, R. R., 1987, Precise U/Pb dating of zircon at the sub-nanogram Pb level: Chemical Geology (Isotope Geoscience Section), v. 66, p. 111–121.

Schouenborg, B. E., Johansson, L., and Gorbatschev, R., 1991, U-Pb zircon ages of basement gneisses and discordant felsic dykes from Vestranden, westernmost Baltic Shield and central Norwegian Caledonides: Geologische Rundschau, v. 80, p. 121–134.

Sinha, A. K. and Frisch, T., 1976, Whole-rock Rb-Sr and zircon U-Pb ages of metamorphic rocks from northern Ellesmere Island, Canadian Arctic Archipelago. II. The Cape Columbia Complex: Canadian Journal of Earth Sciences, v. 13, p. 774–780.

Sloss, L. L., 1988, Tectonic evolution of the craton in Phanerozoic time, in Sloss, L. L., ed., The Geology of North America: Boulder, Geological Society of America, v. D-2, p. 25–51.

Smith, M. P., 1991, Early Ordovician conodonts of East and North Greenland: Meddelelser om Gronland, v. 26, p. 5–18.

Sonderholm, M. and Jepsen, H. F., 1991, Proterozoic basins of North Greenland: Gronlands Geologiske Undersogelse Bulletin 160, p. 49–69.

Steiger, R. H., Hansen, B. T., Schuler, C., Bar, M. T., and Henriksen, N., 1979, Polyorogenic nature of the southern Caledonian fold belt in East Germany: Journal of Geology, v. 87, p. 475–495.

Stephens, M. B., Furnes, H., Robins, B., and Sturt, B. A., 1985, Igneous activity within the Scandinavian Caledonides, in Gee, D. G., and Sturt, B. A., eds., The Caledonide Orogen-Scandinavia and Related Areas: Chichester, John Wiley and Sons, p. 623–656.

Stouge, S. and Peel, J. S., 1979, Ordovician conodonts from the Precambrian Shield of southern West Greenland: Rapport Gronlands Geologiske Undersogelse, v. 91, p. 105–109.

Surlyk, F., 1991, Tectonostratigraphy of North Greenland: Gronlands Geologiske Undersogelse Bulletin 160, p. 25–47.

Taylor, P. N. and Kalsbeek, F., 1990, Dating the metamorphism of Precambrian marbles: examples from Proterozoic mobile belts in Greenland: Chemical Geology (Isotope Geology section), v. 86, p. 21–28.

Thorsteinsson, R., 1960, Lower Paleozoic stratigraphy of the Canadian Arctic Archipelago (abs.), in Raasch, G. O., ed., Geology of the Arctic, Proceedings of the First International Symposium on Arctic Geology: University of Toronto Press, v. 1 , p. 380.

Thorsteinsson, R. and Tozer, E. T., 1970, Geology of the Arctic Archipelago, in Douglas, R. J. W., ed., Geology and Economic Minerals of Canada: Ottawa, Geological Survey of Canada Economic Geology Report 1, p. 547–590.

Till, A. B., 1983, Granulite, peridotite, and blueschist; Precambrian to Mesozoic history of the Seward peninsula: Journal of the Alaska Geological Society, v. 3, p. 59–66.

Tozer, E. T. and Thorsteinsson, R., 1964, Western Queen Elizabeth Islands, Arctic Archipelago: Ottawa, Geological Survey of Canada Memoir 332, 242 p.

Trettin, H. P., 1971, Geology of Lower Paleozoic formations, Hazen Plateau and Southern Grant Land Mountains, Ellesmere Island, Arctic Archipelago: Geological Survey of Canada Bulletin, v. 203, 134 p.

Trettin, H. P., 1978, Devonian Stratigraphy, west-central Ellesmere Island, Arctic Archipelago: Geological Survey of Canada Bulletin, v. 302, 119 p.

Trettin, H. P., 1987, Pearya: a composite terrane with Caledonian affinities in northern Ellesmere Island: Canadian Journal of Earth Sciences, v. 24, p. 224–245.

Trettin, H. P., 1991a, Silurian-Early Carboniferous Deformational Phases and Associated Metamorphism and Plutonism, Arctic Islands, in Trettin, H. P., ed., Geology of the Innuitian Orogen and Arctic Platform of Canada and Greenland: Ottawa, Geological Survey of Canada, The Geology of North America, v. E, p. 293–342.

Trettin, H. P., 1991b, Late Silurian-Early Devonian Deformation, Metamorphism, and Granitic Plutonism, Northern Ellesmere and Axel Heiberg Islands, in Trettin, H. P., ed., Geology of the Innuitian Orogen and Arctic Platform of Canada and Greenland: Ottawa, Geological Survey of Canada, The Geology of North America, v. E, p. 295–301.

Trettin, H. P. (compiler), 1991c, Geotectonic correlation chart, Figure 4, Sheets 1–3, in Trettin, H. P., ed., Geology of the Innuitian Orogen and Arctic Platform of Canada and Greenland: Ottawa, Geological Survey of Canada, The Geology of North America, v. E.

Trettin, H. P., Mayr, U., Long, G. D. F., and Packard, J. J., 1991, Cambrian to Early Devonian Basin Development, Sedimentation, and Volcanism, Arctic Islands, in Trettin, H. P., ed., Geology of the Innuitian Orogen and Arctic Platform of Canada and Greenland: Geological Society of America, The Geology of North America, v. E, p. 163–238.

Trettin, H. P., Parrish, R., and Loveridge, W. D., 1987, U-Pb age determinations on Proterozoic to Devonian rocks from northern Ellesmere Island, Arctic Canada: Canadian Journal of Earth Sciences, v. 24, p. 246–256.

Trettin, H. P., Parrish, R. R., and Roddick, J. C., 1992, New U-Pb and ^{40}Ar-^{39}Ar age determinations from northern Ellesmere and Axel Heiberg islands, Northwest Territories and their tectonic significance: Ottawa, Geological Survey of Canada Paper 92-2, p. 3–30.

Tucker, R. D., Boyd, R., and Barnes, S.-J., 1990a, A U-Pb zircon age for the Rana intrusion, N. Norway: New evidence of basin magmatism in the Scandinavian Caledonides in Early Silurian time: Norsk Geologisk Tidsskrift, v. 70, p. 229–240.

Tucker, R. D., Dallmeyer, R. D., and Strachan, R. A., 1993, Age and Tectonothermal Record of Laurentian Basement, north-east Greenland Caledonides: Journal of the Geological Society (London), v. 150, p. 371–379.

Tucker, R. D., Krogh, T. E., and Raheim, A., 1990b, Proterozoic Evolution and Age-Province Boundaries in the Central Part of the Western Gneiss Region, Norway: Results of U-Pb Dating of Accessory Minerals from Trondheimsfjord to Geiranger, in Gower, C. F., Rivers, T., and Ryan, B., eds., Mid-Proterozoic Laurentia-Baltica: Montreal, Geological Association of Canada, Special Paper 38, p. 149–173.

Villeneuve, M. E., Ross, G. M., Thériault, R. J., Miles, W., Parrish, R. R., and Broome, J., 1993, Tectonic subdivision and U-Pb geochronology of the crystalline basement of the Alberta Basin, western Canada: Geological Survey of Canada Bulletin, v. 447, 86 p.

Young, G. M., 1979, Correlation of middle and upper Proterozoic strata of the northern rim of the North American craton: Transactions of the Royal Society of Edinburgh, v. 70, p. 323–336.

Ziegler, P. A., 1988, Evolution of the Arctic-North Atlantic and the Western Tethys: Tulsa, The American Association of Petroleum Geologists Memoir 43, 198 p.

APPENDIX A: SAMPLE LOCATIONS AND DESCRIPTIONS

Sample A, Bird Fiord Formation (TC-791B)

Location: collected from stream bank exposures of the Bird Fiord Formation on Crozier Island (NTS 68H; UTM zone 14X, 568350 E - 8412150 N).

Description: very fine-grained, green weathering, calcareous, fossiliferous sandstone. Sandy and calcareous components are complexly intercalated and bioturbated. Macrofauna include brachiopods, trilobite pygidia, and assorted shell hash. Framework minerals are calcite, quartz, detrital chlorite, and biotite. Accessory and heavy fraction minerals include apatite, rutile, rutilated quartz, zircon, ilmenite, chloritoid, authigenic pyrite and hematite, minor staurolite, and very minor sillimanite. Heavy and accessory minerals from a separate sample of the Bird Fiord Formation, collected from the Schei Syncline on southern Ellesmere Island, include kyanite, a greater concentration of staurolite, almandine, muscovite, biotite, chlorite, and dravite tourmaline.

Sample B, Hecla Bay Formation (TC-Hecla Bay)

Location: collected from the medial part of the Hecla Bay Formation on the east-dipping, upright limb of the Schei Syncline 20.5 km northeast of Baumann Fiord and 1.9 km west of Vendom Fiord on southern Ellesmere Island (NTS 49D; UTM zone 17X, 441900 E - 8619700 N).

Description: off-white to pale brown, weakly cemented quartz sandstone with 98% quartz, 1% potash feldspar, 1% muscovite, and accessory rutile, rutilated quartz, dravite tourmaline, zircon, spinel, and authigenic hematite. In thin section, the quartz is fine grained (0.1 to 0.4 mm), subrounded, and almost entirely monocrystalline. Polycrystalline quartz and chert each comprise less than 1% of the mineral mode. Five to 10% of the quartz grains display undulatory extinction or internal deformation fractures. Cements include quartz, kaolinite, and minor hematite. The potash feldspar is variably kaolinized, and there are rare grains of unaltered microcline.

Sample C, Fram Formation (TC-1105A)

Location: collected from an exposure of the Fram Formation 28 km south of the head of Vendom Fiord on the east-dipping limb of the Schei Syncline (NTS 49D; UTM zone 17X, 448000 E - 8645000 N).

Description: pale pink weathering, calcite-cemented, fine-grained (0.1–0.2 mm) quartz sandstone containing 58% mostly monocrystalline quartz, 32% calcite, 6% chert, 2% kaolinite, 1% dolomite, and traces of chlorite, muscovite, and authigenic hematite. Heavy fraction mineralogy is dominated by garnet with lesser ilmenite, staurolite, rutile, rutilated quartz, zircon, and authigenic(?) pyrite. The quartz is subrounded to subangular; the more angular grains often feature silica overgrowths and crystal facets against interstitial calcite. About half the quartz grains, including the overgrowths, possess undulatory extinction. Internally fractured grains are rare.

Sample D, Okse Bay Formation (TM-132D)

Location: collected from weakly-deformed, upper Givetian-Frasnian sandstone and conglomerate outliers located in the Yelverton Pass area of northern Ellesmere Island (NTS 340D; UTM zone 17X, 539400E - 9070100N).

Description: well sorted, quartz-cemented, medium-grained sandstone collected from the middle part of the upper member on the southeastern side of Yelverton Pass. The sample is composed of monocrystalline quartz (77%), chert (21%), and semicomposite quartz grains that probably represent vein material and recrystallized chert (2%). In addition, the rock contains subordinate feldspar, muscovite, phyllite rock fragments, chalcedony, zircon, tourmaline, and unidentified opaque minerals. The quartz commonly shows undulatory extinction. The chert is pure to slightly argillaceous and ghosts of entirely recrystallized radiolaria (?) are present in at least one fragment. The quartz and chert appear to have been rounded originally but have been modified by pressure dissolution.

PART III
REGIONAL STUDIES OF FORELAND BASINS

CHRONOSTRATIGRAPHY AND TECTONIC SIGNIFICANCE OF LOWER CRETACEOUS CONGLOMERATES IN THE FORELAND OF CENTRAL WYOMING

MICHAEL T. MAY, LLOYD C. FURER, ERIK P. KVALE, AND LEE J. SUTTNER
Department of Geological Sciences, Indiana University, Bloomington, IN 47405
GARY D. JOHNSON
Department of Earth Sciences, Dartmouth College, Hanover, NH 03755
AND
JAMES H. MEYERS
Department of Geology, Winona State University, Winona, MN 55987

ABSTRACT: Intra- and inter-basinal correlations between outcrop and subsurface over most of northern and central Wyoming indicate that chert-bearing conglomerates in the lower Cretaceous Cloverly Formation in the foreland of central Wyoming occupy three distinct stratigraphic levels. The two older conglomerates are in the lower Cloverly Formation in the western Wind River Basin and reflect northerly to northeasterly dispersal. The youngest conglomerate is in the upper Cloverly Formation in the eastern portion of the basin; gravels in this interval also were transported to the north and northeast.

The two older conglomerates are separated from the youngest conglomerate by up to 35 m of purple to gray, smectite-rich mudstones that contain distinctive 10 to 90 cm-thick layers of white to dark green devitrified tuff, as well as silica and carbonate nodular beds. Fission-track ages of 125–128 Ma have been obtained from three samples of tuff in the Wind River Basin. These tuffs can be correlated to prominent tuffs further north in the Bighorn Basin where a paleomagnetic stratigraphy has been established. Fission-track ages of zircons from devitrified tuff layers and magnetostratigraphy of mudstones suggest that the older two conglomerates in the Wind River Basin were deposited between 133 and 128 Ma and the youngest conglomerate at about 118 to 115 Ma.

Three-dimensional, spatially controlled and temporally constrained reconstructions of paleodrainage systems for Cloverly conglomerates illustrate the complexity of fluvial drainage networks within the evolving Early Cretaceous foreland basin. Sand-body geometry and dispersal patterns within these fluvial networks were partially controlled by tectonic activity, which created a series of northeast-oriented horsts and grabens in the Wind River Basin. Location of trunk rivers was controlled by the positions of grabens within the basin.

INTRODUCTION

The goal of this paper is to document the spatial and temporal distribution of fluvial sandstones and conglomerates in the lower Cretaceous Cloverly Formation in central Wyoming in order to better understand the tectonic and depositional history of the early stage of development of the Rocky Mountain foreland basin. Delimiting areas of abrupt thickness changes of conglomerates and sandstones in the earliest Cretaceous foreland basin fill has made it possible to identify areas of increased accommodation space for fluvial sediment deposition and preservation.

Fluvial depositional systems are easily influenced by subtle tectonism (Russ, 1982; Burnett and Schumm, 1983). Recurrent movement along linear structural weaknesses in the upper lithosphere often controls the location of major rivers in sedimentary basins and, therefore, the temporal and spatial distribution of nonmarine facies within a basin. Thus, fluvial systems can be used to define intraforeland-basin structural features that actively influenced distribution of sedimentary facies. In the Rocky Mountain foreland basin, paleodispersal indicators and conglomerate thickness trends indicate possible fault locations and assist in recognizing intrabasinal tectonic motion which influenced differential subsidence and paleodisperal (DeCelles, 1984, 1986; Schwartz and DeCelles, 1988; Meyers and others, 1992; Kvale and Vondra, 1993).

Outcrop and subsurface mapping of fluvial lithostratigraphic units constrained by chronostratigraphic horizons (ash beds) in the fine-grained facies can define spatial and temporal migration of subsidence in nonmarine foreland basins. Most previous depositional models for foreland basins have not been based on all available subsurface and chronostratigraphic data. Consequently, previous foreland basin models which have synthesized paleocurrent and thickness data obtained mostly from outcrop have failed to fully identify stratigraphic complexities that reflect structural and geomorphic complexities within the basin. This paper integrates data from both the outcrop and subsurface for better understanding the three-dimensional complexity and early subsidence and depositional history of the Rocky Mountain foreland basin.

The Rocky Mountain foreland basin formed east of the Sevier fold-thrust belt as part of the Cordilleran mobile belt that extends from Alaska to Mexico. The timing of the onset of deformation in the Sevier belt in Idaho, Wyoming, and Utah in part has been interpreted on the basis of lateral and vertical lithofacies relations of Lower Cretaceous strata in Wyoming and Montana (Curry, 1960; Heller and others, 1986; Heller and Paola, 1989). These strata have been interpreted as early foreland basin-fill and include the mostly nonmarine upper Jurassic Morrison and lower Cretaceous Cloverly and Lakota Formations and their western equivalents in the Gannett Group along the Wyoming-Idaho border (Fig. 1).

In order to better understand timing of thrust movement and subsequent subsidence history of the basin, several studies of Lower Cretaceous chert-bearing conglomerates have been conducted throughout the Western Interior of the U.S.A. and Canada (Heller and Paola, 1989; Schultheis, 1970; Corman, 1972; Knight, 1978; Varley, 1982). However, because conglomeratic intervals previously have been correlated by use of lithostratigraphy alone, and because most studies have been conducted in relatively local areas independent of each other, interpretations of Lower Cretaceous stratigraphy and its tectonic significance are quite varied. For example, deposition of Lower Cretaceous chert

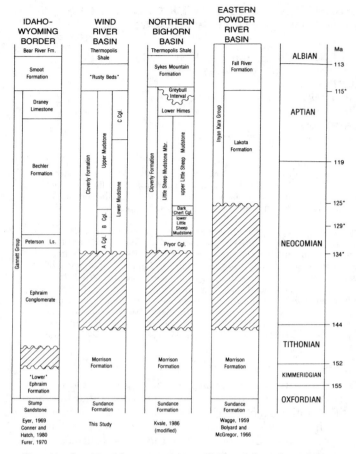

Fig. 1.—Stratigraphic nomenclature of Upper Jurassic and Lower Cretaceous strata of north-central Rocky Mountain region. A and B conglomeratic intervals within the Cloverly Formation in the western Wind River Basin are defined as coarse clastic units with intercalated mudstones overlain by a thick mudstone representing C-interval deposition. In the eastern part of the Basin, the C interval can be subdivided into a mudstone overlain by a conglomerate. Asterisks with ages represent nonstandard boundaries based on Swierc (1990); all other age boundaries based on Palmer (1983).

gravels in Wyoming is considered to be either (1) synchronous with an initial episode of thrusting (e.g., Armstrong and Oriel, 1965; Wiltschko and Dorr, 1983; Heller and others, 1986), or (2) post-orogenic and indicative of a period of tectonic quiescence (Kvale and Beck, 1985; Heller and others, 1988), or (3) pre-orogenic, having been generated by regional uplift or thermal doming that occurred just prior to Sevier thrusting (Heller and Paola, 1989).

Data presented in this study permit an estimate of the absolute age ranges of three distinct stratigraphic intervals in the Cloverly Formation, designated informally from oldest to youngest as A, B, and C (Fig. 1). These letter-designated intervals contain chert-bearing conglomerates and conglomeratic sandstones in the Wind River and Bighorn Basins. Our findings contradict earlier studies that suggest the existence of only one significant conglomerate in the Cloverly Formation (e.g., Moberly, 1960; Mirsky, 1962; Hooper, 1962). Moreover, we have identified at least two regionally extensive devitrified tuff beds and distinctive pedogenic horizons and lacustrine beds between the B and C conglomerates in outcrop and subsurface sections. The tuff beds, pedogenic horizons and lacustrine beds can be correlated from the Bighorn Basin into the Wind River Basin and farther east into the Hanna Basin (DeCelles and Burden, 1992). Dating the conglomerate-bearing intervals in the Cloverly Formation permits tectono-stratigraphic interpretation of the history of drainage evolution in the Sevier foreland basin. This interpretation suggests that structural partitioning within the early foreland basin was significant in controlling development of the drainage system.

FIELD METHODS

Outcrops of the Morrison and Cloverly Formations were studied along the margins of the Wind River and southern Bighorn Basins of central and west-central Wyoming (Fig. 2). Areal distribution of Cloverly conglomerate was mapped along approximately 100 km of outcrop in the two areas. Over 1000 paleocurrent measurements were made mainly from trough cross beds and were categorized according to lithofacies and stratigraphic position. Inter- and intra-basinal correlation was based on a database that includes 75 measured stratigraphic sections and correlation of 300 wireline and sample logs. Detailed results of the areal mapping of conglomerates and of the paleocurrent and petrographic studies for the western part of the Wind River Basin are found in Meyers and others (1992) and May (1992a, b). Absolute-age ranges for regional chronostratigraphic correlations were established by zircon fission-track age-dating of irradiated euhedral zircons obtained from devitrified tuffs.

MORRISON-CLOVERLY STRATIGRAPHY

Introduction

The upper Jurassic Morrison and lower Cretaceous Cloverly Formations are part of a regionally extensive interval of fluvial and lacustrine sediments deposited during a break in marine sedimentation within the Western Interior Seaway of North America. The Morrison Formation conformably overlies the glauconitic sandstones of the marine upper Jurassic Sundance Formation and consists of a succession of alternating calcite-cemented, quartzo-feldspathic fluvial and eolian sandstones and caliche or lacustrine carbonate-bearing, green and red mudstones.

An unconformity is present between the Morrison and the overlying Cloverly Formations. Its presence is identified by magnetostratigraphic constraints established for mudstones in the Morrison-Cloverly section (Swierc, 1990) and is supported by regional subsurface-outcrop correlation (May, 1992b). The Cloverly Formation contains chert-bearing conglomerate and medium to coarse-grained sandstone, intraformational conglomerate made up mostly of reworked lacustrine and pedogenic carbonate clasts, and variegated smectitic mudstone. In central Wyoming, the Cloverly Formation is overlain by the "rusty beds" sandstone of the Thermopolis Formation which is equivalent to the Sykes Mountain Formation of the Bighorn Basin. The "rusty beds" mark the initiation of Albian marine transgression into this area (e.g., Young, 1970).

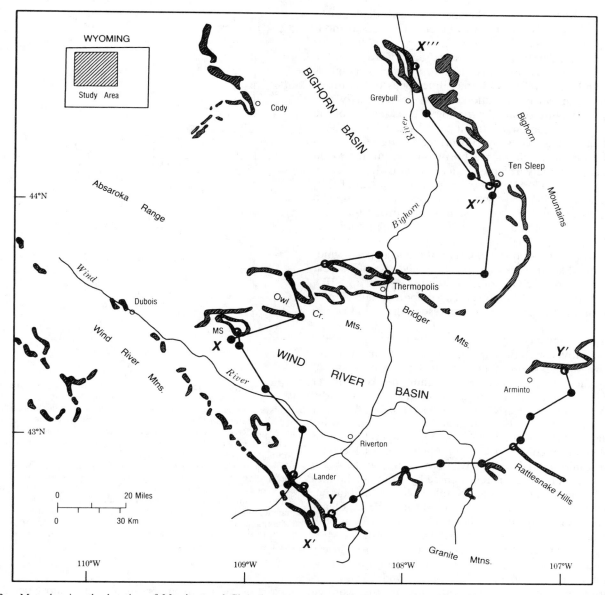

FIG. 2.—Map showing the location of Morrison and Cloverly outcrop (dark bands) in north-central Wyoming. Location of cross sections in Figures 4 and 5 is shown by lines X-X', Y-Y' and X-X"-X'". Open circles on cross section lines represent outcrop sections and solid circles subsurface control. Location of additional outcrop and subsurface control for this study is given in Figure 7.

Methods of Correlation of the Morrison and Cloverly Formations

Time correlation in nonmarine rocks is difficult and heretofore has not been accomplished in the Morrison-Cloverly Formations of Wyoming. Instead, the contact between the Morrison and Cloverly Formations traditionally has been placed at the base of the first chert-bearing sandstone or conglomerate (Mirsky, 1962; Furer, 1970; Winslow and Heller, 1987). This practice, as illustrated in Figure 3A, is particularly common in subsurface studies because the contrast in log signature between sandstones and conglomerates and the encasing mudstones is easily discernible (e.g., Wilson, 1958). However, such correlation is lithostrati-

graphic and obscures important tectono-stratigraphic relationships whose interpretation requires chronostratigraphic correlation. Furthermore, where a conglomerate or chert-bearing sandstone is not present in the lower conglomertic interval of the Cloverly Formation (Kcv (lc)) (e.g., Hudson-Lander outcrop section and South Lander subsurface section in Fig. 4A) the contact cannot be defined in this manner.

We have established a regional chronostratigraphy of the Morrison-Cloverly nonmarine rocks by (i) correlating measured surface sections to the nearest site of subsurface control, (ii) correlating electrical logs throughout the basins, and (iii) using paleomagnetic and zircon fission-track age constraints. Marine time-marker beds, clearly distinguishable on electric logs, are used to bracket the ages of nonmarine Morrison and Cloverly rocks (Fig. 3B). In all localities, our measured sections extend downward from the lower part of the Thermopolis Shale to the top of the glauconitic sandstone and fossiliferous limestones of the upper part of the Sundance Formation, which has an easily recognizable signature on an electric log. The signature of the Sundance Formation on electric logs in Figure 3B has been verified in several wells by subsurface stratigraphic samples. The top of the Sundance Formation within the Wind River Basin is widely accepted as a time-equivalent horizon (Peterson, 1972). Correlation of spontaneous potential (SP) or gamma ray (GR)-resistivity subsurface logs, combined with micropaleontologic studies, has established that several time-equivalent stratigraphic intervals are present in the Albian Thermopolis Shale (Bally, 1987). Moreover, the base of the lower Thermopolis Shale is a reliable upper-bounding time marker for rocks in this study because of its near parallelism with overlying biostratigraphic zones (Fig. 3B). Therefore, the base of the lower Thermopolis Shale is most likely a marine flooding surface representing a rapid rise in sealevel.

Criteria established by Moberly (1960) to define the Morrison-Cloverly contact in the Bighorn Basin in north-central Wyoming seem to be the most reliable means of selecting the contact in outcrop throughout the Wind River Basin. This contact can then be easily correlated to a nearby subsurface section. Moberly (1960, p. 1145) placed the lower contact of the Cloverly Formation at the "base of the lowest beds which either show evidence of significant additions of volcanic debris or contain pebbles or granules of black chert." The "volcanic debris" in the Bighorn Basin, as Moberly indicated, most typically is represented by montmorillonite-rich mudstones and nodules and veinlets of chert and barite. In both the Bighorn and Wind River Basins, these volcanogenic deposits may reflect the onset of Cretaceous volcanism farther west. As noted by Moberly (1960), the Morrison-Cloverly contact also is marked by a change from nearly ubiquitous, calcrete-bearing red and green mudstones of the Morrison Formation to silica/carbonate- and barite-bearing, variegated purple, orange, red, gray and light-brown mudstones of the Cloverly Formation. Associated with this difference in color between the fine-grained siliciclastic rocks of the Morrison and Cloverly Formations is a subtle difference in the detrital composition of the sandstones. Throughout the Wind River Basin, Morrison sandstone contains 8–15 percent feldspar and less than three percent chert, whereas Cloverly sandstone contains 2–4 percent feldspar and 10–90 percent detrital chert (Furer, 1970; Meyers and others, 1992; May, 1992b).

Our study in the Wind River Basin has revealed at least two stratigraphic levels containing white to dark-green siliceous mudstones. The mudstones occur as discontinuous 10- to 90- cm-thick beds and contain angular mudstone rip-up clasts and euhedral zircon microphenocrysts. The stratigraphically lower white-green bed occurs in some localities just above the "B" conglomerate of the lower Cloverly Formation and correlates with a similar green mudstone described by DeCelles and Burden (1992) from the Alcova

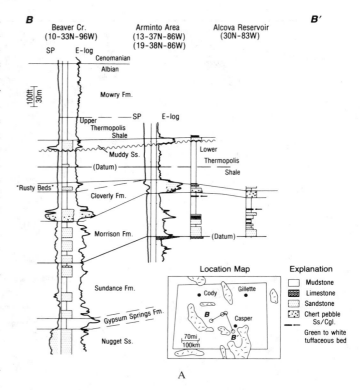

FIG. 3.—Subsurface-outcrop cross sections showing differences between traditional correlation of the Morrison-Cloverly section and equivalents versus correlations in this study based on paleontologic, magnetostratigraphic and fission-track age constraints. (A) Cross section from the Wind River Basin to the Hanna Basin showing traditional method of picking the contact between the Morrison and Cloverly Formations. This correlation is based on lithostratigraphic characteristics. (B) Cross section from Wind River Basin to Shirley and Powder River Basins showing paleontological zonations in the Thermopolis Shale and typical lithologic and geophysical characteristics of the Morrison-Cloverly section. (C) Cross section from the Wind River and Hanna basins showing chronostratigraphic correlation of the Morrison-Cloverly sections as used in this study. Paleomagnetic, paleontologic and lithologic correlation of fine-grained facies (i.e., lavender beds) defines the disparate nature in time and space of Cloverly conglomerates. Dashed lines are approximate time lines within mudstones.

area (Fig. 3C). An upper green bed is most persistent in the Wind River Basin and occurs no more than seven meters (22 ft) below the base of the "rusty beds" in the western Wind River Basin. This green bed is correlated to a similar bed below the "C" conglomerate in the Arminto area in the eastern Wind River Basin (Fig. 3C). We have not been able to detect a signature for this bed on electric logs, but have documented many localities where it exists in outcrop near a subsurface data point, such as in the Arminto area.

We interpret the white to green mudstone beds as the diagenetically-altered remains of reworked ash deposits or tuffs. The beds do not appear to represent a single depositional event. Instead, based on our mapping and regional correlation, the younger reworked ash beds appear to occupy a relatively thin stratigraphic interval at approximately the same level as a thick devitrified-ash bed in the upper part of the Little Sheep Mudstone Member of the Cloverly Formation throughout the Bighorn Basin (Fig. 1; cf. Mob-

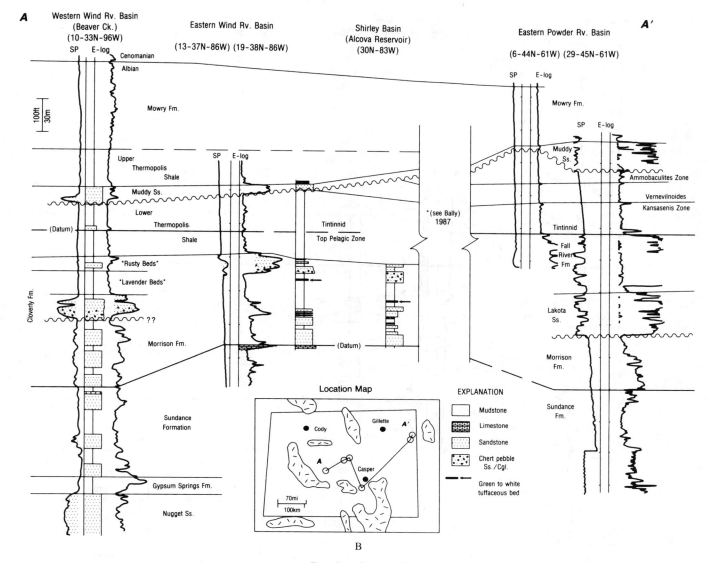

Fig. 3.—Continued.

erly, 1960; Heady, 1992). The upper white to green tuff bed always is underlain by barite and silica/carbonate pedogenic horizons or lacustrine beds, and in the western half of the Wind River Basin, always is associated with smectitic mudstones directly beneath the transitional marine "rusty beds" of the overlying Thermopolis Shale. Consequently, we suggest that the interval containing the two major white to green tuff beds can be used as a chronostratigraphic marker zone reflecting a segment of time characterized by eruptive volcanic events that deposited ash atop well developed, laterally-extensive pedogenic horizons and lacustrine beds throughout the Bighorn and Wind River Basins. This interval is referred to as the "lavender beds" (Love, 1948).

Although the white to green tuffs are too thin to be seen on electric logs, the mudstones encasing them are distinctive and their stratigraphic position can be correlated from measured outcrop sections to subsurface logs by correlating the bounding marine time-marker beds in the Sundance and Thermopolis Formations as illustrated in Figure 3B. For example, in the Arminto area and in several other localities in the Wind River Basin, the stratigraphic thickness from the top of the marine Sundance Formation to the base of the Thermopolis Shale is nearly the same (Fig. 4A). Thus, the low resistivity signature shown on subsurface logs 3 to 30 m (10–100 ft) below the C chert-pebble conglomerate in Figure 3C represents the mudstone interval containing the upper white to green tuffaceous mudstone in the nearby Arminto outcrop. The lower green bed occurs at a consistent distance above the top of the Sundance Formation allowing it be correlated from the Alcova Reservoir to the Arminto area (Fig. 3C). The white to green tuffaceous marker beds within the "lavender beds" interval, when correlated as illustrated by the short dashed lines in Figure 3C, are sub-parallel to the biostratigraphic zonations established in both the Thermopolis Shale and Sundance Formation in the eastern Wind River Basin. The low resistivity signature of the nonmarine "lavender beds" in the Arminto area correlates in Figure 3C across the Wind River Basin to the low

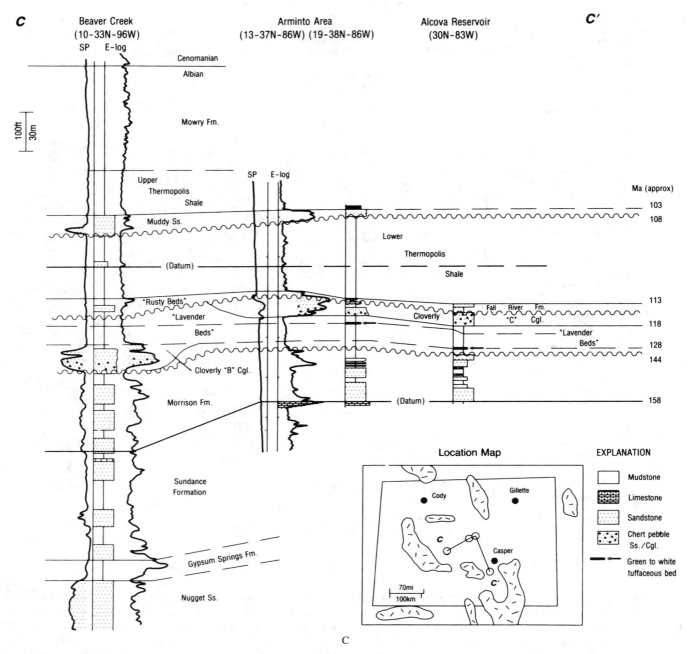

Fig. 3.—Continued.

resistivity mudstones below the "rusty beds" in the Beaver Creek area. This correlation demonstrates that the chert-pebble sandstone and conglomerate of the Cloverly Formation in the western Wind River Basin are older than the "lavender beds," whereas the opposite relationship exists in the eastern part of the Basin.

REGIONAL STRATIGRAPHIC RELATIONS
OF CLOVERLY CONGLOMERATES

Two compositionally distinct types of conglomerates exist within the Cloverly Formation in the Wind River Basin. One variety consists almost entirely of extra-basinally derived chert clasts, which are dominantly pebble-sized. The second consists primarily of intra-formationally derived, granule-to-cobble-sized limestone, siltstone, and mudstone clasts reworked from overbank facies. This conglomerate commonly contains less than 15 percent chert clasts. These two conglomerates collectively constitute what we term the lower Cloverly conglomerate (Kcv (lc) in Figs. 4, 5). Typically, when both types of conglomerates occur in the same section, the chert-pebble conglomerate lies beneath the intra-formational-pebble conglomerate. An identical stratigraphic relationship involving similar conglomerates in the lower Cretaceous Kootenai Formation has been docu-

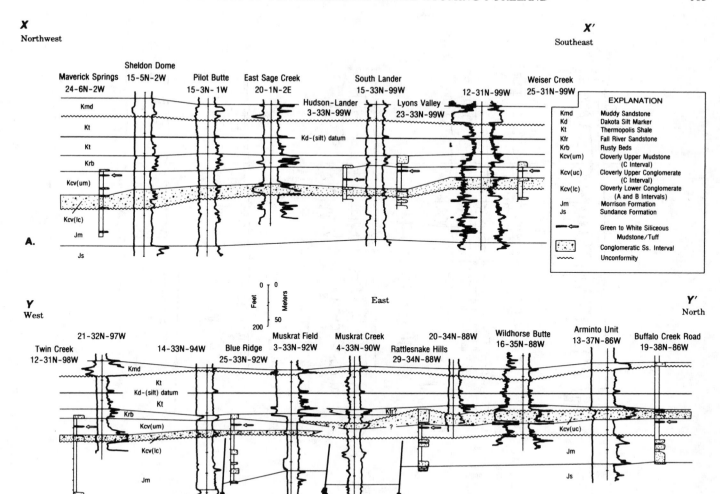

FIG. 4.—(A) X-X'—Northwest to southeast cross section on the west side of the Wind River Basin. Kcv (lc) denotes the combination of A- and B- conglomeratic intervals of the Cloverly Formation of Figure 1. (B) Y-Y'—West to east cross section across the southern margin of the Wind River Basin. Kcv (uc) denotes the uppermost Cloverly conglomerate or C conglomerate of Figure 1, whereas Kcv (lc) is as in Figure 4A. Location of both cross-section lines is given in Figure 2.

mented throughout much of southwestern Montana by DeCelles (1984) and Thompson (1984).

In the western and west-central Wind River Basin, conglomerates are present only near the base of the Cloverly Formation (Cloverly lower conglomeratic A- and B-intervals in Fig. 4A), at least 50 m below the "rusty beds" of the overlying Thermopolis Shale. The lower conglomerate bearing A- and B-interval in the western part of the basin is overlain by the C-interval (designated Kcv (um) or upper mudstone in Figs. 4, 5). The C-interval is comprised of up to 35 m of red, orange, light brown, and purple smectitic mudstones ("lavender" or "lilac" beds of Love, 1948; maroon and purple beds of the Little Sheep Mudstone Member of Swierc, 1990), and gray smectitic mudstones. Most mudstones in the C-interval, particularly the purple or orange/red ones, are intercalated with nodular silica and barite beds and nodular and bedded nonmarine micritic limestone.

These rock types are rarely present in the A- and B-intervals. Discontinuous sandstones ranging in thickness from 0.1–0.5 m occur near the base of the C-interval.

In contrast, in the eastern part of the basin, conglomerates are confined to the upper part of the Cloverly Formation, nearly lateral to or approximately 1–5 m below the base of the "rusty beds" (Cloverly upper conglomeratic C-interval in Fig. 4B). Our on-going investigations suggest that a similar stratigraphic relationship of upper Cloverly conglomerate above smectitic mudstone exists in the Bighorn Basin, as illustrated in Figure 5, and that this conglomerate correlates to the upper C-interval mudstones farther west.

In much of the western part of the Wind River Basin, a chert-pebble conglomerate is the basal Cloverly unit (A-interval) and lies unconformably on the Morrison Formation. This contact has local relief ranging from 6 to 15 m

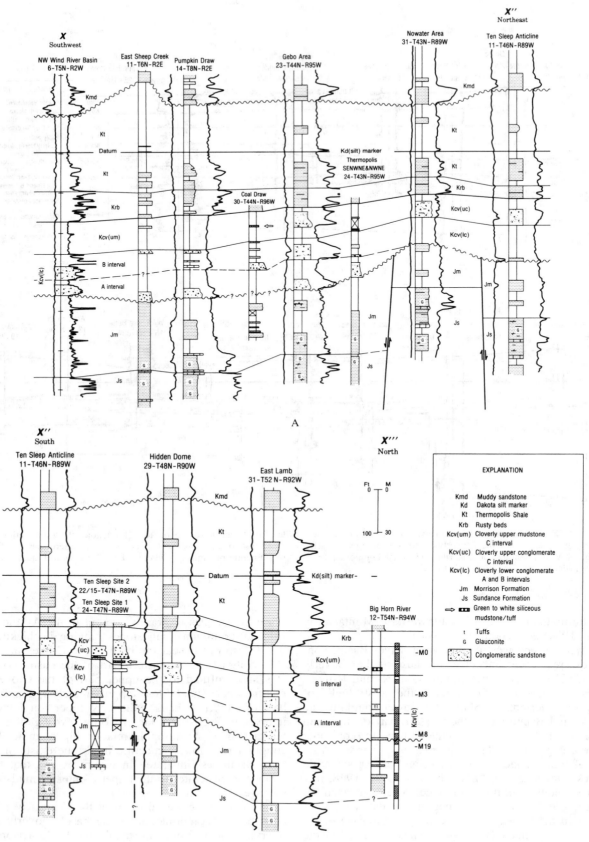

(Figs. 4, 5A). The intra-formational-pebble conglomerate (B-interval) overlies and is locally incised into the A-interval chert-pebble conglomerate (May, 1992a, Fig. 11). More frequently, however, the B-interval conglomerate is separated from the A-interval conglomerate by 1–10 m of mudstone as illustrated by the NW Wind River Basin section at the southwest end of section X-X" in Figure 5A.

ESTIMATES OF THE ABSOLUTE AGE OF FACIES
WITHIN THE CLOVERLY FORMATION

General Constraints

Although the absolute age of the Cloverly Formation remains equivocal, fission-track age determinations on zircons from pyroclastic facies and magnetostratigraphy of fine-grained facies permit tentative interpretations. Samples of the white to green tuffs in the upper Cloverly Formation have been collected from outcrops at three sites within the Wind River Basin—Maverick Springs Dome (MS), Arminto, and in the Rattlesnake Hills (Fig. 2). Fission-track dates on these samples range from 125–128 Ma. Our correlations indicate that the mudstone interval containing these tuff beds is approximately equivalent to the upper Little Sheep Mudstone Member of the Cloverly Formation in the Bighorn Basin (Fig. 5B). Several separate but concordant fission-track ages on zircons from tuff beds in the variegated and purple mudstone interval within the Little Sheep Member have been obtained (Chen, 1989a, b; Heady, 1992; and Tabott, 1992, pers. commun.). These dates suggest that an episode of maximum pyroclastic deposition in the upper Cloverly Formation occurred between 118 and 128 Ma, with principal events around 120 and 127 Ma (Heady, 1992). These two events are represented by the two tuffaceous mudstones in the Arminto and Alcova Reservoir sections in Figure 3C. Consistent with these dates is the fact that the purple and variegated beds in the Little Sheep Mudstone Member record a long period of magnetic polarity reversal correlated by Swierc (1990) to the M3 reversal polarity zone of the Global Magnetic Polarity scale (Fig. 6). This period of reversal spans approximately 123–126 Ma. Based on his paleomagnetic work, Swierc (1990) concluded that the base of the Little Sheep Mudstone Member is about 128 Ma and the top is about 116 Ma.

Our correlations suggest that the "Pryor Conglomerate Member" of the Cloverly Formation, as identified by Winslow and Heller (1987), is approximately time equivalent to the B-interval conglomerate in the Cloverly Formation in the Wind River Basin (Fig. 5B). We conclude that their

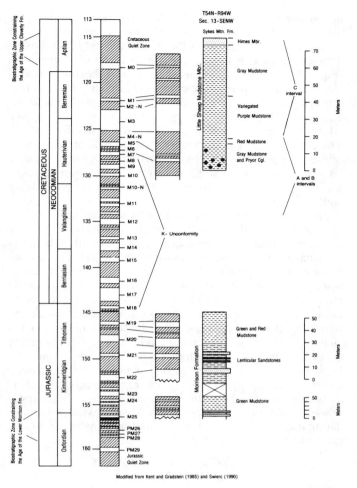

Fig. 6.—Geochronology of Morrison and Cloverly Formations in the Bighorn Basin (Modified from Swierc, 1990). The left column is the geomagnetic time scale from Kent and Gradstein (1985) showing normal (cross-hatched) and reverse polarity. The center column shows magnetic reversal stratigraphy for the Cloverly and Morrison Formations at a site in the northeastern Big Horn Basin.

"Pryor Conglomerate" actually is a chert-pebble conglomerate within the lower part of the Little Sheep Mudstone Member (Fig. 6). The base of the Cloverly Formation is about 20 m below this conglomerate, where an older conglomerate is occasionally present in the subsurface (the Pryor Conglomerate in Figure 1). This lower conglomerate crops

---

Fig. 5.—(A) Southwest to northeast cross section X-X" from the northwestern Wind River Basin to the southeastern Bighorn Basin. Note that the chert-bearing conglomerates are in the Kcv (uc) or C-interval in the southeastern Bighorn Basin. Cross section location shown on Figure 2. Explanation of symbols is shown on 5B. (B) South to north cross section X"-X"' from the southeastern Bighorn Basin to near the Cloverly type section in the northeastern Bighorn Basin. Chert-bearing conglomerates are in the Kcv (lc) or A- and B-intervals near the type section as in the northwestern Wind River Basin. Their relationship to Swierc's (1990) magneostratigraphy is shown in the column to the right of the Big Horn River section. On this column periods of normal polarity are represented by cross-hatched pattern. Lithology on geophysical logs from American Stratigraphic Company and Mills (1956). The Pryor Conglomerate of Moberly (1960) corresponds to the A interval; Winslow and Heller's (1987) Pryor Conglomerate corresponds to the B interval in the East Lamb well. The upper Little Sheep Mudstone of Moberly (1960) is between the B conglomerate and the "rusty beds." The "rusty beds" of the East Lamb well correlate with the Himes Member of Figure 6. The thick sandstone above the "rusty beds" in the East Lamb well is referred to as the Sykes Mountain Formation in the northeast Bighorn Basin.

out at the type section of the Cloverly Formation in the northern end of the Bighorn Basin and is equivalent to the A-interval conglomerate of the western Wind River Basin. This conglomerate is also present in the East Lamb well (Fig. 5B). Along the northeast margin of the Bighorn Basin, the B-interval conglomerate occurs lateral to a whitish tuff bed that Heady (1992) concluded was deposited at about 128 Ma. We correlate this pyroclastic event to the lower white to green tuff bed in the Wind River Basin (Fig. 4). The date of 128 Ma represents the best available estimate of the minimum age of the youngest "Pryor Conglomerate" (B-interval conglomerate) and very likely is a minimum age for the B-interval conglomerate in the Wind River Basin as well.

Fission-track ages of tuffs and magnetostratigraphy of fine-grained facies in the Cloverly Formation aid in constraining the ages of multiple conglomeratic intervals within the foreland basin. Such temporal constraints are relevant to our tectono-stratigraphic and paleogeographic reconstruction associated with early subsidence and infilling of the foreland basin. We are confident in our basin-to-basin correlations of sandstones and conglomerates because paleomagnetic and absolute age-dating of non-channelized facies show that distinctive pedogenic and lacustrine lithofacies occur at similar chronostratigraphic positions from the Bighorn to the Wind River Basins as previously discussed. Channelized sandy/gravelly facies are inherently discontinuous, therefore a benchmark for identifying their stratigraphic position and age is attainable only by mapping and dating the lateral and vertical distribution of associated fine-grained facies and ash beds.

Absolute Age Differences Among A-, B-, and C-Interval Conglomerates

Estimates of the differences in absolute age of the Cloverly conglomerates are important to calculation of the rate at which the major Cloverly trunk rivers migrated from west to east through the foreland basin. The chronostratigraphy and regional correlations we have presented permit a first approximation of these age differences.

Although the A- and B-interval conglomerates are difficult to separate temporally, deposition of both may have ended by 128 Ma, but no later than 122 Ma. This is based on correlations of the conglomerates and mudstones from the Wind River Basin to the section in the Bighorn Basin where Swierc (1990) has established a paleomagnetic stratigraphy (Fig. 5B). A minimum age of the B-interval conglomerate appears to be 128 Ma, based on the youngest fission-track ages for white to green tuff beds. In contrast, according to our correlations in Figures 4 and 5, the C-interval conglomerate in the eastern Wind River Basin was deposited during the final stage of C-interval deposition, roughly corresponding to the 118–115 Ma interval of time from MO, or the Cretaceous quiet zone, to near the base of the Sykes Mountain Formation or "rusty beds" (Fig. 6). Consequently, a minimum of four and a maximum of 13 million years may have elapsed between the time of deposition of B-interval conglomerates in the western Wind River Basin and C-interval conglomerates in the eastern Wind River Basin (Fig. 3C).

PALEOGEOGRAPHIC RECONSTRUCTION—TECTONIC AND PROVENANCE IMPLICATIONS

Cloverly A- and B-interval conglomerates in the western Wind River Basin were transported generally northeastward as were C-interval conglomerates in the eastern Wind River Basin (Fig. 7; cf. May, 1992a, b; Meyers and others, 1992; Curry, 1960). The C-interval trunk river system extended into the southeastern Bighorn Basin near Ten Sleep as reflected by the relatively thick conglomerate in that area (Fig. 5). Major tributaries of this younger trunk system probably did not exist in the western part of the Wind River Basin. Evidence for this is that in both the subsurface and in extensive outcrop in the westernmost part of the basin, no channelized sandstone or conglomerates exist in the C-interval except for a single locality along the Little Popo Agie River southeast of Lander, Wyoming. Instead of widespread deposition of chert-pebble conglomerate, sedimentation in the western part of the basin during the time of deposition of the eastern C-interval chert-pebble gravels is represented mainly by a gray and light-brown smectitic mudstone interval below the "rusty beds." These mudstones probably accumulated in extensive playa lakes.

Our previous studies of the lowermost fluvial systems associated with deposition of the chert gravels in the western Wind River Basin suggest deposition on a broad braid plain, with locally thick channel belts (A-interval deposition in Fig. 8) to meandering systems (B-interval deposition in Fig. 8) (May, 1992b; Meyers and others, 1992). Both systems apparently were structurally controlled (May, 1992b; Meyers and others, 1992). Northward to northeastward dispersal of sediment occurred about 10 million years later, approximately 50 to 150-km to the east (C-interval deposition). Transport was via a complex system of mostly braided rivers. The courses of these rivers also may have been structurally controlled. Syndepositional faults revealed by our subsurface correlation (Figs. 4B, 5A) and shown on Figure 7, regional lineament trends (Thomas, 1971), and dikes in the basement rocks of the Wind River Range (Worl and others, 1986) all are subparallel and coincident with the margins of the major trunk systems (i.e., A- and B-interval conglomerates) throughout the Wind River Basin (May, 1992a, b; Meyers and others, 1992) suggesting structural control on these systems.

Several workers (e.g., Stokes, 1944, p. 980; McGookey and others, 1972; Heller and others, 1988; Heller and Paola, 1989) have concluded that chert gravels in the Cloverly Formation were derived from basin-margin sources located directly west or southwest of the Wind River Basin. However, chert-bearing conglomerates in the Ephraim Formation of the Gannett Group in far western Wyoming differ in composition from the conglomerates of the Cloverly Formation in the Wind River Basin (Lammers, 1939; May, 1991, 1992b, 1993). Red chert grains and pebbles are abundant in the Ephraim conglomerates, whereas light-and dark-gray chert and a near absence of red chert typify the extra-basinally derived conglomerates of the Cloverly Formation of the Wind River Basin. Consequently, both the Ephraim and lower Cloverly conglomerates were not derived from the same source. However, conglomerates in the

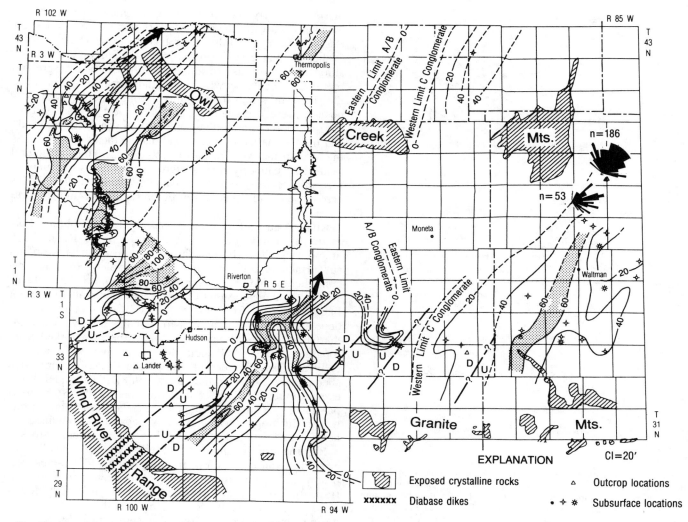

Fig. 7.—Isopach map of Cloverly conglomerates/sandstones in the Wind River Basin. Solid arrows represent mean paleoflow direction as intepreted by May (1992a, b) for area in northwestern corner of map and by Meyers and others (1992) for area southeast of Riverton. Rose diagrams north of Waltman in northeast corner of the map represent dispersion of trough axis orientations in Cloverly C conglomerate as documented in this study. Relative dip-slip movement on steep faults is shown by (U) for up and (D) for down.

Bechler Formation in western Wyoming resemble those of the Cloverly Formation in composition and possibly could have shared a common provenance.

Although time equivalence of conglomeratic units in the Gannett Group from near the Wyoming-Idaho border and those in the Cloverly Formation in the Wind River Basin has not yet been firmly established, we believe the major fluvial systems in the Ephraim Formation and those in the Cloverly Formation in the Wind River Basin represent separate drainage systems, at least through much of Neocomian time (i.e., until about 130 Ma). In fact, the Ephraim Formation may be older than the lower Cloverly conglomerates in the Wind River Basin. Ephraim fluvial systems directly west of the Wind River Basin have been interpreted by Derman and others (1984) as northward-flowing trunk river systems which is consistent with our interpretation.

DISCUSSION AND CONCLUSIONS

The Cloverly Formation in central Wyoming contains chert-pebble conglomerates at three distinct stratigraphic intervals. Each Cloverly conglomerate represents a separate belt of gravel that accumulated in central Wyoming at different times. Both the lower (A- and B-interval) and upper (C-interval) conglomeratic deposits reflect a northerly-to-northeasterly-dipping paleoslope in central Wyoming. Paleogeographic reconstruction based on paleocurrent and subsurface to outcrop mapping of thickness of sandstones and conglomerates (May, 1992a, b; Meyers and others, 1992) indicates northward to northeastward transport of the A- and B-interval gravels into the southwestern Bighorn Basin. The C-interval gravels were transported northeastward into the southeastern Bighorn Basin.

Correlation of surface and subsurface sections of the non-marine Morrison-Cloverly interval throughout Wyoming

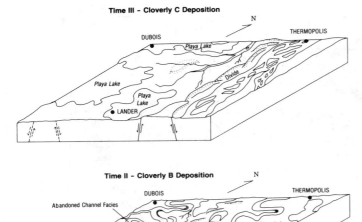

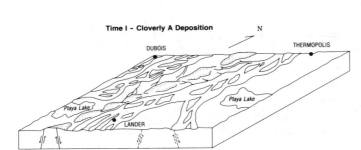

Fig. 8.—Schematic block diagrams showing the paleogeography of the central to western Wind River Basin region during Cloverly A-, B-, and C-interval deposition. Inferred location of horst and graben systems is based on cross-section interpretation (e.g., Figs. 4, 5) and the correspondence between isopach and paleocurrent trends as inferred from Figure 7. Active faults shown with solid lines and inactive are dashed. Presumably not all fault systems in the east were active as early as those in the west.

indicates that early structural partitioning in the Sevier foreland basin controlled formation thicknesses and the location of thick sandstone bodies deposited in trunk rivers. Pre-Cloverly erosion of Morrison sediments occurred over NE-trending horst blocks, and the axes of thick channel deposits in the Cloverly Formation parallel nearby NE-trending faults mapped in underlying rocks. Recurrent movement along these NE-trending faults controlled dispersal of coarse detritus in Cloverly channels in the Wind River Basin (May, 1992a, b; Meyers and others, 1992).

The elapsed time between A-interval conglomerate deposition in the western Wind River Basin and C-interval conglomerate deposition in the eastern Wind River Basin implies a complex history of early basin subsidence, sediment-filling, and local reworking of fluvially deposited coarse-grained clastic sediment. Depocenters associated with trunk river systems appear to have migrated from west to east through time. "Rusty beds" transgression over the non-marine Cloverly section probably began in the northeast, then spread toward the southwestern part of the Wind River Basin. Assuming that the major axis of subsidence had shifted farthest to the east during C-interval conglomerate deposition, this area was presumably the first to be transgressed by the Cretaceous seaway in our study area. However, once the rivers in which the C-interval conglomerates were deposited were drowned, a relatively short period of time elapsed (perhaps as little as one million years) before sandy and muddy marine to transitional marine facies spread across the area to the western part of Wyoming. Thus, structural control on Cloverly river systems probably also influenced the initial location of estuarine and eventually marine facies across the Wind River Basin region (cf. Kvale and Vondra, 1993).

Rivers transporting gravels making up the Ephraim Formation probably did not flow into the Wind River Basin area during Cloverly deposition. Based on our outcrop and subsurface correlations and regional paleogeographic reconstruction and previous provenance studies (May, 1992b, 1993), we conclude that Cloverly sediments also were not likely reworked from sources directly west of the site of deposition in the foreland basin. For this to have occurred, the Cloverly would have had to have been reworked from foredeep deposits (Gannett Group), as they were isostatically uplifted after denudation of the thrust-belt and then shed eastward (Heller and others, 1988). This model, however, implies elasticity of the lithosphere with no structural partitioning. Our work suggests that the lithosphere in this region did not behave elastically because it contained inherent structural weaknesses deeply seated in basement rocks.

Significantly, channels filled with Cloverly gravels have a nearly axial orientation, trending subparallel rather than orthogonal to the structural grain of the Sevier orogenic belt (Fig. 9). Northeasterly-directed drainage apparently developed during Cloverly deposition or perhaps even earlier during upper Morrison deposition; unfortunately, regional paleocurrent data for the channelized deposits in the Morrison Formation are scarce. Presumably, the early Sevier foredeep was filled by the time A- and B-interval Cloverly conglomerates accumulated in the western Wind River Basin. The younger Cloverly conglomerates (C-interval), above the lavender mudstone beds and the white to dark-green tuff-bearing interval in the eastern part of the study area, were perhaps derived more from the south than the older, western conglomerates (A- and B-intervals). Progressively more southeasterly development of fluvial trunk systems carrying chert gravels evolved in Wyoming for one or more of the following reasons: (1) extra-basinal tectonism may have migrated temporally from north to south along the basin margin progressively elevating more southerly-located source regions; (2) movement along intra-basinal basement faults might have been recurrent and migrated through time to the east; or (3) a decrease in overall basin subsidence, coupled with an increase in sediment supply could have filled the foreland basin, causing avulsion of major systems preferentially in a direction parallel to and down paleoslope (cf. Bridge and Leeder, 1979; Seni, 1980). Whatever the cause, we believe that regional intra-basinal structural partitioning, manifested as a series of subtle horsts and gra-

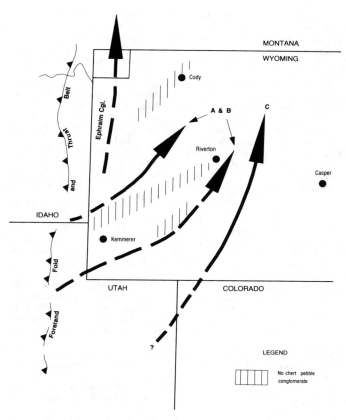

FIG. 9.—Regional paleogeographic reconstruction for Neocomian-Aptian time. Thrusting along the basin margin is inferred to become younger to the south. Arrows indicate major fluvial systems. Source areas for easternmost A- and B-interval conglomerates and C-interval conglomerates probably were in Utah or Nevada based on paleocurrent data and the "no chert-pebble conglomerate" region in southwestern Wyoming. Ephraim conglomerate dispersal data are from Derman and others (1984). Dispersal data in extreme southwestern Wyoming are from H. T. Pile (pers. commun.).

bens, was mainly responsible for controlling the geographic distribution of Cloverly chert-pebble gravels and their geometry. This interpretation is supported by isopach trends of channel sand bodies and regional correlations.

Evaluation of stratigraphic modeling of foreland basins is limited by an understanding of complex basin stratigraphies. Without proper understanding of the chronostratigraphy in a basin, the applicability of quantitative tectonic, climatic, or eustatic models in predicting detailed stratigraphic sequences within basins cannot be rigorously tested. Broad, two-dimensional crustal rheological models, or even two-dimensional sedimentary basin-fill models have traditionally formed a satisfactory basis for collecting and analyzing data from within foreland basins. Our study suggests, however, that such two-dimensional basin models should be modified to accommodate the concept of recurrent structural partitioning of foreland basins and the complexity generated in the resulting stratigraphic sequences.

ACKNOWLEDGEMENTS

We express appreciation to Peter Schwanns, Peter DeCelles, Steve Dorobek and Rod DeBruin for their rigorous and constructive reviews of this manuscript. Peter DeCelles also has graciously provided us measured sections and other outcrop data from the Rattlesnake Hills and Alcova Reservoir areas. Research on foreland-basin sedimentation in Wyoming is supported by NSF Grants EAR-8709039 and EAR-9017775 to Lee Suttner and Gary Johnson. We are indebted to Kim Schulte for typing multiple versions of this manuscript and to members of the Graphics, Cartography and Photography Division of the Indiana University Department of Geological Sciences and the Indiana Geological Survey.

REFERENCES

ARMSTRONG, F. C. AND ORIEL, S. S., 1965, Tectonic development of Idaho-Wyoming thrust belt: American Association of Petroleum Geologists Bulletin, v. 49, p. 1847–1866.

BALLY, A. W., 1987, Atlas of Seismic Stratigraphy: Tulsa, American Association of Petroleum Geologists Studies in Geology 27, v. 1, 125 p.

BOLYARD, D. W. AND MCGREGOR, A. A., 1966, Stratigraphy and petroleum potential of Lower Cretaceous Inyan Kara Group in northeastern Wyoming, southeastern Montana and western South Dakota: American Association of Petroleum Geologists Bulletin, v. 50, p. 2221–2244.

BRIDGE, J. S. AND LEEDER, M. R., 1979, A simulation model of alluvial stratigraphy: Sedimentology, v. 26, p. 617–644.

BURNETT, A. W. AND SCHUMM, S. A., 1983, Active tectonics and river response in Louisiana and Mississippi: Science, v. 222, p. 49–50.

CHEN, Z., 1989a, A fission-track study of the terrigenous sedimentary sequences of the Morrison and Cloverly Formations in the northeastern Bighorn Basin, Wyoming: Unpublished M.S. Thesis, Dartmouth College, Hanover, 66 p.

CHEN, Z., 1989b, A fission-track study of the terrigenous sedimentary sequences of the Morrison and Cloverly Formation in the northeastern Bighorn Basin, Wyoming (abs.): Geological Society of America Abstracts with Programs, v. 26, p. 206.

CONNER, J. L. AND HATCH, J. E., 1980, Stratigraphy of the Sage Valley and Elk Valley quadrangles, western Wyoming and southeastern Idaho, in Hollis, S., Stratigraphy of Wyoming: Casper, Wyoming Geological Association 31st Annual Field Conference Guidebook, p. 263–277.

CORMAN, D., 1972, Formation of a mature siliceous chert conglomerate (Lower Cretaceous) of the Rocky Mountain foothills, Alberta: Unpublished M.S. Thesis, University of Delaware, Newark, 73 p.

CURRY, W. H., III, 1960, Stratigraphy and paleogeography of Upper Jurassic and Lower Cretaceous rocks of central Wyoming: Unpublished Ph.D. Dissertation, Princeton University, Princeton, 306 p.

DECELLES, P. G., 1984, Sedimentation and diagenesis in a tectonically partitioned, nonmarine foreland basin: The Lower Cretaceous Kootenai Formation, southwestern Montana: Unpublished Ph.D. Dissertation, Indiana University, Bloomington, 422 p.

DECELLES, P. G., 1986, Sedimentation in a tectonically partitioned nonmarine foreland basin: the Lower Cretaceous Kootenai Formation, southwestern Montana: Geological Society of America, v. 97, p. 911–931.

DECELLES, P. G. AND BURDEN, E., 1992, Nonmarine sedimentation in the overfilled part of the Jurassic-Cretaceous Cordilleran foreland basin: Morrison and Cloverly Formations, central Wyoming, U.S.A.: Basin Research, v. 4, p. 291–313.

DERMAN, A. S., WILKINSON, B. H., AND DORR, J. A., Jr., 1984, Jurassic-Cretaceous nonmarine foreland basin sedimentation in the western United States (abs.): American Association of Petroleum Geologists Bulletin, v. 68, p. 470–471.

EYER, J. A., 1969, Gannett Group of western Wyoming and southeastern Idaho: American Association of Petroleum Geologists Bulletin, v. 3, p. 1368–1391.

FURER, L. C., 1970, Petrology and stratigraphy of nonmarine Upper Jurassic-Lower Cretaceous rocks of western Wyoming and southeastern Idaho: American Association of Petroleum Geologists Bulletin, v. 54, p. 2282–2302.

HEADY, E., 1992, Stratigraphic and fission track study of Morrison and Cloverly sediments, Bighorn County, Wyoming: Unpublished M.S. Thesis, Dartmouth College, Hanover, 52 p.

HELLER, P. L., BOWDLER, S. S., CHAMBERS, H. P., COOGAN, J. C., HAGEN, W. S., SHUSTER, M. W., WINSLOW, N. S., AND LAWTON, T. F., 1986, Time of initial thrusting in the Sevier orogenic belt, Idaho-Wyoming and Utah: Geology, v. 14, p. 388–391.

HELLER, P. L., ANGEVINE, C. L., AND WINSLOW, N. S., 1988, Two-phase stratigraphic model of foreland-basin sequences: Geology, v. 16, p. 501–504.

HELLER, P. L. AND PAOLA, C., 1989, The paradox of Lower Cretaceous gravels and the initiation of thrusting in the Sevier orogenic belt, United States Western Interior: Geological Society of America Bulletin, v. 101, p. 864–875.

HOOPER, W. F., 1962, Lower Cretaceous stratigraphy of the Casper Arch, Wyoming, in Enyert, R. L., and Curry, W. H., eds., Symposium on Early Cretaceous Rocks of Wyoming and Adjacent Areas: Casper, Wyoming Geological Association 17th Annual Field Conference Guidebook, p. 141–147.

KENT, D. V. AND GRADSTEIN, F. M., 1985, A Cretaceous and Jurassic geochronology: Geological Society of America Bulletin, v. 96, p. 1419–1427.

KNIGHT, R., 1978, Deposition on the pre-Cadomin unconformity surface in north-central Alberta: Unpublished M.S. Thesis, Queen's University, Kingston, 92 p.

KVALE, E. P., 1986, Paleoenvironments and tectonic significance of the Upper Jurassic Morrison/Lower Cretaceous Cloverly Formations, Bighorn Basin, Wyoming: Unpublished Ph.D. Dissertation, Iowa State University, Ames, 191 p.

KVALE, E. P. AND BECK, R. A., 1985, Thrust-controlled sedimentation patterns of the earliest Cretaceous nonmarine sequence of the Montana-Idaho-Wyoming Sevier foreland basin (abs.): Geological Society of America Abstracts with Programs, v. 17, p. 636.

KVALE, E. P. AND VONDRA, C. F., 1993, Effects of relative sea-level changes and local tectonics on a Lower Cretaceous fluvial to transitional marine sequence, Bighorn Basin, Wyoming, USA, in Marzo, M. and Puigdefabregas, C., eds., Alluvial Sedimentation: Oxford, International Association Sedimentologists Special Publication 17, p. 383–399.

LAMMERS, E. C. H., 1939, The origin and correlation of the Cloverly conglomerate: Journal of Geology, v. 47, p. 113–132.

LOVE, J. D., 1948, Mesozoic stratigraphy of the Wind River Basin, central Wyoming, in Maebius, J. B., ed., Wind River Basin: Casper, Wyoming Geological Association Third Annual Field Conference Guidebook, p. 96–111 and plate in pocket.

MAY, M. T., 1991, Architectural analysis of Lower Cretaceous fluvial systems, an aid in tectonostratigraphic study in the central portion of a developing foreland, northwestern Wind River Basin, Wyoming (abs.): Geological Society of America Abstracts with Programs, v. 23, p. 285–286.

MAY, M. T., 1992a, Intra- and extra-basinal tectonism, climate and intrinsic threshold cycles as possible controls on Early Cretaceous fluvial architecture, Wind River Basin, Wyoming, in Mullen, C. E., ed., Rediscover the Rockies: Cheyenne, Wyoming Geological Association 43rd Annual Field Conference Guidebook, p. 61–74.

MAY, M. T., 1992b, A regional tectono-stratigraphic analysis of the Late Jurassic-Early Cretaceous Cordilleran foreland basin, Wind River basin region, Wyoming: Unpublished Ph.D. Dissertation, Indiana University, Bloomington, 308 p.

MAY, M. T., 1993, Petrographic characteristics of Morrison-Cloverly Formations and equivalent rocks in west central and central Wyoming-Implications for tectonic complexity in the early Sevier foreland, in Keefer, W. R., Metzger, W. J., and Goodwin, L. H., eds., Oil and Gas and Other Resources of the Wind River Basin: Cheyenne, Wyoming Geological Association Special Symposium, p. 49–70.

MCGOOKEY, D. P., HAUN, J. D., HALE, L. A., GOODELL, H. G., MCCUBBIN, D. G., WEIMER, R. J., AND WULF, G. R., 1972, Cretaceous System, in Mallory, W. W., ed., Geological atlas of the Rocky Mountain region: Denver, Rocky Mountain Association of Geologists, p. 189–228.

MEYERS, J. H., SUTTNER, L. J., FURER, L. C., MAY, M. T., AND SOREGHAN, M., 1992, Intrabasinal tectonic control on fluvial sandstone bodies in the Cloverly Formation (Early Cretaceous), west-central Wyoming, U. S. A.: Basin Research, v. 4, p. 315–333.

MILLS, N. K., 1956, Subsurface Stratigraphy of the Pre-Niobrara Formations in the Big Horn Basin Wyoming, in Nomenclature Committee, eds., Wyoming Stratigraphy: Casper, Wyoming Geological Association, 98 p.

MIRSKY, A., 1962, Stratigraphy of non-marine Upper Jurassic and Lower Cretaceous rocks, southern Big Horn Mountains, Wyoming: American Association of Petroleum Geologists Bulletin, v. 46, p. 1633–1680.

MOBERLY, R., JR., 1960, Morrison, Cloverly, and Sykes Mountain Formations, northern Bighorn Basin, Wyoming and Montana: Geological Society of America Bulletin, v. 71, p. 1136–1176.

PALMER, A. R., 1983, The decade of North American geology 1983 geologic time scale: Geology, v. 11, p. 503–504.

PETERSON, J. A., 1972, The Jurassic System, in Mallory, W. W., ed., Geologic Atlas of the Rocky Mountain Region: Denver, Rocky Mountain Association of Geologists, p. 176–185.

RUSS, D. P., 1982, Style and significance of surface deformation in the vicinity of New Madrid, Missouri: Washington, D. C., United States Geological Survey Professional Paper 1236-H, p. 95–114.

SCHULTHEIS, N. H., 1970, Petrography, source and origin of the Cadomin conglomerate (Cretaceous) between the North Saskatchewan and Athabasca rivers, Alberta: Unpublished M.S. Thesis, McGill University, Montreal, 133 p.

SCHWARTZ, R. K. AND DECELLES, P. G., 1988, Cordilleran foreland basin evolution and synorogenic sedimentation in response to interactive Cretaceous thrusting and reactivated foreland partitioning, in Schmidt, C. J., and Perry, R. W. J., eds., Interaction of the Rocky Mountain Foreland and the Cordilleran Thrust Belt: Boulder, Geological Society of America Memoir., 171, p. 489–514.

SENI, S. J., 1980, Sand-body geometry and depositional systems, Ogallaha Formation, Texas: Austin, Texas Bureau of Economic Geology Report of Investigations No. 105, 35 p.

STOKES, W. L., 1944, Morrison Formation and related deposits in and adjacent to the Colorado Plateau: Geological Society of America Bulletin, v. 55, p. 951–992.

SWIERC, J. E., 1990, The timing of Latest Jurassic to Earliest Cretaceous nonmarine sedimentation, northeastern Bighorn Basin, Wyoming: Magneto-stratigraphic implications: Unpublished M.S. Thesis, Dartmouth College, Hanover, 156 p.

THOMAS, G. E., 1971, Continental plate tectonics: southwest Wyoming in Renfro, A. R., ed., Symposium on Wyoming Tectonics and Their Economic Significance: Casper, Wyoming Geological Association 23rd Annual Field Conference Guidebook, p. 103–123.

THOMPSON, T. A., 1984, Limestone-clast conglomerate in the Early Cretaceous foreland basin in southwestern Montana: Origin and significance: Unpublished M.S. Thesis, Indiana University, Bloomington, 73 p.

WAAGE, K. M., 1959, Stratigraphy of the Inyan Kara Group in the Black Hills: United States Geological Survey Bulletin 1081-B, p. 11–90.

WILSON, J. M., 1958, Stratigraphic relations of non-marine Jurassic and pre-Thermopolis Lower Cretaceous strata of north central and northeastern Wyoming in Strickland, J., ed., Powder River Basin: Casper, Wyoming Geological Association 13th Annual Field Conference Guidebook, p. 77–78.

WILTSCHKO, D. B. AND DORR, J. A., 1983, Timing of deformation in overthrust belt and foreland of Idaho, Wyoming and Utah: American Association of Petroleum Geologists Bulletin, v. 67, p. 1304–1322.

WINSLOW, N. S. AND HELLER, P. L., 1987, Evaluation of unconformities in Upper Jurassic and Lower Cretaceous nonmarine deposits, Bighorn Basin, Wyoming and Montana, U. S. A.: Sedimentary Geology, v. 53, p. 181–202.

WORL, R. G., KOESTERER, M. E., AND HULSEBOSCH, T. P., 1986, Geologic map of the Bridger Wilderness and the Green-Sweetwater roadless area, Sublette and Fremont Counties, Wyoming: Reston, United States Geological Survey Miscellaneous Field Studies Map (MF-1636-B).

VARLEY, C. J., 1982, The sedimentology and diagenesis of the Cadomin Formation, Elmworth area, northwestern Alberta: Unpublished M.S. Thesis, University of Calgary, Calgary, 173 p.

YOUNG, R. G., 1970, Lower Cretaceous of Wyoming and the southern Rockies, in Enyert, R. L., ed., Symposium on Wyoming Sandstones: Casper, Wyoming Geological Association 22nd Annual Field Conference Guidebook, p. 147–160.

DIACHRONOUS THRUST LOADING AND FAULT PARTITIONING OF THE BLACK WARRIOR FORELAND BASIN WITHIN THE ALABAMA RECESS OF THE LATE PALEOZOIC APPALACHIAN—OUACHITA THRUST BELT

WILLIAM A. THOMAS
Department of Geological Sciences, University of Kentucky, Lexington, KY 40506-0053

ABSTRACT: The triangular outline of the late Paleozoic Black Warrior foreland basin on the southern edge of the North American craton in Alabama and Mississippi is framed on the southwest by the northwest-striking Ouachita thrust front and on the southeast by northeast-striking Appalachian thrust-belt structures. The nearly orthogonal intersection of the Ouachita and Appalachian thrust belts implies a composite history of flexural subsidence of the foreland. A long homocline that dips southwest beneath the Ouachita thrust front defines the structure of the basin, and a southwestward-thickening, northeastward-prograding synorogenic clastic wedge of Mississippian and Pennsylvanian rocks fills the basin, indicating a thrust load and sediment source (Ouachita thrust belt) on the southwest. A synorogenic clastic wedge in the Appalachian thrust belt (Cahaba synclinorium) is similar in provenance and dispersal to that in the Black Warrior basin, indicating that the palinspastic site of Appalachian thrust sheets was also part of the original Ouachita foreland. Greater thickness of the clastic wedge in the Cahaba synclinorium reflects partitioning of the Ouachita foreland by reactivation of the down-to-southeast Birmingham basement fault system. Addition of northwest-prograding clastic sediment during the Early Pennsylvanian records initiation of Appalachian orogenesis on the southeast. Subsequently, the southeastern part of the southwest-dipping Black Warrior foreland basin was displaced by northwest-propagating Appalachian thrusts, and part of the older, northeastward-prograding, Ouachita-derived clastic wedge was imbricated.

INTRODUCTION

The late Paleozoic Black Warrior foreland basin in Alabama and Mississippi has a triangular outline that is bounded on the southwest by the northwest-striking Ouachita thrust front, on the southeast by northeast-striking Appalachian thrust-belt structures, and on the north by the North American craton (Fig. 1). The nearly orthogonal intersection of the late Paleozoic Ouachita and Appalachian thrust belts defines the Alabama structural recess (cratonward-concave bend of the thrust belt) and indicates diachronous translation of thrust loads from two directions onto North American continental crust (Thomas, 1989). Convergence of thrust belts framing two sides of the Black Warrior basin implies a composite history of foreland flexural subsidence. Interactions of thrust loads and foreland basins have been described primarily in terms of a thrust belt parallel with structural strike of the foreland basin and thrust translation perpendicular to structural strike (for example, Jordan, 1981; Speed and Sleep, 1982; Quinlan and Beaumont, 1984; Schedl and Wiltschko, 1984; Flemings and Jordan, 1989). Parallel thrust belts and foreland structure generally characterize structural salients (broad, cratonward-convex curves of the thrust belt) in contrast to angular structural intersections in recesses. The distinctive setting of the Black Warrior basin provides a model for interactions of thrust belts and a foreland basin within a large-scale structural recess.

The stratigraphic succession in the Appalachian-Ouachita foreland includes a Cambrian to Lower Mississippian passive-margin, carbonate-shelf succession and an Upper Mississippian and Pennsylvanian synorogenic clastic wedge (Thomas, 1988a). The clastic wedge consists mainly of southwestward-thickening, northeastward-prograding, deltaic to shallow-marine sediment dispersed from an orogenic source on the southwest (Thomas, 1972a, b, 1988a; Mack and others, 1983). A comprehensive interpretation of facies and thickness distributions, dispersal directions, and sandstone petrography includes an arc-continent collision (Ouachita thrust belt) and thrusting of a subduction complex onto continental crust along the southwest side of the Alabama promontory (Figs. 1, 2) accompanied by flexural downwarp and sedimentary filling of the Black Warrior foreland basin (Thomas, 1976, 1988a, 1989; Mack and others, 1983; Hines, 1988). In addition to the southwesterly derived clastic sediment, the upper part of the clastic wedge in the southeastern part of the Black Warrior basin includes some southeasterly derived components (Horsey, 1981; Mack and others, 1983; Sestak, 1984; Pashin and others, 1990), indicating initiation of sediment dispersal from an orogenic source terrane along the southeast (Appalachian) side of the foreland basin.

The stratigraphic succession imbricated in the Appalachian thrust belt (Figs. 1, 3) is similar to that in the Black Warrior foreland basin, consisting of Cambrian-Mississippian passive-margin, carbonate-shelf strata and a Mississippian-Pennsylvanian southwestward-thickening, northeastward-prograding, deltaic to shallow-marine synorogenic clastic wedge (Thomas, 1972a, 1985; Thomas and others, 1991). Like that in the Black Warrior basin, the upper part of the synorogenic clastic wedge in the Appalachian thrust belt includes southeasterly derived components. Although facies distribution patterns and vertical succession persist regionally, the Mississippian-Pennsylvanian synorogenic clastic wedge in the Appalachian thrust belt is more than twice as thick as that in the Black Warrior foreland basin. Palinspastically, the thicker clastic wedge in the Appalachian thrust belt is southeast of a large-scale, down-to-southeast basement fault system (Birmingham basement fault system, Figs. 1, 3) which underlies and parallels the thrust-belt structures (Thomas, 1985). The difference in thickness is interpreted to be a result of synsedimentary reactivation of the basement fault system (Thomas, 1986).

Available data may be assembled into a general interpretation of the interaction of the Ouachita and Appalachian thrust belts with the Black Warrior foreland basin, and the purpose of this article is to test the following specific interpretations: (1) prior to northwest-directed Appalachian thrusting, the Ouachita foreland encompassed not only the present Black Warrior foreland basin but also the palin-

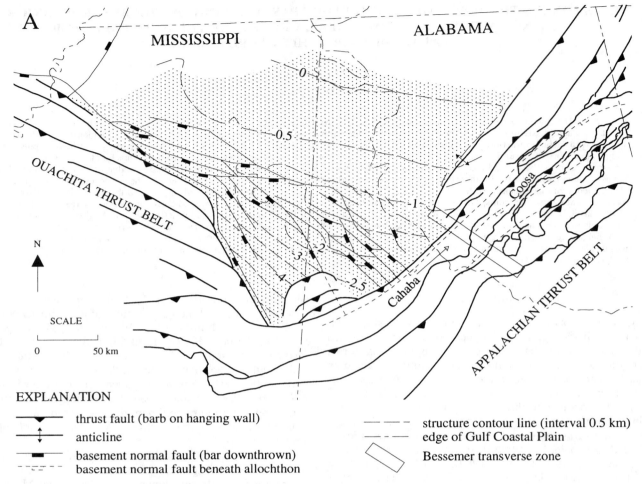

Fig. 1.—Outline structural geology map (A) of the Black Warrior foreland basin (area of stipple) and the Appalachian and Ouachita thrust belts (pre-Mesozoic subcrop map beneath Gulf Coastal Plain based on well data shown in Thomas and others, 1989). Structure contour map of top of Mississippian Tuscumbia Limestone (stratigraphic level of oldest part of synorogenic clastic wedge) based on geophysical logs of ~1,400 wells within the stippled area and structural data from outcrops around the eastern part of the Black Warrior basin (from Thomas, 1988a). Map of Birmingham basement fault system (dotted lines) in area beneath Appalachian thrust belt, interpreted from six regional seismic reflection profiles across the Appalachian thrust belt in Alabama. Location map (B) of lines of structural cross sections identified by letters A-B (Fig. 2) and B-C (Fig. 3); lines of stratigraphic cross sections identified by letters D-E (Fig. 4), F-G (Fig. 6), and H-I (Fig. 7); wells and measured stratigraphic sections shown in the cross sections; and seismic reflection profiles. Inset map (C) shows spatial relationship of Alabama recess (R) of the late Paleozoic Appalachian-Ouachita thrust belt to Alabama promontory (P) of the late Precambrian-Cambrian rifted margin of North American continental crust (from Thomas, 1991).

spastic site of deposition of the strata now in the Appalachian thrust belt; (2) the abrupt large increase in thickness of the northeastward-prograding synorogenic clastic wedge from the Black Warrior basin southeastward into the Appalachian thrust belt reflects partitioning of the foreland by a northeast-striking system of synsedimentary basement faults; (3) late addition of northwest-prograding components of the synorogenic clastic wedge records initiation of orogenesis along the southeastern side of the foreland basin; and (4) late northwest-directed Appalachian thrusting imbricated part of the northeastward-prograding, Ouachita-derived clastic wedge along the southeastern side of the Black Warrior basin. Data from geophysical logs of approximately 1,400 wells, sample descriptions of more than 70 wells, 60 measured stratigraphic sections at outcrops around the eastern side of the Black Warrior basin and in the Appalachian thrust belt, and seismic reflection profiles are summarized here in maps and cross sections (Figs. 1–7; Table 1); the specific data base is described in the figure captions.

TECTONIC SETTING

The Appalachian-Ouachita thrust belt and the Black Warrior foreland basin record late Paleozoic orogenic events, but the present configuration and exposure of the late Paleozoic structures also reflect both pre-orogenic and post-orogenic processes (Thomas, 1989). Synorogenic late Paleozoic rocks in the eastern part of the basin are exposed in northeastern Alabama. The foreland basin and bordering

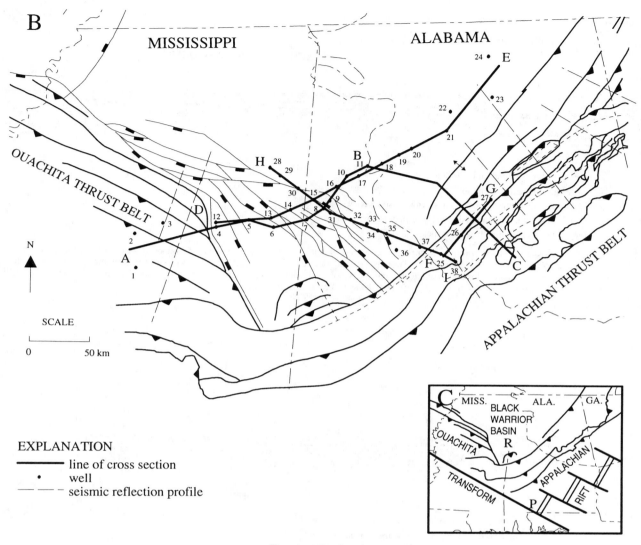

Fig. 1.—Continued.

thrust belts are unconformably overlapped by southwest-dipping, post-orogenic Mesozoic-Cenozoic strata of the Gulf Coastal Plain in northwestern Alabama and northern Mississippi, reflecting subsidence associated with Mesozoic opening of the Gulf of Mexico (Fig. 1; Thomas, 1988b). Deep wells and seismic reflection profiles document late Paleozoic stratigraphy and structure beneath the Coastal Plain cover. Around the southern apex of the triangular foreland basin beneath the Gulf Coastal Plain in eastern Mississippi, Appalachian structures curve into a westerly strike and truncate the northwest-striking Ouachita structures (Fig. 1; Thomas, 1989). The large-scale bend in the late Paleozoic thrust belt (the Alabama recess) mimics the shape of the Alabama promontory of North American continental crust that was originally formed by late Precambrian-Cambrian rifting (Thomas, 1991). Northeast-striking rift segments and northwest-striking transform faults outlined the continental margin around the Alabama promontory, and rift-related extensional faults (for example, the Birmingham basement fault system) propagated into continental crust on the promontory during the late stages of rifting and opening of the Iapetus Ocean (Fig. 1C; Thomas, 1991).

PRE-OROGENIC STRATIGRAPHY ON THE ALABAMA PROMONTORY

Syn-rift strata include Lower Cambrian units of sandstone, dolostone, and fine-grained clastic rocks that pinch out northwestward as a result of transgression onto the North American craton (Sauk sequence; Sloss, 1963, 1988) and overlap of the down-to-southeast Birmingham basement fault system (Figs. 1, 3; Thomas, 1991). Middle to Upper Cambrian fine-grained clastic rocks and carbonate rocks reflect final filling of rift-related downthrown blocks. Paleozoic post-rift reactivation of the Birmingham basement fault system is indicated by stratigraphic variations in thickness and facies, sedimentary slump structures, and unconformities (Thomas, 1986; Ferrill, 1989), as well as fault-rock fabrics (Wu, 1989).

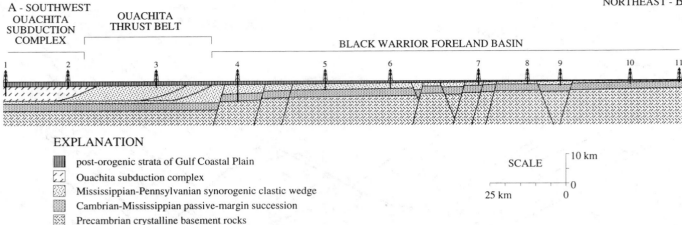

Fig. 2.—Structural cross section of southwest-dipping homocline of Black Warrior foreland basin and northwest-striking Ouachita thrust belt. Cross section of Black Warrior basin (modified from Thomas, 1988a) derived from all well data incorporated in structure contour map in Figure 1A; representative wells identified by number, located in Figure 1B, and listed in Table 1. Cross section of Ouachita thrust belt based on well data and two seismic reflection profiles (traces shown in Fig. 1B). Line of cross section shown in Figure 1B.

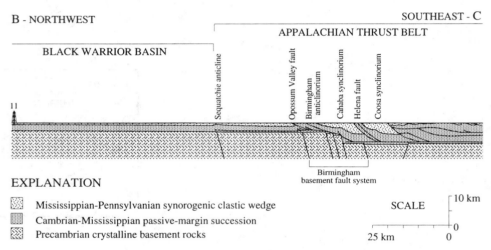

Fig. 3.—Structural cross section of Black Warrior foreland basin and northeast-striking Appalachian thrust belt. Cross section of Black Warrior basin derived from all well data incorporated in structure contour map in Figure 1A and from outcrop data along Sequatchie anticline; well identified by number, located in Figure 1B, and listed in Table 1. Balanced, restorable cross section of the Appalachian thrust belt based on outcrop geology (including bedding attitudes and measured stratigraphic sections) and regional seismic reflection profiles perpendicular to strike (traces shown in Fig. 1B). Line of cross section shown in Figure 1B.

A thick (~1,200 m), Upper Cambrian-Lower Ordovician carbonate unit characterizes a passive-margin shelf on the Alabama promontory of continental crust and marks maximum transgression onto the continent during deposition of the Sauk sequence (Sloss, 1963, 1988). Middle Ordovician through Lower Mississippian carbonate rocks and chert indicate persistence of a passive-margin shelf on the western part of the Alabama promontory in the present location of the Black Warrior foreland basin (Thomas, 1988a). Thin, westward-pinching clastic units in the eastern part of the Black Warrior foreland basin and adjacent Appalachian thrust belt in Alabama indicate distal effects of the Taconic and Acadian orogenies farther northeast along the Appalachian orogenic belt (Thomas, 1977, 1988a), but no record of these events is evident in the stratigraphy farther west on the passive-margin shelf. The uppermost part of the passive-margin succession is a Lower Mississippian carbonate-shelf unit (Fort Payne Chert and Tuscumbia Limestone) that extends widely across the Alabama promontory throughout the foreland and the Appalachian thrust belt.

STRUCTURE OF THE OUACHITA THRUST BELT

Samples from deep wells in Mississippi indicate that the frontal thrust sheets of the Ouachita thrust belt beneath the Gulf Coastal Plain consist of Mississippian-Pennsylvanian mudstone and sandstone, and that the more internal thrust sheets are characterized by slaty cleavage and quartz veins

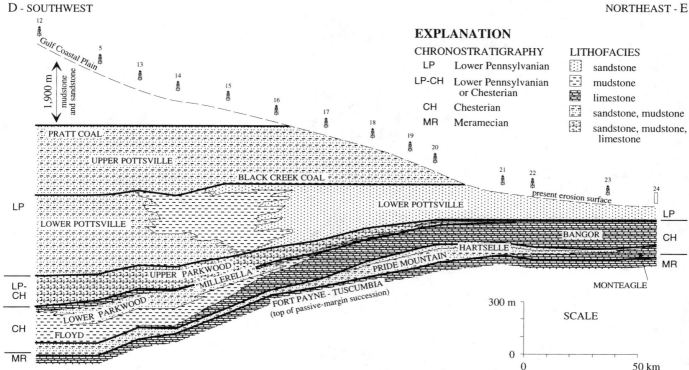

FIG. 4.—Stratigraphic cross section of lithostratigraphic, lithofacies, and chronostratigraphic subdivisions of Mississippian-Pennsylvanian rocks approximately perpendicular to structural and stratigraphic strike in the Black Warrior foreland basin. Cross section based on wells within a corridor across Black Warrior basin and measured stratigraphic sections in eastern part of basin. Representative wells and measured stratigraphic sections identified by number, located in Figure 1B, and listed in Table 1. Each well shown in the cross section represents several wells in a correlation network along the corridor, except that wells 22 and 23 are correlated by mapping to nearby measured stratigraphic sections. Preserved Pennsylvanian clastic-wedge rocks above Pratt coal shown schematically (not to scale). Correlation markers shown by bold lines. Chronostratigraphic correlations based on biostratigraphic data from references cited in text. Line of cross section shown in Figure 1B.

in a mudstone-dominated succession (Thomas, 1972b, 1973, 1989). By analogy with the exposed Ouachita orogen in Arkansas and Oklahoma, the subsurface Ouachita orogen is interpreted to consist of a frontal belt of thrust-imbricated, synorogenic clastic-wedge rocks and an interior belt that includes an accretionary prism of pre-orogenic off-shelf rocks and trench-fill deposits (Ouachita thrust belt and subduction complex, Fig. 2; Viele and Thomas, 1989). Distribution of rock types mapped from well data and seismic reflection profiles document northwest strike of the Ouachita frontal structures in east-central Mississippi (Fig. 1; Thomas, 1973, 1989; Thomas and others, 1989).

STRUCTURE OF THE BLACK WARRIOR BASIN

Structurally, the Black Warrior foreland basin is a homocline which has an average dip of less than 2° southwestward (Figs. 1, 2; Thomas, 1988a). The foreland basin is asymmetric, and the single long homoclinal limb dips southwestward beneath northwest-striking frontal structures of the Ouachita thrust belt (Fig. 2). The southwest-dipping homocline is broken by a system of northwest-striking, predominantly down-to-southwest normal faults. The northwesterly strike of the homocline persists as far southeast as the northeast-striking structural front of the Appalachian thrust belt where the structure of the foreland basin is truncated by the frontal Appalachian faults (Figs. 1, 3).

STRUCTURE OF THE APPALACHIAN THRUST BELT

The Appalachian thrust belt along the southeastern side of the Black Warrior basin is characterized by large-scale, internally coherent thrust sheets detached near the base of the Paleozoic sedimentary sequence (Fig. 3). The thrust sheets contain the pre-orogenic shelf succession and overlying synorogenic clastic-wedge rocks (Fig. 3). The frontal (northwestern) structures are broad, flat-bottomed synclines and narrow, asymmetric anticlines having relief of <3,000 m (Fig. 3). Structures on the southeast include folds associated with large frontal ramps having structural relief of >4,000 m and low-angle, broad, multiple-level thrust sheets. The difference in scale of structures reflects the difference in depth to basement across the down-to-southeast Birmingham basement fault system (Figs. 1, 3). The most northwesterly of the large-scale frontal ramps (Birmingham anticlinorium, Fig. 3) in the thrust belt are positioned over the basement fault system (Thomas, 1985). The regional décollement is within Lower and Middle Cambrian fine-grained clastic rocks; the structural style of the thrust sheets is controlled by the thick, Upper Cambrian-Lower Ordov-

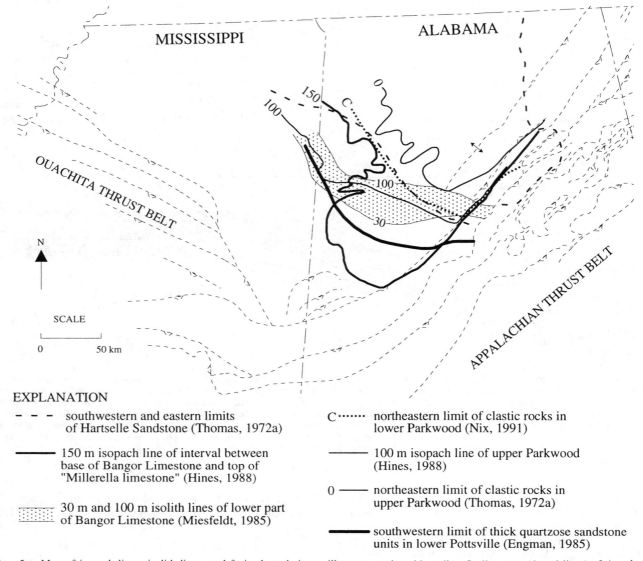

Fig. 5.—Map of isopach lines, isolith lines, and facies boundaries to illustrate stratigraphic strike. Outline map (dotted lines) of Appalachian-Ouachita structure from Figure 1.

ician massive carbonate stiff layer. The contrast in composition between Appalachian and Ouachita thrust sheets is reflected in distinct differences in structural style, and these differences are used to identify the thrust belts in the subsurface in eastern Mississippi where west-striking Appalachian structures truncate northwest-striking Ouachita structures (Fig. 1; Thomas, 1989).

Synorogenic Mississippian-Pennsylvanian clastic-wedge rocks are preserved in two deep synclinoria (Cahaba and Coosa) in the Appalachian thrust belt (Figs. 1, 3); the strike length of the outcrop belt along the Cahaba synclinorium offers better exposure for comparing stratigraphy to that of the Black Warrior foreland basin. The Cahaba synclinorium persists along strike approximately 100 km from the edge of Coastal Plain cover and ends abruptly northeastward in steep southwest plunge at a lateral ramp. Northwest-directed Appalachian thrusting post-dated the youngest preserved synorogenic strata and shortened the distance between rocks in the Cahaba synclinorium and those in the foreland by at least 4 km (Thomas, 1985; Ferrill, 1989). The northwest limb of the Cahaba synclinorium is formed by a large-scale, complex frontal ramp/ramp anticline (Birmingham anticlinorium), and the trailing edge of the synclinorium is truncated by the Helena fault (Figs. 1, 3).

The structural profile of each large-scale structure in the Appalachian thrust belt in Alabama persists along strike for several tens of kilometers, but beyond that strike length, the structures are characterized by abrupt along-strike terminations or structural changes associated with lateral ramps (Thomas, 1985). Furthermore, many (but not all) sites of abrupt along-strike change are aligned across strike of the thrust belt, defining transverse zones (Thomas, 1985, 1990).

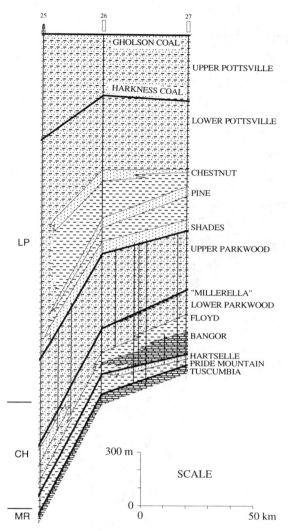

FIG. 6.—Stratigraphic cross section of lithostratigraphic, lithofacies, and chronostratigraphic subdivisions of Mississippian-Pennsylvanian rocks along present structural strike of the Cahaba synclinorium in the Appalachian thrust belt. Cross section based on measured stratigraphic sections and wells identified by number, located in Figure 1B, and listed in Table 1. Details in cross section based on two additional measured stratigraphic sections between composite sections 25 and 26, and four additional measured stratigraphic sections and three drill holes between composite sections 26 and 27 (shown by vertical lines) (descriptions in Thomas, 1972a; Osborne, 1985; Leverett, 1987). Drill holes document otherwise poorly exposed Bangor Limestone in the northeastern part of the Cahaba synclinorium and a mudstone succession that contains very little limestone to the southwest (Butts, 1927; Thomas, 1972a). Correlation markers shown by bold lines. Chronostratigraphic correlations based on biostratigraphic data from references cited in text; correlations supplemented by mapping of sandstone units and coal beds (Butts, 1907, 1911; Thomas, 1972a; Osborne, 1985; Leverett, 1987). Line of cross section shown in Figure 1B. Explanation of lithologic symbols and chronostratigraphic abbreviations same as in Figure 4.

Lateral stratigraphic variations in parts of the Paleozoic succession (for example, changes of thickness, changes of facies, truncations at unconformities) are also localized along the structural transverse zones, suggesting genetic effects on both stratigraphy and later thrust-belt structure by episodically reactivated, northwest-striking basement faults (Ferrill, 1989; Thomas, 1990). An along-strike change in the geometry of the Birmingham anticlinorium (frontal ramp), an abrupt curve in strike of the Helena fault, and up-plunge termination of the Coosa synclinorium in the Helena thrust sheet define the trace of the Bessemer transverse zone (Thomas, 1985); the transverse zone crosses the Cahaba synclinorium (Fig. 1).

MISSISSIPPIAN-PENNSYLVANIAN SYNOROGENIC CLASTIC WEDGE
IN THE BLACK WARRIOR FORELAND BASIN

Large-scale facies architecture of the Mississippian-Pennsylvanian clastic wedge in the Black Warrior foreland basin defines a five-part vertical subdivision (in ascending order, Pride Mountain-Hartselle, lower Parkwood, upper Parkwood, lower Pottsville, and upper Pottsville, Fig. 4). The five stratigraphic subdivisions are based on distinctive correlation markers (Fig. 4) which provide a framework for detailed comparison with the succession in the Cahaba synclinorium. The lower three subdivisions of the clastic wedge intertongue with and grade northeastward into a Mississippian shelf-carbonate facies, and Pennsylvanian siliciclastic rocks in the upper two subdivisions extend northeastward over the Mississippian carbonate facies (Fig. 4; Thomas, 1988a). Stratigraphic limits of siliciclastic and carbonate intertongues and facies boundaries of internal components of the clastic wedge strike northwestward across the foreland basin, approximately parallel with the Ouachita structural front along the southwest side of the basin, and approximately perpendicular to strike of Appalachian thrust-belt structures along the southeast side of the basin (Fig. 5). The synorogenic clastic wedge fills the Black Warrior basin and thickens progressively southwestward (Fig. 4). Deltaic to shallow-marine environments prevailed throughout deposition of the clastic wedge (Thomas, 1988a), indicating that sediment-accumulation rate approximately equaled subsidence rate and that the basin remained filled with sediment throughout subsidence. Previous research has addressed thickness and facies distributions, depositional systems, water-depth indicators, sediment-dispersal indicators, and provenance for various parts of the clastic wedge in the Black Warrior basin; reference citations in the following summary indicate sources of detailed documentation.

The lowest subdivision of the clastic wedge is a tongue of mudstone and sandstone (Pride Mountain Formation and Hartselle Sandstone) that progrades northeastward over the Monteagle Limestone (Fig. 4). The clastic tongue contains four distinct units of quartzose sandstone that represent shallow-marine-bar and barrier-island environments. The entire tongue pinches out northeastward between the Monteagle and Bangor Limestones, a succession of oolitic and bioclastic limestones (Figs. 4, 5; Thomas, 1972a, 1988a; Handford, 1978; Thomas and Mack, 1982; Higginbotham, 1985). The Hartselle Sandstone (Fig. 5) and the middle Pride Mountain sandstone pinch out southwestward into back-

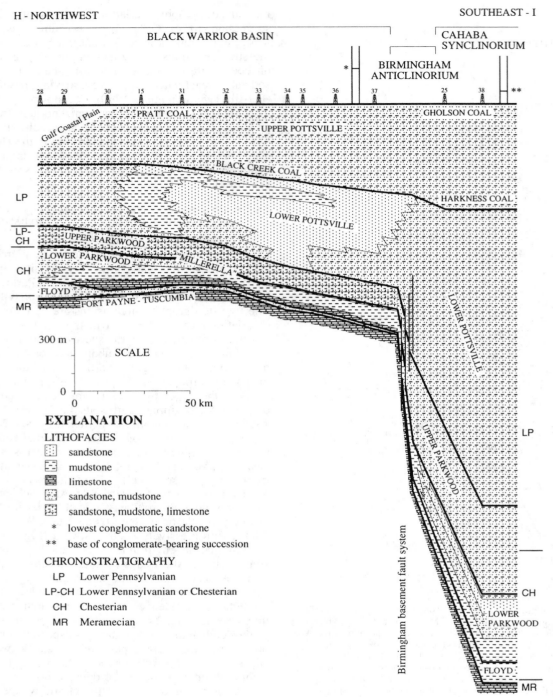

Fig. 7.—Stratigraphic cross section of lithostratigraphic, lithofacies, and chronostratigraphic subdivisions of Mississippian-Pennsylvanian rocks across the Black Warrior foreland basin approximately parallel with structural strike, extending to palinspastically restored locations of wells and stratigraphic sections in the southwestern part of the Cahaba synclinorium. Palinspastic restoration of locations in Cahaba synclinorium derived from balanced structural cross section (Fig. 3). Cross section based on wells and measured stratigraphic sections identified by number, located in Figure 1B, and listed in Table 1. The section in well 38 possibly includes structural thickening in the extraordinarily thick lower Pottsville; however, no direct indication of structural thickening is available. Thickness of the upper Pottsville in well 38 is the same as that in a nearby core hole and measured stratigraphic section, and the section below the Pottsville is similar in thickness to that in well 25 and to nearby measured outcrop sections (Fig. 6). Interpretation of drape fold and small faults over the Birmingham basement fault system based on indicated times of synsedimentary fault movement, geometry of fault system in Paleozoic rocks, and palinspastic restoration of southeast limb of Birmingham anticlinorium (Thomas, 1985, 1986; Ferrill, 1989; Wu, 1989). Mississippian-Pennsylvanian strata have been eroded from the crest of the Birmingham anticlinorium between wells 25 and 37. Stratigraphic levels of lowest conglomeratic beds above Pratt and Gholson coals shown in vertical columns to scale. Correlation markers shown by bold lines. Chronostratigraphic correlations based on biostratigraphic data from references cited in text. Line of cross section shown in Figure 1B.

TABLE 1.—LIST OF WELLS AND MEASURED STRATIGRAPHIC SECTIONS IDENTIFIED BY NUMBERS IN FIGURES 1–7

1. Rudman No. 1 Wigley [g,i]
   sec. 25, T 15 N, R 5 E, Attala County, Mississippi
2. Texaco No. 1 Whitehead [g,a]
   sec. 22, T 18 N, R 5 E, Carroll County, Mississippi
3. Gulf No. 1 Parker [g,a]
   sec. 22, T 19 N, R 7 E, Montgomery County, Mississippi
4. Exxon No. 1 Fulgham [g,i]
   sec. 33, T 19 N, R 12 E, Oktibbeha County, Mississippi
5. McAlester No. A-1 Sudduth [g]
   sec. 6, T 19 N, R 15 E, Oktibbeha County, Mississippi
6. Shell No. 1 Gearhiser [g]
   sec. 1, T 18 N, R 17 E, Lowndes County, Mississippi
7. Bow Valley No. 1 Bain [g]
   sec. 7, T 18 S, R 16 W, Pickens County, Alabama
8. Pruet No. 1 Vann [g]
   sec. 15, T 17 S, R 15 W, Lamar County, Alabama
9. Burns No. 20-1 Tomlin [g]
   sec. 20, T 16 S, R 14 W, Lamar County, Alabama
10. Warrior No. 23–1 Bynum [g]
    sec. 23, T 14 S, R 13 W, Fayette County, Alabama
11. Pine No. 1 Baldwin [g,a]
    sec. 20, T 13 S, R 11 W, Fayette County, Alabama
12. Texaco No. 1 Sheely [g]
    sec. 28, T 19 N, R 12 E, Oktibbeha County, Mississippi
13. composite of Atlantic No. 1 Dunning [g,a]
    sec. 12, T 19 N, R 16 E, Clay County, Mississippi; and
    Shell No. 1 Chism [g]
    sec. 24, T 19 N, R 16 E, Lowndes County, Mississippi
14. composite of Pruet and Hughes and Amerada Hess No. 1 Livingston [g]
    sec. 11, T 17 S, R 18 W, Lowndes County, Mississippi; and
    Pruet and Baria Mason No. 1 Gilmer [g]
    sec. 14, T 17 S, R 18 W, Lowndes County, Mississippi
15. Hughes and Hughes and Warrior No. 1 Bardon [g]
    sec. 14, T 16 S, R 16 W, Lamar County, Alabama
16. Terra No. 1 Glasgow [g]
    sec. 19, T 15 S, R 14 W, Lamar County, Alabama
17. Warrior No. 20–1 Breitling [g]
    sec. 20, T 14 S, R 12 W, Fayette County, Alabama
18. Champlin No. 1 Galloway [McGlamery, 1955]
    sec. 12, T 13 S, R 10 W, Walker County, Alabama
19. Energy No. 17–12 Deason [g]
    sec. 17, T 12 S, R 8 W, Winston County, Alabama
20. Kelton No. 1 Grant [g]
    sec. 32, T 11 S, R 7 W, Winston County, Alabama
21. Texas Eastern No. 1 Horton [g,a,o]
    sec. 19, T 10 S, R 4 W, Cullman County, Alabama
22. composite of measured stratigraphic section [Thomas, 1972a]
    sec. 25, T 8 S, R 4 W, Morgan County, Alabama; and
    Cullman No. 1 Collum [McGlamery, 1955]
    sec. 36, T 8 S, R 4 W, Cullman County, Alabama
23. Young No. 1 Black [a]
    sec. 22, T 7 S, R 1 E, Marshall County, Alabama
24. measured stratigraphic section [Thomas, 1972a]
    sec. 4, 9, 10, 15, T 4 S, R 1 E, Madison County, Alabama
25. composite of Meridian No. 1 Gulf States Paper [g,i] (below Harkness coal)
    sec. 27, T 21 S, R 5 W, Bibb County, Alabama; and
    data from composite section 38 (above Harkness coal)
26. composite of measured stratigraphic section [Womack, 1983]
    sec. 18, 30, T 19 S, R 2 W, sec. 24, T 19 S, R 3 W, sec. 18, T 20 S, R 3 W, sec. 12, 13, T 20 S, R 4 W, Shelby and Jefferson Counties, Alabama; and
    measured stratigraphic section [Thomas, 1972a; Leverett, 1987]
    sec. 1, 12, T 20 S, R 4 W, Jefferson County, Alabama; and
    Woodward Iron Company core hole RO-25 [Thomas, 1972a]
    sec. 19, T 19 S, R 3 W, Jefferson County, Alabama
27. composite of measured stratigraphic section [Butts, 1907; Womack, 1983]
    sec. 18, T 17 S, R 1 E, sec. 13, 14, 15, T 17 S, R 1 W, Jefferson County, Alabama; and
    measured stratigraphic section [Thomas, 1972a]
    sec. 8, 17, 18, 19, 20, T 17 S, R 1 W, sec. 13, 24, T 17 S, R 2 W, Jefferson County, Alabama; and
    Sloss-Sheffield core hole K [Thomas, 1972a]
    sec. 9, T 17 S, R 1 W, Jefferson County, Alabama
28. Magnolia No. 1 Pierce [g,a]
    sec. 22, T 13 S, R 7 E, Monroe County, Mississippi
29. Burns No. 6–1 Hilliard [g]
    sec. 6, T 14 S, R 18 W, Monroe County, Mississippi
30. Dalton-Buie No. 1 Rye [g]
    sec. 15, T 15 S, R 17 W, Monroe County, Mississippi
31. Pruet No. 1 Williamson [g]
    sec. 18, T 17 S, R 14 W, Lamar County, Alabama
32. Hunt No. 1 Brown Foundation [g]
    sec. 1, T 18 S, R 13 W, Pickens County, Alabama
33. Carless No. 1 Patterson [g]
    sec. 20, T 18 S, R 11 W, Tuscaloosa County, Alabama
34. American Quasar and Conwood No. 1 Wright [g]
    sec. 18, T 19 S, R 10 W, Tuscaloosa County, Alabama
35. Terra No. 1 Friedman [g,o]
    sec. 36, T 19 S, R 10 W, Tuscaloosa County, Alabama
36. waste disposal well, Reichhold Chemicals [g,o,a]
    sec. 3, T 21 S, R 9 W, Tuscaloosa County, Alabama
37. Terra No. 1 Durrett [g]
    sec. 36, T 20 S, R 7 W, Tuscaloosa County, Alabama
38. Warrior No. 1 Kimberly Clark [g,s-D. E. Leverett, T. E. Osborne]
    sec. 23, T 22 S, R 4 W, Shelby County, Alabama;
    section above Harkness coal correlated to measured stratigraphic section [Womack, 1983] sec. 2, 3, T 24 N, R 10 E, sec. 28, T 22 S, R 5 W, Bibb County, Alabama; and
    School of Mines and Energy Development core hole [g,s-T. E. Osborne]
    sec. 15, T 22 S, R 4 W, Shelby County, Alabama

Sources of data shown in brackets
g: geophysical well log;
a: sample description by author;
o: sample description in Alabama Geological Survey open file;
i: petroleum industry sample description;
s: sample description by Graduate Research Assistant with author's supervision.

barrier and coastal-marsh mudstones, but deltaic facies of the upper and lower Pride Mountain sandstones extend southwest beyond the limit of well data (Higginbotham, 1985). Gradual progradation of the Pride Mountain-Hartselle tongue over the Monteagle Limestone is indicated across a wide area of regressive facies, where successively higher clastic rocks extend progressively farther northeast (Fig. 4; Thomas, 1972a). A subsequent southwestward transgression was very rapid as indicated by the wide extent and persistent stratigraphic level of the basal Bangor Limestone (correlation marker in Fig. 4). The transgressive lower part of the Bangor Limestone extends farther southwest than the southwestern limit of the Hartselle Sandstone (Fig. 5), but it thins and grades farther southwest into mudstone (Floyd Shale, Fig. 4).

The lower tongue of Bangor Limestone is overlapped by a northeastward-prograding cyclic succession of deltaic sandstones and mudstones of the lower part of the Parkwood Formation (Miesfeldt, 1985; Nix, 1991). Successively higher sandstone units extend progressively farther northeast; but toward the northeast, the entire clastic tongue pinches out between the regressive top of the lower tongue of the Bangor Limestone and a very extensive, thin tongue of the middle Bangor Limestone, informally called the "Millerella limestone" (Figs. 4, 5). The wide extent of the "Millerella limestone" tongue at a persistent stratigraphic level indicates rapid transgression over the lower Parkwood deltaic sandstones. The "Millerella limestone" is a distinctive correlation marker throughout most of the Black Warrior basin (Fig. 4), and it is commonly associated with distinctive maroon and green mudstones in contrast to the gray mudstones that dominate the succession (Thomas, 1972a, b).

The upper part of the Parkwood Formation is a succession of sandstone and mudstone that grades northeastward into the upper part of the Bangor Limestone (Figs. 4, 5). Several thin, discontinuous limestone tongues extend into the clastic facies (Miesfeldt, 1985). Sandstone distribution suggests deltaic deposition on the southwest and marine bars on the northeast.

The lower part of the Pottsville Formation consists of three, distinct, laterally equivalent facies in the Black Warrior basin: stacked, barrier-island sandstones in the northeastern part of the basin; a thick mudstone succession representing a back-barrier lagoon in the central part; and a succession of fining-upward, northeastward-prograding deltaic sandstones and mudstones in the southwestern part of the basin (Fig. 4; Engman, 1985; Dismukes, 1989). The facies boundary between the barrier sandstones and the lagoonal mudstone strikes northwestward across the basin (Fig. 5).

The upper part of the Pottsville Formation is a cyclic succession of marine and non-marine mudstones, sandstones, and coals (Culbertson, 1964; Sestak, 1984; Shadroui, 1986). The Black Creek coal marks the base of the upper part of the Pottsville Formation, and the Pratt coal defines a useful correlation marker within the upper Pottsville succession (correlation markers in Fig. 4). In the downdip, southwestern part of the Black Warrior basin, a mudstone-sandstone-coal succession as much as 1,900 m thick is preserved above the level of the Pratt coal (shown schematically in Fig. 4 because the limited area of preservation precludes regional correlation). In the southeastern part of the Black Warrior basin, the upper part of the Pottsville Formation locally thickens southeastward and includes some northwestwardly prograding sandstones and conglomeratic sandstones, indicating initiation of sediment supply from the southeast as a result of Appalachian orogenesis (Horsey, 1981; Mack and others, 1983; Sestak, 1984; Pashin and others, 1990). The stratigraphically lowest conglomeratic sandstones, which contain chert pebbles, are approximately 200 m above the Pratt coal.

MISSISSIPPIAN-PENNSYLVANIAN SYNOROGENIC CLASTIC WEDGE IN THE CAHABA SYNCLINORIUM

The five subdivisions of the Mississippian-Pennsylvanian synorogenic clastic wedge and the correlation markers defined in the Black Warrior basin also are recognizable in the Cahaba synclinorium (Thomas and others, 1991). The general vertical succession and distribution of facies are similar; however, the succession in the Cahaba synclinorium is more than twice as thick as that in the Black Warrior foreland basin (compare Figs. 4, 6). Despite the great difference in thickness, depositional systems in the Cahaba synclinorium indicate persistent deltaic to shallow-marine environments and a consistently filled basin.

As in the Black Warrior basin, the Pride Mountain and Hartselle formations in the Cahaba synclinorium constitute the lowermost subdivision of the clastic wedge and represent shallow-marine to barrier-island environments (Fig. 4; Thomas and Mack, 1982). Northeastward along Appalachian structural strike (northeast of the up-plunge northeast end of the Cahaba synclinorium) in stratigraphic relationships like those in the Black Warrior basin, the Pride Mountain-Hartselle tongue prograges northeastward over the Monteagle Limestone and pinches out between the Monteagle and Bangor Limestones (Fig. 5; Thomas, 1972a). The southwest limit of the Hartselle Sandstone in the Cahaba synclinorium is aligned with the linear southwest limit of the northwest-striking, elongate sandstone in the Black Warrior basin (Fig. 5), and both the Hartselle and Pride Mountain sandstones pinch out southeastward across structural strike into a marine mudstone succession in the southeastern structures of the Appalachian thrust belt. Maximum thickness of the lower clastic tongue in the Cahaba synclinorium is similar to that in the Black Warrior basin (compare Figs. 4, 6).

In the northeastern part of the Cahaba synclinorium, the Hartselle Sandstone is overlain by the Bangor Limestone (Fig. 6); southwestward transgression of the lower Bangor Limestone is further indicated by marine-reworked sandstones at the top of the Hartselle Sandstone (Thomas and Mack, 1982). As in the Black Warrior basin, the lower tongue of the Bangor Limestone thins and grades southwestward to marine mudstones of the Floyd Shale (Fig. 5). Facies distributions are similar to those in the Black Warrior basin, except that the Hartselle Sandstone extends farther southwest than the basal Bangor Limestone in the Cahaba synclinorium (Figs. 4, 5, 6). Isopach lines representing the lower tongue of the Bangor Limestone strike northwestward across the Black Warrior basin and in the Cahaba synclinorium (Fig. 5).

Southwestward transgression reflected in the lower Bangor tongue was succeeded by northeastward progradation of another tongue of clastic sediment, including the Floyd Shale and an overlying cyclic succession of deltaic to shallow-marine sandstones and mudstones of the Parkwood Formation (Fig. 6; Thomas, 1972a; Osborne, 1985; Leverett, 1987; Osborne and others, 1991). Within one of the mudstones in the middle part of the Parkwood Formation in the Cahaba synclinorium, a locally exposed bioclastic limestone unit is associated with maroon and green mudstones and marine-reworked sandstone (Thomas, 1972a; Thomas and others, 1982). The limestone unit in the Cahaba synclinorium is in the same stratigraphic position as the "Millerella limestone" in the Black Warrior basin, has similar stratigraphic associations, and is probably part of the same widespread transgressive tongue, thereby providing an important correlation marker (Figs. 4, 6). In the Parkwood Formation below the "Millerella limestone" in the Cahaba synclinorium, as in the Black Warrior basin, successively higher sandstone units extend progressively farther northeast (Osborne and others, 1991), but that part of the clastic succession is significantly thicker than the equivalent section in the Black Warrior basin (compare Figs. 4, 6).

The upper part of the Parkwood Formation (above the "Millerella limestone") generally shows progressively less marine influence upward; however, the succession of depositional systems is nonsystematic in detail, ranging from upper delta plain to marine bay (Thomas and others, 1982; Osborne, 1985; Leverett, 1987; Osborne and others, 1991). Storm deposits, containing marine invertebrate fossils in poorly sorted sandstone lenses, provide biostratigraphic data for the upper part of the Parkwood Formation (T. W. Henry, pers. commun.). In contrast to the distribution in the Black Warrior basin, the upper part of the Parkwood Formation extends far to the northeast along the Appalachian thrust belt and does not grade northeastward into an equivalent carbonate facies (Bangor Limestone); however, the clastic succession includes more marine components northeast of the Cahaba synclinorium along Appalachian strike (Thomas, 1972a; Mintz, 1987). The upper part of the Parkwood Formation is substantially thicker than the equivalent section in the Black Warrior basin (compare Figs. 4, 6), and the greater northeastward extent of the clastic facies records greater northeastward progradation in the area southeast of the Birmingham basement fault system (Fig. 5).

In the Cahaba synclinorium, the lower part of the Pottsville Formation is characterized by thick, quartzose sandstones and local quartz-pebble conglomerates that reflect barrier-island systems comparable to the lower Pottsville sandstones in the eastern part of the Black Warrior basin (Hobday, 1974). Three quartzose sandstones (Shades, Pine, and Chestnut Sandstone Members) in the Cahaba synclinorium are separated by thicker mudstone units (Fig. 6), unlike the succession of stacked quartzose sandstone units (Engman, 1985) in the eastern part of the Black Warrior basin (Fig. 4). Paleobotanical data are interpreted to show that the Harkness coal (Fig. 6) in the Cahaba synclinorium is the likely biostratigraphic equivalent of the Black Creek coal (Butts, 1926) which marks the top of the lower part of the Pottsville Formation in the Black Warrior basin (Fig. 4). A succession of mudstones, coals, and thinner sandstones between the Chestnut Sandstone Member and the Harkness coal is generally similar to the overlying upper part of the Pottsville Formation, but differs from the evidently equivalent upper part of the lower Pottsville succession in the eastern part of the Black Warrior basin (compare Figs. 4, 6). The lower part of the Pottsville Formation in the Cahaba synclinorium is more than twice as thick as that in the Black Warrior basin, and the differences in the vertical succession probably reflect more rapid sediment accumulation on the southeast.

Correlations within the Parkwood Formation and lower part of the Pottsville Formation are somewhat problematic in the southwestern part of the Cahaba synclinorium. Stratigraphic sections penetrated in wells in the southwestern part of the synclinorium do not contain distinctive, thick, quartzose sandstones at a stratigraphic level equivalent to those in the lower part of the Pottsville Formation of the Cahaba outcrops (Fig. 6), although some of the wells are not far from the outcrops. Instead, a relatively thick sandstone has been penetrated at a stratigraphic level (measured up from the Tuscumbia Limestone) that suggests a position within the lower part of the Parkwood Formation of the nearby outcrop sections; the sandstone in the wells is probably continuous with marine-reworked sandstone that includes a bioclastic limestone interbed (Leverett, 1987) in the southwestern part of the outcrop belt (Fig. 6). In the Black Warrior basin, lower Parkwood sandstones pinch out northeastward (Fig. 4), and data from the southwestern part of the Cahaba synclinorium suggest a similar sandstone with similar distribution. In the Black Warrior basin, the characteristic thick, quartzose sandstones of the lower part of the Pottsville Formation pinch out abruptly southwestward into mudstone (Figs. 4, 5); the lack of quartzose sandstones at the well locations in the southwestern part of the Cahaba synclinorium suggests similar facies distributions (Figs. 5, 6). The relationships between outcrop sections and the wells in the southwestern part of the Cahaba synclinorium suggest facies distributions that are similar to those in the Black Warrior basin and that are consistent with northeastward progradation; however, poor exposure, sparse wells, and somewhat complex local structures prohibit detailed correlations.

The upper part of the Pottsville Formation is a cyclic succession of sandstone, mudstone, coal, and conglomerate. Correlations along the synclinorium are particularly reliable where surface mines in the southeast-dipping coal beds may be traced along strike. The Harkness and Gholson coals are useful stratigraphic markers (Fig. 6) and are correlated with the Black Creek and Pratt coals, respectively, in the Black Warrior basin (Butts, 1926). A general northeastward decrease in the amount of sandstone and northeastward thinning of the section are evident (Fig. 5; Womack, 1983). Thick polymictic conglomerates (Straven Conglomerate Member) are distinctive deposits of the upper part of the Pottsville Formation (Osborne, 1991). The stratigraphically lowest polymictic conglomerate is approximately 75 m above the Gholson coal and is near the stratigraphic level of the lowest chert-pebble-bearing conglomeratic sandstones in the Black Warrior basin (Fig. 7). The preserved section above the Gholson coal (Fig. 7) is approximately 275 m thick and contains conglomerates throughout the upper part. The coarse conglomerates reflect distal-fan and braided-fluvial deposits characterized by northwesterly paleocurrents (Osborne, 1991). Distribution and sedimentology of the coarse polymictic conglomerates and associated sandstones indicate a sediment source on the southeast and progradation toward the northwest (Mack and others, 1983; Womack, 1983; Osborne, 1991). The upper Pottsville succession includes the record of an important change in tectonic framework. Earlier components of the clastic wedge indicate an exclusive source on the southwest and progradation toward the northeast; the upper part of the Pottsville Formation reflects both continuation of sediment dispersal from the southwest and the addition of a supply of clastic sediment from the southeast.

CORRELATION OF CHRONOSTRATIGRAPHIC HORIZONS BETWEEN SECTIONS IN THE BLACK WARRIOR BASIN AND CAHABA SYNCLINORIUM

Two elements of the physical stratigraphy define approximate chronostratigraphic horizons. Cyclicity in the clastic-wedge stratigraphy records repeated transgression and regression, and the maximum extent of each transgressive and progradational tongue provides a marker that approximates a synchronous surface. More importantly, the bases of both the lower tongue of the Bangor Limestone and the "Millerella limestone" tongue of the Bangor indicate very rapid transgression; each of these horizons is a useful chronostratigraphic marker (Figs. 4, 6). Tracing of the extensive coal beds in the upper part of the Pottsville Formation is highly reliable within the Cahaba synclinorium and within the Black Warrior foreland basin, but correlations between the two areas are somewhat less certain.

General biostratigraphic similarities link the succession in the Cahaba synclinorium to that in the Black Warrior basin. The lowest part of the clastic wedge is Meramecian in age, and the Meramecian-Chesterian boundary is within the lower part of the Pride Mountain Formation (Figs. 4, 6; Butts, 1926; Drahovzal, 1967). The Bangor Limestone in the eastern part of the Black Warrior basin contains Chesterian faunas (Butts, 1926; Drahovzal, 1967). The extensive "Millerella limestone" tongue within the Parkwood Formation contains Chesterian invertebrate fossils, both in the Black Warrior basin (Gordon, 1953) and in the Cahaba synclinorium (T. W. Henry, pers. commun.). Beds between the "Millerella limestone" and the base of the Pottsville Formation in the Black Warrior basin contain a transitional fauna that has both Mississippian and Pennsylvanian elements (Gordon, 1953), and the Mississippian-Pennsylvanian boundary is placed within the Parkwood Formation above the "Millerella limestone" in the Cahaba synclinorium on the basis of marine invertebrate faunas (Figs. 4, 6; Butts, 1926; T. W. Henry, pers. commun.). Palynological data (Eble and others, 1991) indicate that all of the strata assigned to the exposed Pottsville Formation in the Black Warrior basin and in the Cahaba synclinorium are Early Pennsylvanian in age and that the Mississippian-Pennsylvanian boundary is within the Parkwood Formation in the Cahaba synclinorium. Paleobotanical data also document the Mississippian-Pennsylvanian boundary within the Parkwood Formation in the Cahaba synclinorium (Jennings and Thomas, 1987). Available data do not resolve the age of the Parkwood-Pottsville contact; possibly some beds assigned to the lower part of the Pottsville Formation in the Black Warrior basin are temporally equivalent to beds assigned to the upper part of the Parkwood Formation in the Cahaba synclinorium. Paleobotanical data indicate that the Black Creek coal at the base of the upper part of the Pottsville Formation in the Black Warrior basin is correlative with the Harkness coal in the Cahaba synclinorium and, similarly, that the Pratt coal in the Black Warrior basin is correlative with the Gholson coal in the Cahaba synclinorium (Figs. 4, 6, 7; Butts, 1926). Palynomorphs indicate that the youngest preserved beds in the subsurface part of the Black Warrior basin are as young as Middle Pennsylvanian (Upshaw, 1967), younger than any strata preserved in the Cahaba synclinorium.

THICKNESS DISTRIBUTION AND BASIN SUBSIDENCE

The synorogenic clastic wedge thickens southwestward across structural strike of the Black Warrior foreland basin (Fig. 4). The minimum thickness on the northeast includes the Mississippian carbonate facies, and southwestward thickening is associated with southwestward change to the clastic facies. For most of the clastic wedge, thickness distribution conforms generally to southwestward thickening and northwest-striking isopach lines and facies boundaries (Fig. 5; Thomas, 1972a, b; Cleaves, 1981; Thomas and Womack, 1983; Sestak, 1984; Engman, 1985; Higginbotham, 1985; Miesfeldt, 1985; Hines, 1988; Dismukes, 1989; Nix, 1991). Subdivisions within the upper part of the Pottsville Formation thicken southwestward in the southwestern part of the basin and thicken southward or southeastward in the southeastern part of the basin (Figs. 4, 7; Cleaves, 1981; Sestak, 1984; Pashin and others, 1990).

In the Cahaba synclinorium, the thickness of the Mississippian-Pennsylvanian clastic wedge, as well as the thicknesses of recognizable subdivisions, increases toward the southwest (Fig. 6). The pattern of southwestward thickening in the Cahaba synclinorium is similar to that in the Black Warrior basin (Fig. 4); however, the succession in the Cahaba synclinorium is more than twice as thick as that in the Black Warrior basin (compare Figs. 4, 6, 7). The gradient of southwestward thickening along the Cahaba synclinorium is interrupted by an abrupt southwestward increase in thickness across the Bessemer transverse zone, which trends northwestward across the Appalachian thrust belt (Figs. 1, 6). Stratigraphic variations in pre-Mississippian units across the Bessemer transverse zone indicate episodic synsedimentary movement of an inferred northwest-striking basement fault (Ferrill, 1989), which is further suggested by greater depth to basement southwest of the transverse zone as interpreted from seismic reflection profiles. The abrupt southwestward increase in thickness of the Mississippian-Pennsylvanian synorogenic clastic wedge suggests an additional episode of movement on the structure that ultimately localized the Bessemer transverse zone.

The contrast in thickness between sections in the Black Warrior basin and those in the Cahaba synclinorium indicates greater rates of subsidence and sediment accumulation southeast of the Birmingham basement fault system (Fig. 7). Except for the upper part of the Pottsville Formation in the southeastern part of the basin, components of the clastic wedge show no regional southeastward gradient of thickening across the Black Warrior basin toward the Cahaba synclinorium. The present apparent gradient of southeastward thickening into the Appalachian thrust belt has been exaggerated by northwestward translation of strata in the Cahaba synclinorium; however, the palinspastically restored thickness gradient (Fig. 7) is much steeper than any southeastward thickening component in the Black Warrior basin. The local abrupt southeastward thickening into the Cahaba synclinorium suggests that the Birmingham basement fault system was reactivated with down-to-southeast displacement during deposition of the northeastward-prograding synorogenic clastic wedge (Thomas, 1986). The persistence of deltaic to shallow-marine environments on both sides of the basement fault system indicates that sediment-accumulation rate approximately equaled subsidence rate throughout the foreland.

The mechanism of fault movement to accommodate greater sedimentary accumulation southeast of the Birmingham basement fault system is suggested by a distinct system of northeast-striking, mostly down-to-southeast normal faults in Paleozoic strata on the southeast limb of the Birmingham anticlinorium (Simpson, 1965). The fault system is as much as 7 km wide, and a typical cross section of the system includes six or more faults. Individual faults have separation as great as 130 m (Simpson, 1965). Stratigraphic variations indicate episodic fault movement from Ordovician to Mississippian time (Thomas, 1986). Detailed studies (Wu, 1989) of fault-rock fabrics in Ordovician limestones indicate at least two episodes of deformation, an early episode of normal faulting under low confining pressure (thin cover) and a later episode of subhorizontal compression associated with Appalachian thrusting under greater confining pressure (thicker cover). The early episode (as identified by Wu, 1989) is consistent in timing and cover thickness with a Middle Silurian episode of fault movement that is indicated by stratigraphic thickness variations and sedimentary slump structures (Thomas, 1986; Ferrill, 1989). Rotational, listric faults in the Hartselle Sandstone and upper Pride Mountain Formation indicate down-to-southeast slumping on unstable depositional slopes in the area of the normal-fault system (Thomas, 1968); they do not represent upward propagation of specific basement faults into the Mississippian strata. The lack of a large-scale normal fault in the Paleozoic strata comparable in separation to that documented by seismic reflection profiles for the Birmingham basement fault system, along with the mapped width and geometry of the system of smaller down-to-southeast faults in Paleozoic strata, suggests that the cover strata responded to vertical basement-fault movement by drape folding (Thomas, 1986). The system of normal faults in the cover strata probably reflects extension associated with drape folding. Continued drape folding of the cover strata is adequate to account for the difference in sediment-accumulation rates on opposite sides of the Birmingham basement fault system during deposition of the Mississippian-Pennsylvanian clastic wedge; alternatively, one or more faults may have contributed to the vertical separation. Rocks and structures in the pre-thrust drape fold subsequently were translated on the Opossum Valley thrust fault into the southeast limb of the Birmingham anticlinorium (Fig. 3), and any faults that propagated upward into the cover strata from the basement fault system would presently be in the structurally high part of the Birmingham ramp anticlinorium where Mississippian-Pennsylvanian strata have been eroded. No direct evidence of either a large-scale Mississippian-Pennsylvanian synsedimentary fault or proximal facies to a large-separation fault is preserved, and no large normal faults have been recognized in the preserved pre-Mississippian strata.

In both the Black Warrior basin and Cahaba synclinorium, subsidence rates increased through time during deposition of the Mississippian-Pennsylvanian synorogenic clastic wedge, and subsidence rates on the southwest exceeded those on the northeast (Table 2; Thomas and others, 1991). Greater subsidence on the southwest in the Black Warrior basin is interpreted to indicate thrust loading by northeastward-advancing Ouachita thrust sheets, and similarity of subsidence history indicates that the Cahaba synclinorium was part of the Ouachita foreland during deposition of most of the Mississippian-Pennsylvanian clastic wedge (Fig. 8). Relatively greater subsidence at the palinspastic site of the Cahaba synclinorium than within the present Black Warrior basin is interpreted to indicate that the Ouachita foreland was partitioned by the reactivated, northeast-striking Birmingham basement fault system and that down-to-southeast displacement resulted in greater subsidence rates southeast of the basement fault system. Lack of paleobathymetric contrasts demonstrates that sediment accumulation kept pace with subsidence on both sides of the basement fault system. Introduction of a new sediment source on the southeast during deposition of the upper part of the Pottsville Formation is represented by northwestward progradation of a clastic wedge (Fig. 8), which presaged the northwestward advance of Appalachian thrust sheets and ultimate thrust imbrication of the southeastern side of the older, Ouachita-related, southwest-dipping Black Warrior foreland basin (Fig. 1).

TABLE 2.—COMPARISON OF SUBSIDENCE RATES VERTICALLY AND LATERALLY ACROSS THE BLACK WARRIOR FORELAND BASIN AND ALONG THE CAHABA SYNCLINORIUM IN THE APPALACHIAN THRUST BELT

	Black Warrior Basin		Cahaba Synclinorium	
	Southwest	Northeast	Southwest	Northeast
Lower part of Pottsville Formation	7.8 cm/ky	6.0 cm/ky	20.5 cm/ky	10.8 cm/ky
Lower part of clastic wedge (Pride Mountain-Hartselle)	1.8 cm/ky	1.2 cm/ky	3.2 cm/ky	1.6 cm/ky

Subsidence rates were computed from decompaction and back stripping, using a Macintosh program written by M. S. Wilkerson and A. T. Hsui. Correlation markers are shown in Figures 4, 5, and 7; age of each correlation marker was estimated from the time scale of the Decade of North American Geology extrapolated to smaller subdivisions. Water depth was considered negligible on the basis of depositional systems identified within the clastic wedge (references cited in text).

SUMMARY AND CONCLUSIONS

A change from passive margin to active convergent margin during Meramecian time (middle Mississippian) resulted from Ouachita arc-continent collision along the southwest side of the Alabama promontory of North American continental crust. Initial down-to-southwest flexural subsidence and northeastward progradation of synorogenic clastic sediment into the original (greater) Black Warrior foreland basin encompassed the area of the present Black Warrior foreland basin and also the palinspastic location of strata now in the Appalachian thrust belt. Greater thickness of the synorogenic clastic wedge in the Cahaba synclinorium in the Appalachian thrust belt reflects synsedimentary down-to-southeast reactivation of the northeast-striking Birmingham basement fault system that partitioned the Ouachita foreland. Progradation of synorogenic clastic sediment from the southeast was added to the northeastward-prograding foreland sediment dispersal system during the Early Pennsylvanian, recording the initiation of northwest-directed Appalachian thrusting. Subsequently, northeast-striking, northwest-advancing Appalachian thrust faults

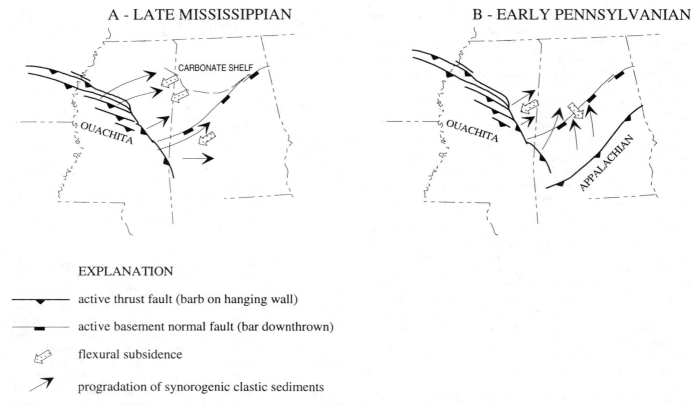

FIG. 8.—Reconstructions of successive stages of thrusting around the Black Warrior foreland basin: (A) Late Mississippian (Parkwood Formation): northeast-directed Ouachita thrusting; down-to-southwest flexural subsidence of foreland basin; northeastward-prograding synorogenic clastic wedge; Ouachita foreland includes area of the present Black Warrior basin and the palinspastic site of Appalachian thrust sheets. (B) Early Pennsylvanian (upper part of Pottsville Formation): northwest-directed Appalachian thrusting in addition to northeast-directed Ouachita thrusting; limited down-to-southeast flexural subsidence in addition to down-to-southwest flexural subsidence; northwestward-prograding synorogenic clastic wedge in addition to northeastward-prograding clastic wedge; subsequently, Appalachian thrust front propagated farther northwest and imbricated the southeastern part of the original Ouachita foreland.

truncated the northwest-striking structure of the Black Warrior foreland basin and imbricated part of the older, southwestward-thickening, northeastward-prograding, Ouachita-derived clastic wedge.

Diachronous loading by thrust belts advancing onto orthogonal continental margins contrasts with emplacement of a thrust load along a single direction of translation. Although a thrust front commonly advances to incorporate the proximal part of a synorogenic clastic wedge into the thrust belt, part of the southwest-dipping Ouachita component of the Black Warrior basin was imbricated in the independently acting Appalachian thrust belt, the strike of which is nearly perpendicular to strike of the the Ouachita-related Black Warrior foreland basin. The distinctive, two-component subsidence history and fault partitioning of the Black Warrior foreland basin reflect the location within a large-scale, thrust-belt recess, where diachronous, two-directional thrusting framed the foreland basin.

ACKNOWLEDGMENTS

This research has been supported by grants from the National Science Foundation (EAR-8218604, EAR-8905229), the Petroleum Research Fund of the American Chemical Society (17069-AC2), and the University of Alabama School of Mines and Energy Development. In addition to the work reported in theses and dissertations cited herein, Graduate Research Assistants participated in various parts of the research, as follows: S. H. Womack, W. E. Osborne, and D. E. Leverett measured stratigraphic sections and described cores and cuttings; T. E. Osborne and R. A. Hines described cores and cuttings; B. A. Ferrill compiled stratigraphic sections and field checked correlations; and J. L. Allen calculated subsidence rates. A draft of the manuscript was reviewed by J. C. Mars, W. J. Sims, and B. M. Whiting. J. C. Mars assisted in design of the illustrations. Seismic reflection profiles were provided by Grant-Norpac, Digicon, and Geophysical Field Surveys.

REFERENCES

BUTTS, C., 1907, The northern part of the Cahaba coal field, Alabama: United States Geological Survey Bulletin 316, p. 76–115.
BUTTS, C., 1911, The southern part of the Cahaba coal field, Alabama: United States Geological Survey Bulletin 431, p. 89–146.
BUTTS, C., 1926, The Paleozoic rocks, in Geology of Alabama: Tuscaloosa, Alabama Geological Survey Special Report 14, p. 41–230.
BUTTS, C., 1927, Bessemer-Vandiver folio, Alabama: Washington, D. C., United States Geological Survey Geologic Atlas of the United States No. 221, 22 p.

CLEAVES, A. W., 1981, Resource evaluation of Lower Pennsylvanian (Pottsville) depositional systems of the western Warrior coal field, Alabama and Mississippi: University, Mississippi Mineral Resources Institute Technical Report No. 81–1, 125 p.

CULBERTSON, W. C., 1964, Geology and coal resources of the coal-bearing rocks of Alabama: United States Geological Survey Bulletin 1182-B, 79 p.

DISMUKES, M. B., 1989, Stratigraphy and environments of deposition of the lower Pottsville Formation in Mississippi: Unpublished M.S. Thesis, University of Alabama, Tuscaloosa, 165 p.

DRAHOVZAL, J. A., 1967, The biostratigraphy of Mississippian rocks in the Tennessee Valley, in Smith, W. E., ed., A Field Guide to Mississippian Sediments in Northern Alabama and South-Central Tennessee: Tuscaloosa, Alabama Geological Society Guidebook, 5th Annual Field Trip, p. 10–24.

EBLE, C. F., GILLESPIE, W. H., AND HENRY, T. W., 1991, Palynology, paleobotany, and invertebrate paleontology of Pennsylvanian coal beds and associated strata in the Warrior and Cahaba coal fields, in Thomas, W. A., and Osborne, W. E., eds., Mississippian-Pennsylvanian Tectonic History of the Cahaba Synclinorium: Tuscaloosa, Alabama Geological Society Guidebook, 28th Annual Field Trip, p. 119–132.

ENGMAN, M. A., 1985, Depositional systems in the lower part of the Pottsville Formation, Black Warrior basin, Alabama: Unpublished M.S. Thesis, University of Alabama, Tuscaloosa, 250 p.

FERRILL, B. A., 1989, Middle Cambrian to Lower Mississippian synsedimentary structures in the Appalachian fold-thrust belt in Alabama and Georgia: Unpublished Ph.D. Dissertation, University of Alabama, Tuscaloosa, 270 p.

FLEMINGS, P. B. AND JORDAN, T. E., 1989, A synthetic stratigraphic model of foreland basin development: Journal of Geophysical Research, v. 94, p. 3851–3866.

GORDON, M., JR., 1953, Report on referred fossils [unpublished]: Washington, D.C., United States Geological Survey.

HANDFORD, C. R., 1978, Monteagle Limestone (Upper Mississippian)—Oolitic tidal-bar sedimentation in southern Cumberland Plateau: American Association of Petroleum Geologists Bulletin, v. 62, p. 644–656.

HIGGINBOTHAM, D. R., 1985, Regional stratigraphy, environments of deposition, and tectonic framework of Mississippian clastic rocks between the Tuscumbia and Bangor Limestones in the Black Warrior basin of Alabama and Mississippi: Unpublished M.S. Thesis, University of Alabama, Tuscaloosa, 177 p.

HINES, R. A., JR., 1988, Carboniferous evolution of the Black Warrior foreland basin, Alabama and Mississippi: Unpublished Ph.D. Dissertation, University of Alabama, Tuscaloosa, 231 p.

HOBDAY, D. K., 1974, Beach- and barrier-island facies in the upper Carboniferous of northern Alabama: Boulder, Geological Society of America Special Paper 148, p. 209–223.

HORSEY, C. A., 1981, Depositional environments of the Pennsylvanian Pottsville Formation in the Black Warrior basin of Alabama: Journal of Sedimentary Petrology, v. 51, p. 799–806.

JENNINGS, J. R. AND THOMAS, W. A., 1987, Fossil plants from Mississippian-Pennsylvanian transition strata in the southern Appalachians: Southeastern Geology, v. 27, p. 207–217.

JORDAN, T. E., 1981, Thrust loads and foreland basin evolution, Cretaceous, western United States: American Association of Petroleum Geologists Bulletin, v. 65, p. 2506–2520.

LEVERETT, D. E., 1987, Stratigraphy and depositional environments of the Mississippian-Pennsylvanian Parkwood Formation on a part of the northwest limb of the Cahaba synclinorium, Jefferson and Bibb Counties, Alabama: Unpublished M.S. Thesis, University of Alabama, Tuscaloosa, 135 p.

MACK, G. H., THOMAS, W. A., AND HORSEY, C. A., 1983, Composition of Carboniferous sandstones and tectonic framework of southern Appalachian-Ouachita orogen: Journal of Sedimentary Petrology, v. 53, p. 931–946.

McGLAMERY, W., 1955, Subsurface stratigraphy of northwest Alabama: Tuscaloosa, Alabama Geological Survey Bulletin 64, 503 p.

MIESFELDT, M. A., 1985, Facies relationships between the Parkwood and Bangor formations in the Black Warrior basin: Unpublished M.S. Thesis, University of Alabama, Tuscaloosa, 149 p.

MINTZ, J. L., 1987, Stratigraphy and depositional environments of the Parkwood and Pottsville Formations at Blount and Chandler Mountains, central Alabama: Unpublished M.S. Thesis, University of Alabama, Tuscaloosa, 206 p.

NIX, M. A., 1991, Facies and facies relationships of the lower part of the Parkwood Formation in the Black Warrior basin of Mississippi and Alabama: Unpublished M.S. Thesis, University of Alabama, Tuscaloosa, 362 p.

OSBORNE, T. E., 1991, The depositional environment and provenance of the Straven Conglomerate Member of the Pottsville Formation, in Thomas, W. A., and Osborne, W. E., eds., Mississippian-Pennsylvanian Tectonic History of the Cahaba Synclinorium: Tuscaloosa, Alabama Geological Society Guidebook, 28th Annual Field Trip, p. 73–98.

OSBORNE, W. E., 1985, Depositional environments and dispersal system of the Parkwood Formation (Carboniferous) on the northwest limb of the Cahaba syncline, Jefferson County, Alabama: Unpublished M.S. Thesis, University of Alabama, Tuscaloosa, 160 p.

OSBORNE, W. E., LEVERETT, D. E., AND THOMAS, W. A., 1991, Depositional environments and sediment dispersal of the Parkwood Formation (Mississippian-Pennsylvanian), northwest limb of Cahaba synclinorium, Appalachian thrust belt in Alabama, in Thomas, W. A., and Osborne, W. E., eds., Mississippian-Pennsylvanian Tectonic History of the Cahaba Synclinorium: Tuscaloosa, Alabama Geological Society Guidebook, 28th Annual Field Trip, p. 53–72.

PASHIN, J. C., WARD, W. E., II, WINSTON, R. B., CHANDLER, R. V., BOLIN, D. E., HAMILTON, R. P., AND MINK, R. M., 1990, Geologic evaluation of critical production parameters for coalbed methane resources, Part II, Black Warrior basin: Tuscaloosa, Alabama Geological Survey, Annual Report, Gas Research Institute Contract No. 5087-214-1594, 177 p.

QUINLAN, G. M. AND BEAUMONT, C., 1984, Appalachian thrusting, lithospheric flexure, and the Paleozoic stratigraphy of the eastern interior of North America: Canadian Journal of Earth Sciences, v. 21, p. 973–996.

SCHEDL, A. AND WILTSCHKO, D. V., 1984, Sedimentological effects of a moving terrain: Journal of Geology, v. 92, p. 273–287.

SESTAK, H. M., 1984, Stratigraphy and depositional environments of part of the Pennsylvanian Pottsville Formation in the Black Warrior basin: Alabama and Mississippi: Unpublished M.S. Thesis, University of Alabama, Tuscaloosa, 184 p.

SHADROUI, J. M., 1986, Depositional environments of the Pennsylvanian Bremen Sandstone Member and associated strata, Pottsville Formation, north-central Alabama: Unpublished M.S. Thesis, University of Alabama, Tuscaloosa, 172 p.

SIMPSON, T. A., 1965, Geologic and hydrologic studies in the Birmingham red-iron-ore district, Alabama: Washington, D. C., United States Geological Survey Professional Paper 473-C, 47 p.

SLOSS, L. L., 1963, Sequences in the cratonic interior of North America: Geological Society of America Bulletin, v. 74, p. 93–114.

SLOSS, L. L., 1988, Tectonic evolution of the craton in Phanerozoic time, in Sloss, L. L., ed., Sedimentary Cover—North American Craton: U. S.: Boulder, Geological Society of America, The Geology of North America, v. D-2, p. 25–51.

SPEED, R. C. AND SLEEP, N. H., 1982, Antler orogeny and foreland basin: A model: Geological Society of America Bulletin, v. 93, p. 815–828.

THOMAS, W. A., 1968, Contemporaneous normal faults on flanks of Birmingham anticlinorium, central Alabama: American Association of Petroleum Geologists Bulletin, v. 52, p. 2123–2136.

THOMAS, W. A., 1972a, Mississippian stratigraphy of Alabama: Tuscaloosa, Alabama Geological Survey Monograph 12, 121 p.

THOMAS, W. A., 1972b, Regional Paleozoic stratigraphy in Mississippi between Ouachita and Appalachian Mountains: American Association of Petroleum Geologists Bulletin, v. 56, p. 81–106.

THOMAS, W. A., 1973, Southwestern Appalachian structural system beneath the Gulf Coastal Plain: American Journal of Science, v. 273-A, p. 372–390.

THOMAS, W. A., 1976, Evolution of Ouachita-Appalachian continental margin: Journal of Geology, v. 84, p. 323–342.

THOMAS, W. A., 1977, Evolution of Appalachian-Ouachita salients and recesses from reentrants and promontories in the continental margin: American Journal of Science, v. 277, p. 1233–1278.

THOMAS, W. A., 1985, Northern Alabama sections, in Woodward, N. B., ed., Valley and Ridge Thrust Belt: Balanced Structural Sections, Pennsylvania to Alabama (Appalachian Basin Industrial Associates):

Knoxville, University of Tennessee Department of Geological Sciences Studies in Geology 12, p. 54–61.

Thomas, W. A., 1986, A Paleozoic synsedimentary structure in the Appalachian fold-thrust belt in Alabama, *in* McDowell, R. C., and Glover, L., III, eds., The Lowry Volume: Studies in Appalachian Geology: Blacksburg, Virginia Tech Department of Geological Sciences Memoir 3, p. 1–12.

Thomas, W. A., 1988a, The Black Warrior basin, *in* Sloss, L. L., ed., Sedimentary Cover—North American Craton: U. S.: Boulder, Geological Society of America, The Geology of North America, v. D-2, p. 471–492, Plate 8.

Thomas, W. A., 1988b, Early Mesozoic faults of the northern Gulf Coastal Plain in the context of opening of the Atlantic Ocean, *in* Manspeizer, W., ed., Triassic-Jurassic Rifting: Amsterdam, Elsevier, Developments in Geotectonics 22, p. 463–476.

Thomas, W. A., 1989, The Appalachian-Ouachita orogen beneath the Gulf Coastal Plain between the outcrops in the Appalachian and Ouachita Mountains, *in* Hatcher, R. D., Jr., Thomas, W. A., and Viele, G. W., eds., The Appalachian-Ouachita Orogen in the United States: Boulder, Geological Society of America, The Geology of North America, v. F-2, p. 537–553.

Thomas, W. A., 1990, Controls on locations of transverse zones in thrust belts: Eclogae Geologicae Helvetiae, v. 83, p. 727–744.

Thomas, W. A., 1991, The Appalachian-Ouachita rifted margin of southeastern North America: Geological Society of America Bulletin, v. 103, p. 415–431.

Thomas, W. A., Ferrill, B. A., Allen, J. L., Osborne, W. E., and Leverett, D. E., 1991, Synorogenic clastic-wedge stratigraphy and subsidence history of the Cahaba synclinorium and the Black Warrior foreland basin, *in* Thomas, W. A., and Osborne, W. E., eds., Mississippian-Pennsylvanian Tectonic History of the Cahaba Synclinorium: Tuscaloosa, Alabama Geological Society Guidebook, 28th Annual Field Trip, p. 37–52.

Thomas, W. A. and Mack, G. H., 1982, Paleogeographic relationship of a Mississippian barrier-island and shelf-bar system (Hartselle Sandstone) in Alabama to the Appalachian-Ouachita orogenic belt: Geological Society of America Bulletin, v. 93, p. 6–19.

Thomas, W. A., Tull, J. F., Neathery, T. L., Mack, G. H., and Ferrill, B. A., 1982, A field guide to the Appalachian thrust belt in Alabama, *in* Thomas, W. A., and Neathery, T. L., eds., Appalachian Thrust Belt in Alabama: Tectonics and Sedimentation (Field Trip Guidebook, Geological Society of America 1982 Annual Meeting, New Orleans, Louisiana): Tuscaloosa, Alabama Geological Society, p. 5–40.

Thomas, W. A., Viele, G. W., Arbenz, J. K., Nicholas, R. L., Denison, R. E., Muehlberger, W. R., and Tauvers, P. R., 1989, Tectonic map of the Ouachita orogen, *in* Hatcher, R. D., Jr., Thomas, W. A., and Viele, G. W., eds., The Appalachian-Ouachita Orogen in the United States: Boulder, Geological Society of America, The Geology of North America, v. F-2, Plate 9.

Thomas, W. A. and Womack, S. H., 1983, Coal stratigraphy of the deeper part of the Black Warrior basin in Alabama: Gulf Coast Association of Geological Societies Transactions, v. 33, p. 439–446.

Upshaw, C. F., 1967, Pennsylvanian palynology and age relationships in the Warrior basin, Alabama, *in* Ferm, J. C., Ehrlich, R., and Neathery, T. L., A Field Guide to Carboniferous Detrital Rocks in Northern Alabama (Geological Society of America 1967 Annual Meeting): Tuscaloosa, Alabama Geological Society, p. 16–20.

Viele, G. W. and Thomas, W. A., 1989, Tectonic synthesis of the Ouachita orogenic belt, *in* Hatcher, R. D., Jr., Thomas, W. A., and Viele, G. W., eds., The Appalachian-Ouachita Orogen in the United States: Boulder, Geological Society of America, The Geology of North America, v. F-2, p. 695–728.

Womack, S. H., 1983, Provenance of the Pottsville Formation in the Cahaba syncline: Unpublished M.S. Thesis, University of Alabama, Tuscaloosa, 119 p.

Wu, S., 1989, Strain partitioning and deformation mode analysis of the normal faults at Red Mountain, Birmingham, Alabama: Tectonophysics, v. 170, p. 171–182.

SYNOROGENIC CARBONATE PLATFORMS AND REEFS IN FORELAND BASINS: CONTROLS ON STRATIGRAPHIC EVOLUTION AND PLATFORM/REEF MORPHOLOGY

STEVEN L. DOROBEK
Department of Geology, Texas A&M University, College Station, TX 77843

ABSTRACT: Carbonate platforms and reefs are more common in foreland basins than is generally appreciated and may provide a better record of basin evolution and relative sea-level change than siliciclastic strata. Carbonate platform and reefal facies may develop in the proximal foredeep on a variety of topographic highs, in the distal foreland area far from terrigenous influx, or across the entire foreland basin during tectonically quiescent stages of basin development. Basin geometry, dispersal of siliciclastic sediment, subsidence patterns, and deformation structures across the foreland affect carbonate facies distribution and platform morphology during synorogenic stages of foreland basin evolution.

Synorogenic foreland carbonate platforms typically have ramp profiles that mimic the flexural profile produced by tectonic loading. During active convergence, the flexural profile is driven toward the foreland by the advancing orogenic wedge. Synorogenic carbonate ramps are forced to onlap and/or backstep cratonward. Basinward parts of some foreland carbonate platforms may be drowned (*sensu stricto*) during active convergence, especially when: (1) the underlying lithosphere has low rigidity; (2) the orogenic wedge advances rapidly; (3) a eustatic sea-level rise occurs at the same time as migration of the flexural profile; or (4) some other environmental stress affects carbonate-producing benthos. Two-dimensional forward models show that flexural drowning of some Phanerozoic carbonate platforms, even in the absence of a coeval eustatic sea-level rise or other environmental stress, is possible in less than 250,000 yr.

In some foreland areas, complex patterns of synorogenic differential subsidence and foreland deformation can affect carbonate facies tracts many hundreds of kilometers cratonward of the proximal foredeep. These patterns of differential subsidence and deformation probably reflect the response of preexisting basement structures or rheological anisotropies in the foreland area to tectonic loading along the plate margin or the response to sublithospheric processes. Quantitative subsidence analyses from some foreland areas suggest that differential subsidence in the distal foreland is related temporally to tectonic loading along the continental margin, but cratonward limits of the differential subsidence are beyond reasonable limits of flexurally produced subsidence. In addition, patterns of differential subsidence in the distal foreland do not have "normal" flexural wavelengths, amplitudes, or orientations with respect to the orogenic wedge. Therefore, while subsidence in the distal foreland is temporally related to convergence along the plate margin, alternative models for lithospheric deformation are necessary to explain the differential subsidence in the distal foreland.

Siliciclastic sediment dispersal is another first-order control on carbonate sedimentation in foreland basins. Coarse-grained, siliciclastic sediment may have less affect on suspension-feeding, carbonate-producing benthic organisms than clay and silt. Hence, bedrock geology, paleoclimate, and depositional gradients in the hinterland and foreland sides of the basin indirectly affect carbonate sedimentation. Effects of siliciclastic sedimentation on foreland carbonates also depend on the evolutionary stage of a foredeep. During "underfilled" stages of basin evolution, siliciclastic sediment is trapped in the proximal foredeep and will not affect carbonate-producing benthos on the distal side of the basin. The distal parts of clastic wedges may fill accommodation during "intermediate" stages of basin evolution and provide a substrate for carbonate platforms that prograde from the peripheral bulge. Progradation of siliciclastics during later "overfilled" stages of basin evolution may terminate carbonate platforms even in the distal foreland.

INTRODUCTION

Studies of foreland basin stratigraphy generally focus on the thick packages of siliciclastic sediment that fill many foreland basins. The bulk of these sediments are eroded from the orogenic wedge that comprises the load responsible for initial subsidence in the basin. Sophisticated forward models have been developed recently that link the evolution of the orogenic wedge and adjacent foreland basin with basin-filling processes (e.g., Flemings and Jordan, 1989, 1990; Beaumont and others, 1992; Sinclair and others, 1991; Johnson and Beaumont, this volume; Peper and others, this volume).

Carbonate platform strata, however, have been largely ignored in forward models of foreland basin stratigraphy, even though they may comprise a significant part of the sedimentary fill in many foreland basins. Marine carbonate platform strata in foreland basins may actually be better indicators of relative sea-level change and basin evolution than siliciclastic strata because carbonate facies respond more rapidly to changes in relative sea level and to changes in depositional gradients. Foreland carbonate platforms also are economically important in that they are common hosts for ore deposits and are reservoirs, seals, or source rocks for major hydrocarbon accumulations. Lastly, carbonate strata form important groundwater aquifers in many foreland basin settings.

This paper discusses the controls on synorogenic carbonate platform deposition in foreland basins. "Synorogenic" refers to carbonate platforms that are deposited during major orogenic events when active, flexurally driven subsidence occurs. Some temporal and conceptual overlap necessarily exists between synorogenic platforms and both "pre-orogenic" carbonate platforms (i.e., extant platforms that were suddenly subjected to foreland deformation and flexural subsidence) and "post-orogenic" foreland carbonate platforms (i.e., platforms that were constructed on basin topography inherited from earlier synorogenic stages of foreland basin evolution).

The main intent of this paper is to examine how foreland basin evolution affects shallow water, carbonate platform and reef development. Deep-water and lacustrine carbonate facies are not considered here although they may comprise a large part of the basin fill in some foredeeps (e.g., Eugster and Hardie, 1975; Johnson, 1985; Hattin, 1986; Lafferriere and others, 1987).

The large-scale morphology and stratigraphic development of synorogenic carbonate platforms are emphasized. Case examples of carbonate platforms are used to illustrate conceptual models for various types and stages of foreland basin evolution. It should be emphasized, however, that these case examples demonstrate that every foreland area has a unique subsidence and kinematic history. The structural evolution and patterns of differential subsidence within a particular foreland basin impose unique boundary conditions on stratigraphic development. Thus, marine carbonate strata from any specific foreland basin should be approached with this in mind.

GEODYNAMICS OF FORELAND BASINS

A brief review of geodynamic models for foreland basins is necessary because these models provide a basis for predicting carbonate platform evolution in "normal" foreland basins, where subsidence is driven by simple lithospheric flexure and sediment loading. Flexural models, while they are a good first approximation of deformation across the foreland, do not account for the complex styles of deformation observed in many foreland areas. Several case examples are described later which illustrate how basement-involved structures can dramatically affect carbonate sedimentation in complexly deformed foreland areas.

Overview of Flexural Models for Foreland Basins

Foreland basins (or foredeeps) largely form as a flexural response to vertical loading of continental lithosphere by orogenic belts (Beaumont, 1981; Jordan, 1981; Turcotte and Schubert, 1982). Foreland basins typically have asymmetric cross-sectional profiles; the deepest or proximal part of the basin is located adjacent to the orogenic wedge that borders one side of the basin. The basin floor progressively shallows away from the orogenic wedge toward the distal[1] side of the foreland basin. On the distal side of the foreland basin, a low relief, uplifted area called the *peripheral bulge* (or *forebulge*) develops due to isostatic compensation for the vertical loading of the foreland lithosphere.

Theoretically, foreland basin profiles are determined by the magnitude of applied loads, both vertical and horizontal, and the flexural rigidity of the foreland lithosphere. Typically, only vertical loads (i.e., thrust loads, subsurface loads, and sedimentary fill in the basin) are considered in attempts to model foreland basin geometries. Foreland basin geometry can be described by: (1) the maximum *amplitude* of the deflection of the foreland lithosphere as measured in the proximal part of the foredeep; and (2) the *wavelength* or width of the deflection, which is dependent on the flexural rigidity of the foreland lithosphere.

Foreland basin evolution is directly coupled to deformation in the adjacent orogenic wedge. Migration of foreland basin depocenters typically is attributed to migration of the flexural profile as the orogenic wedge advances toward the foreland. Most theoretical models of flexure in foreland areas assume that the foreland lithosphere is mechanically homogeneous (at least in the horizontal direction). There is, however, growing recognition that heterogeneities in foreland lithosphere can affect basin geometry, subsidence history, and deformation patterns (cf. Stockmal and Beaumont, 1987; Royden and others, 1987; Wu, 1991; Waschbusch and Royden, 1992). For example, the Apennine foreland area is compartmentalized into sub-basins because the foreland lithosphere apparently is segmented by tear faults that trend at high angles to the orogenic belt (Royden and others, 1987). Flexure of each foreland segment is more or less independent of adjacent segments, producing offset foreland depocenters with separate peripheral bulges. In other foreland areas, preexisting zones of weakness in the foreland plate may affect the flexural profile produced by the orogenic wedge (Wu, 1991; Waschbusch and Royden, 1992) or subsidence patterns and styles of deformation many hundreds of kilometers cratonward of the foreland basin depocenter (Ziegler, 1987; Dorobek and others, 1991a, b).

Quantitative models of foreland basin evolution are dependent on assumptions of lithospheric rheology. Rheologies used in various approaches include uniform elastic, uniform viscoelastic, and temperature-dependent viscosity models.

In uniform elastic lithosphere models, the foreland lithosphere responds instantaneously to loading. The form and position of the deflection produced by thrust-loading does not change with time as long as the scale and position of the thrust load do not change.

In uniform viscoelastic models, the form of the deflection changes with time because viscous flow in the lower part of the foreland lithosphere relaxes some of the stress produced by bending. Theoretically, if the foreland lithosphere behaves viscoelastically and with unchanging loads, the proximal foredeep should deepen and the crest of the peripheral bulge should increase in height and migrate toward the thrust load over time.

In temperature-dependent viscoelastic models, viscous flow first occurs in the lower lithosphere which is hotter and less viscous than upper parts of the lithosphere. Relaxation propagates progressively upward into cooler, more viscous lithosphere (Quinlan and Beaumont, 1984; Beaumont and others, 1987). The time required for relaxation also increases progressively upward through the foreland lithosphere because of the corresponding increase in temperature-dependent viscosity. The amount of time necessary to achieve local isostatic equilibrium in the uppermost lithosphere by viscous flow is greater than the age of the Earth. Thus, the temperature-dependent viscosity model can account for the flexural response of the lithosphere as well as any apparent viscous relaxation after loading.

Non-flexural Response to In-plane Stress

In addition to the flexural response of the plate margin, convergence also produces in-plane stresses that can be transmitted into the interior of a foreland plate (Zoback, 1992). These horizontal, in-plane stresses may affect foreland lithosphere many hundreds of kilometers inboard of the actual thrust load and proximal foredeep, far beyond any observable effects of flexure on continental lithosphere (Lambeck and others, 1984; Cloetingh, 1988). As a result, in-plane stresses might enhance deflections of the lithosphere in distal foreland or cratonic areas and can affect sedimentation far from the zone of active convergence.

In-plane stresses generally are ignored in quantitative or conceptual models of foreland basin evolution or foreland stratigraphy (cf. Peper and others, this volume). This as-

---

[1]"Distal" refers to the side of a foreland basin that is farthest from the orogenic wedge.

sumption also is not valid for many foreland areas because in-plane stress produced during convergence can affect preexisting flexural deflections of the lithosphere, or it can cause reactivation of preexisting structures in the foreland plate (Ziegler, 1987; Karner and others, 1993). Deformation and reactivation of preexisting basement faults may occur over hundreds of kilometers inboard from the proximal foredeep, far beyond the effects of plate flexure (Ziegler, 1987). The deformation also is commonly at high angles to the trend of the orogenic wedge and foredeep, further evidence that it is not related entirely to flexure (Dorobek and others, 1991a, b). Finally, if oblique convergence occurs or if there are prominent salients along the plate boundary, collision will be diachronous and the stress field within the foreland plate will evolve over time. Complex tectonic inversion may result across a foreland area as reactivated faults respond to the evolving stress field (Dorobek and others, 1991a, b). Non-flexural deformation can significantly affect carbonate facies patterns over broad regions of the foreland plate, as will be shown later in this paper.

TECTONIC CONTROLS ON SYNOROGENIC FORELAND CARBONATE
PLATFORMS & REEFS

In foreland basins, depositional topography constantly evolves because of tectonically induced differential subsidence and deformation of the foreland plate. Carbonate facies patterns and platform morphology are strongly affected by this differential subsidence and by the kinematic evolution of structural features across the foreland. In this section, comparisons are made between carbonate platform facies that develop in foreland basins formed by simple, flexural subsidence and those that develop in foreland basins where at least some deformation of the foreland is not related to flexure.

Synorogenic Carbonate Ramps: Response to Flexure

With few exceptions, carbonate platforms in active foredeeps form on the distal, cratonward side of the basin and typically have homoclinal ramp profiles. This is true regardless of whether or not reef-building organisms were extant at the time that the foreland carbonate ramps developed. The predominance of carbonate ramps over other platform morphologies points to some intrinsic character of foreland basins that apparently prevents other carbonate platform morphologies from developing.

The most important factor that probably controls carbonate platform morphology in foreland basins is the mechanism for basin subsidence, namely, lithospheric flexure. During active convergence, lithospheric flexure normally produces a gentle, basinward dipping surface on the cratonward side of foreland basins. Synorogenic carbonate ramp profiles on the distal side of foreland basins mimic this flexural profile because flexure influences three important controls on carbonate sedimentation in foreland areas: depositional gradient, subsidence rate, and water depth along the flexural profile.

During active convergence, depositional gradients and subsidence rates within a foreland basin are lowest along the flanks of the peripheral bulge, and both progressively increase toward the orogenic wedge (cf. Burchette and Wright, 1992). As an orogenic wedge advances cratonward, the flexural profile produced by the wedge advances with it (Fig. 1). Thus for an orogenic wedge that advances at a constant rate, depositional gradient, subsidence rate, and water depth for most points along the distal side of the basin will progressively increase non-linearly over time, and foreland carbonate platforms must aggrade at ever increasing rates to keep pace with these changes. Large volumes of siliciclastic sediment eroded from the orogenic wedge and excessive water depth exclude carbonate sedimentation in the proximal foredeep (with important exceptions as described below), so it is irrelevant that siliciclastic sedimentation rates might outpace increases in relative water depth in the proximal foredeep. Changes in depositional gradient, subsidence rate, and water depth are only considered for the distal side of foreland basins, where most carbonate platforms develop. The effects of flexure on each of these parameters are discussed below.

Flexure and its Effects on Relative Water Depth.—

In situ benthic carbonate sedimentation for Phanerozoic platforms is insignificant at water depths greater than 100 m (Hallock and Schlager, 1986). Using this as a conservative limit for autochthonous benthic carbonate sedimentation during the first stages of foreland basin development, the basinward limit of a carbonate ramp can be estimated by determining where the 100-m isobath intersects the flexural profile. Flexural models of elastic plates indicate that productive carbonate ramps (i.e., ramps that are not drowned) should not be much greater than 100 km wide, given a wide range of values for eustatic sea-level position, topographic elevation of the orogenic wedge, and flexural rigidity of the foreland lithosphere (Fig. 2). This will hold true as long as the crest of the peripheral bulge is subaerially exposed. If sea level is very high or some mantle process adds an additional component of plate-scale subsidence that is unrelated to flexural subsidence, the crest of the peripheral bulge and the "back-bulge basin" (cf. Giles and Dickinson, this volume) may be submerged to shallow depths. Aggrading carbonate platforms may extend over several hundred kilometers (dip direction) in these cases (cf. Reid and Dorobek, 1991, 1993). These simple models that relate flexure to carbonate platform width remain to be tested by detailed, high-resolution, chronostratigraphic studies of carbonate platforms that develop on the distal side of foreland basins.

Flexure and its Effects on Depositional Gradient.—

In many sedimentary basins, antecedent topography greatly affects the morphology of carbonate platforms. Local antecedent topographic highs may become preferred sites for accumulation of shoal water facies; topographic lows must be filled with pelagic or allochthonous sediment before carbonate platform facies can prograde over them. While we do not fully understand the causal relationships between antecedent topography and carbonate platform morphology (cf. Read, 1985; Schlager, 1992), it is apparent that carbonate ramps probably cannot form where regional, antecedent de-

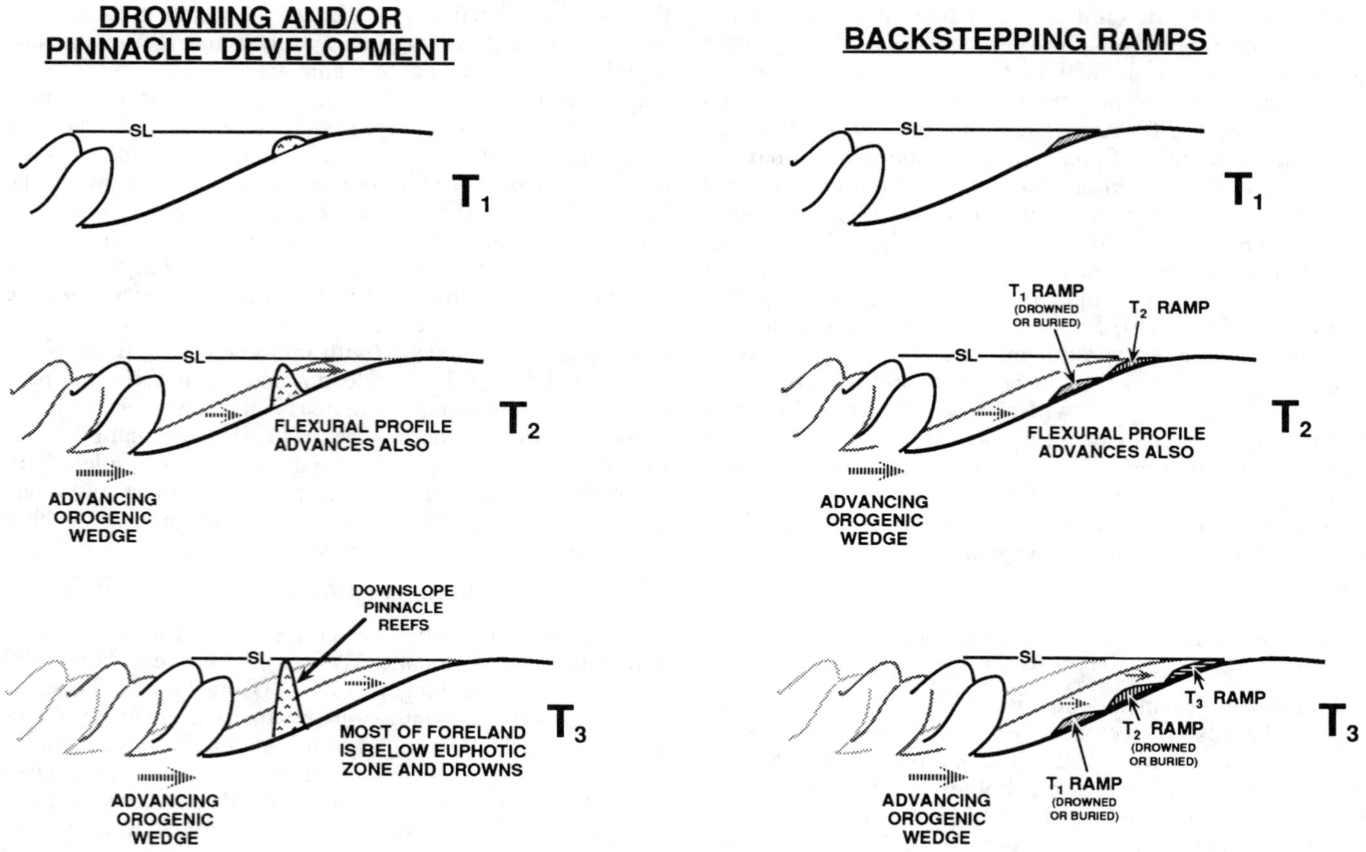

DROWNING AND/OR PINNACLE DEVELOPMENT

BACKSTEPPING RAMPS

FIG. 1.—Conceptual models for foreland basin carbonate ramps and buildups during active convergence. As an orogenic wedge advances cratonward, the flexural profile produced by this load also migrates cratonward; eustatic sea level does not change during migration of the flexural profile in these conceptual diagrams. Carbonate ramps on the distal side of the basin respond to ever increasing subsidence rates by backstepping or drowning. Carbonate reefs/mounds respond by forming pinnacle geometries or by drowning.

positional gradients are greater than 1°. This is true simply because carbonate ramps, by definition, have no significant break in slope along their profiles. A preexisting break in slope produced by normal faulting or submergence of an older platform with a steep margin would result in a carbonate platform with much steeper margin-to-basin profiles (e.g., distally steepened ramp, rimmed shelf, or isolated platform; cf. Read, 1985), not the homoclinal ramps that typify foreland basins.

Figure 2 shows that for foreland lithosphere with moderate to high flexural rigidity, initial depositional gradients are less than 1° along most or all of the basin's profile. Thus, if depositional gradient is a dominant control on carbonate platform morphology, ramps should typify foreland basins that are underlain by lithosphere with high flexural rigidity. Where foreland lithosphere has very low flexural rigidity, depositional gradients exceed 1° within a few tens of kilometers basinward of the crest of the peripheral bulge. In fact, the 100-m isobath may intersect the flexural profile where initial depositional gradients are greater than 1°, especially if the foreland plate initially was *above* sea level, not *at* sea level as shown for the pre-loading position of sea level in Figure 2. Thus, steeper depositional gradients in these foreland basins might exclude broad carbonate ramps from developing.

The predominance of homoclinal ramps in foreland basins is even more notable given that basinward-dipping, normal faults (with local antithetic normal faults) with up to 1 km of dip-slip displacement have been identified on the distal side of some foreland basins (Bradley and Kidd, 1991). These faults may be the result of extension of the crust in the vicinity of the peripheral bulge where bending stress and extensional strain are greatest. Development of homoclinal ramps in these settings suggests several possibilities: (1) significant dip-slip displacement on normal faults only develops far basinward of the 100-m isobath and has little effect on shallow carbonate deposition on the peripheral bulge, or (2) structural topography produced by dip-slip displacement is constantly "smoothed out" by carbonate sedimentation. The first case is more likely, as suggested by the location of extensional faults in several modern foredeeps (Bradley and Kidd, 1991). In addition, stratigraphic evidence for synorogenic extensional faulting (e.g., abrupt thickness or facies changes across faults) are not commonly observed in carbonate platforms that develop on the distal side of foreland basins.

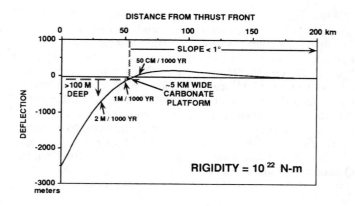
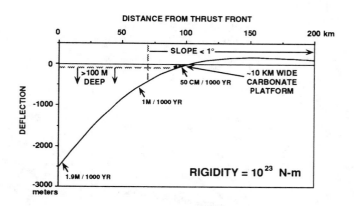
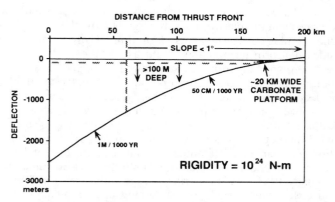
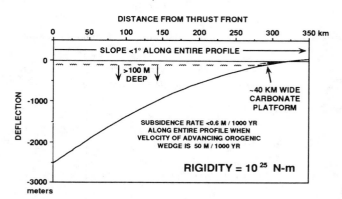

Fig. 2.—Initial flexural profiles for water-filled foredeep, showing area of optimal carbonate deposition. Model flexural profiles were produced using a 5 km high orogenic wedge and low (10^{22} N-m) to high (10^{25} N-m) flexural rigidities; flexural equation used to construct these profiles is from Speed and Sleep (1982, Equation 3a). Note differences in form of individual flexural profiles. For purposes of discussion (see text), the foreland plate is assumed to be exactly at sea level prior to loading; the horizontal line at 0 meters represents the original position of plate before loading by the orogenic wedge. Depositional gradients, 100-m isobath, and predicted widths of foreland carbonate platforms are shown for each flexural rigidity. "Instantaneous" subsidence rates for several points along each profile are indicated by arrows and are expressed as m/1000 yr or cm/1000 yr. Figure 3 shows how subsidence rate changes over time at the point where sea level intersects the initial flexural profile.

Effects of a Migrating Flexural Profile on Subsidence Rates.—

As an orogenic wedge advances cratonward, the flexural profile of the foreland plate advances with it, and the zone of optimum carbonate production must also "backstep" or migrate cratonward. Backstepping, synorogenic ramp successions record the response of carbonate facies to a migrating flexural profile, and many synorogenic ramp successions exhibit this behavior (Fig. 1).

Tectonically forced backstepping can be examined more quantitatively by looking at the change in subsidence rate for any fixed point located basinward of the peripheral bulge as an orogenic wedge advances. Figure 3 shows the change in subsidence rate for the point where sea level intersects the initial flexural profiles from Figure 2; model results for three different rates of wedge advancement (constrained by measured rates of convergence in modern convergent margins; cf. Covey, 1986) and four different plate rigidities are shown.

These figures illustrate how carbonate aggradation rates must keep pace with ever increasing subsidence rates in order for a carbonate platform to maintain a fixed position within the basin. Increasing subsidence rate probably is the dominant control on synorogenic carbonate sedimentation in foreland basins and explains why most synorogenic platforms and/or reefs either drown (*sensu stricto*, Schlager, 1981) or backstep during convergence (Fig. 1). It also probably explains why carbonate ramp morphologies are maintained on the distal side of the foredeep. Aggradation and progradation across ramp facies belts cannot outpace the ever increasing subsidence rates during active convergence, thus preventing ramp profiles from steepening and evolving into rimmed shelves. Examples of backstepping, synorogenic, foreland carbonate ramps include: Middle Ordovician carbonates, Virginia and Tennessee (Read, 1980; Ruppel and Walker, 1984); Mississippian Lodgepole Formation, Montana (Elrick and Read, 1991); Mississippian Fayetteville Limestone, Arkansas (Handford, 1993); Upper Pennsylvanian to Wolfcampian strata, Val Verde Basin and

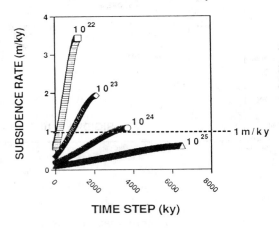

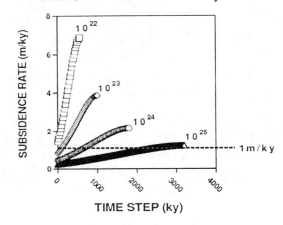

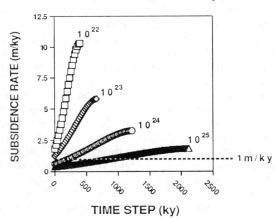

Fig. 3.—Change in subsidence rate over time at the point where sea level intersects the initial flexural profile (see point where sea level intersects each flexural profile shown in Fig. 2). As an orogenic wedge advances cratonward, the subsidence rate at this point increases over time. The curves show model results for three different velocities of wedge advance and four different rigidities of the foreland plate; the orogenic wedge is assumed to be 5 km high in all cases. Upper right end of all curves indicates when the frontal thrust of the orogenic wedge arrives at the initial crossover point of sea level and the initial flexural profile. The dotted horizontal line indicates the probable minimum subsidence rate needed to cause "flexural drowning" (1 m/1000 yr). For low to moderate plate rigidities, flexural drowning is generally possible within one million years of initial convergence, even with low to moderate rates of wedge advancement (uppermost and middle graphs). For high plate rigidities, termination by siliciclastic pollution probably occurs before flexural drowning, even with very high rates of wedge advancement (lowermost graph).

Marathon region, West Texas (Ross and Ross, 1985; Yang and Dorobek, this volume); Cretaceous ramps, Tremp Basin, Spanish Pyrenees (Simo, 1989); Paleogene strata, South Pyrenean foreland basin (Luterbacher and others, 1991); Eocene Nummulitic Limestone, Switzerland (Allen and others, 1991); and Miocene carbonate platforms, northwest Sabah area (Hinz and others, 1989). Examples of drowned synorogenic platforms, reefs, and mounds in foreland settings include: the Proterozoic Kimerot[2] platform, Northwest Territories, Canada (Grotzinger and McCormick, 1988); Middle Ordovician mounds, Virginia (Read, 1982); and Quaternary pinnacle reefs, Sahul Shelf, northwest Australia (Purser, 1983). "Flexural drowning" is discussed in more detail below.

Flexural Drowning of Foreland Carbonate Platforms

Drowning (*sensu stricto*) of carbonate platforms occurs when relative sea-level rise outpaces carbonate accumulation rates and the platform is submerged below the euphotic zone (Schlager, 1981). This definition of drowning excludes platform termination by influx of siliciclastic sediment. Drowned platforms are recognized in the record where shallow carbonate platform deposits are abruptly overlain by deep marine facies. Marine hardgrounds or corroded/mineralized surfaces typically separate drowned, shallow platform carbonates from overlying, deep water facies. Thus, true drowning often is recorded as a surface of non-deposition and can be differentiated from a platform that has been terminated by siliciclastic "pollution" (Kendall and Schlager, 1981). This criterion is especially useful in foreland basins where both true drowning and termination of carbonate platforms by siliciclastic influx are possible.

Schlager (1981) described drowned carbonate platforms as "paradoxical" in that carbonate sedimentation rates typically outpace eustatic sea-level rise and long-term subsidence rates for thermally subsiding basins (i.e., passive

---

[2]The Kimerot platform was a pre-orogenic, passive margin carbonate platform that was drowned during earliest stages of flexural subsidence in the Kilohigok Basin, Northwest Territories, Canada. Continued convergence ultimately resulted in development of the synorogenic Beechey carbonate ramp on the distal side of the Bear Creek Foredeep.

margins, rift basins). Foreland basins, however, may have subsidence rates that are up to two orders of magnitude faster than thermally subsiding basins (Fig. 3; Angevine and others, 1990). Foreland basin subsidence rates also may be several orders of magnitude greater than either accommodation produced by glacio-eustatic rise or aggradation rates of Holocene reefs (cf. Bosscher and Schlager, 1993).

Schlager (1992) estimated that the average growth potential of carbonate platforms in the 10^3 to 10^5 yr time range is on the order of 30 cm to 1m/ky. If growth potential is exceeded by an increase in accommodation over this time range, drowning is likely to occur. Thus, when subsidence in a foreland basin exceeds 1 m/ky (indicated by the dashed lines on Fig. 3) for 10^3 to 10^5 yr, most foreland carbonate platforms and reefs will be drowned or forced to backstep. If eustatic sea level falls at the same time as active convergence in a foreland basin and counteracts the flexurally driven subsidence, it must fall fast enough to outpace the increase in accommodation produced by subsidence. During active convergence, foreland basin subsidence at any point basinward of the peripheral bulge continues to increase even after the critical drowning rate of 1 m/ky has been reached. This readily explains why most synorogenic, foreland carbonate successions backstep or are drowned.

It is important to note that "flexural drowning" is possible without any accompanying eustatic rise in sea level or in the absence of some other environmental stress on carbonate-producing benthic organisms. If even a low amplitude eustatic sea-level rise or moderate change in basin-wide salinity or nutrient level occurs at the same time as active convergence, drowning is even more likely.

Interaction of Flexure and Eustatic Sea Level

As a first approximation, flexure and eustatic sea level combine to determine the areal distribution of foreland carbonate platforms. The initial flexural response of the lithosphere to vertical loading, regardless of the geodynamic model chosen, is instantaneous. Thus, lithospheric flexure and eustatic sea level determine water depth at any point across a foreland basin at the time of initial loading. Thereafter, further convergence and synorogenic sedimentation in the foredeep will constantly modify the basin profile (cf. Flemings and Jordan, 1989, 1990; Sinclair and others, 1991).

Ignoring any other external factors (e.g., siliciclastic/nutrient influx, climatic change, salinity variations) that might affect carbonate sedimentation, theoretical foreland basin profiles show how flexural rigidity of the foreland plate and eustatic sea level may determine the initial width of synorogenic carbonate ramps (Fig. 2). If we first assume that prior to flexural loading, the foreland plate is exactly at sea level, we can see how the geometry of the peripheral bulge determines where the initial strandline position is relative to the front of the orogenic wedge. Higher rigidities for the foreland lithosphere cause the strandline on the distal side of the foredeep to be located farther from the orogenic wedge. Initial depositional gradients also will be less than 1° for all of the distal part of the flexural profile, producing a wide zone where shallow water (<100 m water depth) carbonate sedimentation is possible.

If eustatic sea level is high, the peripheral bulge may be completely submerged and no subaerial unconformity will develop across the bulge. If eustatic sea level is high enough that the crest of the peripheral bulge is at water depths greater than 100 m, the probable maximum water depth for *in situ* carbonate sedimentation (Schlager, 1981; Hallock and Schlager, 1986), no carbonate sediments will be deposited, and the entire foreland area will be drowned (*sensu stricto*). The Sahul Platform off northwestern Australia (Fig. 15) is an example of a modern, drowned foreland ramp where even the crest of the peripheral bulge is below water depths in excess of 100 m (Purser, 1983).

If eustatic sea level is high enough to submerge the crest of the peripheral bulge to only a few-meters water depth, broad carbonate ramps may develop across the entire, shallow "back-bulge" basin on the cratonward side of the peripheral bulge (Pelechaty and Grotzinger, 1988; Flemings and Jordan, 1989; Goebel, 1991; Giles and Dickinson, this volume), as well as extend into the foredeep. The basinward limit of carbonate ramps in the foredeep can be determined by locating the point along the flexural profile where water depths exceed 100 m (this maximum water depth was chosen for reasons discussed earlier). Carbonate sediment derived from shallow water, inner ramp environments may be transported basinward into water depths over 100 m by storms, storm-generated turbidity currents, or other sediment gravity flow mechanisms, but this seems to be a reasonable upper limit for maximum water depth where autochthonous carbonate sedimentation occurs on most carbonate ramps (cf. Burchette and Wright, 1992). Finally, flexural subsidence in the back-bulge basin will be slow enough that there will not be significant depositional gradients produced by flexure. Carbonate facies probably will keep pace with subsidence in the submerged back-bulge basin, and depth-controlled differentiation of carbonate facies tracts will be minimal. Thus, identification of back-bulge basins may only be recognizable on isopach maps.

In contrast, if eustatic sea level is low, the peripheral bulge may be subjected to significant subaerial erosion (cf. Mussman and Read, 1986; Robertson, 1987). The shallow "back-bulge" basin may be completely emergent and the strandline on the foredeep side of the bulge may be located far basinward along the flexural profile. This will result in a more areally limited zone of carbonate sedimentation as indicated by the simple flexural profiles shown in Figure 2.

Viscoelastic Relaxation, the Peripheral Bulge, and Effects on Carbonate Sedimentation

Deciphering chronostratigraphic relationships of carbonate strata on or flanking the peripheral bulge is likely to be difficult. When the orogenic wedge is not advancing, viscoelastic relaxation may occur, resulting in an increase in height and basinward migration of the peripheral bulge (cf. Quinlan and Beaumont, 1984; Tankard, 1986). Carbonate strata may be eroded from the crest of a subaerially exposed peripheral bulge during viscoelastic relaxation, as suggested by stratigraphic relationships in several Paleozoic intervals from the Appalachian Basin (Quinlan and Beaumont, 1984; Tankard, 1986). Assuming sea level is below the crest of the peripheral bulge, erosion on the crest of the bulge should expose progressively older strata as relaxation progresses. Migration of the peripheral bulge toward the orogenic wedge during relaxation also may affect local carbonate facies patterns (cf. Ettensohn, 1993) or force carbonate ramps to prograde basinward because accommodation on the basinward flank of the bulge would decrease. If the crest of the bulge is at or above sea level when relaxation begins, accommodation will progressively decrease along the basinward side of the peripheral bulge during relaxation, forcing ramp progradation toward the basin. An important stratigraphic signature of this "tectonically forced progradation" would be a time-equivalent, basinward shift in the updip limit of progressively younger ramps. Again, the behavior of the peripheral bulge during relaxation and its effects on carbonate sedimentation remain to be tested by high-resolution studies of carbonate strata.

There are many caveats to this generalized view of "bulge behavior" that have important implications for interpreting the stratigraphy on the peripheral bulge. Eustatic sea-level changes during relaxation might greatly complicate chronostratigraphic relationships on the peripheral bulge. If the peripheral bulge remains submerged during viscoelastic relaxation, carbonate sedimentation is likely to keep pace with any relative sea-level changes on the crest of the bulge (except during a high amplitude, short-term eustatic rise that might drown carbonate facies on the bulge). Unconformities may not develop, and changes in relative sea level caused by relaxation may be recorded instead by subtle, diachronous changes in lithofacies. In addition, the time scales at which viscoelastic relaxation operates, if it even does occur, are poorly understood. Estimates for time constants of lithospheric relaxation are on order of 10^6–10^7 yr (Quinlan and Beaumont, 1984; Schedl and Wiltschko, 1984), which may be longer than the actual amount of time represented by shallow water carbonate strata on the peripheral bulge. Detailed sedimentologic and high resolution chronostratigraphic studies are needed to test the actual behavior of the peripheral bulge during foreland basin evolution. Carbonate facies on the peripheral bulge, because they respond almost instantaneously to relative sea-level changes, may be ideally suited to test these concepts.

Non-flexural Deformation in Foreland Areas and Its Effects on Synorogenic Carbonates

Non-flexural deformation can significantly affect carbonate facies patterns over broad regions of the foreland. Structures of all scales can develop in foreland areas at the same time as shortening in the orogenic wedge. Styles of deformation vary from broad, regional folding to fault-bounded uplifts with over 5 km of structural relief from the top of the uplift to the adjacent basinal area. Sea floor topography produced by non-flexural foreland deformation has a great effect on carbonate sedimentation patterns, as discussed below.

Non-flexural deformation across foreland areas may be related to either forces applied at the base of the foreland lithosphere or compressional forces that are transmitted from plate boundaries into interior parts of the foreland plate. It is important to remember, however, that we still do not have a complete understanding of non-flexural deformation of foreland areas. Identifying the process(es) responsible for non-flexural deformation is a difficult problem, and similar stratigraphic relationships and structural features may result from very different processes. Several examples of non-flexural deformation and its effects on carbonate sedimentation are described later and illustrate how difficult it is to identify the sources and orientation of the stresses responsible for deformation.

Sublithospheric Processes.—

Sublithospheric processes that occur beneath a foreland plate during convergence may result in broad deformation across the foreland. In some retroarc foreland basins (*sensu* Dickinson, 1977), the dip angle of the subducted slab may affect deformation patterns in the overlying foreland plate. Shallowing of the subducting Nazca plate beneath Chile and Argentina at approximately 17 Ma may be responsible for broad areas of basement deformation across the Andean foreland (Jordan and others, 1983; Reynolds and others, 1990); the basement deformation is interpreted to be the result of the interaction between flat segments of the subducting Nazca plate with the overlying South American plate. The long-term dynamics of mantle convection beneath a foreland plate and viscous drag at the base of the foreland lithosphere also might affect foreland deformation (Mitrovica and Jarvis, 1985; Mitrovica and others, 1989), but the stratigraphic signature of these processes will be difficult to identify in the geologic record.

In-plane Stresses and Foreland Deformation.—

Convergence along a plate margin produces horizontal in-plane stresses that may be transmitted many hundreds of kilometers inboard of the actual thrust load and proximal foredeep (Zoback, 1992). Deformation of the foreland as a result of in-plane stress may occur at distances that are far beyond those that are normally attributable to simple flexure of continental lithosphere (Lambeck and others, 1984; Ziegler, 1987; Cloetingh, 1988). There basically are two types of foreland deformation that may result from the transmission of in-plane stresses across a foreland plate: (1) changes in preexisting flexural deflections, and (2) reactivation of preexisting structures in the foreland plate.

For simplification, in-plane stresses generally are ignored in quantitative or conceptual models of foreland basin evolution or foreland stratigraphy. Recent quantitative models, however, indicate that fluctuations in the magnitude of in-plane stress can enhance or subdue flexural deflections of the lithosphere, which in turn can affect sedimentation far from the zone of active convergence (e.g., Karner, 1986; Cloetingh, 1988; Peper, 1994; Peper and others, this volume). In general, an increase in compressive in-plane stress should induce greater foreland basin subsidence and greater uplift of the peripheral bulge, causing shoreline regression along the flank of a subaerial peripheral bulge. Conversely, tensile in-plane stress will result in basin uplift and subsidence of the peripheral bulge. Accurately assigning transgressive-regressive stratigraphic sequences in foreland basins to fluctuations in the magnitude of in-plane stress, however, requires that the inverse subsidence relationships between the basin center and peripheral bulge can be identified in well-preserved strata that contain high-resolution isochrons. The level of chronostratigraphic resolution needed to test this model generally does not exist for most foreland basin stratigraphies.

In-plane stresses produced during convergence also might reactivate preexisting structures in the foreland plate (Ziegler, 1987; Heller and others, 1993; Karner and others, 1993). Deformation and reactivation of preexisting basement faults may occur over hundreds of kilometers inboard from the proximal foredeep, far beyond the effects of plate flexure (Ziegler, 1987). The deformation may develop at high angles to the orientation of the orogenic wedge and foredeep, further evidence that it is not related entirely to flexure (cf. Dorobek and others, 1991a, b). If oblique convergence occurs or if there are prominent salients along the plate boundary, collision will be diachronous, and the stress field within the foreland plate will evolve over time (Bradley, 1989). Complex tectonic inversion may result across a foreland area as reactivated faults respond to the evolving stress field (Letouzey, 1986; Dorobek and others, 1991a, b; Karner and others, 1993). Detailed isopach maps, facies analysis, and seismic reflection data can constrain the timing and geometry of foreland deformation related to reactivation of preexisting basement structures, but resolving the stress field(s) responsible for the deformation requires construction of very detailed kinematic histories. Even with detailed kinematic analyses, unique solutions for the stress field(s) may not be possible.

Examples of Non-flexural Deformation and Contemporaneous Carbonate Sedimentation.—

The following examples of non-flexural deformation illustrate the highly variable structures that can form in foreland areas and, in turn, influence sedimentation. Non-flexural deformation may be related to propagation of the orogenic wedge, strike-slip deformation across the foreland, or reactivation of preexisting structures, and may be either thin-skinned or involve basement rocks. Shale/salt diapirs that form in response to loading by progradational siliciclastic strata are a more localized form of non-flexural deformation and can form topographic highs for shallow water carbonate sedimentation.

Reactivation of Basement Structures in the Antler Foreland, Western North America.—

There are many examples in the geologic record where deformation across the foreland is related to reactivation of ancient structures in the foreland plate and apparently is not associated with flexure. As long as some part of the foreland plate is below sea level, carbonate sedimentation is possible, and even subtle differential subsidence can affect carbonate facies distribution over broad areas of the foreland.

An episode of Late Devonian-Early Mississippian convergence, known regionally as the Antler orogeny, affected much of the western margin of North America (Fig. 4; Burchfiel and Davis, 1972; Dickinson, 1977; Speed and Sleep, 1982; Gordey, 1988; Morrow and Geldsetzer, 1988;

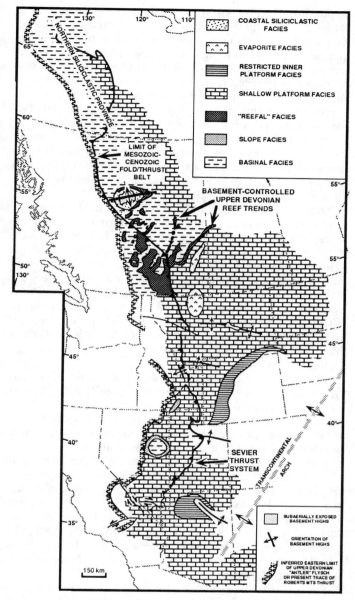

FIG. 4.—Regional Late Devonian paleogeographic relationships of western North America during initial stages of the Antler orogeny (Late Devonian–Mississippian). Note numerous paleotopographic highs and basement uplifts that trend at high angles to the Antler orogenic belt. These reactivated basement structures affected Upper Devonian carbonate facies distribution hundreds of kilometers inboard of the proximal foredeep (e.g., Upper Devonian reef trends in Alberta). After Moore (1988) and Sandberg and others (1988).

Oldow and others, 1989). Devonian and Mississippian carbonate rocks from the western Montana and east-central Idaho portion of the Antler foreland were deposited on a cratonic platform that faced the Antler foredeep to the west (Sandberg and others, 1983; Dorobek and others, 1991a, b; Reid, 1991; Reid and Dorobek, 1991, 1993). Subsidence analyses of the Devonian-Mississippian strata in this area indicate episodic subsidence contemporaneously affected the proximal foredeep and adjacent cratonic platform, an area approximately 800 km wide (palinspastic distance measured perpendicular to the axis of the foredeep; Fig. 5). Isopach maps for this sequence also show that many depocenters and paleohighs were geographically coincident across the foreland although some of these structures were tectonically inverted (i.e., paleohighs became depocenters and vice versa) several times during the 50–60 million years represented by this stratigraphic sequence (Fig. 6). The Devonian-Mississippian foreland structures had cross-sectional wavelengths of 50–200 km and amplitudes of about 50–350 m. Many of these generally east-west trending paleostructures were oriented at high angles to the north-south trending axis of the Antler foredeep and the inferred strike of the Antler orogenic belt.

Structures across the Antler foreland coincide geographically with structural trends produced during either Proterozoic extension and development of the Belt Basin (Winston, 1986) or the much older Great Falls Tectonic Zone (Hoffman, 1989; G. Ross, pers. commun., 1994). This geographic coincidence suggests that ancient faults were reactivated during Antler convergence. Subtle tectonic inversions across the distal Antler foreland in Montana greatly affected carbonate sedimentation by controlling regional onlap patterns onto topographic highs (Dorobek, 1991) and by influencing the location of deep water carbonate mounds in topographic lows (Smith, 1977).

The complex differential subsidence across the Montana-Idaho foreland is not easily attributable to simple flexural models for vertical loading of unbroken elastic plates. Instead, differential subsidence of the foreland may be related to: (1) flexure of mechanically independent, fault-bounded segments of the foreland caused by diachronous, subregional vertical loads, similar to the Apennine foreland (Royden and others, 1987); (2) transmission of compressive in-plane stresses through the foreland lithosphere which reactivated Proterozoic fault systems; (3) waxing and waning of in-plane compressive stresses due to the episodic nature of Antler convergence; and (4) possible sublithospheric processes beneath the Antler foreland plate that involved mantle convection or interaction with the subducted slab (see earlier discussion).

Reactivation of basement structures and tectonic inversion also occurred in western Canada during Antler-age convergence. Late Devonian differential subsidence patterns across western Canada may be the major control on the location of Upper Devonian carbonate platform margins and linear reef trends (Mountjoy, 1980; Ross, 1992; Edwards and Brown, 1993) and isopach/facies patterns in Mississippian carbonates (Brandley and others, 1993). Reactivated basement faults also may have provided conduits for deep, basement-derived(?) brines that caused mineralization of overlying Devonian and Mississippian carbonate strata during later episodes of convergence along the western margin of North America (Mountjoy and Halim-Dihardja, 1991; Dix, 1993) although this remains a controversial topic (Muehlenbachs and others, 1993).

Devonian-Mississippian carbonate strata from the Antler foreland suggest that in settings where the foreland lithosphere is broken by ancient fault systems, the foreland may exhibit complex patterns of differential subsidence that probably reflect a composite response to both vertical and horizontal loads. Similar timing of major subsidence events from both the proximal foredeep and distal foreland provides key evidence that deformation in both regions is genetically related to the same convergence events along the plate margin.

Reactivation of Basement Structures in the Marathon-Ouachita Foreland.—The Marathon-Ouachita orogenic belt developed during Mississippian to early Permian time when Gondwana collided with the southern margin of North America (Kluth and Coney, 1981; Ross, 1986). This complex foreland area consists of sub-basins that are separated by fault-bounded, intraforeland uplifts (Fig. 7). Boundary fault zones consist of complex, steeply dipping faults with variable slip. Uplift of these fault-bounded features occurred at the same time as late Paleozoic shortening in the Marathon-Ouachita orogenic belt, indicating that deformation along the plate margin and in the foreland was mechanically related.

The Central Basin Platform (CBP) is a major uplift in the Permian Basin portion of the Marathon foreland in west Texas (Fig. 8; Yang and Dorobek, this volume). Uplift of the CBP occurred during the late Mississippian to Early Permian when collision apparently reactivated preexisting basement faults (Hills, 1985; Wuellner and others, 1986; Yang and Dorobek, 1992, this volume; Yang, 1993). Right-lateral shear stress caused clockwise rotation of crustal blocks that were bounded by preexisting faults (Fig. 9; Yang and Dorobek, this volume). Clockwise rotation of the blocks, along with some component of east-west directed compression resulted in: (1) formation of steeply dipping reverse faults along the SW and NE corners of blocks and extensional faults at some NW block corners; (2) the large structural relief observed at the SW and NE corners of individual blocks; and (3) decreasing amounts of basement shortening away from thrust-faulted corners that can be documented along both sides of individual blocks.

The variable platform-to-basin relief that was produced during uplift of the CBP resulted in very different patterns of stratigraphic onlap during late Pennsylvanian to early Permian (Wolfcampian) time (Yang, 1993; Yang and Dorobek, this volume). Along extensionally faulted block corners, depositional gradients and topographic relief were much lower than along the SW and NE block corners where there were steep gradients and great topographic relief between the top of the CBP and the adjacent basins. Consequently, upper Pennsylvanian to Lower Permian carbonate ramp facies onlapped the CBP at SE and NW block corners and extended far across the top of the CBP (Fig. 10). In contrast, the lateral extent of onlapping siliciclastic and carbonate facies was much less at the steep, SW and NE block corners (Yang and Dorobek, this volume). By the end of the Wolfcampian, most of the topography on top of the CBP produced during uplift had been buried by onlapping carbonate and siliciclastic strata. The CBP then became a pedestal upon which over 1700 meters of carbonates, evaporites, and siliciclastic platform facies accumulated during the middle to Late Permian (Silver and Todd, 1969; Ward and others, 1986; Frenzel and others, 1988; Hanson and others, 1991).

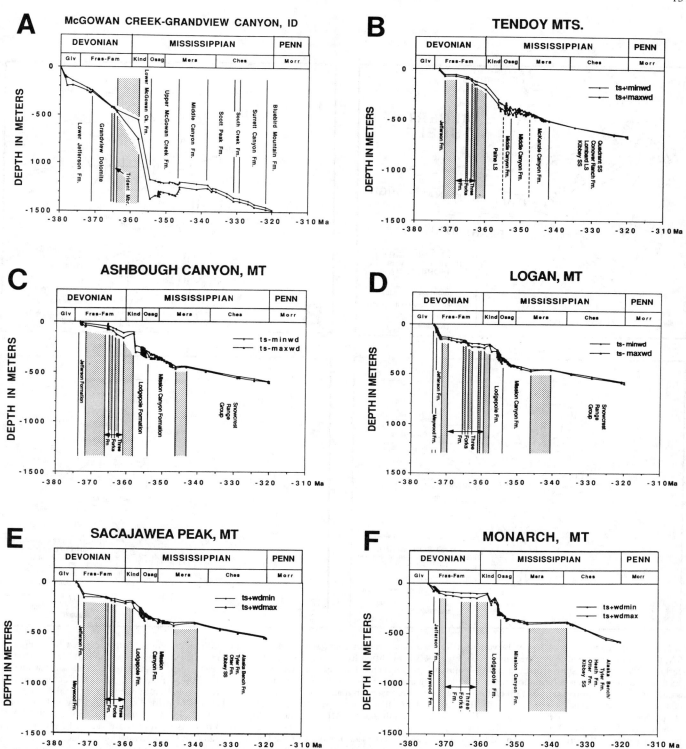

FIG. 5.—Subsidence curves generated for selected localities across the Antler foreland of Montana and Idaho. Unconformities are indicated by stippled intervals. Two subsidence curves are shown and each plot, one generated using corrections for minimum water depth estimates during deposition of each stratigraphic unit ("ts-minwd," upper curve), and the other generated using maximum water depth estimates ("ts-maxwd," lower curve). Most facies were deposited in water depths less than 50 m; therefore, differences in curves due to water depth corrections are negligible. Monarch, Sacajawea Peak, Logan, and Ashbough Canyon are platform localities. Tendoy Mountains section along the Montana-Idaho border is located near the transition from shallow platform to deeper slope environments on the eastern side of the Antler foredeep. McGowan Creek-Grandview Canyon section is the most basinal locality in the Antler foredeep. Note that major inflection points on all curves have similar timing, indicating that deformation across this broad region is mechanically related. See Dorobek and others (1991a, b) for additional discussion of subsidence across the Antler foreland in Montana and Idaho.

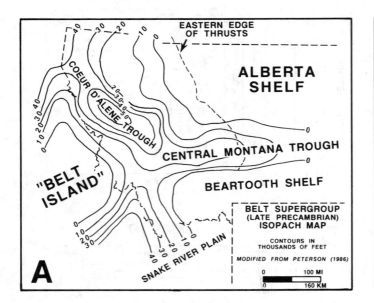
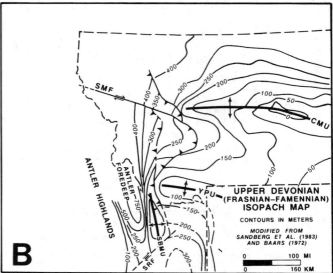
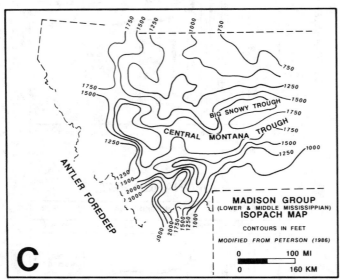
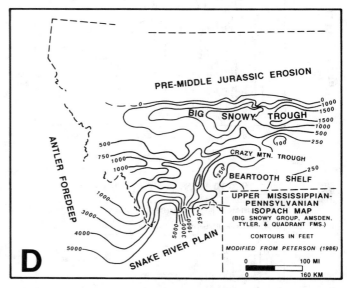

Fig. 6.—Isopach maps of Upper Devonian to lower Pennsylvanian strata across Montana, Idaho, and Wyoming; partly restored base. Figures 6A, 6C, 6D are modified from Peterson (1986); Figure 6B is modified from Sandberg and others (1983). Eastern limit of Laramide (Late Cretaceous-early Tertiary) thrusts shown in Figure 6A. Major paleotopographic features are also indicated on each map. Isopach thicknesses are reasonably accurate because Antler deformation never advanced far enough eastward during Devonian and Mississippian time to significantly deform and/or erode much of the sedimentary section that is contoured in these isopach maps. (A) Isopach map for Proterozoic Belt Supergroup. Several depocenters comprise the "Belt Basin"; note especially the east-west trending Central Montana Trough. (B) Isopach map for Frasnian and lower Famennian strata. SRF = Snake River Fault Zone; SMF = St. Marys Fault Zone; CMU = Central Montana Uplift; YPU = Yellowstone Park Uplift; SBMU = Southern Beaverhead Mountains Uplift. Thrust traces (with sawteeth) delineate major Laramide thrusts; these thrusts are shown here in order to be consistent with the original isopach map of Sandberg and others (1983). Note that the Late Devonian Central Montana Uplift coincides geographically with the Proterozoic Central Montana Trough depocenter shown in Figure 6A. Also note the location of the Late Devonian foredeep depocenter. (C) Isopach map for Lower and Middle Mississippian Madison Group. Note that the Central Montana Uplift which was present across Montana in Late Devonian time (Fig. 6B) underwent tectonic inversion and became the Central Montana Trough in Early to Middle Mississippian time. (D) Isopach map for Upper Mississippian and Pennsylvanian strata. Note the foreland area in Montana still was dissected by numerous paleostructures that were oriented at high angles to the Antler foredeep axis. Also note that the late Mississippian-early Pennsylvanian foredeep depocenter was located much further to the southeast than the Late Devonian foredeep depocenter.

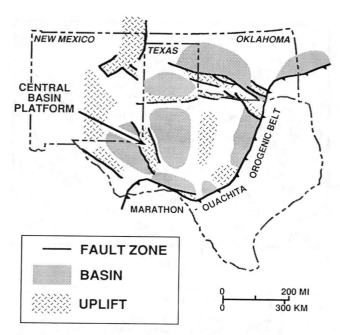

Fig. 7.—Regional base map of the Marathon-Ouachita foreland. Note partitioning of the Marathon-Ouachita foreland into a series of late Paleozoic sub-basins that are separated by high relief, fault-bounded uplifts. Modified from McConnell (1989).

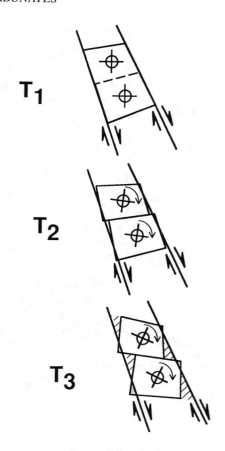

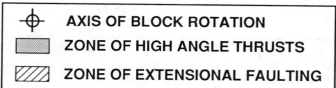

Fig. 9.—Kinematic model for the Central Basin Platform, Permian Basin, West Texas. At T1, preexisting east-west transverse faults might have been cut by a right-lateral shear couple along the boundaries of the CBP. At T2, the CBP split into segmented blocks which then started to rotate in a clockwise direction. Because the boundaries of these blocks were constrained by non-parallel NNW-SSE trending bounding fault zones, shortening and extension due to block rotation occurred at the corners of the blocks. At T3, the blocks continued to rotate. Note that because the CBP is bounded by non-parallel boundaries, the amount of shortening at the SW block corners would be larger than at the NE block corners. See Yang and Dorobek (this volume) for more discussion of this model.

Structural Highs in the Proximal Foredeep.—In some foreland basins, local topographic highs may develop in the proximal foredeep by various deformation mechanisms. These structural highs may become sites for carbonate sedimentation if they are at the proper water depth and if siliciclastic sediment eroded from subaerial parts of the orogenic wedge does not terminate the carbonate benthos.

Diapirs can develop where thick, salt or shale horizons are loaded by siliciclastic sediment shed from the orogenic

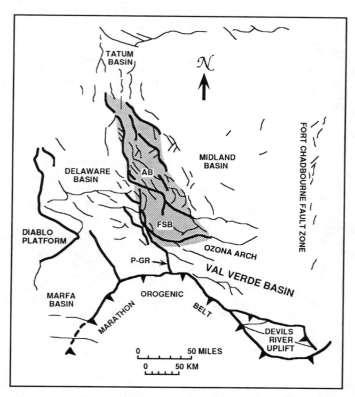

Fig. 8.—Tectonic map for the Central Basin Platform (shaded area) and adjacent basinal areas, Permian Basin, West Texas. Modified from Ewing (1990).

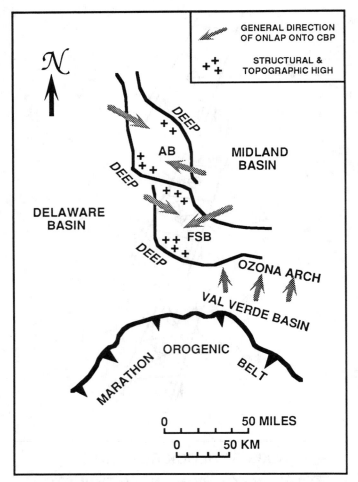

Fig. 10.—Patterns of stratal onlap onto CBP basement topography during Late Pennsylvanian to early Permian time. Note that strata onlap less onto reverse-faulted SW and NE block corners that have the greatest structural and topographic relief.

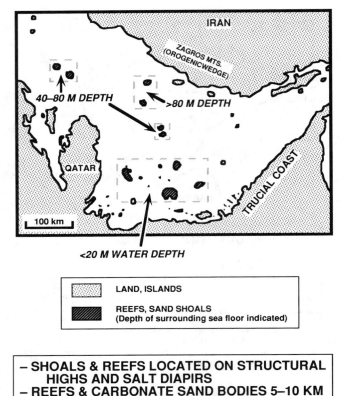

– SHOALS & REEFS LOCATED ON STRUCTURAL HIGHS AND SALT DIAPIRS
– REEFS & CARBONATE SAND BODIES 5–10 KM ACROSS IN *PROXIMAL* FOREDEEP

Fig. 11.—Distribution of carbonate reefs and carbonate shoals in proximal parts of the Persian Gulf foredeep. Depths of the sea floor surrounding these buildups are indicated. Reefs and shoals in the proximal foredeep are locally surrounded by basin floor that is over 80 m below sea level. Most of these buildups are localized on top of salt diapirs that rise from the Hormuz Salt and are 5 to 10 km across. Reef-capped salt or shale diapirs may develop in proximal parts of other foreland basins. Modified from Purser (1973).

wedge. These highs may become preferred sites for carbonate sedimentation, even in the most proximal parts of a foredeep. Recent reefs and skeletal sand shoals in axial parts of the Persian Gulf foredeep locally are over 10 km across and developed on top of salt diapirs that rose from the Hormuz Salt (Fig. 11; Purser, 1973). It is important to note that in ancient stratigraphic sequences from proximal foredeep areas, the record of synorogenic salt diapirism may be poorly preserved if dissolution or subsequent deformation of salt occur. The preserved record of the former reef or shoal-capped salt highs might consist of deformed or brecciated carbonate strata surrounded by thick, deformed proximal foredeep siliciclastic facies.

Local carbonate buildups and isolated platforms also can form where blind thrusts deform overlying strata and produce anticlinal highs on the floor of the proximal foredeep (Fig. 12). Siliciclastic sediment eroded from the orogenic wedge might pond behind the carbonate-capped structural highs or bypass the steep slope on the proximal side of the foredeep and accumulate downdip in axial parts of the ba-

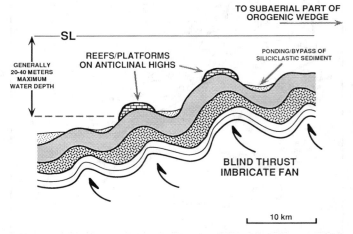

Fig. 12.—Cross-sectional view of a foredeep with carbonate reefs/platforms on top of blind thrust anticlines in proximal parts of the basin. Continued shortening and displacement along the blind thrusts will disrupt stratigraphic relationships that developed in carbonate facies during earlier stages of basin evolution. After Luterbacher and others (1991).

sin. Continued displacement along the blind thrusts can disturb any stratigraphy that develops during periods of limited fault displacement; stratigraphic relationships might be difficult to reconstruct in these cases. Examples of these types of carbonate facies include Paleogene carbonate sand banks and coralgal reefs in proximal parts of the Pyrenean foredeep (Luterbacher and others, 1991), Oligocene carbonate strata in proximal parts of the Taranaki foredeep, New Zealand (King and Thrasher, 1992), and possibly a lower Proterozoic stromatolite reef in the Northwest Territories, Canada (Jackson, 1989).

Localization of carbonate strata on structural highs in proximal parts of foreland basins, regardless of their origin, may produce an ideal juxtaposition of facies for entrapment of hydrocarbons. Permeable carbonate facies surrounded by fine-grained, deep-water siliciclastic sediments of the proximal foredeep might form significant hydrocarbon reservoirs.

OTHER CONSIDERATIONS FOR SYNOROGENIC CARBONATE DEPOSITION

There are other factors that can affect carbonate sedimentation in foreland basins but which are not easily identified from the stratigraphic record.

Orientation of Platform Margin/Reef Facies Relative to an Orogenic Wedge

The direction in which synorogenic foreland carbonate platforms backstep is dependent on the orientation of carbonate platform margin and reef facies tracts with respect to an advancing orogenic wedge. If a platform margin trend is subparallel to the axis of an orogenic wedge and its adjacent foredeep, the entire margin will backstep as the orogenic wedge advances toward the platform (cf. Davies and others, 1989). In contrast, if the platform margin is oriented at a high angle to the orogenic wedge and foredeep axis, backstepping will be highly diachronous (Davies and others, 1989; Pigram and others, 1989).

Synorogenic backstepping will also be highly diachronous if there are prominent salients along the margins of either colliding plate boundary or if oblique subduction occurs along the foreland plate boundary (Fig. 13; cf. Bradley, 1989). This has important implications when interpreting the timing of convergence from a particular foreland basin stratigraphy.

Episodic Convergence

Episodic convergence along a plate margin and attendant, sporadic movement of the orogenic wedge toward the foreland may result in a series of stacked and/or shingled carbonate ramp sequences that prograde from the distal side of the foredeep. During advancement of the orogenic wedge, the migrating flexural profile forces carbonate ramps on the distal side of the foredeep to backstep, but when the orogenic wedge is not advancing, flexural subsidence on the distal side of the foredeep will be driven largely by sedimentation in the basin (Flemings and Jordan, 1989; Sinclair and others, 1991). The amount of flexural subsidence in the distal foreland caused by basin filling is not as great as

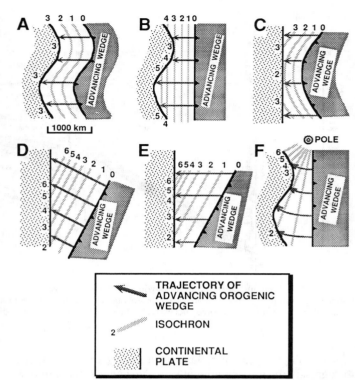

FIG. 13.—Various plate margin geometries and plate trajectories which lead to synchronous or diachronous convergence. Isochron numbers indicate relative timing of collision between the advancing wedge and the foreland plate. This figure suggests that the response of foreland carbonate platforms (not shown on this figure) to flexural subsidence may not be coeval along an entire plate margin, depending on plate geometry and relative plate motions. Modified from Bradley (1989).

that produced by the orogenic wedge. Thus, ramps may prograde basinward from the peripheral bulge when the orogenic wedge is not advancing because flexural subsidence will be less. Basin-filling, siliciclastic sediments also may fill accommodation (*sensu* Van Wagoner and others, 1988) in the basin and provide a substrate for carbonate ramp progradation away from the peripheral bulge. Ramp progradation will be most rapid and will extend further basinward with lesser amounts of accommodation, lower depositional gradients, and higher production of carbonate sediment along the distal side of the foredeep.

A series of backstepping, foreland carbonate ramps formed on the distal side of the Val Verde foreland basin, West Texas, during the Pennsylvanian (Fig. 14; Yang and Dorobek, this volume). The Pennsylvanian Strawn, Canyon, and Cisco formations comprise a succession of carbonate ramp sequences that accumulated on the south-facing side of the Ozona Arch. Each ramp sequence is separated by Pennsylvanian siliciclastic strata derived from the Marathon orogenic belt to the south. Pennsylvanian siliciclastic facies that separate each carbonate interval progressively increase in thickness up-section. In addition, each successive carbonate ramp sequence backsteps and does not extend as far basinward as the underlying ramp sequence (Fig. 14). Each ramp sequence also generally thins updip. The geometric

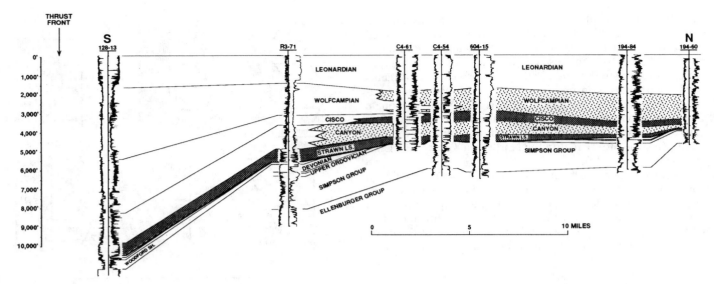

Fig. 14.—Well log cross section showing backstepping, synorogenic carbonate ramps in the Val Verde Basin, West Texas. Note that each successive ramp does not extend as far basinward as the underlying ramp sequence, while at the same time, basinal siliciclastic facies that are equivalent to each ramp become successively thicker. See text for more discussion.

relationships of the siliciclastic and ramp packages suggest progressively greater amounts of *basinward* accommodation during the Pennsylvanian. Given that these units were deposited during a long-term ("second order") fall in sea level (Ross and Ross, 1987), it is even more likely that the Strawn, Canyon, and Cisco ramps record episodic advance of the Marathon orogenic wedge. The successively thicker siliciclastic wedges record increasing basinward accommodation during active convergence, while carbonate ramps prograded into the basin when the orogenic wedge was not advancing (cf. Wuellner and others, 1986).

Effects of Siliciclastic Sedimentation on Foreland Carbonates

Siliciclastic sediments can affect the distribution of carbonate strata in foreland basins during all stages of basin development. The volume, grain size, and dispersal directions of siliciclastic sediment supplied to the foredeep influence carbonate sedimentation. Siliciclastic influx from the orogenic wedge or from cratonic sources can result in complex interfingering of carbonate and siliciclastic mixtures or may terminate carbonate deposition.

Effects of Sediment Type.—

Large volumes of fine-grained siliciclastic sediment probably have a greater effect on carbonate-producing benthos than sandy or coarser grained siliciclastic sediment because fine-grained sediments may remain suspended and might be dispersed over much of a foreland basin. Fine-grained siliciclastic sediment increases the turbidity of the water column and has a more deleterious effect on suspension-feeding organisms than coarse-grained siliciclastic sediment (Adey, 1977; Lighty and others, 1978). Carbonate strata with admixtures of siliciclastic sand and interlayered sandstone are common in some foreland basins because many suspension-feeding organisms are not affected greatly by sand-sized material, regardless of its composition.

Influence of Climate and Bedrock Lithology on Siliciclastic Sediment.—

As the grain size of siliciclastic sediment affects carbonate sedimentation, it follows that climate and bedrock geology in turn affect the type of siliciclastic sediment supplied to foreland basins. Orogenic wedges or cratonic areas that are comprised of crystalline rocks and that are located in humid settings with high weathering rates might supply abundant clay and silt to either side of a foreland basin. This fine-grained sediment is more effective at terminating carbonate sedimentation than coarse, sandy sediment.

In contrast, orogenic wedges consisting of deformed passive margin strata (e.g., platform carbonates and quartz arenites) that are subaerially exposed under arid climatic settings might supply only carbonate clasts or recycled quartz sand to the foredeep. This coarser grained sediment might have little effect on benthic, carbonate-producing organisms.

Siliciclastic Sedimentation and Stage of Foredeep Evolution.—

Siliciclastic sediments will have less of an effect on carbonate platforms on the distal side of foreland basins during early, "underfilled" stages of basin development because the proximal foredeep acts as a trap for siliciclastic sediment. Conversely, during later stages of basin development when foreland basins are "overfilled" with sediment or have reached a "steady state" balance between subsidence, wedge advancement, and basin-filling (cf. Covey, 1986), siliciclastic sediments are dispersed across the entire basin and will exclude carbonate sedimentation even in distal foreland areas.

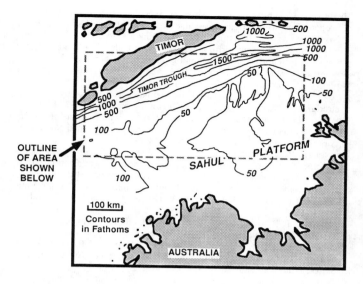

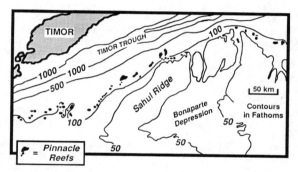

Fig. 15.—Generalized map of the Sahul Platform, northwest Australia (distal foreland basin) and Timor Trough (proximal foredeep). The Sahul Platform is a drowned foreland carbonate platform. The Sahul Ridge is the crest of the peripheral bulge. Immediately downdip from the peripheral bulge, large pinnacle reefs (up to 20 km across) built upwards from sea floor that is now over 100 fathoms (200 m) deep. Although these reefs are now drowned, the cause of the drowning is uncertain and could be related to either post-Pleistocene sea-level rise or flexure. Pinnacle development must have started before the post-Pleistocene sea-level rise because the surrounding sea floor would have been about 50 fathoms (100 m) deep during the Pleistocene lowstand at about 18,000 yr ago. Modified from Purser (1983).

Orographic Effects on Siliciclastic Sediment Dispersal and Foredeep Evolution.—

The position of an orogenic belt relative to trade winds can affect rates of erosional unloading which, in turn, may influence the tectonic evolution of the orogenic wedge, flexure in an adjacent foredeep, and the volume and type of siliciclastic sediment that fills the foredeep (Hoffman and Grotzinger, 1993). Recent geophysical models (Koons, 1989; Beaumont and others, 1992; Johnson and Beaumont, this volume) have attempted to integrate the effects of orographic precipitation with the tectonic evolution of the orogenic wedge and foredeep and mass flux between the orogenic wedge and the foredeep. Hoffman and Grotzinger (1993) suggest that erosion rates at windward mountain fronts may be sufficient to reduce the topographic load (i.e., the orogenic wedge) and thus the amount of flexural subsidence in the adjacent foredeep. Windward mountain fronts might be bordered by an "overfilled" foredeep. In contrast, low erosion rates at leeward mountain fronts might produce "underfilled" foredeeps because erosion does not occur fast enough to denude the topographic load; lithospheric flexure is greater because of the increased topographic load.

Orographic precipitation therefore might indirectly affect carbonate sedimentation in foreland basins by influencing dispersal of siliciclastic sediment across the foredeep. Windward mountain fronts and overfilled foredeeps might exclude carbonate sedimentation on the distal side of the foredeep, whereas leeward mountain fronts might have thick carbonate accumulations because siliciclastic sediments do not prograde across the foredeep for a large part of the basin's history. These concepts have yet to be tested by studies of foreland basin carbonates.

The Persian Gulf: A Case Example.—

The modern Persian Gulf illustrates how climate, bedrock geology, basin bathymetry, and basement structure influence carbonate and siliciclastic sedimentation. The Persian Gulf is an asymmetric foreland basin produced by subduction of the Arabian plate beneath Iran. Siliciclastic sediment is supplied to the basin by the Tigris and Euphrates rivers in the northwestern end of the basin and from the Zagros Mountains along the northeast side of the basin. Sediment from the Tigris and Euphrates rivers is largely trapped in fault-bounded grabens in the northwestern end of the Persian Gulf (Purser and Seibold, 1973). Siliciclastic sediment is also eroded from the Zagros Mountains, but most of this sediment is trapped in the adjacent proximal foredeep and has little effect on the carbonate ramp along the distal (southwestern) side of the Persian Gulf. In contrast, there are few rivers and streams on the Arabian side of the Persian Gulf because of the extremely arid climate of the region. Thus, little siliciclastic sediment is supplied along the Arabian side of the Persian Gulf, and carbonate ramp facies are deposited there.

CONCLUSIONS

Marine carbonate strata from foreland basins may provide a better record of basin evolution than siliciclastic strata because they respond essentially instantaneously to changes in basin floor topography and basin subsidence. This paper is an attempt to relate carbonate stratigraphy and platform evolution to the important variables that are inherent to foreland basins. Detailed aspects of the conceptual models presented here and in other papers on foreland basin evolution, however, remain to be tested by high resolution stratigraphic and sedimentologic studies of foreland stratigraphy.

Nevertheless, some important conclusions are:

1. Synorogenic carbonate platforms typically form on the distal side of foreland basins and have ramp-like profiles. The ramp profile probably mimics the flexural profile produced by an orogenic wedge. Long-term (i.e., 1–10 my time scales) maintenance of the ramp profile requires that there is limited aggradation across facies

belts that comprise foreland ramps. This probably reflects continued convergence, migration of the flexural profile, and the inability of ramp facies tracts to keep pace with increasing subsidence rates during convergence.

2. Synorogenic platforms and reefs are forced to backstep or are ultimately drowned as an orogenic wedge advances cratonward. This probably is the response of carbonate platforms and reefs to ever increasing rates of flexural subsidence although dispersal of siliciclastic sediments, eustatic sea-level rise, or some other environmental stress also may contribute to backstepping/drowning.

3. Facies patterns in synorogenic platforms also can be affected by non-flexural deformation of the foreland plate. Structural styles can be highly variable and may be unique to a particular foreland setting. Reactivation of ancient, preexisting structural grain is common in many foreland areas and can have a dramatic affect on carbonate facies distribution. It is essential to characterize the types of structures found in foreland basins and their kinematic history in order to understand carbonate facies distribution and platform evolution.

ACKNOWLEDGMENTS

Various research projects that have led to this paper were supported by the U.S. Department of Energy, Basic Energy Sciences Program (Grant DE–FG05–87ER13767); The Petroleum Research Fund, administered by the American Chemical Society (Grants #19519–G2 and #23269-AC); the National Science Foundation (Grant #EAR-91–17935); and the Center for Energy and Mineral Resources at Texas A&M University. Numerous oil companies have supported various aspects of studies that generated data for this paper, but I would especially like to acknowledge Ron Genter at Chevron USA for providing significant amounts of seismic data for our studies of the Permian Basin. I would also like to acknowledge Steve Reid and Kenn-Ming Yang, two of my former students whose respective Ph.D. research projects contributed much to this paper. Bob Stanton, Texas A&M University, provided insights into the complex interactions between benthic organisms and siliciclastic sediments. Kenn-Ming Yang also assisted with the flexural modeling. Comments provided by John Grotzinger, Fred Read, and Gerry Ross greatly improved the manuscript.

REFERENCES CITED

ADEY, W. H., 1977, Shallow water Holocene bioherms of the Caribbean Sea and West Indies, in Taylor, D. L., ed., Proceedings: Third International Coral Reef Symposium: Miami, University of Miami, v. 2. p. xxi–xxiv.

ALLEN, P. A., CRAMPTON, S. L., AND SINCLAIR, H. D., 1991, The inception and early evolution of the North Alpine Foreland Basin, Switzerland: Basin Research, v. 3, p. 143–163.

ANGEVINE, C. L., HELLER, P. L., AND PAOLA, C., 1990, Quantitative Sedimentary Basin Modeling: Tulsa, American Association of Petroleum Geologists, Continuing Education Course Note Series #32, 133 p.

BEAUMONT, C., 1981, Foreland basins: Geophysical Journal of the Royal Astronomical Society, v. 65, p. 291–329.

BEAUMONT, C., FULLSACK, P., AND HAMILTON, J., 1992, Erosional control of active compressional orogens, in McClay, K. R., ed., Thrust Tectonics: New York, Chapman and Hall, p. 1–18.

BEAUMONT, C., QUINLAN, G., AND HAMILTON, J., 1987, The Alleghanian orogeny and its relationship to the evolution of the Eastern Interior, North America, in Beaumont, C., and Tankard, A. J., eds., Sedimentary Basins and Basin-forming Mechanisms: Calgary, Canadian Society of Petroleum Geologists Memoir 12, p. 425–445.

BOSSCHER, H. AND SCHLAGER, W., 1993, Accumulation rates of carbonate platforms: Journal of Geology, v. 101, p. 345–355.

BRADLEY, D. C., 1989, Taconic plate kinematics as revealed by foredeep stratigraphy, Appalachian orogen: Tectonics, v. 8, p. 1037–1049.

BRADLEY, D. C. AND KIDD, W. S. F., 1991, Flexural extension of the upper continental crust in collisional foredeeps: Geological Society of America Bulletin, v. 103, p. 1416–1438.

BRANDLEY, R. T., THURSTON, J., AND KRAUSE, F. F., 1993, Basement-tectonic control on carbonate ramp deposition: Lower Carboniferous Mount Head Formation, southwest Alberta, southeast British Columbia, in Ross, G. M., ed., Alberta Basement Transects Workshop: Vancouver, LITHOPROBE Report #31, University of British Columbia, p. 98–118.

BURCHETTE, T. P. AND WRIGHT, V. P., 1992, Carbonate ramp depositional systems: Sedimentary Geology, v. 79, p. 3–57.

BURCHFIEL, B. C. AND DAVIS, G. A., 1972, Structural framework and evolution of the southern part of the Cordilleran orogen, western United States: American Journal of Science, v. 272, p. 97–118.

CLOETINGH, S., 1988, Intraplate stresses: a new element in basin analysis, in Kleinspehn, K. L., and Paola, C., eds., New Perspectives in Basin Analysis: New York, Springer-Verlag, p. 205–230.

COVEY, M., 1986, The evolution of foreland basins to steady state: evidence from the western Taiwan foreland basin, in Allen, P., and Homewood, P., eds., Foreland Basins: Oxford, International Association Sedimentologists Special Publication 8, p. 77–90.

DAVIES, P. J., SYMONDS, P. A., FEARY, D. A., AND PIGRAM, C. J., 1989, The evolution of the carbonate platforms of northeast Australia, in Crevello, P. D., Wilson, J. L., Sarg, J. F., and Read, J. F., eds., Controls on Carbonate Platform and Basin Development: Tulsa, Society of Economic Paleontologists and Mineralogists Special Publication 44, p. 233–258.

DICKINSON, W. R., 1977, Paleozoic plate tectonics and the evolution of the Cordilleran continental margin, in Stewart, J. H., Stevens, C. H., and Fritsche, A. E., eds., Paleozoic Paleogeography of the Western United States: Los Angeles, Pacific Section, Society of Economic Paleontologists and Mineralogists, Pacific Coast Paleogeography Symposium 1, p. 137–155.

DIX, G. R., 1993, Patterns of burial- and tectonically controlled dolomitization in an Upper Devonian fringing-reef complex: Leduc Formation, Peace River Arch area, Alberta, Canada: Journal of Sedimentary Petrology, v. 63, p. 628–640.

DOROBEK, S. L., 1991, Cyclic platform carbonates of the Devonian Jefferson Formation, southwestern Montana, in Cooper, J. D., and Stevens, C. H., eds., Paleozoic Paleogeography of the Western United States–II: Los Angeles, Pacific Section, Society of Economic Paleontologists and Mineralogists, v. 67, p. 509–526.

DOROBEK, S. L., REID, S. K., AND ELRICK, M., 1991a, Foreland response to episodic Antler convergence events, Devonian-Mississippian stratigraphy of Montana and Idaho: effects of horizontal and vertical loads, in Cooper, J. D., and Stevens, C. H., eds., Paleozoic Paleogeography of the Western United States–II: Los Angeles, Pacific Section, Society of Economic Paleontologists and Mineralogists, v. 67, p. 487–508.

DOROBEK, S. L., REID, S. K., ELRICK, M., BOND, G. C., AND KOMINZ, M. A., 1991b, Subsidence across the Antler foreland of Montana and Idaho: Tectonic versus eustatic effects, in Franseen, E., and Watney, L., eds., Sedimentary Modeling: Lawrence, Computer Simulation and Methods for Improved Parameter Definition: Kansas Geological Society Special Volume 233, p. 231–251.

EDWARDS, D. J. AND BROWN, R. J., 1993, A geophysical investigation of basement control on reef growth, in Ross, G. M., ed., Alberta Basement Transects Workshop: Vancouver, LITHOPROBE Report #31, University of British Columbia, p. 18–28.

ELRICK, M. AND READ, J. F., 1991, Cyclic ramp-to-basin carbonate deposits, lower Mississippian, Wyoming and Montana: A combined field and computer modeling study: Journal of Sedimentary Petrology, v. 61, p. 1194–1224.

ETTENSOHN, F. R., 1993, Possible flexural controls on the origins of extensive, ooid-rich, carbonate environmtexttextents in the Mississippian of the United States, *in* Keith, B. D., and Zuppann, C. W., eds., Mississippian Oolites and Modern Analogs: Tulsa, American Association of Petroleum Geologists, Studies in Geology 35, p. 13–30.

EUGSTER, H. P. AND HARDIE, L. A., 1975, Sedimentation in an ancient playa-lake complex: the Wilkins Peak Member of the Green River Formation of Wyoming: Geological Society of America Bulletin, v. 86, p. 319–334.

EWING, T. E., 1990, The tectonic map of Texas: Austin, Bureau of Economic Geology, The University of Texas at Austin.

FLEMINGS, P. B. AND JORDAN, T. E., 1989, A synthetic stratigraphic model of foreland basin development: Journal of Geophysical Research, v. 94, p. 3851–3866.

FLEMINGS, P. B. AND JORDAN, T. E., 1990, Stratigraphic modeling of foreland basins: Interpreting thrust deformation and lithosphere rheology: Geology, v. 18, p. 430–434.

FRENZEL, H. N., BLOOMER, R. R., CLINE, R. B., CYS, J. M., GALLEY, J. E., GIBSON, W. R., HILLS, J. M., KING, W. E., SEAGER, W. R., KOTLOWSKI, F. E., THOMPSON, S., III, LUFF, G. C., PEARSON, B. T., AND VAN SICLEN, D. C., 1988, The Permian Basin region, *in* Sloss, L. L., ed., Sedimentary Cover–North American Craton; U. S.: Boulder, Geological Society of America, The Geology of North America, v. D–2, p. 261–306.

GOEBEL, K. A., 1991, Paleogeographic setting of Late Devonian to Early Mississippian transition from passive to collisional margin, Antler Foreland, eastern Nevada and western Utah, *in* Cooper, J. D., and Stevens, C. H., eds., Paleozoic Paleogeography of the Western United States – II: Los Angeles, Pacific Section, Society of Economic Paleontologists and Mineralogists, v. 67, p. 401–418.

GORDEY, S. P., 1988, Devono-Mississippian clastic sedimentation and tectonism in the Canadian Cordilleran miogeocline, *in* McMillan, N. J., Embry, A. F., and Glass, D. J., eds., Devonian of the World: Volume 2: Sedimentation: Calgary, Canadian Society of Petroleum Geologists Memoir 14, p. 1–14.

GROTZINGER, J. AND MCCORMICK, D. S., 1988, Flexure of the early Proterozoic lithosphere and the evolution of the Kilohigok Basin (1.9 Ga), Northwest Canadian Shield, *in* Kleinspehn, K. L., and Paola, C., eds., New Perspectives in Basin Analysis: New York, Springer-Verlag, p. 405–430.

HALLOCK, P. AND SCHLAGER, W., 1986, Nutrient excess and the demise of coral reefs and carbonate platforms: Palaios, v. 1, p. 389–398.

HANDFORD, C. R., 1993, Sequence Stratigraphy of a Mississippian Carbonate Ramp, North Arkansas and Southwestern Missouri: Tulsa, Field Guidebook for American Association of Petroleum Geologists Annual Convention, April 25–28, 1993, 13 p.

HANSON, B. M., POWERS, B. K., GARRETT, C. M., JR., MCGOOKEY, D. E., MCGLASSON, E. H., HORAK, R. L., MAZZULLO, S. J., REID, A. M., CALHOUN, G. G., CLENDENING, J., AND CLAXTON, B., 1991, The Permian basin, *in* Gluskoter, H. J., Rice, D. D., and Taylor, R. B., eds., Economic Geology, U. S.: Boulder, Geological Society of America, The Geology of North America, v. P–2, p. 339–356.

HATTIN, D. E., 1986, Interregional model for deposition of Upper Cretaceous pelagic rhythmites, U. S. Western Interior: Paleoceanography, v. 1, p. 483–494.

HELLER, P. L., BEEKMAN, F., ANGEVINE, C. L., AND CLOETINGH, S. A. P. L., 1993, Cause of tectonic reactivation and subtle uplifts in the Rocky Mountain region and its effect on the stratigraphic record: Geology, v. 21, p. 1003–1006.

HILLS, J. M., 1985, Structural evolution of the Permian Basin of west Texas and New Mexico, *in* Dickerson, P. W., and Muehlberger, W. R., eds., Structure and Tectonics of Trans-Pecos Texas: Midland, West Texas Geological Society Publication 85–81, p. 89–99.

HINZ, K., FRITSCH, J., KEMPTER, E. H. K., MOHOMMAD, A. M., MEYER, J., MOHAMED, D., VOSBERG, H., WEBER, J., AND BENAVIDEZ, J., 1989, Thrust tectonics along the north-western continental margin of Sabah/Borneo: Geologische Rundschau, v. 78, p. 705–730.

HOFFMAN, P. F., 1989, Precambrian geology and tectonic history of North America, *in* Bally, A. W., and Palmer, A. R., eds., The Geology of North America – An Overview: Boulder, Geological Society of America, The Geology of North America, v. A, p. 447–512.

HOFFMAN, P. F. AND GROTZINGER, J. P., 1993, Orographic precipitation, erosional unloading, and tectonic style: Geology, v. 21, p. 195–198.

JACKSON, M. J., 1989, Lower Proterozoic Cowles Lake foredeep reef, N. W. T., Canada, *in* Geldsetzer, H. H. J., James, N. P., and Tebbutt, G. E., eds., Reefs, Canada and Adjacent Areas: Calgary, Canadian Society of Petroleum Geologists Memoir 13, p. 64–71.

JOHNSON, R. C., 1985, Early Cenozoic history of the Uinta and Piceance Creek Basins, Utah and Colorado, with special reference to the development of Eocene Lake Uinta, *in* Flores, R. M., and Kaplan, S. S., eds., Cenozoic Paleogeography of the West-Central United States, Rocky Mountain Paleogeography Symposium 3: Denver, Rocky Mountain Section, Society of Economic Paleontologists and Mineralogists, p. 247–276.

JORDAN, T. E., 1981, Thrust loads and foreland basin evolution, Cretaceous, Western United States: American Association of Petroleum Geologists Bulletin, v. 65, p. 2506–2520.

JORDAN, T. E., ISACKS, B. L., ALLMENDINGER, R. W., BREWER, J. A., RAMOS, V. A., AND ANDO, C. J., 1983, Andean tectonics related to geometry of subducted Nazca plate: Geological Society of America Bulletin, v. 94, p. 341–361.

KARNER, G. D., 1986, Effects of lithospheric in-plane stress on sedimentary basin stratigraphy: Tectonics, v. 5, p. 573–588.

KARNER, G. D., DRISCOLL, N. W., AND WEISSEL, J. K., 1993, Response of the lithosphere to in-plane force variations: Earth and Planetary Science Letters, v. 114, p. 397–416.

KENDALL, C. G. ST. C., AND SCHLAGER, W., 1981, Carbonates and relative changes in sea level: Marine Geology, v. 44, p. 181–212.

KING, P. R. AND THRASHER, G. P., 1992, Post-Eocene development of the Taranaki Basin, New Zealand—convergent overprint of a passive margin, *in* Watkins, J. S., Zhiqiang, F., and McMillen, K. J., eds., Geology and Geophysics of Continental Margins: Tulsa, American Association of Petroleum Geologists Memoir 53, p. 93–118.

KLUTH, C. F. AND CONEY, P. J., 1981, Plate tectonics of the Ancestral Rocky Mountains: Geology, v. 9, p. 10–15.

KOONS, P. O., 1989, The topographic evolution of collisional mountain belts: A numerical look at the Southern Alps, New Zealand: American Journal of Science, v. 289, p. 1041–1069.

LAFERRIERE, A. P., HATTIN, D. E., AND ARCHER, A. W., 1987, Effects of climate, tectonics, and sea-level changes on rhythmic bedding patterns in the Niobrara Formation (Upper Cretaceous), U. S. Western Interior: Geology, v. 15, p. 233–236.

LAMBECK, K., MCQUEEN, H. W. S., STEPHENSON, R. A., AND DENHAM, D., 1984, The state of stress within the Australian continent: Annales Geophysicae, v. 2, p. 723–742.

LETOUZEY, J., 1986, Cenozoic paleo-stress pattern in the Alpine Foreland and structural interpretation in a platform basin: Tectonophysics, v. 132, p. 215–231.

LIGHTY, R. G., MACINTYRE, I. G., AND STUCKENRATH, R., 1978, Submerged early Holocene barrier reef, south-east Florida shelf: Nature, v. 275, p. 59–60.

LUTERBACHER, H. P., EICHENSEER, H., BETZLER, C. H., AND VAN DEN HURK, A. M., 1991, Carbonate-siliciclastic depositional systems in the Paleogene of the South Pyrenean foreland basin: a sequence-stratigraphic approach, *in* Macdonald, D. I. M., ed., Sedimentation, Tectonics and Eustasy: Sea-level Changes at Active Margins: Oxford, International Association of Sedimentologists Special Publication Number 12, p. 391–407.

MCCONNELL, D. A., 1989, Determination of offset across the northern margin of the Wichita uplift, southwest Oklahoma: Geological Society of America Bulletin, v. 101, p. 1317–1332.

MITROVICA, J. X., BEAUMONT, C., AND JARVIS, G. T., 1989, Tilting of continental interiors by the dynamical effects of subduction: Tectonics, v. 8, p. 1079–1094.

MITROVICA, J. X. AND JARVIS, G. T., 1985, Surface deflections due to transient subduction in a convecting mantle: Tectonophysics, v. 120, p. 211–237.

MOORE, P. F., 1988, Devonian geohistory of the western interior of Canada, *in* McMillan, N. J., Embry, A. F., and Glass, D. J., eds., Devonian of the World: Volume 1: Regional Syntheses: Calgary, Canadian Society of Petroleum Geologists Memoir 14, p. 67–83.

MORROW, D. W. AND GELDSETZER, H. H. J., 1988, Devonian of the eastern Canadian Cordillera, *in* McMillan, N. J., Embry, A. F., and Glass, D. J., eds., Devonian of the World: Volume 1: Regional Syntheses: Calgary, Canadian Society of Petroleum Geologists Memoir 14, p. 85–121.

MOUNTJOY, E. W., 1980, Some questions about the development of Upper Devonian carbonate buildups (reefs), Western Canada: Bulletin of Canadian Petroleum Geology, v. 28, p. 315–344.

MOUNTJOY, E. W. AND HALIM-DIHARDJA, M. K., 1991, Multiple phase fracture and fault-controlled burial dolomitization, Upper Devonian Wabamun Group, Alberta: Journal of Sedimentary Petrology, v. 61, p. 590–612.

MUEHLENBACHS, K., BURWASH, R. A., AND CHACKO, T., 1993, A major oxygen isotope anomaly in the basement rocks of Alberta, in Ross, G. M., ed., Alberta Basement Transects Workshop: Vancouver, LITHOPROBE Report #31, University of British Columbia, p. 120–124.

MUSSMAN, W. J. AND READ, J. F., 1986, Sedimentology and development of a passive- to convergent-margin unconformity: Middle Ordovician Knox unconformity, Virginia Appalachians: Geological Society of America Bulletin, v. 97, p. 282–295.

OLDOW, J. S., BALLY, A. W., AVÉ LALLEMANT, H. G., AND LEEMAN, W. P., 1989, Phanerozoic evolution of the North American Cordillera; United States and Canada, in Bally, A. W., and Palmer, A. R., eds., The Geology of North America – An Overview: Boulder, Geological Society of America, The Geology of North America, v. A, p. 139–232.

PELECHATY, S. M. AND GROTZINGER, J. P., 1988, Stromatolite bioherms of a 1.9 Ga foreland basin carbonate ramp, Beechey Formation, Kilohigok Basin, Northwest Territories, in Geldsetzer, H. H. J., James, N. P., and Tebbutt, G. E., eds., Reefs, Canada and Adjacent Areas: Calgary, Canadian Society of Petroleum Geologists Memoir 13, p. 93–104.

PEPER, T., 1994, Tectonic and eustatic control on Albian shallowing (Viking and Paddy Formations) in the Western Canada Foreland Basin: Geological Society of America Bulletin, v. 106, p. 253–264.

PETERSON, J. A., 1986, General stratigraphy and regional paleotectonics of the western Montana overthrust belt, in Peterson, J. A., ed., Paleotectonics and Sedimentation in the Rocky Mountain Region, Unites States: Tulsa, American Association of Petroleum Geologists Memoir 41, p. 57–86.

PIGRAM, C. J., DAVIES, P. J., FEARY, D. A., AND SYMONDS, P. A., 1989, Tectonic controls on carbonate platform evolution in southern Papua New Guinea: Passive margin to foreland basin: Geology, v. 17, p. 199–202.

PURSER, B. H., 1973, Sedimentation around bathymetric highs in the southern Persian Gulf, in Purser, B. H., ed., The Persian Gulf—Holocene Carbonate Sedimentation and Diagenesis in a Shallow Water Epicontinental Sea: New York, Springer-Verlag, p. 157–177.

PURSER, B. H., 1983, Sédimentation et Diagenèse des Carbonates Néritiques Récents: Les Domaines de Sédimentation Carbonatée Néritiques Récents; Application à L'interprétation des Calcaires Anciens, Tome 2: Paris, Éditions Technip, 389 p.

PURSER, B. H. AND SEIBOLD, E., 1973, The principal environmental factors influencing Holocene sedimentation and diagenesis in the Persian Gulf, in Purser, B. H., ed., The Persian Gulf—Holocene Carbonate Sedimentation and Diagenesis in a Shallow Water Epicontinental Sea: New York, Springer-Verlag, p. 1–9.

QUINLAN, G. M. AND BEAUMONT, C., 1984, Appalachian thrusting, lithospheric flexure and the Paleozoic stratigraphy of the eastern interior of North America: Canadian Journal of Earth Sciences, v. 21, p. 973–996.

READ, J. F., 1980, Carbonate ramp-to-basin transitions and foreland basin evolution, Middle Ordovician, Virginia Appalachians: American Association of Petroleum Geologists Bulletin, v. 64, p. 1575–1612.

READ, J. F., 1982, Geometry, facies, and development of Middle Ordovician carbonate buildups, Virginia Appalachians: American Association of Petroleum Geologists Bulletin, v. 66, p. 189–209.

READ, J. F., 1985, Carbonate platform facies models: American Association of Petroleum Geologists Bulletin, v. 69, p. 1–21.

REID, S. K., 1991, Evolution of the Lower Mississippian Mission Canyon platform and distal Antler foredeep, Montana and Idaho: Unpublished Ph. D. Dissertation, Texas A&M University, College Station, 115 p.

REID, S. K. AND DOROBEK, S. L., 1991, Controls on development of third- and fourth-order depositional sequences in the Lower Mississippian Mission Canyon Formation and stratigraphic equivalents, Idaho and Montana, in Cooper, J. D., and Stevens, C. H., eds., Paleozoic Paleogeography of the Western United States – II: Los Angeles, Pacific Section, Society of Economic Paleontologists and Mineralogists, v. 67, p. 527–542.

REID, S. K. AND DOROBEK, S. L., 1993, Sequence stratigraphy and evolution of a progradational, foreland carbonate ramp, Lower Mississippian Mission Canyon Formation and stratigraphic equivalents, Montana and Idaho, in Loucks, R. G., and Sarg, J. F., eds., Carbonate Sequence Stratigraphy – Recent Developments and Applications: Tulsa, American Association of Petroleum Geologists Memoir 57, p. 327–352.

REYNOLDS, J. H., JORDAN, T. E., JOHNSON, M. M., DAMANTI, J. F., AND TABBUTT, K. D., 1990, Neogene deformation of the flat-subduction segment of the Argentine-Chilean Andes: Magnetostratigraphic constraints from Las Juntas, La Rioja province, Argentina: Geological Society of America Bulletin, v. 102, p. 1607–1622.

ROBERTSON, A. H. F., 1987, Upper Cretaceous Muti Formation: transition of a Mesozoic nate (sic) platform to a foreland basin in the Oman Mountains: Sedimentology, v. 34, p. 1123–1142.

ROSS, C. A., 1986, Paleozoic evolution of southern margin of Permian basin: Geological Society of America Bulletin, v. 97, p. 536–554.

ROSS, C. A. AND ROSS, J. R. P., 1985, Paleozoic tectonics and sedimentation in west Texas, southern New Mexico, and southern Arizona, in Dickerson, P. W., and Muehlberger, W. R., eds., Structure and Tectonics of Trans-Pecos Texas: Midland, West Texas Geological Society Publication 85–81, p. 221–230.

ROSS, C. A. AND ROSS, J. R. P., 1987, Late Paleozoic sea levels and depositional sequences, in, Ross, C. A., and Haman, D., eds., Timing and Depositional History of Eustatic Sequences: Constraints on Seismic Stratigraphy: Washington, D.C., Cushman Foundation for Foraminiferal Research, Special Publication No. 24, p. 137–149.

ROSS, G. M., 1992, Basement structure, in-plane stresses and the stratigraphic evolution of cratonic ramps in foreland basins (abs.): American Association of Petroleum Geologists, 1992 Annual Convention Official Program, p. 111.

ROYDEN, L., PATACCA, E., AND SCANDONE, P., 1987, Segmentation and configuration of subducted lithosphere in Italy: An important control on thrust-belt and foredeep-basin evolution: Geology, v. 15, p. 714–717.

RUPPEL, S. C. AND WALKER, K. R., 1984, Petrology and depositional history of a Middle Ordovician carbonate platform: Chickamauga Group, northeastern Tennessee: Geological Society of America Bulletin, v. 95, p. 568–583.

SANDBERG, C. A., GUTSCHICK, R. C., JOHNSON, J. G., POOLE, F. G., AND SANDO, W. J., 1983, Middle Devonian to Late Mississippian geologic history of the Overthrust Belt region, western United States: Denver, Rocky Mountain Association of Geologists, Geologic Studies of the Cordilleran Thrust Belt, v. 2, p. 691–719.

SANDBERG, C. A., POOLE, F. G., AND JOHNSON, J. G., 1988, Upper Devonian of western United States, in McMillan, N. J., Embry, A. F., and Glass, D. J., eds., Devonian of the World: Volume 1: Regional Syntheses: Calgary, Canadian Society of Petroleum Geologists Memoir 14, p. 183–220.

SCHEDL, A. AND WILTSCHKO, D. V., 1984, Sedimentologic effects of a moving terrain: Journal of Geology, v. 92, p. 273–287.

SCHLAGER, W., 1981, The paradox of drowned reefs and carbonate platforms: Geological Society of America Bulletin, v. 92, p. 197–211.

SCHLAGER, W., 1992, Sedimentology and Sequence Stratigraphy of Reefs and Carbonate Platforms: A Short Course: Tulsa, American Association of Petroleum Geologists Continuing Education Course Note Series #34, 71 p.

SILVER, B. A. AND TODD, R. G., 1969, Permian cyclic strata, northern Midland and Delaware Basins, west Texas and southeastern New Mexico: American Association of Petroleum Geologists Bulletin, v. 53, p. 2223–2251.

SIMO, A., 1989, Upper Cretaceous platform-to-basin depositional-sequence development, Tremp Basin, south-central Pyrenees, Spain, in Crevello, P. D., Wilson, J. L., Sarg, J. F., and Read, J. F., eds., Controls on Carbonate Platform and Basin Development: Tulsa, Society of Economic Paleontologists and Mineralogists Special Publication 44, p. 365–378.

SINCLAIR, H. D., COAKLEY, B. J., ALLEN, P. A., AND WATTS, A. B., 1991, Simulation of foreland basin stratigraphy using a diffusion model of mountain belt uplift and erosion: an example from the Central Alps, Switzerland: Tectonics, v. 10, p. 599–620.

SMITH, D. L., 1977, Transition from deep- to shallow-water carbonates, Paine Member, Lodgepole Formation, central Montana, in Cook, H. E., and Enos, P., eds., Deep-water Carbonate Environments: Tulsa, Society of Economic Paleontologists and Mineralogists Special Publication No. 25, p. 187–201.

SPEED, R. C. AND SLEEP, N. H., 1982, Antler orogeny and foreland basin: A model: Geological Society of America Bulletin, v. 93, p. 815–828.

STOCKMAL, G. S. AND BEAUMONT, C., 1987, Geodynamic models of convergent margin tectonics: the southern Canadian Cordillera and the Swiss Alps, *in* Beaumont, C., and Tankard, A. J., eds., Sedimentary Basins and Basin-Forming Mechanisms: Calgary, Canadian Society of Petroleum Geologists Memoir 12, p. 393–411.

TANKARD, A. J., 1986, On the depositional response to thrusting and lithospheric flexure: examples from the Appalachian and Rocky Mountain basins, *in* Allen, P. A., and Homewood, P., eds., Foreland Basins: Oxford, International Association of Sedimentologists Special Publication 8, p. 369–394.

TURCOTTE, D. L. AND SCHUBERT, G., 1982, Geodynamics—Applications of Continuum Physics to Geological Problems: New York, John Wiley and Sons, 450 p.

VAN WAGONER, J. C., POSAMENTIER, H. W., MITCHUM, R. M., VAIL, P. R., SARG, J. F., LOUTIT, T. S., AND HARDENBOL, J., 1988, An overview of the fundamentals of sequence stratigraphy and key definitions, *in* Wilgus, C. K., Hastings, B. S., Kendall, C. G. St. C., Posamentier, H. W., Ross, C. A., and Van Wagoner, J. C., eds., Sea-level Changes: An Integrated Approach: Tulsa, Society of Economic Paleontologists and Mineralogists Special Publication 42, p. 39–45.

WARD, R. F., KENDALL, C. G. ST. C., AND HARRIS, P. M., 1986, Upper Permian (Guadalupian) facies and their association with hydrocarbons—Permian Basin, west Texas and New Mexico: American Association of Petroleum Geologists Bulletin, v. 70, p. 239–262.

WASCHBUSCH, P. J. AND ROYDEN, L. H., 1992, Episodicity in foredeep basins: Geology, v. 20, p. 915–918.

WINSTON, D., 1986, Sedimentation and tectonics of the Middle Proterozoic Belt Basin and their influence on Phanerozoic compression and extension in western Montana and northern Idaho, *in* Peterson, J. A., ed., Paleotectonics and Sedimentation: Tulsa, American Association of Petroleum Geologists Memoir 41, p. 87–118.

WU, P., 1991, Flexure of lithosphere beneath the Alberta foreland basin: Evidence of an eastward stiffening continental lithosphere: Geophysical Research Letters, v. 18, p. 451–454.

WUELLNER, D. E., LEHTONEN, L. R., AND JAMES, W. C., 1986, Sedimentary-tectonic development of the Marathon and Val Verde basins, West Texas, U. S. A.: a Permo-Carboniferous migrating foredeep, *in* Allen, P., and Homewood, P., eds., Foreland Basins: Oxford, International Association of Sedimentologists Special Publication 8, p 15–39.

YANG, K.-M., 1993, Late Paleozoic stratigraphy, tectonic evolution, and flexural modeling of the Permian Basin, West Texas and New Mexico: Unpublished Ph.D. Dissertation, Texas A&M University, College Station, 142 p.

YANG, K.-M. AND DOROBEK, S. L., 1992, Mechanisms for late Paleozoic synorogenic subsidence of the Midland and Delaware Basins, Permian Basin, Texas and New Mexico, *in* Mruk, D., and Curran, B., eds., Permian Basin Exploration and Production Strategies–Applications of Sequence Stratigraphic and Reservoir Characterization Concepts: Midland, West Texas Geological Society, 1992 Fall Symposium, p. 45–60.

ZIEGLER, P. A., 1987, Late Cretaceous and Cenozoic intra-plate compressional deformations in the Alpine foreland—a geodynamic model: Tectonophysics, v. 137, p. 389–420.

ZOBACK, M. L., 1992, First- and second-order patterns of stress in the lithosphere: The World Stress Map Project: Journal of Geophysical Research, v. 97, p. 11703–11728.

THE PERMIAN BASIN OF WEST TEXAS AND NEW MEXICO: TECTONIC HISTORY OF A "COMPOSITE" FORELAND BASIN AND ITS EFFECTS ON STRATIGRAPHIC DEVELOPMENT

KENN-MING YANG[1] AND STEVEN L. DOROBEK
Department of Geology, Texas A&M University, College Station, TX 77843

ABSTRACT: The Permian Basin of West Texas and southern New Mexico is located in the foreland of the Marathon-Ouachita orogenic belt. This complex foreland area consists of several sub-basins that are separated by intraforeland uplifts. This study examined the tectonic, kinematic, and subsidence history of the Permian Basin in order to evaluate how intraforeland deformation affected stratigraphic development. We focused on: (1) the kinematic history of the Central Basin Platform (CBP), an intraforeland uplift that trends at high angles to the frontal thrust of the Marathons and separates the Delaware and Midland Basins; and (2) subsidence and stratigraphic analyses of the Midland, Delaware, and Val Verde Basins.

Structure contour maps, seismic profiles, and balanced structural cross sections show that the CBP can be subdivided into two fault-bounded "blocks," the Fort Stockton and Andector Blocks, which are arranged in a left-stepping, *en echelon* pattern. The distribution of structural features associated with the CBP is best explained by clockwise rotation of these blocks plus an additional component of east-west compression. The rotation model explains: (1) the steeply dipping reverse faults at the SW and NE corners and local extensional faults at NW corners of individual crustal blocks that comprise the CBP; (2) the large structural relief observed at the SW and NE corners of individual blocks; and (3) decreasing amounts of basement shortening away from thrust-faulted corners that can be documented along both sides of the Fort Stockton Block. An additional component of shortening is required to account for imbalances between the amount of shortening versus extension observed along block boundaries.

Subsidence analyses from several points in the Midland, Delaware, and Val Verde Basins indicate that the main phase of tectonic activity probably began during middle Pennsylvanian time. Rapid subsidence in each basin began at that time and continued until Early Permian time. Thereafter, subsidence slowed considerably to the end of the Permian. Late Paleozoic unconformities developed across the CBP and locally in adjacent basinal areas at the same time as rapid subsidence in the basins, suggesting that they are related to the same general episode of tectonic activity.

Upper Pennsylvanian to Lower Permian stratigraphic cross sections show that synorogenic strata generally thicken toward the CBP. These stratal relationships indicate that the CBP acted as an intraforeland load that caused flexure of the adjacent sub-basins that comprise the Permian Basin. The thickest accumulation of upper Pennsylvanian to Lower Permian strata is developed next to the SW and NE corners of the Fort Stockton and Andector Blocks, which also corresponds to the areas where block uplift was greatest.

Finally, the variable platform-to-basin relief that was produced during uplift of the CBP resulted in very different patterns of stratigraphic onlap during late Pennsylvanian to Early Permian time. Onlapping upper Pennsylvanian to Wolfcampian strata extend farthest across the top of the CBP where structural relief was least (i.e., at the NW and SE block corners). Stratal onlap is minimal at the SW and NE block corners because structural relief was greatest there. This study illustrates how patterns of intraforeland deformation can dramatically affect basin stratigraphy during synorogenic stages of basin development.

INTRODUCTION

Foreland basins are produced initially by flexure of lithosphere beneath thrust sheets emplaced along one side of the basin (Beaumont, 1981; Jordan, 1981). During subsequent stages of basin evolution, sediments eroded from the orogenic wedge are deposited in the basin and may redistribute topographic loads. Redistribution of these loads alters the form of the flexural profile (Quinlan and Beaumont, 1984; Stockmal and others, 1986; Flemings and Jordan, 1989). In most two-dimensional flexural models of foreland basins, the loaded lithosphere is regarded as an unbroken, homogeneous plate, and lithospheric heterogeneity across the foreland is not considered. In addition, the orogenic wedge along one side of the basin generally is viewed as the only tectonic load that influences the flexural profile. Several case studies, however, show that synorogenic crustal deformation *within* the foreland may also affect foreland basin geometry and stratigraphic development. In fact, intraforeland uplifts may behave as additional topographic loads and complicate the evolution of foreland basins (Hagen and others, 1985; Flemings and others, 1986; Beck and others, 1988; Shuster and Steidtmann, 1988; McConnell and others, 1990; Yang and Dorobek, 1992).

Mechanisms for this intraforeland deformation, however, are poorly understood. Formation of uplifts in the Laramide foreland has been ascribed to regional compressive stress induced by low-angle subduction of lithosphere beneath the North American craton (Sales, 1968; Lowell, 1974; Dickinson and Snyder, 1978; Blackstone, 1980; Scheevel, 1983; Beck, 1984; Bird, 1984). In other foreland areas where low-angle subduction is absent, complex foreland structures have been attributed to reactivation of pre-existing faults, typically by in-plane transmission of compressive stress (Sengor, 1976; Walper, 1977; Sengor and others, 1978; Kluth and Coney, 1981; Thomas, 1983; Cooper and Williams, 1989; Dorobek and others, 1991). In some cases, foreland structures are repeatedly reactivated or inverted due to progressive changes in the magnitude or orientation of principal stresses in the foreland (Sengor, 1976; Mauch and others, 1984; Ross and Ross, 1985; Letouzey, 1986; Ziegler, 1987; Larroque and Laurent, 1988; Dorobek and others, 1991). One approach toward understanding the effects of intraforeland deformation on synorogenic basin development is to examine restored stratigraphic and structural cross sections, subsidence histories, and lithofacies patterns in the foreland area. Kinematic histories can then be reconstructed based on these detailed stratigraphic and structural analyses.

The Permian Basin of West Texas and southern New Mexico is located in the foreland of the Marathon-Ouachita orogenic belt (Fig. 1). Several sub-basins that are separated by intraforeland uplifted areas comprise this complex foreland region (Fig. 2). Despite an immense subsurface data-

---

[1]Current address: Exploration of Development Research Institute, Chinese Petroleum Corporation, 1 Ta Yuan, Wen Shan, Miaoli, Taiwan 36010.

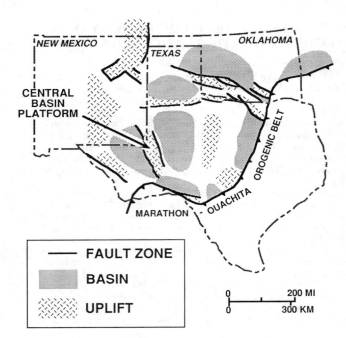

Fig. 1—Regional base map of the Marathon-Ouachita foreland. Note partitioning of the Marathon-Ouachita foreland into a series of late Paleozoic sub-basins that are separated by high relief, fault-bounded uplifts. Modified from McConnell (1989).

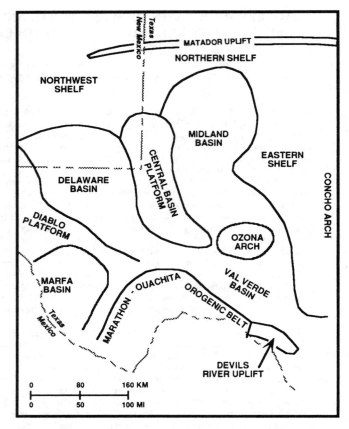

Fig. 2—Regional paleogeographic map of the Permian Basin region showing major paleotopographic features of late Paleozoic age. Modified from Hanson and others (1991).

base that has been compiled after seventy years of hydrocarbon exploration and production in the Permian Basin, the late Paleozoic tectonic history of this region remains enigmatic.

A main objective of this study was to reconstruct the late Paleozoic tectonic evolution and kinematic history of the uplifted areas and sub-basins that comprise most of the Permian Basin region. The kinematic history of one of these uplifted areas, the Central Basin Platform (CBP)[1] was analyzed by using reconstructed structural profiles across its boundary fault zones. We have also examined the late Paleozoic stratigraphic evolution of the sub-basins by documenting regional lithofacies and onlap patterns for Pennsylvanian through Lower Permian strata on stratigraphic cross sections across the basins. In our companion paper on the Permian Basin in this volume, we examine possible flexural effects of the Marathon orogenic belt, the CBP, and other uplifts of the Marathon foreland on subsidence in the adjacent basins.

This study provides an integrated model for the tectonic history of the Permian Basin region. Our observations, reconstructions, and model results provide an explanation for the complex structural evolution and subsidence history of this area. In addition, lateral variability in late Paleozoic stratigraphy in the Permian Basin (Fig. 3) are better understood if foreland structural features and their effects on basin subsidence are documented first. Lastly, the structural styles that we have documented across the Permian Basin may not be unique to this foreland. Grossly similar structures and basin histories might be found in other foreland areas that are underlain by pre-existing zones of lithospheric weakness or where there are significant salients along the orogenic belt that marks the suture between two plates.

TECTONIC FEATURES OF THE PERMIAN BASIN REGION

Several tectonic maps of the Permian Basin have been published recently (GEOMAP, 1983; Gardiner, 1990; Ewing, 1990). The Permian Basin has not been subjected to significant deformation since late Paleozoic time (except for minor tilting along its margins; Ewing, 1993), so the present structural features are essentially the same as those that

---

[1] "Central Basin Platform" is widely used to describe the topographically high area between the Midland and Delaware Basins that became the site of extensive shallow water deposition during Permian time. The term "platform" more appropriately describes the type of Permian sedimentation here, rather than the pre-Permian structure and topography of this uplifted area. Because the Central Basin Platform consists of several crustal blocks with variable topography, Ewing (1993) refers to it as the "Central Basin axis," while Shumaker (1992) prefers "Central Basin uplift." Ewing (1993) also refers to the Permian Basin region as the "West Texas Basin," because this area has a complex geologic history that spans much more than just the Permian. We prefer to retain the terms "Central Basin Platform" and "Permian Basin" because of their precedence and wider usage by geologists working in this area.

SYSTEM	SERIES/STAGE	NORTHWEST SHELF	CENTRAL BASIN PLATFORM	MIDLAND BASIN & EASTERN SHELF	DELAWARE BASIN	VAL VERDE BASIN
PERMIAN	OCHOAN	DEWEY LAKE RUSTLER SALADO	DEWEY LAKE RUSTLER SALADO	DEWEY LAKE RUSTLER SALADO	DEWEY LAKE RUSTLER SALADO CASTILE	RUSTLER SALADO
	GUADALUPIAN	TANSILL YATES SEVEN RIVERS QUEEN GRAYBURG SAN ANDRES —GLORIETA—	TANSILL YATES SEVEN RIVERS QUEEN GRAYBURG SAN ANDRES —GLORIETA—	TANSILL YATES SEVEN RIVERS QUEEN GRAYBURG SAN ANDRES —SAN ANGELO—	DELAWARE MT GROUP BELL CANYON CHERRY CANYON BRUSHY CANYON	TANSILL YATES SEVEN RIVERS QUEEN GRAYBURG SAN ANDRES
	LEONARDIAN	CLEAR FORK YESO WICHITA —ABO—	CLEAR FORK WICHITA	LEONARD SPRABERRY, DEAN	BONE SPRING	LEONARD
	WOLFCAMPIAN	WOLFCAMP	WOLFCAMP	WOLFCAMP	WOLFCAMP	WOLFCAMP
PENNSYLVANIAN	VIRGILIAN	CISCO	CISCO	CISCO	CISCO	CISCO
	MISSOURIAN	CANYON	CANYON	CANYON	CANYON	CANYON
	DESMOINESIAN	STRAWN	STRAWN	STRAWN	STRAWN	STRAWN
	ATOKAN	ATOKA —BEND—	ATOKA —BEND—	ATOKA —BEND—	ATOKA —BEND—	(ABSENT)
	MORROWAN	MORROW	(ABSENT)	(ABSENT ?)	MORROW	(ABSENT)
MISSISSIPPIAN	CHESTERIAN MERAMECIAN OSAGEAN KINDERHOOKIAN	CHESTER MERAMEC OSAGE KINDERHOOK	CHESTER MERAMEC OSAGE "BARNETT" KINDERHOOK	CHESTER MERAMEC OSAGE "BARNETT" KINDERHOOK	CHESTER MERAMEC OSAGE "BARNETT" KINDERHOOK	MERAMEC OSAGE "BARNETT" KINDERHOOK
DEVONIAN		—WOODFORD— DEVONIAN	—WOODFORD— DEVONIAN	—WOODFORD— DEVONIAN	—WOODFORD— DEVONIAN	—WOODFORD— DEVONIAN
SILURIAN		SILURIAN (UNDIFFERENTIATED)	SILURIAN SHALE FUSSELMAN	SILURIAN SHALE FUSSELMAN	MIDDLE SILURIAN FUSSELMAN	MIDDLE SILURIAN FUSSELMAN
ORDOVICIAN	UPPER	MONTOYA	MONTOYA	SYLVAN MONTOYA	SYLVAN MONTOYA	SYLVAN MONTOYA
	MIDDLE	SIMPSON	SIMPSON	SIMPSON	SIMPSON	SIMPSON
	LOWER	ELLENBURGER	ELLENBURGER	ELLENBURGER	ELLENBURGER	ELLENBURGER
CAMBRIAN	UPPER	CAMBRIAN	CAMBRIAN	CAMBRIAN	CAMBRIAN	CAMBRIAN
PRECAMBRIAN						

Fig. 3—Generalized stratigraphic correlation chart for the Permian Basin region. After Hanson and others (1991).

existed at the end of the late Paleozoic (Hills, 1984; Ward and others, 1986; Frenzel and others, 1988).

The Permian Basin consists of several uplifts and sub-basins (Figs. 1, 2). The Delaware Basin is bounded to the west by the Diablo Platform, to the north by the Northern and Northwestern Shelf areas, and to the south by the Marathon orogenic belt, while the north-south trending CBP separates the Delaware Basin from the Midland Basin. Tectonic maps for the Permian Basin show that the CBP is composed of several fault-bounded structural domains that are arranged in a general, left-stepping, *en echelon* pattern (Fig. 4; Gardiner, 1990; Ewing, 1991). These domains are separated from each other by faults and from adjacent basinal areas by complex boundary fault zones.

The Midland Basin is bounded on the east by a complex series of north-south trending fault segments called the Fort Chadbourne Fault Zone (Ewing, 1991). The Fort Chadbourne Fault Zone generally coincides with the transition from marine platform facies on the Eastern Shelf to basinal facies in the Midland Basin. Southward, Midland Basin strata thin onto the Ozona Arch, an easterly extension of the southern CBP that separates the Midland and Val Verde Basins. The Ozona Arch is a broad, east-west trending structural high. The limbs of the Ozona Arch dip northward into the Midland Basin and southward into the Val Verde Basin, although a left-lateral strike-slip fault with vertical displacement is located between the Ozona Arch and the Midland Basin. In contrast to the CBP, no high angle faults with large vertical displacement are observed along the boundaries of the Ozona Arch. The different structural styles associated with the CBP and Ozona Arch suggest different tectonic mechanisms are responsible for their formation.

A few faults are mapped in the Midland and Delaware Basins beyond the boundary fault zones of the CBP. In the Delaware Basin, some of these steeply dipping faults intersect the CBP boundary faults at high angles. These faults have been interpreted as left-lateral strike-slip faults (Ewing, 1991; Shumaker, 1992) and are evidence for late Paleozoic strike-slip deformation in the Permian Basin (Walper, 1977; Goetz and Dickerson, 1985; Horak, 1985).

The thickest accumulations of Pennsylvanian-Permian strata in the Permian Basin region are located in the Val Verde Basin and southernmost part of the Delaware Basin. Both of these areas are located adjacent to the Marathon orogenic belt and the adjoining Devils River Uplift. The Val Verde Basin is separated from the southern Delaware Basin by a steeply dipping fault that extends southward from the western boundary fault zone of the CBP (Fig. 4). We refer to this fault as the "Puckett-Grey Ranch Fault" for several oil fields that are located near it. The Val Verde Basin has a typical flexural geometry with the maximum deflection adjacent to the Marathon thrust front and Devils River Uplift. In contrast, Delaware Basin profiles that are oriented normal to the Marathon orogen do not have typical flexural geometries. A few steeply dipping faults have been documented in the Val Verde Basin and are subparallel to the Marathon orogenic belt (Fig. 4). These faults have been interpreted as left-lateral strike-slip faults (Hills, 1970;

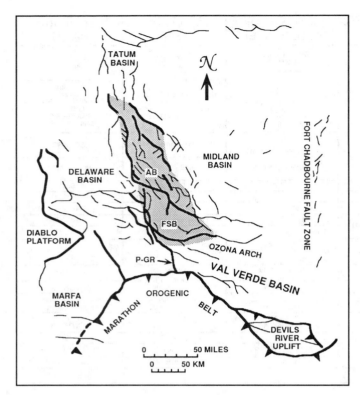

FIG. 4—Generalized tectonic map of the Central Basin Platform (CBP) and adjacent areas. Major boundary faults of the CBP are shown as heavy solid lines; other faults shown as thinner solid lines. Modified from GEO-MAP (1983), Ewing (1990), and Reed and Strickler (1990). The various sub-basins and uplifted areas that comprise the Permian Basin region are labeled. The shaded area represents the Central Basin Platform. AB = Andector Block; FSB = Fort Stockton Block; P–GR = Puckett–Grey Ranch Fault Zone.

Walper, 1977; Goetz and Dickerson, 1985; Horak, 1985; Gardiner, 1990; Ewing, 1991).

STRUCTURAL DOMAINS OF THE CENTRAL BASIN PLATFORM

The CBP consists of a complex series of fault-bounded crustal blocks. Gardiner (1990) identified six structural "subprovinces," with each area having a different structural style from adjacent areas. Ewing (1991) also identified similar, but not identical, structural subprovinces which comprise the CBP. We subdivide the CBP into two dominant crustal "blocks," the Andector and Fort Stockton Blocks (Fig. 4). This simplification is justified because these block boundaries encompass domains that have generally similar senses of vergence, structural relief, and orientations of structural features. Very different structural styles are typically juxtaposed on either side of our block boundaries. Even though smaller, fault-bounded crustal blocks with the same general sense of vergence can be identified within our larger blocks, we prefer this simplification because it facilitates visualization of the large-scale, kinematic history of the CBP.

Description of the Fort Stockton Block

The Fort Stockton Block comprises the southern portion of the CBP. Our designation of the Fort Stockton Block includes the "Fort Stockton and Sand Hills Blocks" of Gardiner (1990) and Shumaker (1992). A regional structure contour map (Fig. 5) shows the magnitude of vertical displacement along the boundary fault zones of the CBP and the differential uplift in its interior. The greatest amounts of vertical displacement along the CBP's boundaries are at the SW and NE corners of the Fort Stockton Block. Vertical displacement also progressively decreases from south to north along the western side of the Fort Stockton Block and from north to south along its eastern side.

Seismic profiles and structural cross sections constructed from well logs (Fig. 6) show that the SW and NE corners of the Fort Stockton Block are characterized by steeply dip-

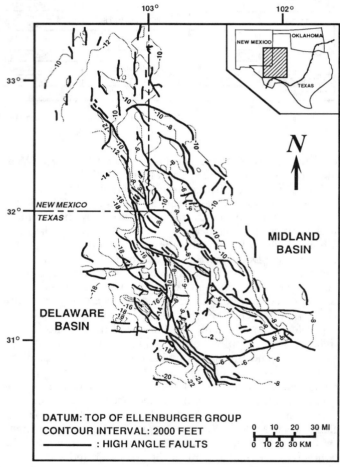

FIG. 5—Regional structure contour map showing the magnitude of vertical displacement along the boundary fault zones of the CBP and the differential uplift in its interior. Modified from Ewing (1990) and Gardiner (1990). The greatest amounts of vertical displacement along the CBP's boundaries are at the southwest and northeast corners of the Fort Stockton and Andector Blocks. Vertical displacement also progressively decreases from south-to-north along the western sides of the Fort Stockton and Andector Blocks and from north-to-south along their eastern sides. Locations of seismic profiles shown in Figure 7 are also indicated.

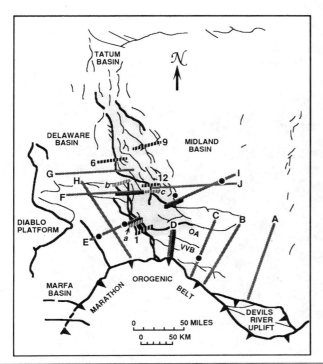

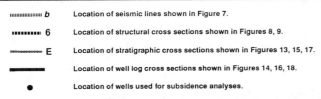

FIG. 6.—Location map for seismic sections (shown in Fig. 7), structural cross sections (shown in Figures 8 and 9), stratigraphic cross sections (shown in Figs. 13, 15, and 17), well-log cross sections (shown in Figs. 14, 16, and 18), and wells used for subsidence analyses.

ping reverse faults that dip toward the interior of the block (Figs. 7A, C, 8). In contrast, normal faults at the NW corner of the block dip westward toward the Delaware Basin (Fig. 7B). We do not have seismic data from the southern boundary of the Fort Stockton Block, but the regional structure on top of the Ellenburger Group indicates that moderate structural relief exists across this fault zone. Strike-slip deformation structures (flower structures, pop-ups, near vertical faults) are located along the east-west fault zone that separates the Fort Stockton Block from the Andector Block, although the seismic data that we have seen for this area are proprietary and cannot be shown.

Description of the Andector Block

The Andector Block comprises the middle portion of the CBP. Our designation of the Andector Block includes: (1) the "Mid-CBP Transition Zone, the western part of the Andrews Shear Zone, and the Lea-Andrews Block" of Gardiner (1990); (2) the "Eunice-Andector ridges and Monahans and Hobbs transverse zones" of Ewing (1991); and (3) the "Eunice, Embar, and Emperor" block uplifts of Shumaker (1992). The regional structure contour map (Fig. 5) shows that the greatest amount of vertical displacement along the boundaries of the Andector Block is located at the SW block corner. Similar to the Fort Stockton Block, vertical displacement also progressively decreases from south to north along the western side of the Andector Block and from north to south along its eastern side.

Seismic data and structural cross sections (Fig. 9) indicate that the SW and NE corners of the Andector Block are characterized by steeply dipping reverse faults that dip toward the interior of the block. Ewing (1991) shows the Andector Block to consist of a series of "ridges" or crustal blocks; these ridges usually are bounded on their eastern flank by steeply dipping reverse faults with eastward vergence and are cut by transverse normal faults. We did not, however, have enough seismic data or well control to characterize the regional variation in structural style along the western and eastern sides of the Andector Block nor details about internal deformation within the Andector Block. Proprietary seismic data, however, show that the fault zone along the northern boundary of the Andector Block is characterized by steeply dipping reverse faults that dip to the south.

Tatum Ridges

The area north of the Andector Block is generally known as the Tatum Basin but has also been called the "Tatum Ridges" by Ewing (1991). Flower structures can be identified on east-west seismic profiles across this area and illustrate the strike-slip style of deformation that characterizes the Northwest Shelf area immediately north of the CBP (Yang, 1993). Other evidence for strike-slip offset along the boundary faults of CBP blocks to the south include *en echelon* folds and faults and offsets in lower Paleozoic subcrop patterns (Walper, 1977; Hills, 1970; Henderson and others, 1984).

TECTONIC MODEL FOR UPLIFT OF THE CBP

Most previous models for the uplift of the CBP are based on the inferred displacements along the CBP's boundary faults (i.e., whether the boundary faults are reverse, strike-slip, or normal faults). Few detailed, balanced structural cross sections, however, have actually been made across the boundary fault zones of the CBP (cf. Shumaker, 1992).

Structure contour maps, seismic profiles, and well data were used in this study to characterize the various structures that comprise the CBP. These data also were used to construct balanced structural cross sections which in turn allowed us to calculate the amount of basement shortening across the boundary fault zones of the CBP. Using a variety of observations, we describe the styles of deformation associated with the two dominant structural domains that comprise the CBP, the Fort Stockton and Andector Blocks, and offer a kinematic model that explains the tectonic origin of the CBP.

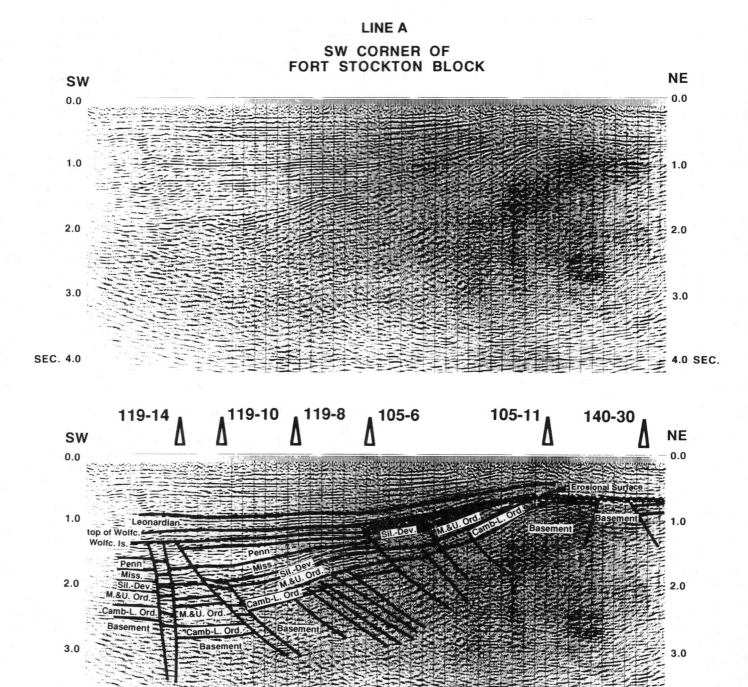

Fig. 7—Seismic lines showing structural styles associated with the boundary fault zones of the Central Basin Platform. Uninterpreted and interpreted seismic profiles are shown. For well locations indicated on interpreted lines, see Yang (1993). *Line A*, migrated seismic profile across SW corner of the Fort Stockton Block, southern end of Central Basin Platform. Note the east–dipping reverse and thrust faults that characterize the boundary fault zone at the SW corner of the Fort Stockton Block. Also note the erosional surface (indicated by an arrow) that truncates lower Paleozoic strata and Precambrian basement on the upper part of the Fort Stockton Block. Wolfcampian limestone along the flank of the Fort Stockton Block pinches out onto this erosional surface. *Line B*, unmigrated seismic profile across NW corner of Fort Stockton Block. Note the west–dipping normal faults that characterize the boundary fault zone along this part of the Fort Stockton Block. *Line C*, migrated seismic profile across NE corner of Fort Stockton Block. Note the west–dipping reverse and thrust faults that characterize the boundary fault zone along this part of the Fort Stockton Block. The erosional surface across the top of the Fort Stockton Block (indicated by an arrow) produces low–angle truncation of lower Paleozoic strata.

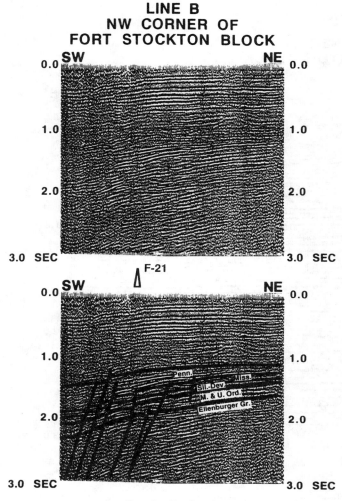

Fig. 7—Continued.

Structural Cross Sections, Structural Relief, and Calculated Basement Shortening along the Boundaries of the CBP

Structural cross sections were constructed using well data tied to seismic profiles across the boundary fault zones of the CBP (Figs. 8, 9). These cross sections were constructed by assuming that pre-Pennsylvanian strata were folded into simple monoclines along the reverse- and thrust-faulted corners of CBP blocks, and then faulted. Monoclinal folds can be produced by passive folding over a series of stepped, planar, and parallel basement faults (Spang and others, 1985; McConnell, 1987). In the case of the boundary fault zones of the CBP, some basement faults cut through overlying pre-Pennsylvanian strata. These structural cross sections were then used to calculate the amount of basement shortening along the faulted boundaries of the CBP. More detailed descriptions of the techniques used to construct balanced structural cross sections and methods for calculating basement shortening are outlined in Yang (1993), but a brief overview is presented here.

Calculation of the amount of basement shortening was done by assuming constant bed length in lower to middle Paleozoic strata before and after uplift of the CBP (Spang and others, 1985; McConnell, 1987; Cook, 1988). Lower to middle Paleozoic strata were eroded from parts of some of the blocks that comprise the CBP. Erosion of pre-Pennsylvanian strata occurred during synorogenic uplift of these blocks as indicated by the angular unconformity that locally separates onlapping Pennsylvanian and Permian strata from underlying, gently dipping Precambrian basement and lower to middle Paleozoic strata. Eroded strata were reconstructed by extrapolating the thicknesses of preserved pre-Pennsylvanian strata from other parts of the CBP. We assume that stratal thicknesses and regional dip did not change from uneroded parts of blocks to areas of maximum erosion and removal of pre-Pennsylvanian strata. We also assume minimal out-of-plane displacement along our cross sections. The reconstructed, deformed pre-Pennsylvanian strata were then palinspastically restored to their original, presumably horizontal attitude, and the amount of basement shortening was measured.

Calculated amounts of basement shortening decrease northward along the western boundary of the Fort Stockton Block, away from its SW corner (Fig. 10). Structural relief across the boundary fault zone also decreases northward (Fig. 5). Some steeply dipping reverse faults at the SW corner of the Fort Stockton Block appear to be traceable along the western side of the block to its NW corner where they become high-angle normal faults (Figs. 7, 8), indicating extension in that area. Detailed structural analysis of the transition zone between areas of shortening to areas of extension was not possible because seismic data were not available to us across the transition zone.

The NE corner of the Fort Stockton Block, sometimes referred to as the "Sand Hill Ridge," is also characterized by thrust faults although they have much less vertical displacement than at the SW corner (Fig. 8, Line 12). Basement shortening across this corner decreases southward, away from the NE corner (Fig. 10). The SE corner of the Fort Stockton Block merges eastward with the Ozona Arch. Characteristics of the Fort Stockton Block's boundary faults in this area are poorly understood because seismic data were not available to us. Steeply dipping boundary faults on the SE side of the Fort Stockton Block can be identified, however, based on well-log correlations that show abrupt changes in the thickness of synorogenic strata and large differences in the structural level of pre-Pennsylvanian strata on either side of the boundary fault zone.

The amount of basement shortening was also calculated along the eastern and western sides of the Andector Block. Like the Fort Stockton Block, basement shortening decreases northward from the SW corner of the Andector Block and southward from the NE block corner (Fig. 10).

Kinematic Model for the CBP

Previous tectonic models have attributed the origin of the CBP either to regional extension (Elam, 1984), compression (Galley, 1958; Ewing, 1984, 1991; Reed and Strickler, 1990), or strike-slip deformation (Hills, 1970; Walper, 1977;

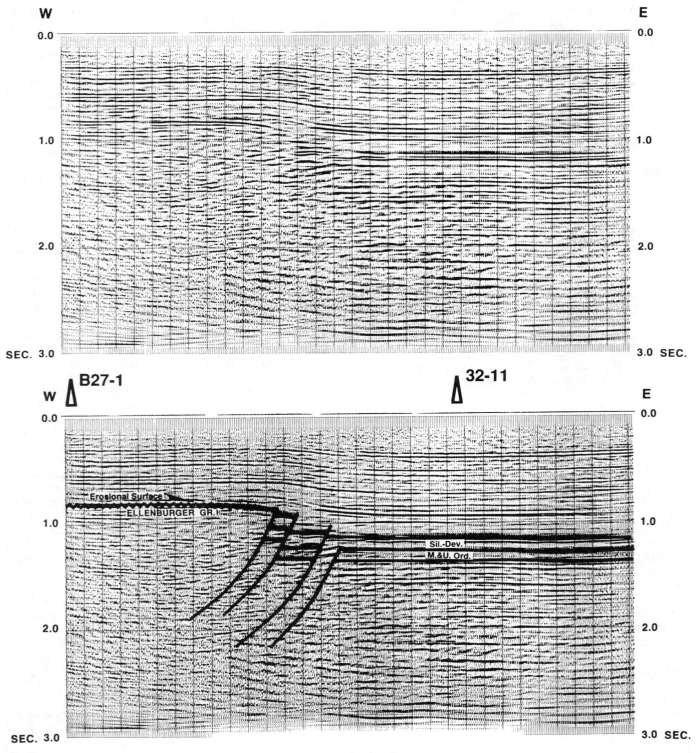

Fig. 7—Continued.

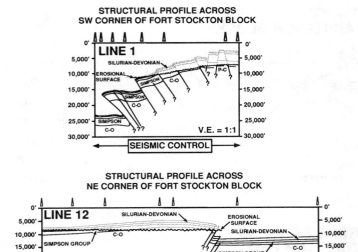

FIG. 8—Reconstructed structural cross sections across the SW and NE corners of the Fort Stockton Block. Small triangles at top of cross section indicate locations of wells used; see Yang (1993) for actual well locations. Note east–dipping reverse faults that characterize the boundary fault zone along the SW corner of the Fort Stockton Block (Line 1) and the west–dipping reverse faults that characterize the NE corner (Line 12). Also note the erosional surface (indicated by an arrow) that truncates lower Paleozoic strata and Precambrian basement on the upper part of the Fort Stockton Block. Reconstructed lower Paleozoic strata above the erosional surface (indicated by an arrow) are shown as shaded lines across the top of the Fort Stockton Block.

Goetz and Dickerson, 1985; Shumaker, 1992). Most of the these previous models, however, have not addressed the differential structural relief documented in interior parts of the CBP and the complex structural styles associated with its boundary fault zones. In addition, many of these earlier interpretations were based on observations from limited areas and without the benefit of seismic profiles. Seismic data are essential for characterizing the types of faults along the boundaries of the CBP.

The structural features used to constrain these previous tectonic models for the CBP and Permian Basin region are not necessarily mutually exclusive. Our regional structural analyses, constrained by seismic sections, lead to an alternative kinematic model that explains the structural evolution of the CBP and incorporates many aspects of the previous models. This kinematic model not only explains the regional variation in structural style along the margins of the CBP but also provides a mechanism for at least some component of its uplift.

We attribute deformation and uplift of the CBP to clockwise rotation of the Fort Stockton and Andector Blocks, the two crustal blocks that comprise the CBP (Fig. 11; Yang and Dorobek, 1991, 1992, 1993). Block rotation occurred during late Pennsylvanian to Early Permian time. Stratigraphic evidence for dating the tectonic activity is discussed below. The Fort Stockton and Andector Blocks are bounded on their eastern and western sides by NNW-SSE trending fault zones that are highly variable along strike. Fault types include steeply dipping reverse/thrust, normal, or strike-slip faults. We propose that these complex boundary fault zones initially were steeply dipping strike-slip faults that formed when the Permian Basin was subjected to regional right-lateral shear stress during late Paleozoic time. These faults may have coincided with preexisting fault zones that were reactivated during late Paleozoic deformation (Wuellner and others, 1986).

Strike-slip motion along the boundary faults would have induced a prominent right-lateral shear component along the trend of the CBP and caused the CBP to split into the Fort Stockton and Andector Blocks (Fig. 11). In our model, the ENE-WSW trending boundary between the Fort Stockton and Andector Blocks is a left-lateral strike-slip fault zone. Proprietary seismic profiles from this area have flower structures and near vertical faults that are typical of strike-slip deformation. The fault zones that define the southern boundary of the Fort Stockton Block and northern boundary of the Andector Block may also have been preexisting zones of weakness that were reactivated during formation of the CBP or they may be cross faults that were produced by the regional shear stress.

As a result of continued right lateral shear, the segmented blocks rotated clockwise simultaneously. Clockwise rotation of the Fort Stockton and Andector Blocks explains: (1) the present left-stepping, *en echelon* pattern of the Fort Stockton and Andector Blocks; (2) the high angle thrusts and reverse faults at SW and NE block corners that dip toward the interior of the CBP and westward-dipping normal faults at the NW corner of the Fort Stockton Block (Figs. 7A–C); (3) the greatest vertical displacement that is observed along the boundary fault zones at the SW and NE corners of these blocks; (4) progressively decreasing amounts of basement shortening from south to north along the western side of the Fort Stockton and Andector Blocks and from north to south along their eastern sides (Fig. 10); and (5) left-lateral strike-slip deformation features (flower structures, high angle faults with variable vertical displacement along strike) along generally E-W trending cross-faults that separate the blocks.

Our model differs from that of Shumaker (1992) in that he attributes the clockwise rotation of crustal blocks which comprise the CBP to left-lateral displacement along east-west trending cross faults that separate the blocks. We prefer instead to attribute clockwise rotation to dominantly right-lateral displacement along NNW-SSE fault zones which bounded the CBP during its initial stages of formation. Evidence that supports our model includes: (1) extensive north-south trending *en echelon* folds and faults (including flower structures) just north of the Andector Block which are a northerly expression of the right-lateral displacement (Yang and Dorobek, 1992; Yang, 1993); and (2) the close match between east-west trending, steeply dipping faults on either side of the CBP (Big Lake and Grisham fault zones of Ewing, 1991), if it is assumed that those faults were once continuous across the CBP and were subsequently offset by right-lateral displacement, as proposed here (Fig. 11). We do not argue that left-lateral displacement did not occur along the Big Lake and Grisham fault zones, but rather that displacement on these east-west faults was a consequence of

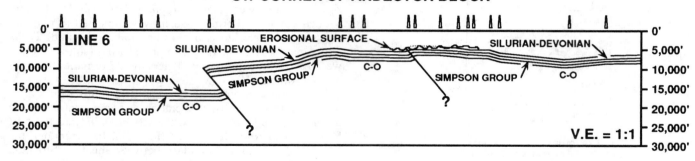

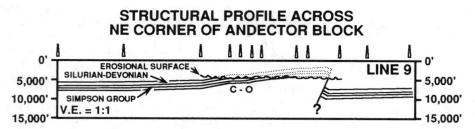

FIG. 9—Reconstructed structural cross sections across the SW and NE corners of the Andector Block. Small triangles at top of cross section indicate locations of wells used; see Yang (1993) for actual well locations. Note east–dipping reverse faults that characterize the boundary fault zone along the SW corner of the Andector Block (Line 6) and the west–dipping reverse faults that characterize the NE corner (Line 9). The interior parts of the Andector Block may be more structurally complex than is shown on these cross sections (cf. Ewing, 1991). Also note the erosional surface (indicated by an arrow) that truncates lower Paleozoic strata and Precambrian basement on the upper part of the Andector Block. Reconstructed lower Paleozoic strata above the erosional surface (indicated by an arrow) are shown as shaded lines across the top of the Andector Block.

right-lateral displacement along the NNW-SSE trending boundary faults of the CBP and concomitant clockwise rotation of the blocks that comprise the CBP.

Our block rotation model can not explain all of the complex structural relationships that characterize the Permian Basin area nor can it explain all of the uplift patterns associated with the CBP. First of all, the Fort Stockton and Andector Blocks are complex features with different areal dimensions and are not the simple polygons shown in map view in Figure 11. Thus, complex patterns of extensional and compressional deformation are to be expected along the boundaries of these blocks. Secondly, the eastern and western boundary fault zones that define the CBP are not parallel. As block rotation proceeded, some local areas may have been subjected to episodes of compression followed by extension, or vice versa (Fig. 11). This is especially true for mid-points along the block boundaries. Present strike-lengths of thrust complexes and amounts of basement shortening in various zones of compression vary at different block corners because the eastern and western boundary fault zones are non-parallel. Lastly, the interior of the Fort Stockton has minor deformation, whereas the interior of the Andector Block contains several zones of faults and folds. This suggests that the Andector Block may not have behaved as rigidly during rotation as the Fort Stockton Block. Further detailed structural analyses are necessary to refine our kinematic model, but clockwise block rotation still explains most of the structural features that we and others have documented.

EFFECTS OF LATE PALEOZOIC TECTONISM ON STRATIGRAPHIC RELATIONSHIPS ACROSS THE CENTRAL BASIN PLATFORM AND ADJACENT BASINAL AREAS

Regional stratigraphic relationships for lower Ordovician to Lower Permian strata in the Permian Basin region are shown on several stratigraphic cross sections (see Fig. 6 for locations of cross sections); more detailed stratigraphic relationships across selected boundary fault zones of the CBP are shown on seismic profiles (Fig. 7). These cross sections illustrate regional differences in basin geometry and the effects of differential uplift of the CBP on the adjacent basins. Patterns of stratal truncation and onlap in the sub-basins and on the CBP are important for constraining the timing of foreland deformation and regional subsidence patterns. In turn, the different structural styles associated with late Paleozoic deformation of the CBP and in the adjacent basins greatly affected stratigraphic evolution and facies distribution in the Permian Basin.

All of our stratigraphic cross sections are hung from the top of Wolfcampian (lowermost Permian) strata. The top of Wolfcampian strata is easily identifiable on well logs. Well-log character was the only criterion used in this study to pick formation tops; no biostratigraphic data were available to us. Even if our picks for formation tops are in error by as much as a few hundred meters, it does not significantly affect any of our general tectonic interpretations or model results. For simplicity, our stratigraphic cross sections also indicate that Wolfcampian strata completely filled the Midland, Delaware, and Val Verde basins *to sea level*.

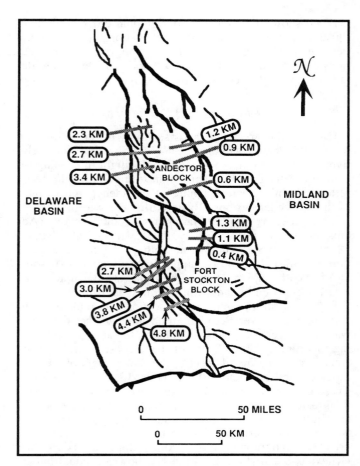

Fig. 10—Map showing the variation in amount of basement shortening along the boundaries of the CBP. The amount of basement shortening decreases toward the north along the western boundaries of the Fort Stockton and Andector Blocks but decreases toward the south along the eastern boundaries of the blocks.

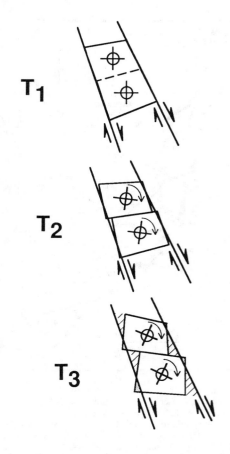

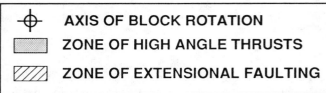

Fig. 11—Kinematic model for the origin of the CBP. At T1, preexisting east-west transverse faults might have been cut by a right-lateral shear couple along the boundaries of the CBP. At T2, the CBP split into segmented blocks which then started to rotate in a clockwise direction (indicated by curved arrows). Because the boundaries of these blocks were constrained by non-parallel NNW-SSE trending bounding fault zones, shortening and extension due to block rotation occurred at the corners of the blocks. At T3, the blocks continued to rotate. Note that because the CBP is bounded by non-parallel boundaries, the amount of shortening at the SW block corners would be larger than at the NE block corners.

This is not true, as these sub-basins may have had water depths of over several hundred meters during Early Permian deposition. Without detailed, regional facies analyses of Wolfcampian strata, we were not able to constrain water depths across the Permian Basin, so we opted for this simplification.

The "top of Wolfcampian" datum was chosen because it is the last prominent stratigraphic horizon that is cut by major faults along the boundaries of the CBP and in the adjacent basins. Thus, uppermost Wolfcampian strata were the last units to be deposited during uplift of the CBP. We identify the earliest phases of late Paleozoic deformation throughout most of the Permian Basin region by the first evidence for major changes in stratal thickness in the basins. The middle Pennsylvanian (Desmoinesian) Strawn Limestone generally has constant thickness across most of the Permian Basin region. Pre-Strawn deformation affected many parts of the Permian Basin (cf. Hills, 1984; Wuellner and others, 1986), but our subsidence analyses and stratigraphic cross sections indicate the major phase of deformation post-dates deposition of the Strawn Limestone. Upper Pennsylvanian (Missourian) units are the oldest strata that display significant changes in original thickness (i.e., not due to post-depositional erosion) on stratigraphic cross sections. Thus, except for the tops of uplifted areas and some local basinal areas, upper Pennsylvanian through Wolfcampian strata were deposited across the Permian Basin region during the most significant phase of deformation.

We make the simplifying assumption that the cross-sectional profile of upper Pennsylvanian through Wolfcampian strata reflects the actual basement profile (i.e., top of pre-upper Pennsylvanian strata) in the sub-basins that comprise

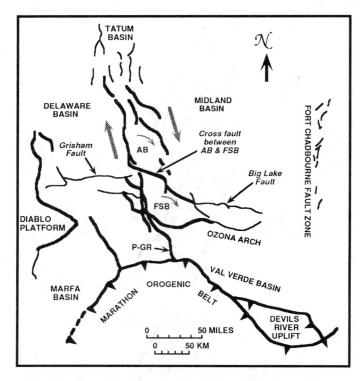

FIG. 12—Generalized tectonic map for Permian Basin region which shows possible right-lateral offset in the Big Lake and Grisham fault zones on either side of the CBP. This offset is additional evidence in support of the kinematic model shown in Figure 11. Shaded arrows indicate right-lateral shear across the CBP's boundary faults and clockwise rotation of the Fort Stockton (FSB) and Andector (AB) Blocks. Cross fault between the Fort Stockton and Andector Blocks can be rotated counter-clockwise to link up the Big Lake and Grisham faults. See text for additional discussion.

the Permian Basin. This assumption is important for understanding the general driving mechanisms for basinal subsidence and for the flexural models presented in our accompanying paper in this volume. This assumption also is reasonable along most of the length of the stratigraphic cross sections because upper Pennsylvanian through Wolfcampian strata in the basins are largely deep water facies (turbidites and hemipelagic facies) that were deposited on a basin floor with minor(?) depositional dip. Hanging stratigraphic cross sections on the top of Wolfcampian strata produces a view of stratigraphic thicknesses that probably closely mimics true depth to basement at the end of Wolfcampian time, if sediment compaction and water depth within the sub-basins are ignored. The only part of the cross sections where this assumption is invalid is along the boundary fault zones of the CBP where depositional dip and topographic relief were significant during late Pennsylvanian to Wolfcampian time (e.g., Figs. 7, 8, 9). Stratigraphy across the boundary fault zones, however, comprise only a small part (<5%) of the stratigraphic cross sections.

Lastly, stratigraphic relationships within the upper Pennsylvanian to Wolfcampian interval were used to document the timing and styles of syntectonic deformation and patterns of onlap across the Permian Basin region. In this paper, we show only selected stratigraphic cross sections to illustrate the general stratigraphic relationships across the Permian Basin region. Additional stratigraphic cross sections can be found in Yang (1993) and provide a more detailed view of the regional stratigraphy.

Delaware Basin Stratigraphy

On all Delaware Basin cross sections, lower Ordovician to middle Pennsylvanian (Desmoinesian) strata are relatively uniform in thickness, with only slight thickening toward the CBP (Fig. 13). This slight stratigraphic thickening reflects slightly greater amounts of subsidence in the center of the antecedent Tobosa Basin during early to middle Paleozoic time, prior to the Marathon-Ouachita orogeny (Adams and others, 1951; Galley, 1958; Hills, 1984; Horak, 1985; Ross, 1986; Wuellner and others, 1986; Frenzel and others, 1988). In contrast, upper Pennsylvanian (Missourian) to Wolfcampian strata in the Delaware Basin are characterized by rapid changes in thickness and lithofacies.

East-west stratigraphic cross sections of middle Pennsylvanian to Lower Permian strata indicate that the Delaware Basin generally was asymmetric toward the CBP during this time interval. Some of the deepest portions of the Delaware Basin are adjacent to the SW corners of the Fort Stockton and Andector Blocks (Fig. 13, Line E and G). These deep parts of the Delaware Basin also coincide with the greatest amount of structural relief between the boundary fault zone of the CBP and adjacent parts of the Delaware Basin (locally up to 6 km). In contrast, north-south oriented stratigraphic cross sections along the axis of the Delaware Basin (Fig. 13, Line H) show a gradual but irregular increase in thickness of upper Pennsylvanian to Wolfcampian strata toward the Marathon thrust front.

Upper Pennsylvanian to Wolfcampian strata in the Delaware Basin generally thicken toward the SW corner of the Fort Stockton Block (Figs. 7A; 13, Line E; 14A). These units then rapidly pinch out onto the crest of the block where younger Leonardian strata directly overlie lower Paleozoic strata or Precambrian basement rocks (Figs. 6A; 13, Line E; 14A). Within the boundary fault zone, middle Pennsylvanian (Desmoinesian) to lower Wolfcampian strata unconformably overlie the Mississippian Barnett Shale. The pre-Desmoinesian unconformity in the boundary fault zone grades westward into a conformable surface in the Delaware Basin. Middle Pennsylvanian to Wolfcampian strata are thickest in the center of the Delaware Basin and then thin toward the western side of the basin (Fig. 13, Line E and F). Stratal thicknesses in the Delaware Basin, especially Wolfcampian strata, also were locally affected by late Paleozoic movement along steeply dipping strike-slip(?) faults within the basin. Thickened strata were deposited on the downthrown side of these faults (Fig. 13, Line F, H).

Depositional gradients across the boundary fault zone at the SW corner of the Fort Stockton Block were very high as indicated by the structural relief shown on seismic profiles and by the rapid facies changes, thickness changes, and significant stratal truncation from the crest of the Fort Stockton Block to the adjacent Delaware Basin (Figs. 7A; 13, Line E; 14A). Carbonate facies characterize most of the

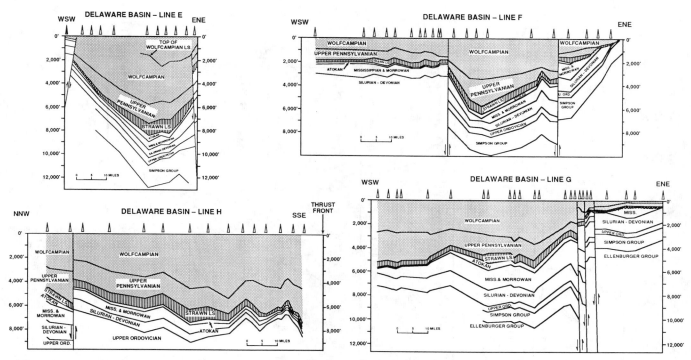

Fig. 13—Delaware Basin stratigraphic cross sections. Small triangles at top of cross section indicate locations of wells used; see Yang (1993) for actual well locations. *Line E*, stratigraphic cross section across the southern part of the Delaware Basin. Cross section extends from the SW corner of the Fort Stockton Block westward into the Delaware Basin. In this and all other stratigraphic cross sections, the shaded area indicates synorogenic strata (upper Pennsylvanian through Wolfcampian strata) and the horizontal datum at the top of each cross section is the top of Wolfcampian strata. Note that the synorogenic strata are thickest in the eastern part of the basin next to the CBP. The transition zone between the platform and the basin is narrow, and the synorogenic strata thin rapidly eastward from the Delaware Basin onto the CBP. *Line F*, east-west stratigraphic cross section across the center of the Delaware Basin. Cross section extends from the NW corner of the Fort Stockton Block westward into the Delaware Basin. Note that the synorogenic strata are thickest in the middle part of the basin. The transition zone between the platform and the basin is wide, and synorogenic strata gradually onlap eroded lower Paleozoic strata toward the NE corner of the Fort Stockton Block. *Line G*, stratigraphic cross section from the northern part of the Delaware Basin. Cross section extends from the SW corner of the Andector Block into the Delaware Basin. *Line H*, stratigraphic cross section from the southern part of the Delaware Basin. Cross section extends perpendicular to the frontal thrust of the Marathon orogenic belt into the southern part of the Delaware Basin. Note that synorogenic strata do not thicken dramatically toward the Marathon orogenic belt.

upper Pennsylvanian(?) to Wolfcampian stratigraphy on the Fort Stockton Block and in the boundary fault zone. However, these carbonate facies abruptly grade into much thicker shale facies in the Delaware Basin west of the boundary fault zone. Thick upper Pennsylvanian to Wolfcampian shales in the deepest part of the Delaware Basin then thin westward toward the Diablo Platform (Fig. 13, Line E, F).

Middle Pennsylvanian to Wolfcampian stratal onlap and facies patterns are much different on the northern part of the Fort Stockton Block than on its southern part because of differences in the structural styles and antecedent topography in both areas (Figs. 7C; 13, Line F). On the northern part of the Fort Stockton Block, middle Pennsylvanian to Wolfcampian strata are thicker and onlap farther east across the top of the CBP than at the southern thrust-faulted corner of the block. Middle Pennsylvanian to Wolfcampian strata, however, also eventually wedge out eastward toward the NE thrust-faulted corner of the Fort Stockton Block where Leonardian strata ultimately overlie the Ordovician Ellenburger Group. The unconformity beneath middle Pennsylvanian to Wolfcampian strata in the boundary fault zone becomes a conformable surface westward in the Delaware Basin.

Middle Pennsylvanian to Wolfcampian strata from the northern part of the Fort Stockton Block mainly consist of dolomitic limestone. These carbonate facies grade westward into siliciclastic or mixed carbonate-siliciclastic turbidite facies that were deposited on the deep basin floor (Fig. 14B). Turbidite facies, in turn, grade westward into basinal shales which reach their maximum thickness in the basin center and then thin gradually toward the western side of the basin. Stratal thicknesses in this part of the Delaware Basin, especially for Wolfcampian strata, also were affected by late Paleozoic movement along steeply dipping strike-slip(?) faults within the basin. Thickened strata occur on the locally downthrown side of these faults.

On the north-south cross section that extends along the axis of the Delaware Basin (Fig. 13, Line H), upper Pennsylvanian through Wolfcampian strata increase in thickness toward the Marathon thrust front. Lithofacies along the entire length of this cross section consist mostly of shale and siltstone. In the southernmost part of the cross section, over

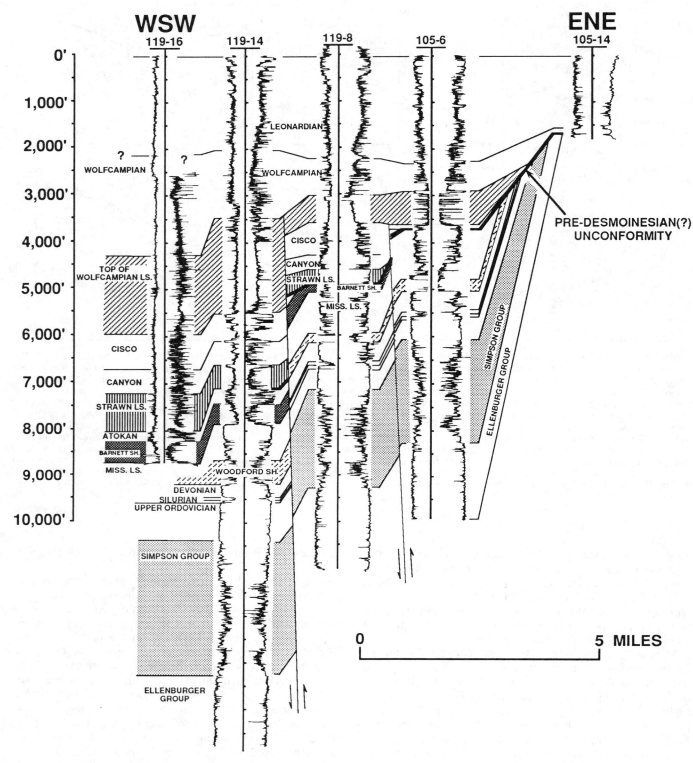

FIG. 14—Well-log cross sections from the corners of the Fort Stockton Block into the adjacent Delaware Basin. See Yang (1993) for exact locations of wells indicated at tops of each log. (A) Well-log cross section across the SW corner of the Fort Stockton Block. Note that lithofacies in the Strawn Limestone and upper Pennsylvanian through Wolfcampian units rapidly change from carbonates at the margin of the Fort Stockton Block to shale beyond the boundary fault zone. (B) Well-log cross section across the NW corner of the Fort Stockton Block. Note that lithofacies changes in the Strawn Limestone and upper Pennsylvanian through Wolfcampian units are more gradual across the boundary fault zone. The unconformity between the Strawn Limestone and overlying strata also becomes a conformable surface in the basin.

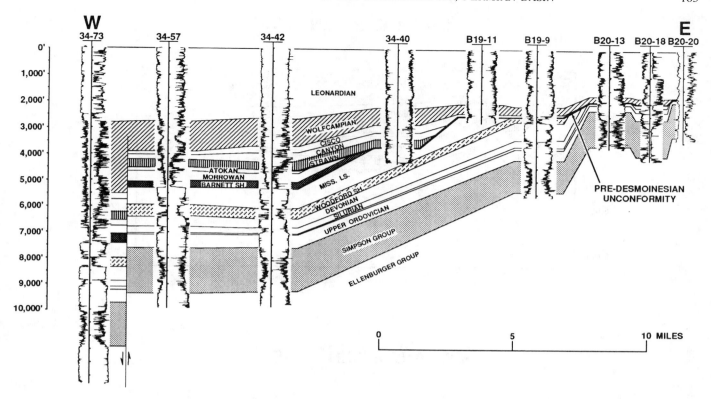

Fig. 14—Continued.

less than 20-km horizontal distance, upper Pennsylvanian through Wolfcampian strata abruptly thicken toward the thrust front. Displacement along high angle faults within the basin apparently also caused local variations in stratigraphic thickness. In contrast to upper Pennsylvanian through Wolfcampian strata, underlying lower Paleozoic through middle Pennsylvanian strata thicken away from the Marathon orogenic belt.

Midland Basin Stratigraphy

Compared to equivalent strata in the Delaware Basin, upper Pennsylvanian to Wolfcampian strata along east-west basin profiles through the Midland Basin are thinner overall, with no dramatic changes in thickness. Two SW-NE stratigraphic cross sections across the Midland Basin, one extending from the NE corner and the other from the SE corner of the Fort Stockton Block (Fig. 15), show that lower Paleozoic to middle Pennsylvanian strata are thickest immediately adjacent to the CBP and thin eastward toward the Eastern Shelf. Lithofacies within these stratigraphic units are also relatively uniform or change gradually across the basin. In contrast, overlying upper Pennsylvanian to Wolfcampian strata are thickest in easternmost parts of the cross sections, although there is some stratigraphic thickening adjacent to the boundary fault zone of the CBP.

Shallow water carbonate lithofacies of the middle Pennsylvanian Strawn Limestone extend across most of the Midland Basin and have relatively constant thickness (Hanson and others, 1991). Lithofacies in upper Pennsylvanian to Wolfcampian strata, however, change abruptly near the eastern and western edges of the Midland Basin. Upper Pennsylvanian to Wolfcampian strata on both the eastern and western sides of the Midland Basin consist dominantly of carbonate platform facies that change basinward into fine-grained siliciclastic facies. The westward limit of onlap onto the eastern side of the CBP is a function of the topographic relief and regional dip of the pre-upper Pennsylvanian surface (see below). In the basin center, thick, upper Pennsylvanian shales overlie shallow water carbonates of the Strawn Limestone. Near the eastern side of the Midland Basin, upper Pennsylvanian to Wolfcampian carbonate strata are thicker than in the basin center. These carbonate strata form a series of westward prograding platforms that built out from the Eastern Shelf and into the Midland Basin (Brown and others, 1973).

The greatest differences between the two east-west Midland Basin cross sections are along the boundary fault zone of the CBP. On the cross section that extends from the SE corner of the Fort Stockton Block, middle Pennsylvanian to Wolfcampian strata onlap westward onto eroded lower Paleozoic strata (Figs. 15, Line I; 16); successively younger strata onlap progressively farther west onto the CBP. On the cross section that extends across the NE corner of the block and into the adjacent basinal area (Fig. 15, Line J), onlap patterns are much different. Lower Leonardian strata unconformably overlie the Ordovician Ellenburger Group along the easternmost edge of the CBP with no intervening onlap of middle Pennsylvanian to Wolfcampian strata. A

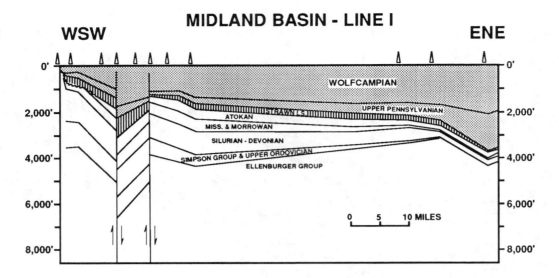

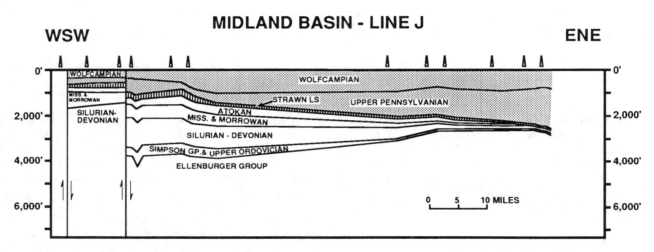

Fig. 15—Midland Basin stratigraphic cross sections. Small triangles at top of cross section indicate locations of wells used; see Yang (1993) for actual well locations. *Line I*, stratigraphic cross section from the southern part of the Midland Basin. Cross section extends from the SE boundary of the Fort Stockton Block into the Midland Basin. Note that the Midland Basin generally is more symmetrical than the Delaware Basin profiles shown in Figure 13, except for the anomalous deflection in the eastern end of this profile. Also note that lower Paleozoic strata thin eastward away from the Fort Stockton Block, while synorogenic upper Pennsylvanian to Lower Permian strata thicken eastward. The Strawn Limestone and synorogenic strata also onlap westward onto the Fort Stockton Block in contrast to the eastward onlap patterns for coeval strata at the NW corner of this block. *Line J*, stratigraphic cross section from the middle part of the Midland Basin. Cross section extends from the NE boundary of the Fort Stockton Block into the Midland Basin. Note essentially no onlap of synorogenic strata onto the CBP.

relatively continuous lower Paleozoic to Wolfcampian stratigraphic section, however, was deposited basinward of the boundary fault at this corner.

Val Verde Basin Stratigraphy

The sedimentary fill in the Val Verde Basin can be divided into lower and upper Paleozoic packages that are separated by a regional unconformity (Fig. 17, Lines A to D). Upper Mississippian to lower Pennsylvanian strata are missing over the entire Val Verde Basin area except in the proximal part of the basin immediately adjacent to the Marathon orogenic wedge (Fig. 17, Lines A to D). On a north-south cross section across the Val Verde Basin, the amount of missing upper Mississippian to lower Pennsylvanian stratigraphy progressively increases northward and reaches a maximum on the crest of the Ozona Arch (Fig. 17, Lines A, B). In the Eastern Shelf area, middle Pennsylvanian strata unconformably overlie the Ordovician Ellenburger Group (Fig. 17, Lines C, D; Brown and others, 1973), indicating significant erosion and/or non-deposition of Ordovician through lower Pennsylvanian strata.

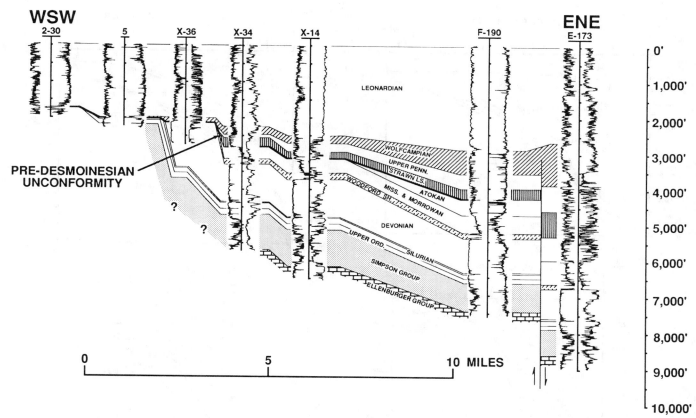

FIG. 16—Well-log cross section across the SE corner of the Fort Stockton Block. See Yang (1993) for exact locations of wells indicated at tops of each log. Note that middle Pennsylvanian to Wolfcampian strata onlap westward onto eroded lower Paleozoic strata. Successively younger strata onlap progressively farther west onto the CBP.

The unconformity on top of lower to middle Paleozoic strata in the Val Verde Basin is overlain by the relatively thin, middle Pennsylvanian Strawn Limestone. Shallow marine carbonate facies of the Strawn Limestone (Hanson and others, 1991) have relatively uniform thickness compared to overlying upper Pennsylvanian to Wolfcampian strata. In western parts of the Val Verde Basin, upper Pennsylvanian to Wolfcampian facies bordering the CBP consist dominantly of coarse-grained sandstone, conglomerate, and limestone(?) (Fig. 18); these coarse-grained facies change southward, toward the axis of the Val Verde Basin, into shale (Hanson and others, 1991).

The thickness of upper Pennsylvanian to Wolfcampian strata increases greatly from north to south across the Val Verde Basin, toward the Marathon orogenic belt (Fig. 17). The thickest part of the stratigraphic section consists mostly of shale. Overlying Leonardian strata, however, have relatively uniform thickness across the Val Verde Basin, Ozona Arch, and Midland Basin, except locally in the southern part of the Midland Basin where this interval was partially eroded during regional uplift.

LATE PALEOZOIC TECTONIC AND KINEMATIC HISTORY OF THE PERMIAN BASIN

The stratigraphic and structural relationships described above constrain the timing of late Paleozoic deformation in the Permian Basin region. Deformation across the Permian Basin region was contemporaneous with deformation in the Marathon-Ouachita orogenic belt (Galley, 1958; Hills, 1970, 1984). Thus, we refer to phases of major faulting, uplift, and subsidence across the Permian Basin region as "synorogenic" and later, less tectonically active intervals as "postorogenic" stages of basin evolution. When basinwide stratigraphic relationships and subsidence histories are examined in detail, a complex history of deposition, differential subsidence, and erosion can be documented (cf. Gardiner, 1990; Hanson and others, 1991).

Early Stages of Synorogenic Deposition across the Permian Basin

Lower Paleozoic strata across the antecedent Tobosa Basin consist mostly of shallow water carbonate facies, with some thin interbedded shales. Deposition of these shallow water facies continued until late Mississippian time when initial collision between Gondwanaland and Laurasia began in the Marathon-Ouachita area to the south and east of the Tobosa Basin (Ross and Ross, 1985; Ross, 1979, 1986; Hills, 1984, 1985; Wuellner and others, 1986; Frenzel and others, 1988). Thick siliciclastic deposits of the upper Mississippian to lower Pennsylvanian Tesnus Formation provide the first stratigraphic record of convergence in the

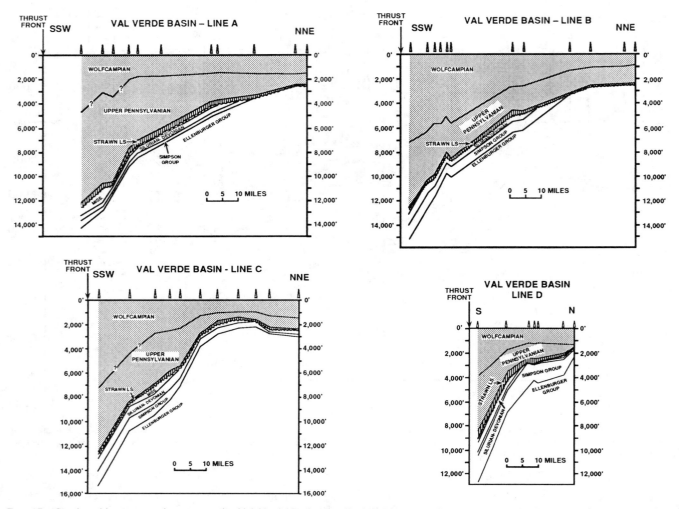

Fig. 17.—Stratigraphic cross sections across the Val Verde Basin. Small triangles at top of cross section indicate locations of wells used; see Yang (1993) for actual well locations. *Line A* extends from the frontal thrust of the Devils River Uplift to southern part of the Eastern Shelf. *Line B* extends from the frontal thrust of the Marathon orogenic belt to the southern part of the Midland Basin. *Line C* extends from the frontal thrust of the Marathon orogenic belt to southern part of the Midland Basin. *Line D* extends from the frontal thrust of the Marathon orogenic belt to southern part of the CBP. Note the regional unconformity that separates lower Paleozoic strata and upper Pennsylvanian to Lower Permian strata on all the cross sections. The Strawn Limestone above this unconformity surface has relatively uniform thickness across each cross section. However, upper Pennsylvanian to Lower Permian strata thicken dramatically toward the Marathon thrust front. Also note on Lines C and D that maximum erosion of pre-Strawn strata is at the same approximate location where synorogenic strata are thinnest.

Marathon orogenic belt (Ross, 1981, 1986; Wuellner and others, 1986; Frenzel and others, 1988). The Tesnus Formation represents deep water flysch deposited in proximal parts of the earliest Marathon foredeep (McBride, 1978, 1989; Hanson and others, 1991). In the Delaware Basin, strata equivalent to the Tesnus Formation include the upper Mississippian Barnett Shale and lower Pennsylvanian "Morrowan sand." These units are thinner than the Tesnus Formation, and their thickness is relatively constant across the basin. This suggests there was no significant differential subsidence over much of the Delaware Basin from late Mississippian to early Pennsylvanian time. The Tesnus Formation was also probably deposited in a basin that was much farther to the south or southeast of the Delaware Basin (McBride, 1978, 1989; Ewing, pers. commun., 1993).

The Barnett Shale is easily picked on well logs from the Midland Basin, but Morrowan strata are more difficult to identify. Accordingly, some workers have suggested that Morrowan strata are absent in the Midland Basin, with the tectonic implication that little subsidence occurred there during early Pennsylvanian time (Hanson and others, 1991). Our regional observations of Morrowan strata around the Permian Basin indicate that characteristic gamma-ray signatures for Morrowan strata in the Delaware Basin can be identified locally in the Midland Basin. Thus, on our stratigraphic cross sections across the Midland Basin, Morrowan(?) strata are tentatively included with upper Mississippian strata (Fig. 15). Regardless, there was no significant differential subsidence in the Midland Basin during late Mississippian to early Pennsylvanian time. The Barnett Shale

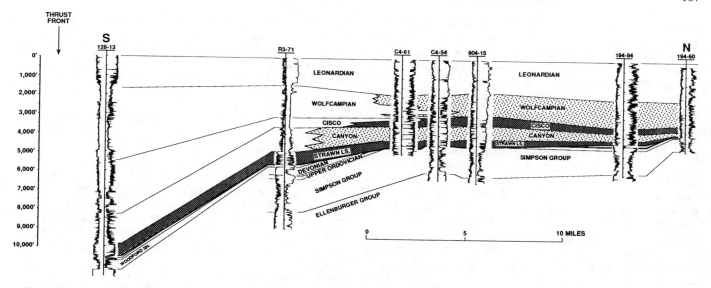

Fig. 18—Well-log cross section across the western part of the Val Verde Basin. See Yang (1993) for exact locations of wells indicated at tops of each log. Note that upper Pennsylvanian to Wolfcampian strata change from carbonate facies in the north to shale in southern parts of the basin. Carbonate facies in upper Pennsylvanian to Wolfcampian strata also progressively backstep northward over time and the thickness of shale facies increases toward the south.

and Morrowan strata are absent over most of the CBP either because of non-deposition of these units on top of the CBP and/or post-Morrowan uplift and erosion.

A regional pre-Desmoinesian unconformity across the Permian Basin also constrains the timing of tectonic activity. Upper Mississippian to lower Pennsylvanian strata in the Val Verde Basin are missing although, within thrust sheets of the Marathon orogenic belt to the south, Mississippian through lower Pennsylvanian strata are preserved (Ross, 1981). Differences in the stratigraphy of the two areas indicate that a foreland basin first formed during the late Mississippian to early Pennsylvanian in the Marathon region, well south of the present Val Verde Basin. Early foredeep facies of the upper Mississippian-lower Pennsylvanian Tesnus, Dimple, and Haymond Formations were then folded and thrust during later Marathon-Ouachita deformation and tectonically transported northward. During late Pennsylvanian to Early Permian time, the orogenic belt moved northward to its present location, and the present Val Verde Basin was the foredeep in front of the orogen.

The basinwide unconformity beneath the middle Pennsylvanian Strawn Limestone indicates that regional, pre-Desmoinesian uplift occurred across the present position of the Val Verde Basin and much of the CBP. The greatest amount of missing, lower to middle Paleozoic section is along the crest of the present Ozona Arch (Fig. 17).

Shallow water carbonate facies of the Strawn Limestone that overlie the pre-Desmoinesian unconformity (Hanson and others, 1991) maintain constant thickness across the Val Verde Basin, right up to the Marathon thrust front. Strawn carbonate strata probably were deposited on a ramp that extended across all of the non-overthrusted parts of the Val Verde Basin.

No significant pre-Strawn uplift occurred in the Delaware Basin and across much of the Midland Basin (excluding the Ozona Arch). The pre-Desmoinesian unconformity across the Val Verde Basin can be traced westward to the Puckett-Grey Ranch fault zone between the Delaware and Val Verde Basins (Fig. 4). West of the Puckett-Grey Ranch fault zone, apparently no Mississippian to lower Pennsylvanian section is missing throughout the Delaware Basin. This indicates that vertical displacement along the Puckett-Grey Ranch fault zone determined the westward limit of the pre-Desmoinesian unconformity in the Val Verde Basin. North of the Ozona Arch, the pre-Desmoinesian unconformity becomes a conformable surface in the center of the Midland Basin. Regional stratigraphic relationships between the sub-basins therefore indicate that the pre-Desmoinesian unconformity is areally limited to the Val Verde Basin/Ozona Arch area and to the top of the CBP.

Thickness of the middle Pennsylvanian Strawn Limestone is relatively uniform across most of the Midland and Val Verde Basins although it is somewhat thicker in the center of the Delaware Basin (Fig. 13). In contrast, strata that overlie the Strawn Limestone in the Midland Basin thicken away from the CBP, the exact opposite of thickness trends in Mississippian through Strawn strata (Fig. 15). This thickening suggests that subsidence patterns in the Midland Basin changed dramatically at the end of the middle Pennsylvanian. In the Val Verde Basin, upper Pennsylvanian through Wolfcampian strata thicken dramatically toward the Marathon orogenic belt (Figs. 17, 18) while the Strawn Limestone maintains relatively constant thickness from the southern part of the basin to the Ozona Arch. These thickness trends indicate that subsidence in the Val Verde Basin also changed dramatically at the end of middle Pennsylvanian deposition. Upper Pennsylvanian strata in the Delaware Basin also thicken toward the Marathon orogenic belt and locally along the western side of the CBP. This contrasts with only minor thickening of Mississippian through

Strawn strata in the center of the Delaware Basin (Fig. 13). These regional variations in stratigraphic thickness indicate that while initial deformation in the Marathon-Ouachita orogenic belt may have begun in the Mississippian, significant differential subsidence in the present Midland, Delaware, and Val Verde Basins did not start until after the middle Pennsylvanian.

Lithofacies variations in middle to upper Pennsylvanian strata across the Midland, Delaware, and Val Verde Basins provide additional evidence for the main episode of differential subsidence in these basins. The middle Pennsylvanian Strawn Limestone is a thin interval of carbonate strata that was deposited in the Midland and Val Verde Basins prior to deposition of overlying deep water siliciclastic facies. The Strawn Limestone in the Delaware Basin, however, is characterized by rapid lateral facies changes. Shallow carbonate platform facies characterize the Strawn Limestone locally along the western side of the CBP (Hanson and others, 1991); these facies grade westward into deep-water shale facies in the adjacent Delaware Basin. Thus, subsidence was more rapid in the Delaware Basin than in the Midland or Val Verde Basins (at least in northern, non-overthrusted parts of the Val Verde Basin) during the Desmoinesian. In contrast to middle Pennsylvanian lithofacies patterns, upper Pennsylvanian strata throughout the Permian Basin region are characterized by rapid lateral facies changes from shallow carbonate platform facies on top of paleotopographic highs to deep-water shale facies in adjacent basinal areas. These lateral facies changes indicate that significant differential subsidence in all basins began in the late Pennsylvanian.

Quantitative subsidence analyses also indicate the timing of tectonically driven subsidence at selected localities in the Delaware (Figs. 19A, B), Midland (Figs. 19C, D), and Val Verde Basins (Fig. 19E). All the subsidence curves have similar forms throughout the Paleozoic. Gradual subsidence occurred across the entire Tobosa Basin/Permian Basin region from early Paleozoic to early Pennsylvanian time. A major pulse of subsidence, however, occurred in all three basins beginning in middle to late Pennsylvanian time (at ~290–300 Ma).

End of Major Late Paleozoic Tectonic Activity in the Permian Basin Region

The precise end of synorogenic deposition across the Permian Basin is difficult to identify from subsurface data. Fault terminations in particular stratigraphic intervals, stratal onlap patterns, and subsidence curves provide some constraints on the end of major tectonic activity in the Marathon-Ouachita foreland.

High-angle faults locally terminate in lower Wolfcampian strata along the boundary fault zones between the CBP and the adjacent Delaware and Midland Basins (Silver and Todd, 1969). A regional unconformity also separates lower and upper Wolfcampian strata across much of the CBP's margins and may mark the end of synorogenic uplift (Silver and Todd, 1969). On the seismic profile across the SW corner of the Fort Stockton Block (Fig. 7A), the top of lower Wolfcampian limestone can be traced eastward where it pinches out toward the crest of the CBP. East of this pinch-out, an angular unconformity separates upper Wolfcampian strata from lower Paleozoic strata. Pinch-out of lower Wolfcampian limestone may indicate the limit of early Wolfcampian onlap onto the CBP or it may record erosional beveling of lower Wolfcampian strata during the last phases of uplift of the CBP.

In the Val Verde Basin and southern Delaware Basin, the end of late Paleozoic deformation is indicated in the frontal part of the Marathon orogenic belt where the youngest thrust faults cut through upper Pennsylvanian to lower Wolfcampian strata (Ross, 1981, 1986; Wuellner and others, 1986; Reed and Strickler, 1990). Upper Wolfcampian strata unconformably overlie the thrust sheets along the leading edge of the orogenic belt (Moore and others, 1981; Ross and Ross, 1985; Reed and Strickler, 1990), which indicates that shortening in the Marathon orogenic belt to the south of the Val Verde Basin ceased by the end of early Wolfcampian time. Well-log correlations across the Val Verde Basin, however, indicate the entire Wolfcampian section is a coarsening-upward sequence that prograded away from the Marathon orogenic belt and filled the basin during late Wolfcampian time (Fig. 18). Thickening of upper Wolfcampian strata toward the Marathons reflects the greater amount of flexural subsidence in the Val Verde Basin caused by the orogenic belt. Other workers have also suggested that the Val Verde Basin formed during final stages of shortening in the Marathon orogenic belt but was not filled with sediments until latest Wolfcampian time (Sanders and others, 1983; Wuellner and others, 1986). Therefore, while steeply dipping reverse faults only cut through lower Wolfcampian strata, it can be argued that the southward thickening wedge of upper Wolfcampian strata in the Val Verde Basin should be included with the synorogenic basin fill. In contrast, the lower Leonardian Spraberry Formation and correlative strata that overlie upper Wolfcampian strata are relatively uniform in thickness, and lithofacies do not vary significantly across the Val Verde Basin. This indicates that subsidence due to flexural loading by the Marathon orogenic belt ceased after the Wolfcampian, and later regional subsidence in the Val Verde Basin was more uniform.

Lastly, subsidence curves from all three basins also constrain the end of major tectonic activity. The interval of rapid subsidence that began at ~300 Ma slowed considerably at the end of Wolfcampian time (~275 Ma; Fig. 19) but continued until the end of the Permian.

GEODYNAMIC MODEL FOR THE LATE PALEOZOIC HISTORY OF THE PERMIAN BASIN REGION

The geodynamic history of the Permian Basin is a complex, three-dimensional problem that can only be understood by documenting regional stratigraphic relationships and the timing of deformation across the Permian Basin region. Quantitative flexural models in a companion paper in this volume provide a test of these conceptual geodynamic models.

Geodynamic Evolution of the Val Verde Basin

Upper Mississippian to lower Pennsylvanian stratigraphic relationships in the Val Verde Basin and Marathon

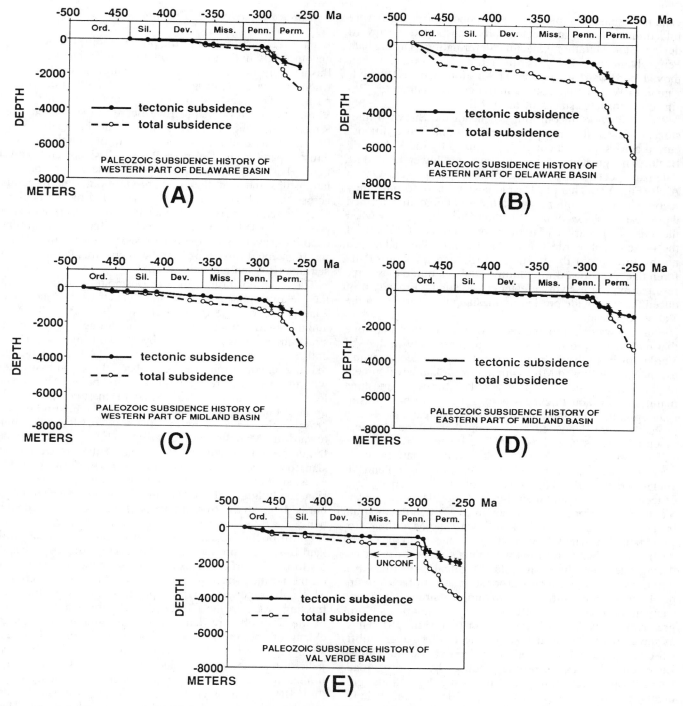

Fig. 19—Subsidence curves for the Delaware (Figs. 19A, B), Midland (Figs. 19C, D), and Val Verde Basins (Fig. 19E). See Figure 6 for well locations used for subsidence analyses. Subsidence in each basin was gradual during early Paleozoic to middle Pennsylvanian time and then increased rapidly during latest middle Pennsylvanian to Early Permian time.

orogenic belt constrain the kinematic history of the *proximal* foreland area during early stages of the Marathon-Ouachita orogeny. Upper Mississippian to lower Pennsylvanian strata in the Val Verde Basin are missing although, within thrust sheets of the Marathon orogenic belt to the south, thick Mississippian through lower Pennsylvanian strata are preserved (Ross, 1981). Differences in the stratigraphy of these two areas indicate that the earliest Marathon foredeep first developed during the late Mississippian to early Pennsylvanian in the Marathon region far south of the present Val Verde Basin (cf. Hanson and others, 1991). Early foredeep facies of the upper Mississippian-lower Pennsylvanian Tesnus, Dimple, and Haymond Formations were folded and thrusted and tectonically transported northward during later Marathon-Ouachita deformation. During late Pennsylvanian to Early Permian time, the orogenic belt moved northward to its present location. Upper Pennsylvanian to Wolfcampian strata in the present Val Verde Basin represent later stages of basin fill.

The basinwide unconformity beneath the Strawn Limestone indicates that regional, pre-Desmoinesian uplift occurred across the area currently occupied by the Val Verde Basin. The greatest amount of missing, lower to middle Paleozoic section is along the crest of the present Ozona Arch (Fig. 17). This pre-Desmoinesian unconformity may record northward migration of the forebulge that formed during earlier phases of the Marathon-Ouachita orogeny when the axis of the earlier Marathon foredeep was far south of the present Val Verde Basin. As the thrust belt advanced northward, so did the flexural profile and position of the forebulge. Northward migration of the forebulge may have resulted in progressively greater erosion and ultimately produced the basinwide pre-Desmoinesian unconformity in the Val Verde Basin.

Shallow-water carbonate facies of the Strawn Limestone that overlie the pre-Desmoinesian unconformity in the Val Verde Basin probably represent a period of tectonic quiescence that followed Mississippian to Atokan development of the initial Marathon foredeep. The Strawn Limestone maintains constant thickness across the Val Verde Basin, right up to the Marathon thrust front. Strawn carbonate strata probably were deposited on a ramp that extended across much of what is now the Val Verde Basin.

The abrupt change from thin Strawn carbonate facies to overlying deep-water siliciclastic facies of upper Pennsylvanian strata marks the beginning of another major phase of tectonic activity and main flexural subsidence in the Val Verde Basin. Upper Pennsylvanian to Wolfcampian strata in the Val Verde Basin thicken toward the Marathon orogenic belt. Estimated depositional gradients from the Ozona Arch to Marathon thrust front were three to four times greater during late Pennsylvanian to late Wolfcampian time than during the Desmoinesian. Carbonate platform facies in upper Pennsylvanian through Wolfcampian strata (Cisco and Canyon Formations, Wolfcampian Limestone) also progressively backstep to the north over time (Fig. 18). Thus, as the Marathon orogen (and/or the Devils River Uplift) episodically moved northward after deposition of the Strawn Limestone, the flexural profile produced by the orogen advanced with it, resulting in the increasing depositional gradients, progressive northward onlap of deep-water shale facies, and backstepping of carbonate facies over time.

Geodynamic Evolution of the Midland and Delaware Basins

Like the Val Verde Basin, the Midland and Delaware Basins apparently were subjected to two major phases of deformation during Pennsylvanian to Early Permian time. The geodynamic history of the Midland and Delaware Basins, however, was affected much more by uplift of the CBP than by the encroaching Marathon orogenic belt.

Middle Pennsylvanian Evolution.—

The areal distribution of the pre-Desmoinesian forebulge unconformity indicates that both the Midland and Delaware Basins were north of the initial flexural profile produced when the Marathon orogenic belt was far south of its final position. The pre-Desmoinesian unconformity in the Val Verde Basin can be mapped westward to the Puckett-Grey Ranch fault zone (Fig. 4). Just west of the Puckett-Grey Ranch fault zone, however, and throughout non-overthrusted parts of the Delaware Basin, Mississippian to lower Pennsylvanian strata are preserved. These stratigraphic relationships indicate that the pre-Desmoinesian forebulge was truncated by the Puckett-Grey Ranch fault zone. The Puckett-Grey Ranch fault zone probably caused segmentation of the initial Marathon foredeep into several sub-basins with separate forebulges. Similar segmented foredeeps have been documented in modern settings (Royden and others, 1987).

The exact position of the pre-Desmoinesian forebulge in the Delaware Basin west of the Puckett-Grey Ranch fault zone is uncertain. Structural cross sections from the Marathon orogenic belt show Ordovician through Mississippian strata unconformably overlain by Strawn Limestone in several thrust sheets (Reed and Strickler, 1990). We suggest that this unconformity records subaerial erosion associated with uplift of the pre-Desmoinesian forebulge. The unconformity surface and overlying Strawn Limestone were caught up in or buried by later Marathon thrusts. Our stratigraphic cross sections also show that the Strawn Limestone thins toward the Marathon orogenic belt. This probably reflects thinning onto the northern backlimb of the pre-Desmoinesian forebulge in the Delaware Basin (Fig. 13, Line H). Thus, strata which record the position of the pre-Desmoinesian forebulge in the Delaware Basin are now within or buried beneath the frontal thrusts of the Marathon orogenic belt.

North of the Ozona Arch toward the center of the Midland Basin, the pre-Desmoinesian unconformity becomes a conformable surface. This represents continuous deposition on the northern backlimb of the early Marathon forebulge. The only other area where pre-Desmoinesian strata are missing is across the CBP, although later phases of uplift of the CBP obscure details about the timing and regional extent of the pre-Desmoinesian unconformity. Regional stratigraphic relationships between the sub-basins therefore indicate that the pre-Desmoinesian unconformity was areally limited to the Val Verde Basin/Ozona Arch area, across the CBP, and areas south of the Delaware Basin that are now occupied by the Marathon orogenic belt. The pre-Desmoinesian unconformity represents subaerial erosion associated with an early Marathon forebulge and probably with early stages of uplift of the CBP.

The thickness and lithofacies composition of the Strawn Limestone are uniform across the Val Verde and Midland Basins, indicating little differential subsidence in those areas during the Desmoinesian. Significant differential subsidence occurred in the Delaware Basin, however, as indicated by the variable thickness of the Strawn Limestone across the basin (Fig. 13). In addition, most of the lower two-thirds of the Strawn Limestone in the Delaware Basin consists of shale. The Strawn Limestone consists entirely of carbonate facies only in the boundary fault zone of the CBP where Strawn carbonates unconformably overlie lower to middle Paleozoic strata (Fig. 14). These stratigraphic relationships provide further evidence that the western boundary fault zone of the CBP, including the Puckett-Grey Ranch fault zone, was tectonically active during middle Pennsylvanian time. It is not clear if crustal shortening occurred along the CBP's western boundary fault zone. There is, however, no obvious indication of flexure near the fault zone because middle Pennsylvanian strata do not thicken dramatically toward the fault zone (Fig. 13). There also is no simple tectonic explanation for the slightly greater subsidence in central parts of the Delaware Basin.

Late Pennsylvanian to Wolfcampian Evolution.—

Uplift of the CBP during late Pennsylvanian to Wolfcampian time was accompanied by basement shortening along its margins (Figs. 7A, C, 8, 9). An important consequence of this shortening is that the CBP then became a topographic load *within* the Marathon-Ouachita foreland that caused flexure in the adjacent Midland and Delaware Basins. The amount of basement shortening and differential uplift across the CBP also is variable, which in turn produced variable basin geometries. For example, the late Pennsylvanian-Wolfcampian geometry of the Delaware Basin is most asymmetric near the SW corner of the Fort Stockton Block where the greatest structural relief and basement shortening along the boundaries of the CBP are located. Basement shortening across the SW corner of the Fort Stockton Block created the largest topographic load that caused flexure and produced the greatest asymmetric basin profile in the adjacent Delaware Basin. East-west Delaware Basin cross sections from the NW corner of the Fort Stockton Block indicate that the basin was relatively symmetric during late Pennsylvanian-Wolfcampian time. This symmetry reflects the eastern vergence of basement shortening across the northern part of the Fort Stockton Block (Figs. 7C; 8, Line 12) and the lesser amount of shortening and uplift at the NW corner of the block.

Throughout late Paleozoic time, the Midland Basin was located north of the flexural profile produced by the Marathon orogenic belt. Upper Pennsylvanian through Wolfcampian strata along north-south Midland Basin cross sections actually thin onto the Ozona Arch, which most likely was the forebulge to the Val Verde foredeep. In contrast, upper Pennsylvanian to Wolfcampian strata in the Delaware Basin thicken somewhat toward the Marathon orogenic belt although not as dramatically as along north-south cross sections through the Val Verde Basin (Figs. 13, Line H; 17). Thus, the Marathon orogenic wedge also affected subsidence in the Delaware Basin during late Pennsylvanian-Wolfcampian time when the wedge had advanced far enough northward to cause flexure in southern parts of the Delaware Basin. The late Pennsylvanian-Wolfcampian, north-south profile of the Delaware Basin, however, is less dramatic than in the Val Verde Basin. This less dramatic indication of flexure in the southern part of the Delaware Basin probably reflects the "yoking" of flexural profiles produced by the Marathon orogenic belt and southern parts of the CBP.

East-west cross sections of upper Pennsylvanian-Wolfcampian strata in the Midland Basin show that the entire basin was relatively symmetrical at that time. These thickness patterns indicate that uplift along the eastern side of the CBP did not produce significant flexure in the Midland Basin, which is consistent with observed structural relief along the eastern side of the CBP. In fact, the greatest amount of late Pennsylvanian-Wolfcampian deposition was in the eastern half of the Midland Basin, far from any flexure caused by the CBP. This suggests that subsidence in the eastern Midland Basin might have been affected more by shortening along the eastern side of the basin, possibly along the Fort Chadbourne fault zone. Unfortunately, we do not have seismic data from this zone of *en echelon* faults and folds, and it is difficult to characterize the structures there.

CONCLUSIONS

We have constructed an integrated model for the Permian Basin region which incorporates aspects of structural, stratigraphic, and kinematic analyses within a large-scale conceptual geodynamic framework. The following conclusions result from this study:

1. The Permian Basin is located in the foreland area of the Marathon-Ouachita orogenic belt. The basin was segmented into several sub-basins by uplifted areas during the late Paleozoic Marathon-Ouachita orogeny. One of these uplifted areas, the Central Basin Platform, strikes at a high angle to the Marathon orogenic belt. The CBP is located near the center of the antecedent Tobosa Basin and separates the Delaware Basin from the Midland Basin.

2. Initial stages of foreland deformation and basin evolution in the Permian Basin region began during Mississippian time. The main phase of basin differentiation and uplift of the CBP, however, began in latest middle Pennsylvanian time, as indicated by the beginning of rapid subsidence in each basin and by sudden changes in thickness and lithofacies of middle or upper Pennsylvanian to Lower Permian strata.

3. The CBP consists of the Fort Stockton and Andector Blocks which are arranged in a left-stepping *en echelon* pattern; the Andector Block actually is comprised of smaller fault-bounded crustal blocks, but most have a consistent eastward sense of vergence. The SW and NE corners of the Fort Stockton Block (the southernmost block of the CBP) are characterized by overthrust faults that dip toward the interior of the block. In contrast, normal faults at the NW corner of the block dip westward toward the Delaware Basin. Similar, steeply dipping, reverse faults characterize the SW and NE corners of the Andector Block, although normal faults have not been identified at the NW corner of this block. The area north of the Andector Block is characterized by strike-slip deformation.

4. Estimated basement shortening across the boundary fault zones of the Fort Stockton Block decreases northward along the western boundary and southward along the

eastern boundary. Similar variations in basement shortening are calculated along the eastern and western boundaries of the Andector Block.

5. Regional variations in structural styles and the amount of basement shortening along the block boundaries indicate that the blocks rotated clockwise during deformation. Block rotation occurred between NNW-SSE trending, right-lateral, strike-slip faults in the Marathon foreland.

6. East-west profiles of restored upper Pennsylvanian-Wolfcampian strata in the Delaware Basin are asymmetric (i.e., thicker) towards the thrusted SW corners of blocks that comprise the CBP. In contrast, the Delaware Basin is relatively symmetric along east-west profiles that extend from the NW, normal-faulted corner of the Fort Stockton Block.

7. Upper Pennsylvanian-Wolfcampian strata in the Midland Basin are thinner overall than in the Delaware Basin, and east-west Midland Basin profiles show thickening toward its eastern side. This thickening suggests that the CBP did not cause prominent flexure in western parts of the Midland Basin and that there is an additional, unknown cause for subsidence in the eastern Midland Basin.

8. Upper Pennsylvanian through Wolfcampian strata in the Val Verde Basin thicken to the south and reflect the flexural profile produced by the Marathon orogenic belt. However, a typical flexural profile is absent in the southern Delaware Basin because of the combined flexural effects of the Marathon orogenic belt and SW corner of the Fort Stockton Block.

9. The foredeep basin was restricted to the Marathon area during Mississippian to early Pennsylvanian time. There was a mild uplift in the CBP and Ozona Arch areas prior to deposition of the middle Pennsylvanian Strawn Limestone. Uplift of the CBP was mainly accomplished by displacement along its western boundary fault zone. The Puckett-Grey Ranch fault zone, a southern extension of the CBP's western boundary fault zone, probably segmented the early Marathon foredeep into sub-basins. The pre-Desmoinesian unconformity across the present Val Verde Basin terminates at the Puckett-Grey Ranch fault zone, indicating that the fault was active during Mississippian to early Pennsylvanian time.

10. At the beginning of late Pennsylvanian time, the Midland and Delaware Basins began to be affected by obvious block uplift and rapid basin subsidence. The present Val Verde Basin became the foredeep basin that was produced by lithospheric flexure from the Marathon orogenic belt. Across the Marathon foreland area, differential uplift of the CBP produced differential subsidence and variable basin geometry in the adjacent Delaware and Midland Basins. This phase of tectonic activity continued till the end of the Wolfcampian time when the rapid subsidence and deformation in the sub-basins ceased. Subsidence continued, however, until the end of the Permian.

ACKNOWLEDGMENTS

We would like to thank Chevron USA and especially Ron Genter for releasing the seismic profiles used in this study and for giving us permission to publish them. This study was supported by the American Chemical Society, Petroleum Research Fund (Grant #23269–AC2 to S. L. Dorobek), the National Science Foundation (Grant #EAR–9117935 to S. L. Dorobek), and the Center for Energy and Mineral Resources at Texas A&M University. Additional support was provided by a Grant-in-Aid to K.-M. Yang from the American Association of Petroleum Geologists. We also thank Tom Ewing and Charlie Kerans for providing extremely useful reviews of the original manuscript.

REFERENCES

ADAMS, J. E., FRENZEL, H. N., RHODES, M. L., AND JOHNSON, D. P., 1951, Starved Pennsylvanian Midland Basin: American Association of Petroleum Geologists Bulletin, v. 35, p. 2600–2607.

BEAUMONT, C., 1981, Foreland basins: Geophysical Journal of the Royal Astronomical Society, v. 65, p. 291–329.

BECK, M. E., Jr., 1984, Introduction to the special issue on correlations between plate motions and Cordilleran tectonics: Tectonics, v. 3, p. 103–106.

BECK, R. B., VONDRA, C. F., FILKINS, J. E., AND OLANDER, J. D., 1988, Syntectonic sedimentation and Laramide basement thrusting, Cordilleran foreland; timing of deformation, in Schmidt, C. J., and Perry, W. J., Jr., eds., Interaction of the Rocky Mountain Foreland and the Cordilleran Thrust Belt: Boulder, Geological Society of America Memoir 171, p.465–487.

BIRD, P., 1984, Laramide crustal thickening event in the Rocky Mountain foreland and Great Plains: Tectonics, v. 3, p. 741–758.

BLACKSTONE, D. L., JR., 1980, Foreland deformation: compression as a cause: University of Wyoming Contributions to Geology, v. 18, p. 83–100.

BROWN, L. F., JR., CLEAVES, A. W., II, AND ERXLEBEN, A. W., 1973, Pennsylvanian depositional systems in North-Central Texas, a guide for interpreting terrigenous clastic facies in a cratonic basin: Austin, The University of Texas at Austin, Bureau of Economic Geology Guidebook 14, 122 p.

COOK, D. G., 1988, Balancing basement-cored folds of the Rocky Mountain foreland, in Schmidt, C. J., and Perry, W. J., Jr., eds., Interaction of the Rocky Mountain Foreland and the Cordilleran Thrust Belt: Boulder, Geological Society of America Memoir 171, p. 53–64.

COOPER, M. A. AND WILLIAMS, G. D., 1989, Inversion Tectonics: Boston, Geological Society Special Publication 44, Blackwell Scientific Publications, 375 p.

DICKINSON, W. R. AND SNYDER, W. S., 1978, Plate tectonics of the Laramide orogeny, in Matthews, V., III, ed., Laramide Folding Associated with Basement Block Faulting in the Western United States: Boulder, Geological Society of America Memoir 151, p. 355–366.

DOROBEK, S. L., REID, S. K., ELRICK, M., BOND, G. C., AND KOMINZ, M. A., 1991, Foreland response to episodic convergence: Subsidence history of the Antler foredeep and adjacent cratonic platform areas, Montana and Idaho, in Franseen, E. K., Watney, W. L., Kendall, C. G. St. C., and Ross, W., eds., Sedimentary Modeling: Lawrence, Computer Simulation and Methods for Improved Parameter Definition, Kansas Geological Survey Bulletin 233, p. 231–252.

ELAM, J. G., 1984, Structural systems in the Permian Basin: West Texas Geological Society Bulletin, v. 24, p. 7–10.

EWING, T. E., 1984, Late Paleozoic structural evolution of the Permian Basin (abs.): American Association of Petroleum Geologists Bulletin, v. 68, p. 474–475.

EWING, T. E., 1990, The tectonic map of Texas: Austin, Bureau of Economic Geology, The University of Texas at Austin.

EWING, T. E., 1991, The tectonic framework of Texas: Text to accompany "The Tectonic Map of Texas": Austin, Bureau of Economic Geology, The University of Texas at Austin, 36 p.

EWING, T. E., 1993, Erosional margins and patterns of subsidence in the late Paleozoic West Texas Basin and adjoining basins of West Texas and New Mexico: Socorro, New Mexico Geological Society Guidebook, 44th Field Conference, p. 155–166.

FLEMINGS, P. B. AND JORDAN, T. E., 1989, A synthetic stratigraphic model of foreland basin development: Journal of Geophysical Research, v. 94. p. 3851–3866.

FLEMINGS, P. B., JORDAN, T. E., AND REYNOLDS, S. A., 1986, Flexural analysis of two broken foreland basins: The Late Cenozoic Bermejo Basin and the Early Cenozoic Green River Basin (abs.): American Association of Petroleum Geologists Bulletin, v. 70, p. 591.

FRENZEL, H. N., BLOOMER, R. R., CLINE, R. B., CYS, J. E., GALLEY, J. E., GIBSON, W. R., HILLS, J. M., KING, W. E., SEAGER, W. R., KOTTLOWSKI, F. E., THOMPSON III, S., LUFF, G. C., PEARSON, B. T., AND VAN SICLEN, D. C., 1988, The Permian Basin region, in Sloss, L. L., ed., Sedimentary Cover-North American Craton: U. S.: Boulder, Geological Society of America, The Geology of North America, v. D-2, p. 261–306.

GALLEY, J. E., 1958, Oil and geology in the Permian Basin of Texas and New Mexico, in Weeks, L. G., ed., Habitat of Oil—A Symposium: Tulsa, American Association of Petroleum Geologists, p. 395–446.

GARDINER, W. B., 1990, Fault fabric and structural subprovinces of the Central Basin Platform: A model for strike-slip movement, in Flis, J. E., and Price, R. C., eds., Permian Basin Oil and Gas Fields: Innovative Ideas in Exploration and Development: Midland, West Texas Geological Society, Publication 90–87, p. 15–27.

GEOMAP, 1983, Pre-Pennsylvanian Subcrop Map of the Permian Basin of West Texas and Southeast New Mexico: Plano, GEOMAP EXECUTIVE REFERENCE MAP 502.

GOETZ, L. K. AND DICKERSON, P. W., 1985, A Paleozoic transform margin in Arizona, New Mexico, west Texas and northern Mexico, in Dickerson, P. W., and Muehlberger, W. R., eds., Structure and Tectonics of Trans-Pecos Texas: Midland, West Texas Geological Society, Publication 85–81, p. 173–184.

HAGEN, E. S., SHUSTER, M. W., AND FURLONG, K. P., 1985, Tectonic loading and subsidence of intermontane basins: Wyoming foreland province: Geology, v. 13, p. 585–588.

HANSON, B. M., POWERS, B. K., GARRETT, C. M., JR., MCGOOKEY, D. E., MCGLASSON, E. H., HORAK, R. L., MAZZULLO, S. J., REID, A. M., CALHOUN, G. G., CLENDENING, J., AND CLAXTON, B., 1991, The Permian basin, in Gluskoter, H. J., Rice, D. D., and Taylor, R. B., eds., Economic Geology, U. S.: Boulder, Geological Society of America, The Geology of North America, v. P-2, p. 339–356.

HENDERSON, G. J., LAKE, E. A., AND DOUGLAS, G., 1984, Langley Deep Field, discovery and interpretation, in Moore, G., and Wilde, G., eds., Transactions Southwest Section, American Association of Petroleum Geologists: Midland, West Texas Geological Society, Publication SWS-84–78, p. 1–10.

HILLS, J. M., 1970, Paleozoic structural directions in southern Permian Basin, west Texas and southeastern New Mexico: American Association of Petroleum Geologists Bulletin, v. 54, p. 1809–1827.

HILLS, J. M., 1984, Sedimentation, tectonism, and hydrocarbon generation in Delaware Basin, west Texas and southeastern New Mexico: American Association of Petroleum Geologists Bulletin, v. 68, p. 250–267.

HILLS, J. M., 1985, Structural evolution of the Permian Basin of west Texas and New Mexico, in Dickerson, P. W., and Muehlberger, W. R., eds., Structure and Tectonics of Trans-Pecos Texas: Midland, West Texas Geological Society, Publication 85–81, p. 89–99.

HORAK, R. L., 1985, Trans-Pecos tectonism and its effect on the Permian Basin, in Dickerson, P. W., and Muehlberger, W. R., eds., Structure and Tectonics of Trans-Pecos Texas: Midland, West Texas Geological Society Publication 85–81, p. 81–87.

JORDAN, T. E., 1981, Thrust loads and foreland basin development, Cretaceous, western United States: American Association of Petroleum Geologists Bulletin, v. 65, p. 2506–2520.

KLUTH, C. F. AND CONEY, P. J., 1981, Plate tectonics of the Ancestral Rocky Mountains: Geology, v. 9, p. 10–15.

LARROQUE, J. M. AND LAURENT, P. H., 1988, Evolution of the stress field pattern in the south of the Rhine Graben from the Eocene to the present: Tectonophysics, v. 148, p. 41–58.

LETOUZEY, J., 1986, Cenozoic paleo-stress pattern in the Alpine Foreland and structural interpretation in a platform basin: Tectonophysics, v. 132, p. 215–231.

LOWELL, J. D., 1974, Plate tectonics and foreland basement deformation: Geology, v. 2, p. 275–278.

MAUCH, J. J., WETTERAUER, R. H., WALPER, J. L., AND MORGAN, K. M., 1984, The Marfa Basin of West Texas: Foreland basin subsidence and depocenter migration: Midland, West Texas Geological Society, Publication SWS-84–78, p. 141–153.

MCBRIDE, E. F., 1978, Tesnus and Haymond Formations—siliciclastic flysch, in Mazzullo, S. J., ed., Tectonics and Paleozoic Facies of the Marathon Geosyncline, West Texas: Midland, Permian Basin Section, Society of Economic Paleontologists and Mineralogists, Publication 78–17, p. 131–148.

MCBRIDE, E. F., 1989, Stratigraphy and sedimentary history of pre-Permian Paleozoic rocks of the Marathon uplift, in Hatcher, R. D., Jr., Thomas, W. A., and Viele, G. W., eds., The Appalachian-Ouachita Orogen in the United States: Boulder, Geological Society of America, The Geology of North America, v. F-2, p. 603–620.

MCCONNELL, D. A., 1987, Paleozoic structural evolution of the Wichita uplift, southwestern Oklahoma: Unpublished Ph. D. Dissertation, Texas A&M University, College Station, 212 p.

MCCONNELL, D. A., 1989, Determination of offset across the northern margin of the Wichita uplift, southwest Oklahoma: Geological Society of America Bulletin, v. 101, p. 1317–1332.

MCCONNELL, D. A., GOYDAS, M., SMITH, G. N., AND CHITWOOD J. P., 1990, Morphology of the Frontal fault zone, southwest Oklahoma: Implications for deformation and deposition in the Wichita uplift and Anadarko basin: Geology, v. 18, p. 634–637.

MOORE, G. E., MENDENHALL, G. V., AND SAULTZ, W. L., 1981, Northern extent of Marathon thrust Elsinore area, Pecos County, Texas, in Jons, R., ed., Marathon-Marfa Region of West Texas Symposium and Guidebook: Midland, Permian Basin Section, Society of Economic Paleontologists and Mineralogists, 81–20, p. 131–133.

QUINLAN, G. M. AND BEAUMONT, C., 1984, Appalachian thrusting, lithospheric flexure and the Paleozoic stratigraphy of the eastern interior of North America: Canadian Journal of Earth Sciences, v. 21, p. 973–996.

REED, T. A. AND STRICKLER, D. L., 1990, Structural geology and petroleum exploration of the Marathon thrust belt, West Texas, in Laroche, T. M., and Higgins, L., eds., Marathon Thrust Belt: Structure, Stratigraphy, and Hydrocarbon Potential: Midland, West Texas Geological Society and Permian Basin Section-Society of Economic Paleontologists and Mineralogists, Field Seminar May 10–12, 1990, p. 39–64.

ROSS, C. A., 1979, Late Paleozoic collision of North and South America: Geology, v. 7, p. 41–44.

ROSS, C. A., 1981, Pennsylvanian and Early Permian history of the Marathon Basin, West Texas, in Jons, R., ed., Marathon-Marfa Region of West Texas Symposium and Guidebook: Midland, Permian Basin Section, Society of Economic Paleontologists and Mineralogists, 81–20, p. 135–144.

ROSS, C. A., 1986, Paleozoic evolution of southern margin of Permian basin: Geological Society of America Bulletin, v. 97, p. 536–554.

ROSS, C. A. AND ROSS, J. R. P., 1985, Paleozoic tectonics and sedimentation in west Texas, southern New Mexico, and southern Arizona, in Dickerson, P. W., and Muehlberger, W. R., eds., Structure and Tectonics of Trans-Pecos Texas: Midland, West Texas Geological Society, Publication 85–81, p. 221–230.

ROYDEN, L. H., PATACCA, E., AND SCANDONE, P., 1987, Segmentation and configuration of subducted lithosphere in Italy: An important control on thrust-belt and foredeep-basin evolution: Geology, v. 15, p. 714–717.

SALES, J. K., 1968, Cordilleran foreland deformation: American Association of Petroleum Geologists Bulletin, v. 52, p. 2000–2015.

SANDERS, D. E., BOYCE, R. G., AND PETERSON, N., 1983, The structural evolution of the Val Verde Basin, West Texas, in Structure and Stratigraphy of the Val Verde Basin-Devils River Uplift, Texas: Midland, West Texas Geological Society, Publication 83–77, p. 123.

SCHEEVEL, J. R., 1983, Horizontal compression and a mechanical interpretation of Rocky mountain foreland deformation, in Boberg, W. W., ed., Geology of the Bighorn Basin: Wyoming Geological Association 34th Annual Field Association Guidebook, p. 53–62.

SENGOR, A. M. C., 1976, Collision of irregular continental margins: Implications for deformation of Alpine-type orogens: Geology, v. 4, p. 779–782.

SENGOR, A. M. C., BURKE, K., AND DEWEY, J. F., 1978, Rift at high angles to orogenic belts: test for their origin and the upper Rhine Graben as an example: American Journal of Science, v. 278, p. 24–40.

SHUMAKER, R. C., 1992, Paleozoic structure of the Central Basin uplift and adjacent Delaware Basin, West Texas: American Association of Petroleum Geologists Bulletin, v. 76, p. 1804–1824.

SHUSTER, M. W. AND STEIDTMANN, J. R., 1988, Tectonic and sedimentary evolution of the northern Green River basin, western Wyoming, in Schmidt, C. J., and Perry, W. J., Jr., eds., Interaction of the Rocky Mountain Foreland and the Cordilleran Thrust Belt: Boulder, Geological Society of America Memoir 171, p. 515–530.

SILVER, B. A. AND TODD, R. G., 1969, Permian cyclic strata, northern Midland and Delaware Basins, west Texas and southeastern New Mexico: American Association of Petroleum Geologists Bulletin, v. 53, p. 2223–2251.

SPANG, J. H., EVANS, J. P., AND BERG, R. R., 1985, Balanced cross sections of small fold-thrust structures: The Mountain Geologist, v. 22, p. 41–46.

STOCKMAL, G. S., BEAUMONT, C., AND BOUTILIER, R., 1986, Geodynamic models of convergent tectonics: the transition from rifted margin to overthrust belt and consequences for foreland-basin development: American Association of Petroleum Geologists Bulletin, v. 70, p. 181–190.

THOMAS, W. A., 1983, Continental margins, orogenic belts, and intracratonic structures: Geology, v. 11, p. 270–272.

WALPER, J. L., 1977, Paleozoic tectonics of the southern margin of North America: Gulf Coast Association of Geological Societies Transactions, v. 27, p. 230–241.

WARD, R. F., KENDALL, C. G. ST. C., AND HARRIS, P. M., 1986, Upper Permian (Guadalupian) facies and their association with hydrocarbons—Permian Basin, West Texas and New Mexico: American Association of Petroleum Geologists Bulletin, v. 70, p. 239–262.

WUELLNER, D. E., LEHTONEN, L. R., AND JAMES, W. C., 1986, Sedimentary-tectonic development of the Marathon and Val Verde basins, West Texas, U. S. A.: a Permo-Carboniferous migrating foredeep, in Allen, P., and Homewood, P., eds., Foreland Basins: International Association of Sedimentologists Special Publication 8, p. 15–39.

YANG, K.-M., 1993, Late Paleozoic Synorogenic Stratigraphy, Tectonic Evolution, and Flexural Modeling of the Permian Basin, West Texas and New Mexico: Unpublished Ph. D. Dissertation, Texas A&M University, College Station, 142 p.

YANG, K.-M. AND DOROBEK, S. L., 1991, The tectonic mechanism for uplift and rotation of crustal blocks in the Central Basin Platform, Permian Basin, Texas and New Mexico (abs.): American Association of Petroleum Geologists Bulletin, v. 75, p. 698.

YANG, K.-M. AND DOROBEK, S. L., 1992, Mechanisms for late Paleozoic synorogenic subsidence of the Midland and Delaware Basins, Permian Basin, Texas and New Mexico, in Mruk, D., and Curran, B., eds., Permian Basin Exploration and Production Strategies–Applications of Sequence Stratigraphic and Reservoir Characterization Concepts: Midland, West Texas Geological Society, 1992 Fall Symposium, p. 45–60.

YANG, K.-M. AND DOROBEK, S. L., 1993, Late Paleozoic synorogenic stratigraphy and tectonic evolution of the Permian Basin, West Texas and New Mexico, in Cromwell, D., and Gibbs, J., eds., New Dimensions in the Permian Basin: 3D and Environmental Geosciences: Midland, West Texas Geological Society, Publication 93-93, p. 8–18.

ZIEGLER, P. A., 1987, Late Cretaceous and Cenozoic intra-plate compressional deformations in the Alpine foreland—a geodynamic model: Tectonophysics, v. 137, p. 389–420.

DEVELOPMENT OF THE MISSISSIPPIAN CARBONATE PLATFORM IN SOUTHERN NEVADA AND EASTERN CALIFORNIA ON THE EASTERN MARGIN OF THE ANTLER FORELAND BASIN

CALVIN H. STEVENS
Department of Geology, San Jose State University, San Jose, CA 95192
DARRELL KLINGMAN
Pacific Gas and Electric Company, San Ramon, CA 94583
AND
PAUL BELASKY
Permian Research Institute, Boise State University, Boise, ID 83725

ABSTRACT: Development of the Mississippian carbonate platform along the eastern margin of the foreland basin of the Antler Orogen was controlled by subsidence due primarily to emplacement of the Roberts Mountains allochthon, but also to variable rates of carbonate production and eustatic sea-level changes. A series of stratigraphic sections in southern Nevada and eastern California, oriented approximately perpendicular to the original depositional trends, has allowed evaluation of the relative influence of these controlling factors and development of a depositional model that may have wide application.

Mississippian carbonate-platform sedimentation began with a rapid transgression in Kinderhookian time, probably due to a sea-level rise and perhaps initial thrust loading of the older Devonian carbonate platform. In early Osagean time, final emplacement of the Roberts Mountains allochthon onto sialic North America depressed the platform, greatly reducing carbonate production. By middle Osagean time, carbonate production rates exceeded the rate of formation of accommodation space, initiating northwestward progradation of a shallow-water carbonate platform. Continued progradation produced relatively steep, unstable slopes, and by early Meramecian time a rimmed platform with coral buildups had formed. These changes in platform morphology are recorded in base-of-slope deposits where slope-derived submarine slides are overlain by sediment-gravity-flow deposits containing debris derived from the platform margin. Carbonate deposition ceased in early Meramecian time when an eustatic sea-level fall exposed the entire platform.

INTRODUCTION

The region east of the Antler Orogen, a major Late Devonian to Early Mississippian tectonic feature in the western United States, commonly is divided into two broad regions: the foreland basin adjacent to the orogenic belt and a carbonate platform to the east (Fig. 1). Mississippian rocks deposited in the central and eastern parts of the foreland basin and on the carbonate platform in southern Nevada and eastern California are assigned to three major facies belts. From southeast to northwest they are: (1) the limestone-facies belt, (2) the quartzose siltstone-facies belt (both of Stevens and others, 1991), and (3) the shale-conglomerate-facies belt (Fig. 1). Because this paper concerns rocks composed of carbonate sediment produced on the platform in southern Nevada and eastern California, the shale-conglomerate-facies belt, which does not contain such material, will not be considered here.

Shallow- to relatively deep-water platform carbonates are the dominant rock types in the limestone-facies belt, and slope to basinal carbonates are prominent in the adjacent eastern part of the quartzose siltstone-facies belt. The limestone-facies belt was subdivided into two subfacies belts by Stevens and others (1991), but further analysis shows that the limestone belt actually can be subdivided into three belts, here called the inner, central, and outer limestone-subfacies belts (Fig. 1). These belts represent the inner, central, and outer parts of the carbonate platform.

The inner limestone-subfacies belt contains mostly limestone with a variety of shallow-water fossils and chert nodules in the middle part of the section. It is distinguished by a substantial amount of dolomite in the upper part of the platform sequence. The central limestone-subfacies belt is similar to the inner belt but lacks dolomite. The outer limestone-subfacies belt differs from the central and inner belts in (1) the much greater thickness (>200 m) of relatively deep-water limestone in the middle part of the section, (2) the occurrence of the Osagean-Meramecian boundary within or close to the deep-water part of the section, (3) the presence of sediment-gravity-flow deposits, and (4) the presence of abundant massive rugose corals of early Meramecian age (Stevens and others, 1991).

Ages of rocks and stratigraphic relationships in southern Nevada and eastern California were documented by Belasky (1988) and Klingman (1987), respectively. The present paper draws heavily on these works as well as data presented by Stevens and others (1991). Correlation of the southern Nevada-eastern California Mississippian section with Series in the type Mississippian (Fig. 2) is based on conodont zones following Poole and Sandberg (1991) because most of the ages reported here are based on conodont faunas. These correlations are in general agreement with the coral- and foraminiferal-based Series correlations of Mamet and Skipp (1970), Sando (1985), and Sando and Bamber (1985), except for the location of the Osagean/Meramecian boundary. Data on the faunas recovered from samples to which reference is made on Figures 3 and 4 are included in a paper submitted for publication as a U. S. Geological Survey Bulletin.

At present the inner part of the carbonate platform (the inner and central limestone-subfacies belts) is better dated in southern Nevada than in eastern California, whereas the platform margin and slope (outer limestone-subfacies and quartzose siltstone-facies belts) are much better preserved and dated in eastern California. Therefore, rocks in both areas will be described and the data integrated into a single depositional model thought to be widely applicable, especially in the southern Great Basin. In both southern Nevada and eastern California, a series of generalized stratigraphic sections (Figs. 3, 4) representing all facies belts present (except for those north of the Las Vegas Valley fault) are shown. More detailed sections showing interpreted depo-

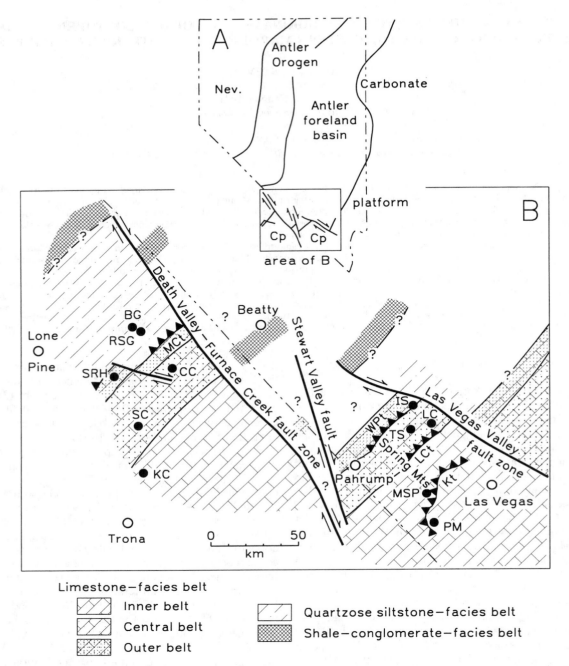

Fig. 1.—Location maps. (A) Major Mississippian tectonic features; (B) Index map of area of this study. BG-Bighorn Gap, CC-Cottonwood Canyon, Cp-carbonate platform, IS-Indian Springs, KC-Knight Canyon, Kt-Keystone thrust fault, LC-Lee Canyon, LCt-Lee Canyon thrust fault, MCt-Marble Canyon thrust fault, MSP-Mountain Springs Pass, PM-Potosi Mine, RSG-Rest Spring Gulch, SC-Stone Canyon, SRH-Santa Rosa Hills, TS-Trough Spring, WPt-Wheeler Pass thrust fault.

sitional environments of post-Tin Mountain rocks of the central and outer limestone-subfacies belts in eastern California (Fig. 1) are presented in Figure 5.

The major purpose of this paper is to report upon new faunal and stratigraphic data concerning Mississippian carbonate rocks in southern Nevada and eastern California, to create a model of the evolving platform geometry, and to interpret the tectonic and eustatic events that led to development of this platform.

MISSISSIPPIAN PLATFORM IN SOUTHERN NEVADA

The carbonate platform in the Spring Mountains south of the Las Vegas Valley fault zone (Fig. 1) is represented in four different plates separated by thrust faults. Rocks of this carbonate platform are referred to as the Monte Cristo Group (Fig. 2), a unit generally overlying shallow-water Devonian limestone and underlying the very thin, latest Mississippian, shallow-water, siliciclastic Indian Springs Formation.

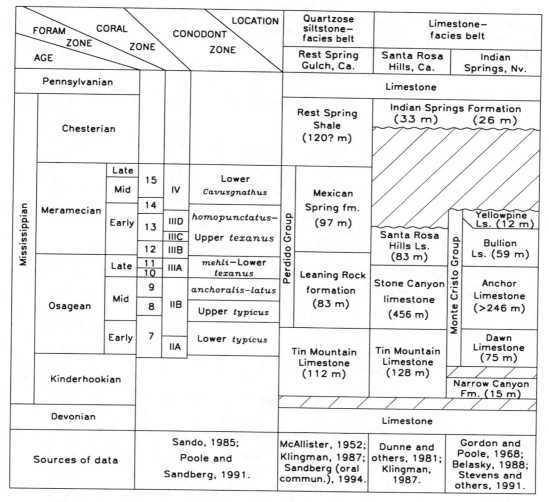

Fig. 2.—Correlation of Mississippian rock units in southern Nevada and eastern California.

The Monte Cristo Group consists of four major formations including, from oldest to youngest: (1) Dawn Limestone, (2) Anchor Limestone, (3) Bullion Limestone, and (4) Yellowpine Limestone. The Arrowhead Limestone, a very thin, locally developed unit between the Bullion and Yellowpine Limestones, will be treated here as the basal member of the Yellowpine Limestone. Another thin Mississippian unit, the Narrow Canyon Formation, which occurs below the Dawn Limestone in the Indian Springs section, will not be considered.

The three younger formations of concern here are lithosomes, each representing a different depositional environment. As shown on Figure 3, these three lithosomes transgress time and thus represent environments that coexisted on the carbonate platform.

Dawn Limestone

The Dawn Limestone generally rests paraconformably on Upper Devonian rocks assigned to the Crystal Pass Limestone in the southern part of the Spring Mountains and the Devils Gate Limestone in the northern part. The character of the Dawn Limestone varies somewhat laterally. In the inner limestone-facies belt, it includes partially dolomitized, dark-gray, medium-grained, pelletal packstone and wackestone with a variety of marine fossils including solitary and colonial corals, brachiopods, bryozoans, molluscs, and pelmatozoan remains. In the outer limestone-subfacies belt to the northwest, the section consists of dark-gray, spiculiferous lime mudstone and massive wackestone with scattered pellets, brachiopods, solitary corals, and pelmatozoan debris. The upper part of the section has yielded a conodont fauna dominated by *Polygnathus*, *Siphonodella*, and *Bispathodus*.

This unit was deposited on a relatively shallow-water, carbonate platform. The fauna in the inner limestone-subfacies belt suggests euphotic, stenohaline marine conditions, and the medium-grained rocks suggest moderate agitation. The platform apparently dipped gently northward so that in the outer limestone-subfacies belt, sediment was deposited below storm-wave base as indicated both by the nature of the sediment (lime mudstone and wackestone) and the conodont biofacies (Sandberg and Gutschick, 1984). Thus, the

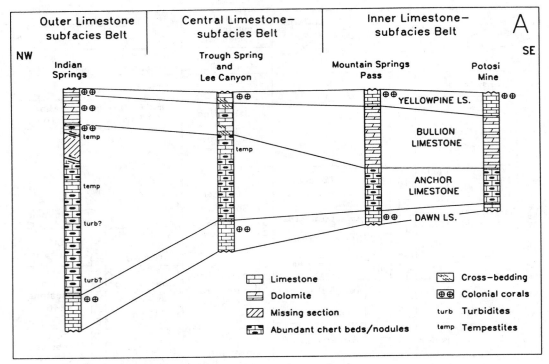

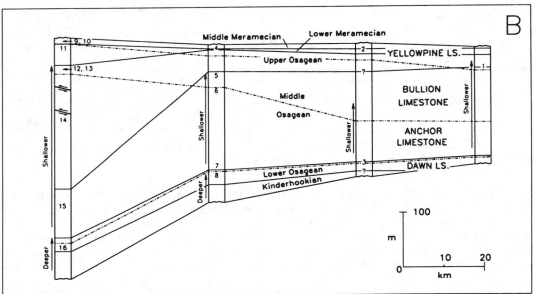

Fig. 3.—Stratigraphic cross sections from Potosi Mine to Indian Springs, southern Nevada. Displacements on Mesozoic thrust faults have been compensated for following Burchfiel and others (1974). (A) Cross section showing formational units, generalized lithology, and some important environmental indicators. (B) Cross section showing interpreted deepening and shallowing trends, time lines, thicknesses, and the position of time-significant fossil samples indicated by the following numbers: 1—PO-5; 2—YPC-4; 3—SM-12; 4—LC-YP-3; 5—LC-BU-8; 6—LC-AN-1; 7—TS-AN-14; 8—DCF-1; 9—ISD-1; 10—IST-BU-1; 11—ISS-31; 12—IST-1; 13—ISS-26; 14—ISB-AN-1B; 15—ISS-12; 16—ISS-11. Paleontologic data on these samples is included in a paper submitted for publication as a U. S. Geological Survey Bulletin.

form of the platform was that of a homoclinal ramp (Read, 1985).

Anchor Limestone

The Anchor Limestone rests conformably on the Dawn Limestone. In the inner limestone-subfacies belt, this unit is composed predominantly of medium-gray, thin- to medium-bedded lime wackestone and packstone with abundant chert lenses and nodules. Fossils include solitary corals, brachiopods, and pelmatozoan parts. A Hindeodid conodont assemblage also was collected from this unit. In the outer limestone-subfacies belt, the lower part of the for-

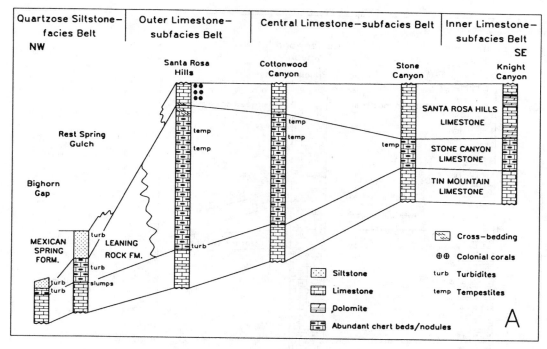

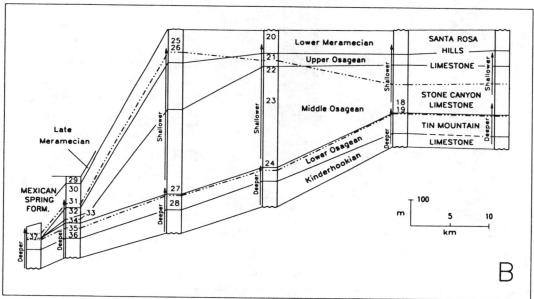

Fig. 4.—Stratigraphic cross section from Knight Canyon to Bighorn Gap, eastern California. The Marble Canyon thrust fault has been compensated for, following Snow (1990). (A) Cross section showing formational units, generalized lithology, and some important environmental indicators. (B) Cross section showing interpreted deepening and shallowing trends, time lines, thicknesses, and the position of time-significant fossil samples indicted by the following numbers: 18—MM-2; 19—MM-1; 20—MC-1; 21—CC-6; 22—CC-4; 23—CC-1; 24—PD-17; 25—SR-31, SR-31A; 26—C-1; 27—SRA-1; 28—SRA-22; 29—QS-94; 30—QS-42, QS-84, QS-85, QS-90; 31—QS-59; 32—QS-37, SJS-1091; 33—QS-25; 34—QS-20; 35—QS-2; 36—QS-1; and 37—BG-108. Paleontologic data on these samples is included in a paper submitted for publication as a U. S. Geological Survey Bulletin.

mation is marked by a 6-m-thick, radiolarian-bearing chert overlain by dark-gray lime mudstone and spiculiferous wackestone with thin beds of dark-brown chert and silicified limestone. The middle part of the section consists of thin- to medium-bedded, dark-gray, spiculiferous wackestone with abundant chert lenses and beds and some crudely graded beds of pelmatozoan-rich packstone with erosional bases and gradational tops. The upper unit contains fine-grained wackestone and packstone interbedded with light-gray, coarse-grained, pelmatozoan-rich grainstone exhibit-

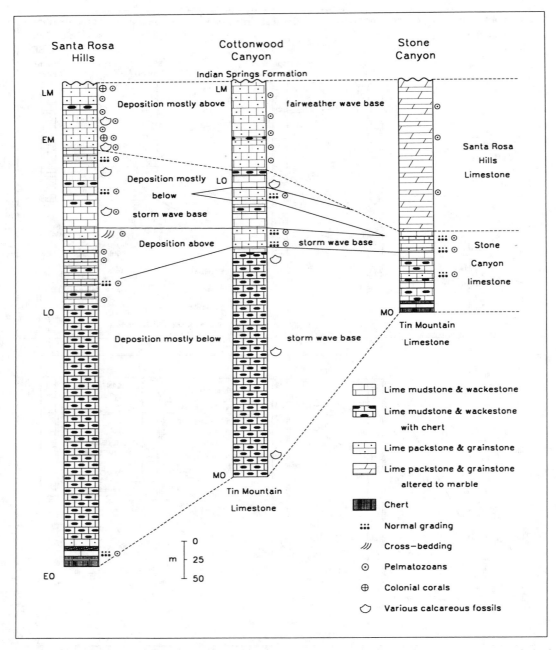

Fig. 5.—Interpreted depositional environments represented in the central to outer platform (central and outer limestone-subfacies belts) in eastern California. EO—early Osagean, MO—middle Osagean, LO—late Osagean, EM—early Meramecian; MM—middle Meramecian.

ing cross-bedding and possible hummocky cross-bedding. A conodont assemblage containing *Cloghergnathus* and *Hindeodus* was collected from this upper unit.

The relatively fine-grained, dark-gray, cherty Anchor Limestone was deposited in a quiescent environment in somewhat deeper water than that of the underlying Dawn Limestone although the Hindeodid conodont biofacies in the inner limestone-subfacies belt still points to deposition in relatively shallow water (Sandberg and Gutschick, 1984). The much thicker section in the outer limestone-subfacies belt was deposited in a considerably deeper marine environment. The lower part of the section at Indian Springs, consisting of unfossiliferous lime mudstone and chert, and lacking bioturbation, was deposited below storm-wave base under dysaerobic conditions. The presence of chert bearing radiolarians and possible turbidites further suggests a relatively deep-water depositional setting. The presence of crudely graded packstones and grainstones, interpreted as tempestites, and a *Cloghergnathus-Hindeodus* conodont assemblage in the upper part of the section shows that shallowing upward to above storm-wave base occurred during deposition of this unit.

Bullion Limestone

The Bullion Limestone is composed of light-gray, thick-bedded, coarse-grained, pelmatozoan-rich packstone and grainstone. In the inner limestone-facies belt this unit has been almost completely dolomitized. In the outer limestone subfacies belt, where colonial rugose corals are abundant and bryozoans, calcareous foraminifera and a *Cloghergnathus-Hindeodus* fauna have been recovered, the section consists entirely of limestone.

The pelmatozoan-rich carbonates were deposited as carbonate-sand shoals or mobile skeletal-sand sheets in an agitated, shallow-water environment. Dolomitization of the unit in the inner limestone-subfacies belt suggests that the shoals may have led to restriction of shallow-water lagoons landward. Farther north the presence of pelmatozoan shoals and sand sheets, the abundance of colonial corals, and the presence of the *Cloghergnathus-Hindeodus* conodont assemblage (Sandberg and Gutschick, 1984) suggest deposition in a well agitated, shallow-water environment near or at the platform margin.

Yellowpine Limestone

This unit is composed primarily of medium- to very dark-gray, medium-grained lime packstone. The diverse faunal assemblage includes pelmatozoans, calcareous foraminifera, solitary and colonial corals, brachiopods, gastropods, and conodonts of the Hindeodid biofacies.

The dark-gray, medium-grained packstones, bearing many fossils and including the Hindeodid conodont biofacies, suggest deposition in moderately quiescent, shallow water (Sandberg and Gutschick, 1984). The corals and other typical marine fossils indicate that normal marine salinity was maintained during deposition.

DEPOSITIONAL MODEL FOR SOUTHERN NEVADA

An understanding of the evolution and original geometry of the Mississippian carbonate platform in southern Nevada at different times during its development can be deduced by aligning the stratigraphic sections along a datum that can be interpreted in terms of water depth. The datum selected for this purpose is the base of the middle Meramecian section (Fig. 3B) because at all localities, these beds and the beds immediately below represent relatively shallow-water conditions. Thus, the base of the middle Meramecian rock sequence represents an essentially paleohorizontal plane.

We have attempted to remove the effects of Mesozoic thrust faulting by removing the shortening estimated by Burchfiel and others (1974), which includes about 20 km of displacement on the Keystone thrust, 7.5 km displacement on the Lee Canyon thrust, and 30 km displacement on the Wheeler Pass thrust.

The Mississippian facies belts in this region probably had an original orientation of about S 40° W (Stevens and others, 1991). Therefore, preparation of the stratigraphic cross sections in Figure 3 involved projection of the sections along the facies belts onto a line drawn perpendicular to that direction (N 50° W) after removal of Mesozoic displacements. Due to faulting in the Indian Springs section, stratigraphic section is missing and a complete reconstruction is precluded. The stratigraphic cross-section constructed on the basis of the existing data (Fig. 3), however, shows a moderate increase in thickness of the Dawn Limestone and a substantial increase in thickness of the Anchor Limestone in a seaward (northwestward) direction. In contrast, the Bullion and Yellowpine Limestones gradually thin seaward (northwestward). Thus, the initial transgressive deposit, the Dawn Limestone, thins shoreward (southeastward) as anticipated, and the shallow-water upper units expected to be progradational (Bullion and Yellowpine Limestones) thin seaward (northwestward). The Anchor Limestone, interpreted as the deepest water deposit, thickens offshore to the northwest.

Time surfaces (Fig. 3B), based on conodonts, colonial rugose corals, and calcareous foraminiferans, although imperfectly constrained, provide an overall picture of the geographic distribution of co-existing depositional environments. The top and bottom of the lower Osagean rocks approximately parallel the Dawn-Anchor contact, but the top of the middle and upper Osagean cross formational boundaries, especially at the outer part of the platform (Indian Springs section). Thus, the upper Osagean rock sequence embraces the upper Bullion and lower Yellowpine Limestones in the Trough Spring section but is mostly within the Anchor Limestone in the Indian Springs section. Because the cross-section is hung on an approximately paleohorizontal surface in the Meramecian, the shape of time surfaces after the last deformational event prior to the middle Meramecian should reflect the morphology of the seafloor at those times. The character of the stratigraphic cross sections suggests that deformation of the platform was minimal except perhaps for the Indian Springs section at the edge of the platform which may have subsided more in post-early Osagean time than the other areas. These cross sections also indicate that at the end of middle Osagean time the platform surface from shore into the Trough Spring area was in relatively shallow water, whereas that at Indian Springs was in considerably deeper water. By early Meramecian time, however, accumulation of vast amounts of carbonate sediment in the Indian Springs area caused aggradation to wave base, and the shallow-water, carbonate platform extended throughout the Spring Mountains.

The available data, therefore, allow the following interpretation of the development of this part of the Mississippian carbonate platform. Deposition of the Dawn Limestone began with a major marine transgression in Kinderhookian time. Rapid deepening of the water column in early Osagean time resulted in initial deposition of the relatively deep-water Anchor Limestone. By early middle Osagean time progradation of the shallow-water Bullion Limestone had begun on the inner part of the platform. Progradation was slow enough, however, that the shallow-shelf carbonates of the Bullion Limestone did not reach Indian Springs until latest Osagean time.

The Yellowpine Limestone presents a separate problem. Paleontologic data indicate that this unit consists of two parts of different ages. A physical break in sedimentation, however, has not been identified. The older (late Osagean and early Meramecian) part of the Yellowpine Limestone apparently belongs to the progradational sequence as indicated by the decreasing age of both the lower part of the Yellowpine and underlying Bullion limestones northwestward (seaward). This lower part of the Yellowpine Limestone is interpreted to represent shallow-platform deposition shoreward of the Bullion carbonate shoals and sand sheets which were formed near the margin of the platform. The upper, middle-Meramecian part of the Yellowpine Limestone occurs throughout the platform, however, and probably represents a later transgression of short duration. After deposition of the Yellowpine Limestone, carbonate deposition over most of the platform ceased. Exposure of the platform is indicated not only by the sharp break between the Monte Cristo Group and the overlying Indian Springs Formation but also by the presence of a forest rooting ground on top of the carbonate-platform rocks in several different areas (Decourten, 1986).

MISSISSIPPIAN PLATFORM IN EAST-CENTRAL CALIFORNIA

Both the limestone-facies and quartzose siltstone-facies belts are well represented in east-central California, and the relationship between these facies belts can be ascertained in the field. Mississippian rocks in all sections in both belts reported here rest upon shallow-water Devonian limestone and are overlain by siliciclastic rocks.

The limestone-facies belt contains a thick limestone sequence that has been divided into three lithologic units (Fig. 2). The lowest is the Tin Mountain Limestone which compares with the Dawn Limestone in southern Nevada. The overlying unit, previously called the limestone facies of the Perdido Formation, will be referred to here informally as the Stone Canyon limestone. It compares with the Anchor Limestone in southern Nevada. The highest unit in the carbonate-platform sequence, the Santa Rosa Hills Limestone, compares lithologically with the Bullion Limestone in southern Nevada. This unit is overlain by the very thin, siliciclastic, shallow-water, Chesterian Indian Springs Formation.

In the quartzose siltstone-facies belt, the Tin Mountain Limestone is overlain by a dominantly limestone unit here informally called the Leaning Rock formation. Both of these units thin and pinch out westward. In the area of this study, the Leaning Rock formation is overlain by a deep-water, dominantly quartzose siltstone unit with interbedded limestone sediment-gravity-flow deposits. These rocks have been called the siltstone facies of the Perdido Formation but here are referred to informally as the Mexican Spring formation. Fossils suggest that this unit is mostly, but not entirely, younger than the limestones in the limestone-facies belt (Stevens and others, 1991). Generalized stratigraphic columns for the sections studied in California are shown in Figure 4. More detailed sections showing interpreted depositional environments in the central and outer limestone-subfacies belts are shown in Figure 5.

Tin Mountain Limestone

The Tin Mountain Limestone is composed of thin- to medium bedded, dark-gray mudstone and wackestone locally with chert lenses and nodules. In the inner and central limestone-subfacies belts, colonial and solitary rugose corals, pelmatozoan debris, brachiopods, bryozoans, and molluscs are present. In the outer limestone-subfacies belt to the north, colonial corals are lacking and the section contains more pelmatozoan-rich beds.

Colonial corals in the inner limestone-subfacies belt and numerous other fossils in the central limestone-subfacies belt suggest deposition in shallow-water, normal-marine environments. In the outer limestone-subfacies belt, the Tin Mountain Limestone contains both lime mudstone, suggesting quiet-water deposition, and a significant number of moderately well-sorted, pelmatozoan-rich beds, indicating temporary agitation. Shallow-water deposition below normal-wave base but above storm-wave base is probable. The platform is interpreted to have deepened gradually northwestward as a homoclinal ramp.

Stone Canyon Limestone

The name Stone Canyon limestone is here employed to describe a cherty limestone sequence of beds between the Tin Mountain Limestone and the Santa Rosa Hills Limestone exposed in the limestone-facies belt. It generally consists of a basal spiculiferous chert overlain by dark-gray lime mudstone with nodular black chert and silicified lime mudstone. A sample bearing the Gnathodid-pseudopolygnathid conodont assemblage was collected near the base of this unit at Cottonwood Canyon; macrofossils are rare to absent. Several thin, sediment-gravity-flow beds are present immediately above the basal chert unit in the outer limestone-subfacies belt, and in the upper part of the section, there are lenses of light- to medium-gray, pelmatozoan-rich limestone with crude graded bedding.

The fine-grained, dark-colored beds, especially the spiculiferous chert at the base of the section, suggest a quiet-water depositional setting. In the Santa Rosa Hills, sediment-gravity-flow deposits, including one debris-flow deposit, suggest a slope or base-of-slope depositional setting. The lack of unequivocal shallow-water macrofossils and the presence of conodonts of the Gnathodid-pseudopolygnathid biofacies (Sandberg and Gutschick, 1984) suggest moderately deep water, and the dark color and lack of bioturbation suggest dysaerobic conditions. The upper part of the unit commonly contains pelmatozoan-rich packstones deposited within the influence of storm-generated waves. This unit was deposited on a relatively deep, carbonate platform that sloped northwestward. Deposition of voluminous amounts of lime mud eventually resulted in the platform aggrading to above storm-wave base.

Leaning Rock Formation

The Leaning Rock formation, a name suggested by J. Kent Snow (pers. commun., 1994), is employed here to describe a dominantly micritic limestone sequence with interbeds of argillaceous mudstone, chert, limestone turbidites, and siliceous, spiculiferous, lime mudstone in the quartzose siltstone-facies belt. It is underlain by the Tin Mountain Limestone and overlain by a dominantly quartzose siltstone and very fine grained sandstone unit here called the Mexican Spring formation. At Rest Spring Gulch, the lower part of this unit consists of thin- to medium-bedded, siliceous, spiculiferous lime mudstone and wackestone interbedded with medium- to dark-gray, argillaceous lime mudstone and chert. The middle part of the sequence consists of intensely folded beds of lime mudstone and chert enclosed between undeformed beds. The upper part of the unit commonly contains trace fossils and well-developed, graded beds (turbidites) containing pelmatozoan debris and fragments of shallow-water fossils including bryozoans, brachiopods, and solitary and colonial rugose corals. Conodonts of the Bispathodid assemblage have been recovered from some of the turbidites.

The spiculiferous, black chert suggests that the lower part of the section was deposited in relatively deep water. The deformed beds in the middle of the section, interpreted as submarine-slide masses consisting of lime mudstone displaced from the slope, suggest a slope or base-of-slope depositional environment and indicate that the platform morphology had become that of a distally-steepened ramp (Read, 1985). The Bispathodid conodont biofacies indicates basinal conditions (Sandberg and Gutschick, 1984). The upper part of the section is replete with turbidites bearing shallow-water fossils that probably were derived from the platform margin after a rimmed shelf (Reed, 1985) had developed.

Santa Rosa Hills Limestone

The Santa Rosa Hills Limestone generally is composed of light-gray, pelmatozoan-rich packstone and grainstone. Fossils, other than pelmatozoan remains, are not common except in the outer limestone-subfacies belt where colonial rugose corals occur in bank-like buildups, and calcareous foraminifera, calcareous algae and brachiopods have been recovered.

Throughout the limestone-facies belt, rocks of this unit represent skeletal sand-sheets and shoals formed in a well-agitated, shallow-water environment. The coral banks in the Santa Rosa Hills probably represent deposition at or near the platform margin and indicate that the platform had evolved into a rimmed shelf (Read, 1985).

Mexican Spring Formation

This unit is dominated by intensely bioturbated, calcareous siltstone with graded, bioclastic limestones containing a variety of calcareous fossils, ooids, and intraclasts of lime mudstone, radiolarian- and spicule-rich chert, and rare phosphatic material.

The calcareous turbidites suggest deposition in a basinal environment. The intense bioturbation indicates that the bottom was well oxygenated.

DEPOSITIONAL MODEL FOR EASTERN CALIFORNIA

Evolution of the Mississippian carbonate platform in eastern California would be most easily interpreted by aligning the stratigraphic sections along a single stratigraphic datum that can be related to original sea level, as was done for the sections in southern Nevada. Because of the local geology, however, a second datum also must be used. The primary datum selected is the top of Meramecian strata (Fig. 4B) because this surface at Knight Canyon, Stone Canyon, Cottonwood Canyon, and the Santa Rosa Hills approximates Meramecian sea level. Post-depositional erosion could have altered this surface somewhat, but the lack of wide variations in thickness of such rocks in southern Nevada suggests that erosion was more or less uniform across the platform.

Northwest of the Santa Rosa Hills, all but the basal Mississippian rocks (i.e., the Tin Mountain Limestone) represent deep-water deposition. Thus, in order to give a reasonable representation of the original morphology of the slope and basin, the Rest Spring Gulch and Bighorn Gap sections must be aligned with the platform sections on the secondary datum. The top of the Tin Mountain Limestone seems the most appropriate because on the platform between Stone Canyon and the Santa Rosa Hills, this unit is close to horizontal, sloping less than 1° after the sections have been aligned along the top of the Meramecian rocks. Therefore, on the assumption that this surface originally continued to slope westward at the same angle (that it constituted a homoclinal ramp), the deeper-water sections were positioned as shown in Figure 4.

The only thrust fault known to separate the sections employed in this stratigraphic cross section is the Marble Canyon thrust (Fig. 1) which telescopes the Rest Spring Gulch and Cottonwood Canyon sections. According to Snow (1990), the displacement is about 3 km, so this amount of shortening was removed in preparation of Figure 4. The sections shown on this figure were projected along depositional strike onto a line drawn perpendicular to the interpreted trend of Mississippian facies belts (i.e., N 50° W).

The Tin Mountain and Santa Rosa Hills Limestones are relatively shallow-water deposits. The Stone Canyon limestone and Leaning Rock formation accumulated in deeper water and water depth increased to the northwest. As in Nevada, the top of lower Osagean strata approximately parallels a lithologic boundary, here the top of the Tin Mountain Limestone, apparently rising slightly in the relatively deep-water section at Rest Spring Gulch. Some of the other age boundaries in eastern California are less well defined, but it is clear that development of the limestone-facies belt here was similar to that in southern Nevada (compare Figs. 3, 4). Seaward of the platform, middle Mississippian rocks thin greatly, and the stratigraphic cross-sectional reconstruction suggests that the water depth in the Rest Spring Gulch area during the early Meramecian was much greater than that to the southeast, perhaps on the order of 600 m.

The depositional history in eastern California apparently was very similar to that in southern Nevada. Deposition of the Tin Mountain Limestone began with a marine trangression in the Kinderhookian. Rapid subsidence, suggested by a large increase in accommodation space in early Osagean time, resulted in deposition of first spiculiferous chert and then relatively deep-water micrite of the Stone Canyon limestone and the Leaning Rock formation. Progradation of the shallow-water Santa Rosa Hills Limestone began in the inner limestone-subfacies belt in middle Osagean time, but apparently it was slow enough that deposition of these shallow-water rocks did not begin in the outer limestone-subfacies belt until early Meramecian time. In the quartzose siltstone-facies belt, submarine slides composed of slope-derived sediment were emplaced in middle Osagean time and sediment-gravity-flow deposits containing shallow-water fossils, probably marking the approach of prograding, shallow-water carbonate-sand sheets, began to accumulate in late Osagean time. The carbonate platform facies, however, never prograded this far seaward. Carbonate deposition ceased in early Meramecian time when the platform was subaerially exposed, as shown by a forest rooting ground on top of the unit in Cottonwood Canyon (Decourten, 1986), but in the quartzose siltstone-facies belt to the northwest, the dominantly siliciclastic Mexican Spring formation continued to accumulate.

COMPARISON OF SOUTHERN NEVADA AND EASTERN CALIFORNIA PORTIONS OF THE PLATFORM

Similarities between the Mississippian carbonate platforms in southern Nevada and eastern California show that they represent parts of the same platform (e.g., Stevens and others, 1991). In both areas the initial transgression of the Mississippian sea was essentially synchronous; relatively shallow-water carbonates occur both below and above cherty, relatively deep-water rocks (i.e., the Anchor Limestone, Stone Canyon limestone, and Leaning Rock formation), and the subsequent late Osagean-early Meramecian progradation of shallow-water carbonates proceeded similarly. In addition, similar facies and subfacies belts also are present in both areas, and thickness patterns in the platform facies in the two areas are similar (Figs. 3, 4). There are, however, significant differences in scale. The greatest thickness in eastern California is about twice that in southern Nevada, and the restored distance between sections representing the inner limestone-subfacies belt (Knight Canyon in California and Mountain Springs Pass in Nevada) and the outer limestone-subfacies belt (Santa Rosa Hills in California and Indian Springs in Nevada) in Nevada is almost twice that in California. Thus, although the similarities show that these carbonate sequences were deposited on the same platform, the nature of the shelf on which the Mississippian carbonate platform developed varied considerably laterally.

PLATFORM EVOLUTION

The developmental history of the Mississippian carbonate platform is shown schematically in Figure 6. Formation of the Mississippian carbonate platform in southern Nevada and eastern California began with a major transgression of the sea in Kinderhookian time over a Late Devonian carbonate platform that probably dipped almost imperceptibly to the northwest. This transgression probably was due to the major eustatic sea-level rise shown by Ross and Ross (1987) in the earliest Mississippian, perhaps combined with regional subsidence due to thrust loading. Relatively shallow-water environments were maintained throughout the area during Kinderhookian and most of early Osagean time.

Probably the most important event in development of the Mississippian carbonate shelf, indicated by all the sedimentologic and paleontologic data, was a rapid deepening of the entire platform at the beginning of deposition of the Anchor Limestone in Nevada and the Stone Canyon lime-

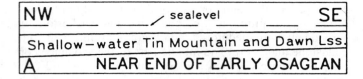
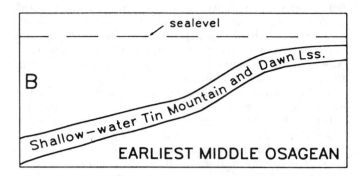
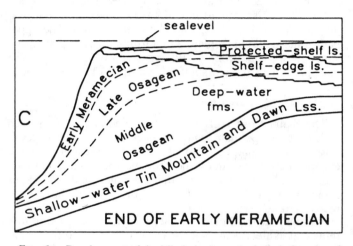

FIG. 6.—Development of the Mississippian carbonate platform based on data from eastern California and southern Nevada. (A) Platform near the end of early Osagean time, (B) Platform during earliest middle Osagean time, (C) Platform at end of early Meramecian time. Vertical exaggeration is about 30x.

stone and Leaning Rock formation in California. This event occurred close to the early-middle Osagean break (within the Lower *typicus* conodont Zone at Rest Spring Gulch), as shown by diagnostic conodonts in both the top of the relatively shallow-water Tin Mountain Limestone and the base of the overlying, relatively deep-water Leaning Rock formation (Figs. 2, 4). This event essentially corresponds to the time of maximum flooding in the western United States shown by Sando (1989), and it is similar to a deepening event recognized throughout eastern Nevada and western Utah by initial deposition of the interpreted deep-water Delle Phosphate (Poole and Sandberg, 1991). A more or less similarly-timed, deepening event also is suggested by subsidence curves shown by Dorobek and others (1991) for sections from central Montana to central Idaho.

This deepening event correlates approximately in time with a sea-level rise, but according to Dorobek and others (1991), this rise was not of great magnitude. This event also may correlate with the final emplacement of the Roberts Mountains allochthon onto the continent during or more likely after the late *isosticha*-Upper *crenulata* Zone (late Kinderhookian). Johnson and Visconti (1992, p. 1217) show the Roberts Mountains allochthon overlying the Dale Canyon Formation which they extended into the *typicus* Zone. They also cited Whitaker (1985) who recovered conodonts of the *typicus* Zone from rocks that depositionally overlie the Roberts Mountains allochthon in the Pinon Range in central Nevada, thus constraining emplacement of the allochthon to this conodont zone (early to early middle Osagean time). The similarity in timing of deepening on the platform and this interpretation of emplacement of the Roberts Mountains allochthon suggest that the deepening may have been due to subsidence in response to thrust loading by the Antler allochthon (Speed and Sleep, 1982). Because thrust activity deforms the lithosphere almost instantaneously geologically, tectonic subsidence history is the best indicator of thrust history in the absence of cross-cutting relations (Jordan and others, 1988). Therefore, subsidence of the platform in eastern California and southern Nevada may date final emplacement of the Roberts Mountains allochthon more closely than the geological relationships in central Nevada. Poole (written commun., 1993) reports, however, that his mapping in the Pinon Range shows that the Roberts Mountains allochthon was emplaced prior to the *isoticha*-Upper *crenulata* Zone. If this is true, final emplacement of the allochthon would correlate better with the beginning of the initial Mississippian transgression.

The early Osagean subsidence event resulted in drowning of the platform and deposition of spiculiferous, locally radiolarian-bearing chert followed by relatively deep-water carbonate mud across most of the homoclinal ramp. In the Santa Rosa Hills on the outer part of the ramp, however, sediment-gravity-flow deposits were emplaced. As the slope here apparently was gentle, seismicity associated with subsidence may have triggered these flows. Prior to the end of middle Osagean time, shallow-water carbonate sediment had prograded into the central limestone-subfacies belt (Fig. 4). Progradation was rapid across the ramp interior, but greater accommodation space offshore to the northwest reduced the progradation rate greatly. Age and sedimentological data (Figs. 3, 4) show that progradation into the outer limestone-subfacies belt was made possible by deposition of vast amounts of lime mud, probably winnowed out from nearshore, shallow-water environments. The mud on the platform slope apparently was unstable and locally became involved in slumping in the latter part of Osagean time. Meanwhile shallow-water environments continued to prograde, and by early Meramecian time the platform had evolved into a rimmed shelf with carbonate sand shoals and colonial-coral buildups along the margin at the outer edge of the outer limestone-subfacies belt. In the inner limestone-subfacies belt in eastern California, shallow-water deposits up to 200 m thick overlie the Stone Canyon limestone which there was deposited in only moderately deep water. Because the overall eustatic sea-level rise aggregated less than 80 m during the time of deposition of these rocks (Dorobek and others, 1991), it is apparent that slow regional subsidence continued throughout development of the platform.

Exposure of most of the carbonate platform in early or middle Meramecian time, marking the end of platform sedimentation, probably was due to a drop in sea level. This event may have been due to the third-order sea-level drop shown by Ross and Ross (1987) after deposition of the lower Salem Limestone in the midcontinent.

REGIONAL RELATIONSHIPS

The history of development of the carbonate platform in southern Nevada and eastern California is similar to that to the northeast in eastern Nevada and western Utah. The initial transgression in both regions began in the latter part of Kinderhookian time (Poole and Sandberg, 1991) and deepening occurred regionally in the early Osagean time. Later, in early Meramecian time, the platform was widely exposed (Sando, 1989; Poole and Sandberg, 1991) so that quartz sand and silt were transported from the exposed transcontinental arch across a karst plain to be deposited, for instance, as the Humbug Formation or deltaic HUM lithosome in Utah (Sando, 1990). Data on the Mexican Spring formation (Klingman, 1987) suggest a similar source for the quartz sand and silt which presumably were carried subaerially to the edge of the carbonate platform and then down the slope in submarine sediment-gravity flows.

There also are similarities between different parts of the carbonate platform. Maps of Sando (1989) show large parts of the platform, including most of Montana, Wyoming, and the Dakotas, exposed during foraminiferal zones 12 and 13. In southern Nevada and eastern California, the platform was exposed in about foraminiferal zones 14 and 13, respectively, and Poole and Sandberg (1991) showed that at the beginning of the *Cavusgnathus* conodont zone (foraminiferal zone 14), the karst plain extended over eastern Utah and most of Arizona. The similarity of timing of exposure throughout the miogeocline suggests an eustatic sea-level drop.

SUMMARY

Development of the carbonate platform in southern Nevada and eastern California began in Kinderhookian time with submergence of the older Devonian carbonate platform, the result of an eustatic sea level rise and/or tectonic loading. In early Osagean time, the platform was depressed and tilted westward, probably in response to the final emplacement of the Roberts Mountains allochthon onto sialic North America. Initially, carbonate production rates decreased, but by the middle Osagean, carbonate production exceeded any increase in accommodation space, initiating northwestward progradation of shallow-water carbonates. By early Meramecian time a rimmed platform developed, and still within early or middle Meramecian time an eustatic sea-level fall resulted in exposure of the entire platform. This history of carbonate platform development in eastern California and southern Nevada is generally similar to that interpreted for other parts of the platform as far north as Montana. Therefore, the model generated here probably can be used to better understand different parts of this and other carbonate platforms.

ACKNOWLEDGEMENTS

We are very grateful to Charles Sandberg (U. S. Geological Survey), who studied all of the conodont samples; Paul Brenckle (Amoco Production Company), who identified the foraminiferans; and William Sando, (U. S. Geological Survey), who identified the corals. Conversations with David Andersen, J. Kent Snow, and especially Paul Stone have added greatly to our understanding of this platform. We also thank David Andersen, Steve Dorobek, Katherine Giles, Forrest Poole, Steven Reid, and William Sando, whose reviews greatly improved the manuscript.

REFERENCES

BELASKY, P., 1988, Stratigraphy and paleogeographic setting of the Mississippian Monte Cristo Group in the Spring Mountains, southern Nevada: Unpublished M. S. Thesis, San Jose State University, San Jose, 152 p.

BURCHFIEL, B. C., FLECK, R. J., SECOR, D. T., VINCELETTE, R. R., AND DAVIS, G. A., 1974, Geology of the Spring Mountains, Nevada: Geological Society of America Bulletin, v. 85, p. 1013–1022.

DECOURTEN, F. L., 1986, *Stigmaria* from the Upper Carboniferous Rest Spring Shale and a possible Lepidodendrale rooting ground, Death Valley region, southeastern California (abs.): Cordilleran Section of Geological Society of America, Abstracts with Programs, v. 18, p. 100.

DOROBEK, S. L., REID, S. K., AND ELRICK, M., 1991, The stratigraphic record of eustatic fluctuations and episodic tectonic events, *in* Cooper, J. D., and Stevens, C. H., eds., Paleozoic Paleogeography of the Western United States—II: Los Angeles, Pacific Section SEPM, Book 67, v. 1, p. 487–507.

DUNNE, G. C., GULLIVER, R. M., AND STEVENS, C. H., 1981, Correlation of Mississippian shelf-to-basin strata, eastern California: Geological Society of America Bulletin, v. 92, p. 1–38.

GORDON, M., JR. AND POOLE, F. G., 1968, Mississippian-Pennsylvanian boundary in southwestern Nevada and southeastern California: Boulder, Geological Society of America Memoir 110, p. 157–168.

JOHNSON, J. G. AND VISCONTI, R., 1992, Roberts Mountains thrust relationships in a critical area, northern Sulphur Spring Range, Nevada: Geological Society of America Bulletin, v. 104, p. 1208–1220.

JORDAN, T. E., FLEMINGS, P. B., AND BEER, J. A., 1988, Dating thrust-fault activity by use of foreland-basin strata, *in* Kleinspehn, K. L., and Paola, C., New Perspectives in Basin Analysis: New York, Springer-Verlag, p. 307–330.

KLINGMAN, D. S., 1987, Depositional environments and paleogeographic setting of the middle Mississippian section in eastern California: Unpublished M. S. Thesis, San Jose State University, San Jose, 231 p.

MAMET, B. L. AND SKIPP, B. A., 1970, Lower Carboniferous calcareous Foraminifera—Preliminary zonation and stratigraphic implications for the Mississippian of North America: Sheffield, 6th International Congress of Carboniferous Stratigraphy and Geology, Compte Rendu, v. 3, p. 1129–1146.

MCALLISTER, J. F., 1952, Rocks and structure of the Quartz Spring area, northern Panamint Range, California: Sacramento, California Division of Mines and Geology Special Report 25, 38 p.

POOLE, F. G. AND SANDBERG, C. A., 1991, Mississippian paleogeography and conodont biostratigraphy of the western United States, *in* Cooper, J. D., and Stevens, C. H., eds., Paleozoic Paleogeography of the Western United States—II: Los Angeles, Pacific Section SEPM, Special Publication 67, (Book 67, v. 1), p. 107–136.

READ, J. F., 1985, Carbonate platform facies models: American Association of Petroleum Geologists Bulletin, v. 69, p. 1–21.

ROSS, C. A. AND ROSS, J. R. P., 1987, Biostratigraphic zonation of Late Paleozoic depositional sequences: Cushman Foundation for Foraminiferal Research Special Publication 24, p. 151–168.

SANDBERG, C. A. AND GUTSCHICK, R. C., 1984, Distribution, microfauna and source-rock potential of Mississippian Delle Phosphatic Member of Woodman Formation and equivalents, Utah and adjacent States, *in* Woodward, J., Meissner, F. F., and Clayton, J. L., eds., Hydrocarbon Source Rocks of the Greater Rocky Mountain Region: Denver, Rocky Mountain Association of Geologists, p. 135–178.

SANDO, W. J., 1985, Revised Mississippian time scale, Western Interior region, conterminous United States: Washington, D. C., United States Geological Survey Bulletin 1605-A, p. A15-A26.

SANDO, W. J., 1989, Dynamics of Carboniferous coral distribution, Western Interior, USA: Memoirs of the Association of Australasian Palaeontologists, v. 8, p. 251–265.

SANDO, W. J., 1990, HUM lithosome: An example of regional stratigraphic synthesis in the Mississippian of the western interior of the United States: Washington, D. C., United States Geological Survey Bulletin 1881, p. E1-E17.

SANDO, W. J. AND BAMBER, E. W., 1985, Coral zonation of the Mississippian System in the Western Interior Province of North America: Washington, D. C., United States Geological Survey Professional Paper 1334, p. 1–61.

SNOW, J. K., 1990, Cordilleran orogenesis, extensional tectonics, and geology of the Cottonwood Mountains area, Death Valley region, California and Nevada: Unpublished Ph. D Thesis, Harvard University, Cambridge, v. 3:1–3:64.

SPEED, R. C. AND SLEEP, N. H., 1982, Antler orogeny and foreland basin: a model: Geological Society of America Bulletin, v. 93, p. 815–828.

STEVENS, C. H., STONE, P., AND BELASKY, P., 1991, Paleogeographic and structural significance of an Upper Mississippian facies boundary in southern Nevada and east-central California: Geological Society of America Bulletin, v. 103, p. 876–885.

WHITAKER, J. H., 1985, Geology of the Willow Creek area, Elko County, Nevada: Unpublished M. S. Thesis, Oregon State University, Corvallis, 116 p.

THE INTERPLAY OF EUSTASY AND LITHOSPHERIC FLEXURE IN FORMING STRATIGRAPHIC SEQUENCES IN FORELAND SETTINGS: AN EXAMPLE FROM THE ANTLER FORELAND, NEVADA AND UTAH

KATHERINE A. GILES
Department of Geological Sciences, New Mexico State University, Las Cruces, NM 88003
AND
WILLIAM R. DICKINSON
Department of Geosciences, University of Arizona, Tucson, AZ 85721

ABSTRACT: Stratigraphic relationships within the Late Devonian to Early Mississippian Antler foreland of eastern Nevada and westernmost Utah suggest that lithospheric flexure in conjunction with eustasy were important primary controls on accommodation trends. Accommodation trends attributed to lithospheric flexure include: (1) sequence-scale opposing relative sea-level trends in time-equivalent strata and (2) regional-scale inversion of topography.

Loading of the lithosphere by eastward thrust emplacement of the Roberts Mountains allochthon (RMA) over the North American craton from Late Devonian to Early Mississippian time resulted in flexural warping of the craton into an asymmetric downwarp (foreland basin), an upwarp at the cratonward edge of this basin (forebulge), and farther cratonward, a broad, shallow downwarp (back-bulge basin). Stratigraphic sequences formed in conjunction with active flexure of the lithosphere display opposing relative sea-level trends. Strata in flexurally subsiding areas reflect increasing trends in accommodation space, and areas of flexural upwarp (forebulge) reflect decreasing trends in accommodation space. During quiescent times, when no new flexural accommodation space was being produced, strata reflect congruous trends in accommodation space predominantly controlled by eustasy.

Eastward migration of the flexural features occurred sporadically in response to eastward transport of the RMA. Migration of the flexural features resulted in regional-scale inversion of topography. Areas formerly occupied by the forebulge were downflexed and incorporated into the foreland basin. Conversely, areas formerly occupied by the back-bulge basin were uplifted as the forebulge migrated cratonward. Stratigraphic evidence of inversion permits delineation of changes in the position and geometry of flexural topography during the transition from a passive to a collisional margin regime.

INTRODUCTION

The nature and development of marine stratigraphic sequences are directly related to changes in accommodation space (space available for deposition) whether in passive margin settings or along tectonically active margins. Accommodation space is a function of subsidence or uplift, sediment supply, and eustasy. Because these factors do not behave independently, determining the relative contributions of each to temporal variations in accommodation space is difficult.

In foreland settings, subsidence and uplift are profoundly affected by lithospheric flexure. Foreland basin subsidence is primarily controlled by downflexing of the lithosphere in response to thrust loading and in some cases, to subsurface loads transmitted from the subduction zone (Jordan, 1981; Beaumont, 1981). Flexural downwarping results in rapid subsidence adjacent to the thrust load. Subsidence rate gradually decreases away from the thrust front producing an asymmetrical depression (Fig. 1A). Flexural uplift (referred to as a peripheral bulge or forebulge; Walcott, 1970; Jacobi, 1981) occurs as an isostatic response to downwarping and forms the distal margin of the foreland basin. At a point on the distal margin of the foreland basin, subsidence decreases to zero (flexural node, Fig. 1A). Cratonward of this point, flexural uplift occurs (forebulge) resulting in a decrease in accommodation space. Though flexurally induced accommodation space decreases, deposition may still occur over the forebulge depending on the amount of accommodation space produced by eustatic sea-level changes. Cratonward of the forebulge flexure, a broad shallow downwarp or intrashelf basin forms, termed a "back-bulge basin" (Turcotte and Schubert, 1982; Quinlan and Beaumont, 1984; Goebel, 1991a). Flexural subsidence results in accommodation space increase within the back-bulge basin (Fig. 1A). Back-bulge basin subsidence is several orders of magnitude less than subsidence in the foreland basin proper.

The dimensions and the amount of flexural subsidence and uplift produced by the flexural features (i.e., foreland basin, forebulge, and back-bulge basin) primarily depend on the geometry and density of the tectonic load, rheology of the lithosphere, density and volume of the sediment infill, and amount of thrust wedge and forebulge erosion (Caldwell and others, 1976; Beaumont, 1981; Jordan, 1981; Goebel, 1991b).

During a single thrust event, flexural uplift of the forebulge and subsidence of the foreland basin and back-bulge basin occur simultaneously. As a result, opposing trends in accommodation space or relative sea-level are produced in laterally equivalent strata. For example, in areas undergoing flexural subsidence (foreland basin and back-bulge basin) accommodation space increases. Stratigraphic patterns associated with accommodation increase are condensed, deepening upward, aggradational, or relatively thick sequences that reflect relative rises in sea level. In contrast, accommodation space decreases in areas undergoing uplift (allochthon and forebulge). Stratigraphic patterns associated with accommodation decrease are shoaling-upward, progradational, or relatively thin sequences separated in part by unconformity surfaces that reflect relative falls in sea level. Coeval, opposing relative sea-level trends produced by flexurally formed accommodation space, is in marked contrast with congruous relative sea-level trends produced by eustatic sea-level changes. The character of accommodation trends in coeval strata allows the relative contributions of flexure-driven accommodation space and eustasy-driven accommodation space to be estimated from depositional facies analysis of stratigraphic sequences.

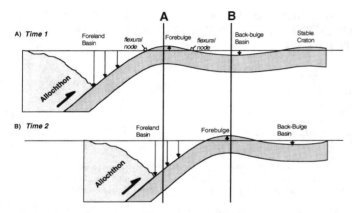

FIG. 1.—Diagrammatic cross sections displaying the relationship of lithospheric flexure to accommodation space in foreland systems. Arrows pointing down indicate an increase in accommodation space produced by lithospheric downwarping and arrows pointing up indicate a decrease in accommodation space due to lithospheric upwarping. (A) *Time 1*: thrust load emplaced, resulting in downwarping of the lithosphere (foreland basin), cratonward upwarping (forebulge), and farther cratonward, gentle downwarping (back-bulge basin). (B) *Time 2*: the thrust load migrates cratonward resulting in cratonward migration of the flexural features; former uplifted area of forebulge (locality A, *Time* 1) is downwarped and incorporated into foreland basin, whereas former back-bulge basin (locality B, *Time* 1) is upwarped over migratory forebulge at *Time* 2.

A second distinctive feature of flexurally induced accommodation trends is the formation of regional-scale inversions of topography. In response to progressive emplacement of thrust loads, flexural features migrate across the foreland downwarping former topographic highs and upwarping former topographic lows (Allen and others, 1986). For example, as the thrust wedge migrates, it flexurally downwarps the area that was previously upwarped over the forebulge and incorporates this area into the foreland basin (Fig. 1B). Conversely, forebulge migration causes uplift of the previously downwarped back-bulge basin.

Erosion of emergent thrusts and local forebulge erosion may increase sediment supply into the foreland and back-bulge basins. Increased sediment supply affects accommodation space primarily in two ways: (1) redistributes the tectonic load and (2) reduces the amount of accommodation space by infill. Sedimentary deposits act as a lithostatic load and produce flexural subsidence. During active tectonism, the amount of subsidence due to sediment loading is relatively minor compared to subsidence due to thrust loading. During quiescent tectonic phases, sediment derived from emergent thrusts or forebulge uplifts effectively redistributes preexisting loads and tends to widen flexurally subsiding basins (Flemings and Jordon, 1989). High sediment influx may allow progradation to fill existing accommodation space. Once the basin is filled, sediment bypasses the basin until additional accommodation space is created.

We have documented accommodation trends for a series of biostratigraphically correlated stratigraphic sections which lie along a regional transect across the Antler foreland in Nevada and Utah. By comparing regional accommodation trends in syntectonic stratigraphic sequences to published third-order eustatic sea-level curves (Ross and Ross, 1987; Johnson and others, 1991), we have attempted to gauge the relative effects of lithospheric flexure and eustasy on the stratigraphic evolution of the Antler foreland region.

GEOLOGIC SETTING OF ANTLER FORELAND

The Antler orogenic belt is an eastward vergent thrust system extending over 2300 km from southern California (terminating at the San Andreas fault system) northward through Nevada and Idaho into British Columbia, Canada. The exact tectonic cause of the orogeny is still enigmatic, and many different models for its origin have been proposed (Nilsen and Stewart, 1980; Dickinson and others, 1983; Burchfiel and Royden, 1991). Most of the models invoke some form of arc-continent interaction, either arc-continent collision or back-arc thrusting.

The onset of the Antler orogeny resulted in eastward thrust emplacement of the Roberts Mountains allochthon (Fig. 2) over the former passive margin of the North American craton (Roberts and others, 1958; Stewart and Poole, 1974). The Roberts Mountains allochthon consists of a structurally complex succession of early to middle Paleozoic deep-water siliciclastic, pelagic, and volcanic rocks that were thrust imbricated during Late Devonian (latest Frasnian) through Early Mississippian (mid-Osage) time (Roberts and others, 1958; Madrid 1987; Johnson and Pendergast, 1981; Johnson and Visconti, 1992; Carpenter and others, 1993a, 1993b). Estimated eastward transport of the Roberts Mountains allochthon was 140 km (Roberts and others, 1958; Nilsen and Stewart, 1980; Murphy and others, 1984). Therefore, flexural features associated with overthrusting are estimated to have migrated eastward a similar distance during Antler orogenesis.

The general stratigraphy in the foreland east of the Roberts Mountains allochthon consists of a thick, passive-margin carbonate platform that developed following upper Precambrian rifting (Figs. 3, 4). The extensive, west facing carbonate platform is overlain by uppermost Devonian through Lower Mississippian siliciclastic and carbonate strata deposited during Antler orogenesis (syntectonic strata). The syntectonic strata are overlain by Middle to Upper Mississippian siliciclastic strata containing submarine-fan and basin slope deposits shoaling to deltaic and fluvial facies (Harbaugh and Dickinson, 1981), which filled the Antler foreland basin (post-tectonic strata). Uppermost Mississippian and Lower Pennsylvanian carbonate strata (Ely Limestone) overlie the foreland basin fill and onlap the Roberts Mountains allochthon (overlap assemblage).

STRATIGRAPHIC EVOLUTION OF ANTLER SYNTECTONIC STRATA

Stratigraphic sequences are classically defined by the relationships of "seismic-scale" depositional geometries (Vail and others, 1977). Because Antler foreland strata are exposed along a series of roughly strike-parallel, Tertiary horst blocks, seismic-scale, dip-oriented profiles are not exposed. Therefore, the sequence stratigraphic framework for this study was inferred from depositional facies analysis of time-correlative stratigraphic packages bounded by sub-regional unconformities and their correlative conformities.

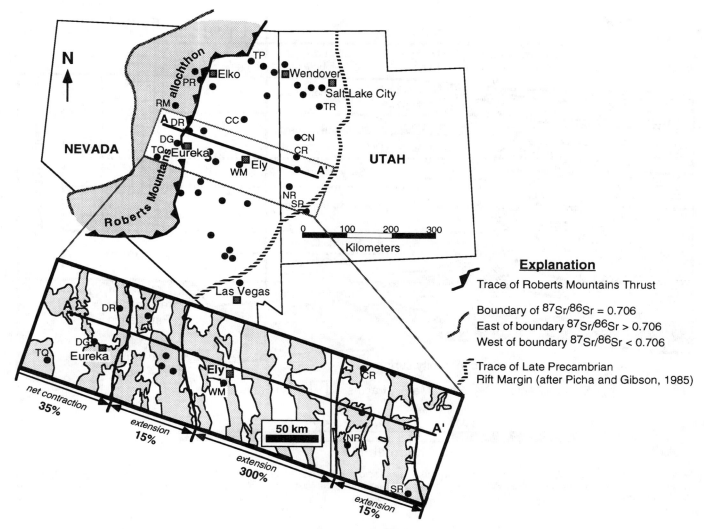

Fig. 2.—Map of the Antler foreland region in eastern Nevada and western Utah. Black dots mark the position of stratigraphic sections used in this study. Stippled squares mark the location of major cities and towns in the area. Sections referred to in this paper are indicated as follows: TQ = Toquima Range, DG = Devils Gate, DR = Diamond Range, PR = Pinon Range, TP = Tripon Pass, CC = Cherry Creek Range, WM = Ward Mountain, NR = Needle Range, SR = Star Range, CR = Confusion Range, CN = Coyote Knolls, TR = Tintic Mountains, TP = Tripon Pass, and ST = Stansbury Mountains. See Goebel, 1991b for detailed description of localities. Within the inset map stippled areas outline surface exposure of Paleozoic rocks and correspond approximately to the outline of mountain ranges. Extension and contraction percentages listed for areas on the inset are derived from Gans (1987) and Dobbs and others (1993). The Upper Devonian and Early Mississippian position of localities on cross section A to A' were determined using these values.

A composite third-order (1 my to 10 my) global coastal onlap curve for latest Devonian and Mississippian time was compiled (Fig. 4) from the Devonian curve of Johnson and others (1991) and the Mississippian curve of Ross and Ross (1987). The composite global coastal onlap curve approximates a eustatic sea-level curve for this time interval. Within the syntectonic time interval, eight third-order global sea-level cycles have been identified: five in the Late Devonian (latest Frasnian through Famennian) and three in the Early Mississippian (Kinderhook through early Osage).

The third-order eustatic sea-level cycles are superimposed on second-order (10 my to 80 my) sea-level cycles (c.f. Goldhammer and others, 1991). Second-order eustatic sea level rose from the Middle Devonian (Emsian) through Late Devonian (Frasnian) time ("Taghanic onlap" of Johnson, 1970). The Late Devonian rise in sea level reached a maximum in latest Frasnian to earliest Famennian time, which corresponds to initiation of thrusting in the Roberts Mountains allochthon (Fig. 4). Subsequent sea-level fall throughout the Late Devonian (Famennian) ended near the Devonian-Mississippian boundary. The following Early Mississippian transgression reached a maximum during early Osagean time near the culmination of Antler thrusting.

Within the syntectonic phase (late Frasnian through mid-Osagean time), we recognize 8 third-order depositional sequences (Figs. 4, 5). Third-order sequence boundaries (Fig. 5) were chosen based primarily on regionally correlatable, missing biostratigraphic zones in conjunction with subaerial

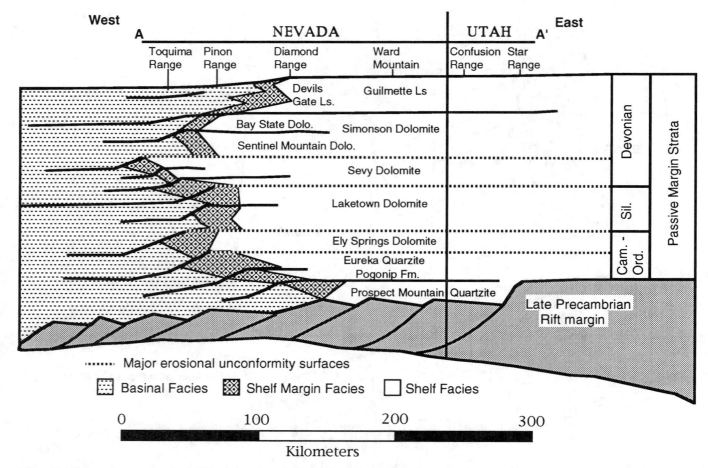

FIG. 3.—Schematic cross section of Cambrian through Upper Devonian passive margin stratigraphy from western Utah into eastern Nevada. Modified from Cook and others (1983). Position of stratigraphic localities palinspastically corrected for post-Antler tectonic events (Gans, 1987; Dobbs and others, 1993). Vertical relationships not to scale.

exposure of subtidal facies, regionally extensive erosion surfaces, and non-Waltherian facies shifts. The regional correlatability (over 100's of kilometers) of these unconformity surfaces implies generation by a fall in eustatic sea level. As a result, it is no coincidence that the regionally correlatable unconformity surfaces used to identify "sequence boundaries" in the Antler syntectonic strata match closely the timing of major eustatic sea-level falls proposed by Johnson and others (1991) and Ross and Ross (1987) for this time period. The sequences were correlated regionally using detailed conodont biostratigraphy derived from the literature and augmented by work completed for this study. Data sources for each locality are listed in the appendix.

To display the lithologic succession within each sequence, we use stratigraphic cross-section A to A' (Figs. 2, 5) which runs approximately perpendicular to the trend of the Roberts Mountains thrust and crosses the Antler foreland to the presumably stable, flexurally unaffected portion of the North American craton. On the cross sections we have attempted to restore the stratigraphic sections to their Devonian/Mississippian relative positions based on the palinspastic restorations of Bogen and Schweikert (1985), Gans (1987), Levy and Christie-Blick (1989), and Dobbs and others (1993). Stratigraphic localities not directly on cross section A to A' were projected on to the line of section.

Sequence 1 (Lower Palmatolepis gigas—Lower Palmatolepis triangularis)

Sequence 1 is Late Devonian (latest Frasnian to early Famennian) in age and spans portions of three conodont zones (Figs. 4, 6) estimated to represent 1.2 my (Sandberg and others, 1988). Deposition of Sequence 1 strata was approximately coincident with initiation of eastward thrusting of the Roberts Mountains allochthon (Poole, 1974; Johnson and Pendergast, 1981; Johnson and Visconti, 1992). The basal sequence boundary is a sub-regionally traceable, sub-aerial erosion surface that exposed portions of underlying subtidal carbonate strata (lower Devils Gate Limestone and Guilmette Limestone). The upper boundary of Sequence 1 is also a sub-regionally traceable, erosion surface.

Sequence 1 Depositional Facies Trends.—

The westernmost preserved strata of Sequence 1 age (Fig. 6, Toquima and Pinon ranges) consist of thin, rhythmically-

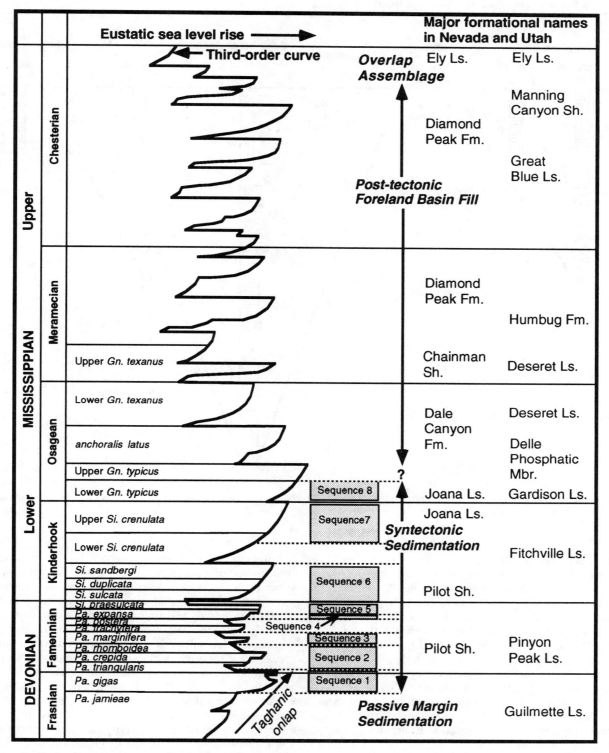

FIG. 4.—Qualitative coastal onlap curve as a general indication of third-order (solid line) eustatic sea-level changes for the Upper Devonian and Mississippian. The composite coastal onlap curve was derived from Johnson and others (1991) for the Upper Devonian and Ross and Ross (1987) for the Mississippian. Eustatic sea-level rise is indicated by shifts to the right on the curve. The Upper Devonian through Lower Mississippian standard conodont zonation shown to the left of the coastal onlap curve is derived from Sandberg and others (1988), Johnson and others (1991), and Poole and Sandberg (1991). The relative positions of the eight syntectonic stratigraphic sequences defined in this study are shown directly to the right of the coastal onlap curve. Formation names of Upper Devonian through Mississippian strata deposited in eastern Nevada and western Utah are shown to the far right of the coastal onlap curve.

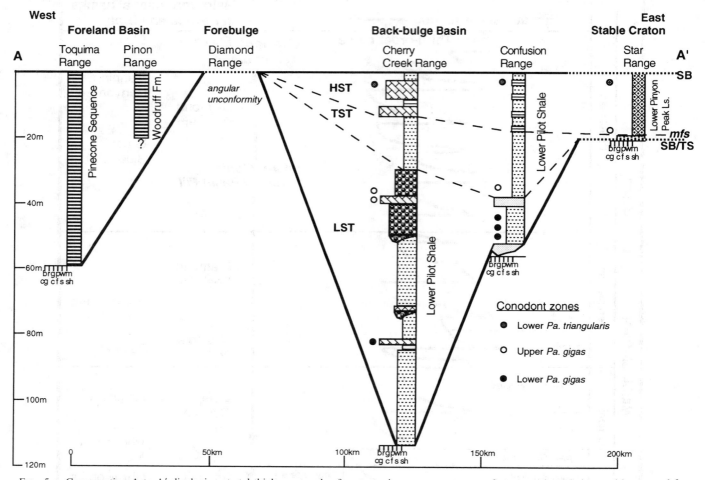

Fig. 5.—Cross section A to A' displaying stratal thickness trends of syntectonic sequences, nature of sequence boundaries, and interpreted forebulge position for each sequence.

bedded chert and hemipelagic claystone (Pinecone Sequence and Woodruff Formation) deposited at bathyal water depths below the oxygen minimum zone (Coles, 1988). Depositional setting and sedimentation rate appear to have remained relatively constant in this area from the Upper Devonian passive margin ramp to Sequence 1 time.

In the Diamond Range, Sequence 2 age strata rest unconformably on tilted passive margin strata (lower Devils Gate Limestone). Strata of Sequence 1 age were either not deposited or subsequently removed from this area. Sedimentation patterns indicating western sourcing of sediments in the Cherry Creek Range suggest that Sequence 1 age strata were eroded from the Diamond Range area prior to Sequence 2 deposition.

The Cherry Creek Range and Confusion Range sections will be discussed together because they contain strata displaying similar depositional facies trends (lower Pilot Shale). The lower portion of these sections consists of a relatively thick succession of dominantly deep-water siliciclastic strata (Gutschick and Rodriguez, 1979) that disconformably overlie subtidal shelf carbonates of the Guilmette Limestone (Larsen and others, 1989). The siliciclastic succession consists of planar-bedded, laminated, calcareous siltstone and shale with planar-bedded and ripple cross-laminated, fine- to coarse-grained, texturally and compositionally mature, quartz sandstone. The sandstone horizons locally thicken and become channelized, containing graded beds of clast-supported, limestone pebble conglomerate grading to fine-grained sandstone. The sandstone horizons are interpreted as turbidites (Jones, 1990). In the Cherry Creek Range, paleocurrent measurements from current ripples in the turbidite horizons show consistent east/southeast paleocurrent directions, suggesting a western source of detritus (Jones, 1990; Goebel, 1991b). In the Confusion Range, the sandstone horizons form relatively continuous sheets and are not organized into a channelized turbidite fan geometry as they are in the Cherry Creek Range (Jones, 1990). Paleocurrent directions in the Confusion Range are more variable and range from southwestward to southeastward. Associated with the turbidite beds in the Cherry Creek Range are medium-

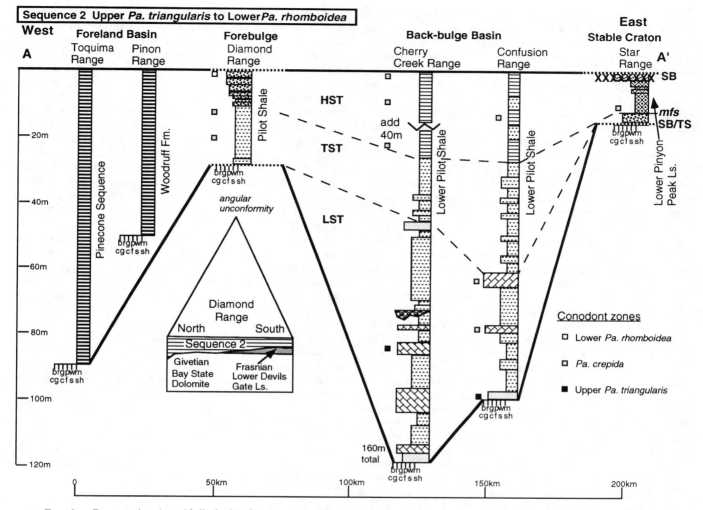

FIG. 6.—Cross section A to A' displaying Sequence 1 (Lower *Pa. gigas* to Lower *Pa. triangularis* conodont zones) foreland strata.

to thick-bedded, matrix-supported, limestone cobble conglomerates interpreted as submarine debris flow deposits (Jones, 1990). The carbonate clasts are composed of peloidal mudstone and are commonly plastically deformed.

The turbiditic units in the Cherry Creek and Confusion ranges are overlain by a thick, monotonous shale sequence (approximately 20m). Laminated shale deposition changes abruptly upward into planar-bedded, calcareous mudstone and wackestone intercalated with calcareous siltstone and shale. Locally (Fig. 2, locality DG and CC) the mudstones are interbedded with thick beds of clast-supported, carbonate breccia (interpreted as intraformational translation breccia), contorted mudstone beds (interpreted as slump deposits), and matrix-supported carbonate conglomerate (interpreted as debris flows). Measurements of vergence directions of folded strata in carbonate slump deposits indicate an east facing paleoslope. Paleocurrent directions taken from climbing ripples in intercalated bioclastic turbidites are east/southeast (consistent with the carbonate slump directions) indicating a western source for carbonate material. Depositional facies in the Cherry Creek and Confusion ranges suggest moderately deep-water conditions with carbonate debris being shed predominantly from paleohighs to the west (Diamond Range area?) of these stratigraphic sections.

Sequence 1 strata in the Star Range consist of a thin (<1.5 m thick), planar- and cross-bedded, basal, quartz sandstone horizon deposited above an erosional, scour surface in a shallow, subtidal depositional setting. The sandstone contains chert fragments and abraded fish teeth possibly eroded from the underlying passive margin carbonate strata. The quartz sandstone horizon grades abruptly upward into peloidal wackestone (lower Pinyon Peak Limestone) interpreted to represent shallow, subtidal carbonate deposition. The sequence is capped by an erosional unconformity surface. Strata of the Lower *Pa. gigas* conodont zone have not been identified in the Star Range.

Interpretation of Sequence 1 Accommodation Trends.—

Accommodation trends in Sequence 1 strata display dramatic changes when compared to those of the underlying passive margin carbonate platform. Upper Devonian, passive-margin carbonate strata were deposited on a distally steepened, west facing ramp in which accommodation space increased gradually in a basinward direction (Fig. 3). The overlying strata of Sequence 1 display highly variable accommodation trends across the former uniformly dipping ramp (Figs. 5, 6).

Depositional facies in the Toquima and Pinon ranges are similar to those of the underlying basinal passive margin strata, suggesting that the depositional setting did not change dramatically. The facies reflect minimal detrital influx into a deep subsiding basin, resulting in a proportionately thin (condensed) stratigraphic section of deep-water deposits. Absence of Sequence 1 age strata in the Diamond Range, due either to non-deposition or erosion, reflects relative accommodation decrease. In contrast, time-correlative stratigraphic sections to the east (e.g., Fig. 6, Cherry Creek Range and Confusion Range) contain thick sequences of deep-water facies reflecting major accommodation increase. In the Star Range, deposition of a thin sequence of shallow-water strata suggests accommodation decrease in comparison to sections directly to the west.

We interpret the large scale, differential accommodation trends described for Sequence 1 to reflect Upper Devonian flexural warping of the former passive-margin carbonate platform into a north-northeast to south-southwest trending foreland basin (Toquima Range and Pinon Range) with concomitant eastward forebulge upwarp (Diamond Range) and still farther east, a broad, downwarped, back-bulge basin (Devils Gate, Cherry Creek Range, and Confusion Range). Strata in the Star Range do not show the same trends in accommodation space as those interpreted to represent back-bulge basin deposits and are interpreted as representing deposition on the relatively stable craton, where subsidence was not profoundly affected by lithospheric flexure.

We have identified smaller scale accommodation trends within Sequence 1, which are apparently synchronous and congruous, allowing us to identify depositional systems tracts. Strata of the lower *Pa. gigas* conodont zone are not present in stratigraphic sections directly to the west of the Cherry Creek and Confusion ranges (e.g., Diamond Range) or to the east (e.g., Star Range). The stratigraphic position, nature, and geometry of these strata (isolated turbidite fan complex directly overlying a sequence boundary) are indicative of the lowstand systems tract (Van Wagoner and others, 1990). The upward transition from coarse turbidite fan deposition to laminated shale indicates temporary termination of coarse detrital material into the basin, reflecting a relative rise in sea level. Coeval shallow-water strata in the Star Range represents initial transgression of the shelf (TST) which reworked stranded lowstand sand and incorporated material eroded from the underlying limestone. Deepening continued to the point where siliciclastic detritus was trapped shoreward (central or eastern Utah), allowing shallow-water carbonate deposition to proceed in the Star Range area. With subsequent relative sea-level fall (HST), carbonate material was delivered into the basin as calciturbidites, debris flows, and slumps derived from shallow-water carbonate complexes to the west and east.

Sediment starvation in the foreland basin may have been due either to lack of significant detrital influx derived from the thrust imbricated Roberts Mountains allochthon to the west or to entrapment of coarse sediment in the deepest, axial portion of the foreland basin resulting in sediment starvation of the distal margin of the foreland basin (Heller and others, 1988). Sediments derived from the allochthon may have not reached the foreland basin because a through-going drainage system had not yet been established. Another possibility is that the Roberts Mountains allochthon may have suffered little erosion due either to flooding of the allochthon during the Late Devonian, second-order sea-level maximum (Fig. 4) or to lack of sufficient structural relief to expose the allochthon subaerially. A volumetrically small amount of clastic material within the foreland basin may have been derived from erosion of the forebulge area to the east. Because of the homogenous, deep-water nature of these deposits, we were unable to differentiate depositional systems tracts.

The unconformity surface in the Diamond Range reflects major accommodation decrease interpreted to represent flexural uplift associated with the forebulge. Terrigenous detritus derived from the craton during sea-level lowstand did not reach the foreland basin, but instead was apparently ponded in the subsiding back-bulge basin behind the forebulge uplift (Cherry Creek and Confusion ranges). The relatively thick succession of clastic strata preserved within the back-bulge basin reflects increased accommodation space due to flexural downwarping in conjunction with increased sedimentation rate from sediment ponding in the intrashelf, back-bulge basin. The presence of western derived, allochthonous, shallow-water carbonate debris in the back-bulge basin requires that some sort of isolated carbonate platform was present to the west (Diamond Range area?), possibly forming on horst blocks associated with forebulge uplift. These deposits were subsequently completely removed at the next sequence boundary.

Sequence 2 (Upper Palmatolepis triangularis—Lower Palmatolepis rhomboidea)

Sequence 2 is Late Devonian (early Famennian) in age and spans portions of three conodont zones corresponding to approximately 1.5 my (Sandberg and others, 1988). The basal sequence boundary is a sub-regionally traceable erosion surface, which locally forms a subaerially exposed, erosion surface (Fig. 7, Diamond Range). The basal sequence boundary in other areas along cross-section A to A' is apparently conformable. The upper sequence boundary forms a sub-regionally traceable, subaerial exposure surface that becomes conformable in other areas.

Sequence 2 Depositional Facies Trends.—

Depositional facies trends in Sequence 2 are similar to those of Sequence 1. Deposition of thin, rhythmically bedded chert and hemipelagic claystone (Pinecone Sequence and Woodruff Formation) continued in the westernmost preserved stratigraphic sections (Fig. 7, Toquima and Pinon ranges).

Sequence 2 age strata in the Diamond Range consist of a comparatively thin siliciclastic sequence (lower Pilot Shale) that unconformably overlie tilted passive margin strata. Near the southern end of the Devonian outcrop belt in the Dia-

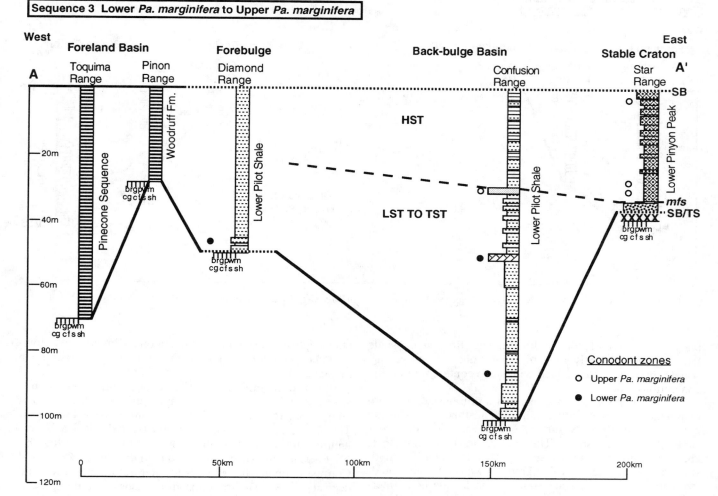

FIG. 7.—Cross section A to A' displaying Sequence 2 (Upper *Pa triangularis* to Lower *Pa rhomboidea* conodont zones) foreland strata.

mond Range (refer to inset on Fig. 7), Sequence 2 strata rest unconformably on lower Devils Gate Limestone of Frasnian age, but northward Sequence 2 strata rest unconformably on Bay State Dolomite of Givetian age. The uppermost surface of the lower Devils Gate Limestone in the Diamond Range is pitted and contains mottled, red staining suggestive of subaerial exposure. A minimum of 205 m of Devils Gate Limestone is estimated to have been removed at this unconformity (thickness from Larson and Riva, 1963). Strata overlying the unconformity surface consist of thin-bedded, platy shale with local thin lenses of siltstone and fine-grained sandstone that grade upward into fissile shale. The fissile shale grades abruptly upward into tabular and trough cross-bedded, quartz sandstone in which grain size coarsens upward in conjunction with increasing bed thickness towards the top of Sequence 2. The coarsening upward and thickening upward trend in the quartz sandstone depositional facies is interpreted to represent shallow-water bar migration.

Sequence 2 dramatically increases in thickness east of the Diamond Range (Fig. 7, Cherry Creek and Confusion ranges), where it consists of predominantly planar-laminated siltstone and shale (lower Pilot Shale). Locally, massive and cross-laminated, fine-grained quartz sandstone beds (interpreted as turbidites; Jones, 1990) are intercalated with thick, massive beds of matrix-supported, carbonate conglomerates (interpreted as debris flows; Jones, 1990). Shale and siltstone horizons below the sandstone turbidites commonly display soft-sediment deformation structures and syndepositional faulting. Paleocurrent data from climbing-ripple sets in the quartz sandstone turbidite intervals display consistent east/southeast directions indicating a western source of detritus. Strata of this age at Devils Gate (Fig. 2, locality DG) contain allochthonous carbonate deposits interpreted as slump deposits, translation breccias, and carbonate turbidites (Goebel, 1991b). East to southeast vergence directions of folds in slumped strata and current ripple directions in turbidite horizons indicate derivation of carbonate material at this locality from an unstable paleohigh to the west.

The turbiditic and debris flow deposits in the Cherry Creek and Confusion ranges are overlain by a thick sequence of

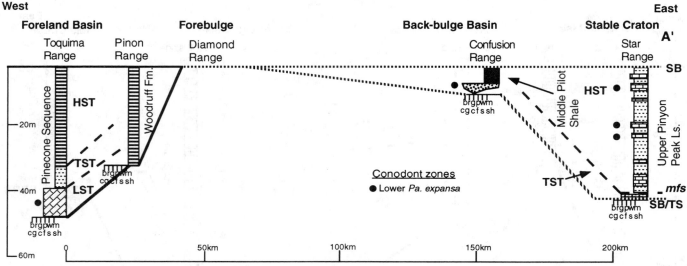

Fig. 8.—Cross section A to A' displaying Sequence 3 (Lower *Pa. marginifera* to Upper *Pa. marginifera* conodont zone) foreland strata.

platy, calcareous siltstone and shale. The siltstone and shale facies grades abruptly upward into laminated, silty, carbonate mudstone interbedded with shale containing micritic concretions.

Strata in the Star Range consist of a thin horizon (<3 m) of planar and cross-bedded quartz sandstone overlying an erosional unconformity surface cut down into Sequence 1 subtidal carbonate strata. The quartz sandstone horizon is abruptly overlain by thin to medium bedded, peloidal mudstone and wackestone, which grade upward into laminated, stromatolitic peloidal wackestone to packstone beds interpreted to represent peritidal carbonate deposition (lower Pinyon Peak Limestone). The sequence is capped by a limestone breccia horizon, indicating subaerial exposure and karsting of the peritidal carbonate sequence.

Interpretation of Sequence 2 Accommodation Trends.—

The position and trends of accommodation displayed by Sequence 2 strata are very similar to those of Sequence 1. For this reason the position of the flexural features is interpreted to have remained stationary during Sequences 1 and 2.

The foreland basin (Toquima and Pinon ranges) was the area of greatest flexural accommodation space yet contains a relatively thin, sedimentary sequence of deep-water strata indicating the foreland basin remained sediment-starved throughout Sequence 2. The nature of the sequence boundaries and the relatively thin sequence of shallow-water facies in the Diamond Range indicate accommodation space remained small compared to adjacent stratigraphic sections. Therefore, the Diamond Range section is interpreted as representing deposition associated with the uplifted forebulge. The relatively thick succession of basinal clastic strata preserved in the Cherry Creek and Confusion ranges reflects deposition within the subsiding back-bulge basin. Terrigenous detritus derived from the craton during sea-level lowstand continued to be ponded in the back-bulge basin as in Sequence 1.

We also identified small-scale accommodation trends within Sequence 2, which are apparently synchronous and congruous, reflecting probable eustatic sea-level changes. The stratigraphic position, geometry (spatially isolated), and nature (turbidites and debris flows) of the strata of the Upper *Pa. triangularis* conodont zone in the Cherry Creek and Confusion ranges have led us to interpret them as lowstand deposits (LST). The transition from sandstone turbidites and carbonate debris flows into siltstone and shale indicates termination of coarse detrital influx into the back-bulge basin, possibly related to a relative rise in sea level (TST). The correlative basal siltstone and shale sequence in the Diamond Range that becomes dominated by fissile shale upward is interpreted to represent transgressive onlap over the exposed angular unconformity surface (TST). The correlative quartz sandstone horizon in the Star Range is interpreted to represent transgression over an erosional scour surface on the stable craton.

Strata above the quartz sandstone horizon in the Star Range display a shoaling-upward sequence with an exposure cap indicating decreasing accommodation space characteristic of the highstand systems tract (HST). The time correlative, carbonate mudstone units in the Cherry Creek and Confusion ranges may represent fine-grained carbonate detritus carried into the basin as turbidites derived from the shoal-water carbonate bank to the east (Star Range) during late highstand progradation. The migrating or prograding quartz sandstone bar system capping Sequence 2 in the Diamond Range reflects shoaling of the sedimentary sequence relative to the underlying fissile shale and is interpreted as representing the highstand systems tract.

Sequence 3 (Lower Palmatolepis marginifera to Upper Palmatolepis marginifera)

Sequence 3 is Late Devonian (mid-Famennian) in age and spans two conodont zones, corresponding to approximately 1 my (Sandberg and others, 1988). The basal sequence boundary is represented by a subaerial exposure surface which resulted in karstification of underlying Sequence 2 carbonate strata in the Star Range. In the Diamond Range, the basal sequence boundary is represented by a disconformity surface and is apparently conformable within the other stratigraphic sections on cross-section A to A'. The upper sequence boundary is a sub-regional, erosional unconformity surface, which becomes conformable only in the westernmost stratigraphic sections (Toquima and Pinon ranges).

Sequence 3 Depositional Facies Trends.—

Depositional facies trends displayed by Sequence 3 strata are similar to those of the underlying syntectonic sequences. Deposition of bedded, phosphatic chert and hemipelagic claystone (Pinecone sequence and Woodruff Formation) continued in the westernmost sections (Toquima and Pinon ranges) indicating deep-water, sediment starved conditions remained throughout Sequence 3. The Diamond Range contains a monotonous section of platy shale containing local, thin-bedded siltstone and fine-grained sandstone horizons at the base. Strata of Sequences 3 through 6 are interpreted to have been removed by erosion during uplift prior to Sequence 7 deposition.

In the Confusion Range, Sequence 3 thickens and consists of laminated siltstone and very fine-grained sandstone at the base grading upward into dominantly silty shale (lower Pilot Shale). The silty shale contains thin, carbonate mudstone horizons and rare, matrix-supported, carbonate clast conglomerates and planar-bedded, coarse-grained, quartz sandstone beds (interpreted as debris flows and turbidites; Sandberg and others, 1988). The upper portion of Sequence 3 in the Confusion Range becomes increasingly more calcareous where it is dominated by thin-bedded, carbonate mudstone indicating an increased influx of fine-grained carbonate material into the area.

In the Star Range, deposition initiated with conglomeratic, quartz sandstone deposited over a subaerially exposed, erosion surface. The conglomerate clasts are composed of peloidal wackestone and packstone and were apparently locally derived from erosion of underlying Sequence 2 carbonate beds. The conglomeratic quartz sandstone horizon grades abruptly upward into subtidal, peloidal wackestone and packstone (lower Pinyon Peak Limestone). Peloidal packstone beds thicken and become more abundant upward, indicating westward progradation of the peloidal packstone facies.

Interpretation of Sequence 3 Accommodation Trends.—

We interpret the topographic profile present during deposition of the first two sequences to have remained the same during Sequence 3 deposition as well. The thin sequence of deep-water strata in the westernmost sections (Toquima and Pinon ranges) represents sediment-starved conditions within the foreland basin. The erosional nature of the sequence boundaries in the Diamond Range indicates this area had significantly less accommodation space than adjacent areas and is interpreted as representing deposition associated with temporary flooding of the uplifted forebulge.

The Confusion Range contains a comparatively thick sequence of moderately deep-water, turbiditic siliciclastic strata representing deposition within the downflexed back-bulge basin. The lower portion of Sequence 3 in this area is interpreted as siliciclastic influx into the back-bulge basin during the lowstand systems tract (LST) and during the subsequent transgression (TST). The conglomeratic quartz sandstone horizon in the Star Range represents the correlative transgression over an erosional scour surface on the stable craton. The shallow, subtidal carbonates overlying the sandstone horizon become more peloid-rich upward indicating current winnowing of mud-sized material possibly related to shoaling (HST). The transition from siliciclastic turbidites and debris flows into carbonate mudstone interbedded with shale is interpreted to represent fine-grained, carbonate turbidites shed from the shoal-water carbonate platform to the east during the highstand systems tract. Equivalent age (Upper *Pa. marginifera*) highstand deposits have not been documented from the inferred forebulge and may have been subsequently removed by extensive uplift and erosion prior to Sequence 7 deposition.

Sequence 4 (Lower Palmatolepis expansa)

Sequence 4 is Late Devonian (late Famennian) in age and spans one conodont zone corresponding to approximately 1 my (Sandberg and others, 1988). The basal sequence boundary is a sub-regional, subaerially exposed(?), erosional unconformity surface and only becomes conformable in the westernmost stratigraphic sections (Fig. 9, Toquima and Pinon ranges). Older strata from the *Pa. trachytera* and *Pa. postera* conodonts zones have not been identified in any of the stratigraphic sections used in this study (Fig. 4). The absence of these zones may represent regional erosion or non-deposition prior to Sequence 4 deposition. The upper sequence boundary is also a sub-regional, erosional unconformity surface that becomes conformable in the westernmost stratigraphic sections (Toquima and Pinon ranges).

Sequence 4 Depositional Facies Trends.—

The basal portion of Sequence 4 in the Toquima Range (Fig. 9) is composed of cross-laminated, crinoidal wackestone and packstone interpreted as calciturbidite horizons (Pinecone Sequence). Isolated crinoidal carbonate beds less than 5 m thick encased within bedded chert locally overlie the calciturbidites and have been interpreted as possible slump blocks (Coles, 1988). Detrital limestone units are abruptly overlain by calcareous argillite that grades upward into bedded chert. Detrital limestone horizons have not been reported in the Pinon Range which contains dominantly bedded phosphatic chert and hemipelagic claystone.

There are no strata of Sequence 4 age preserved in the Diamond Range indicating either non-deposition or sub-

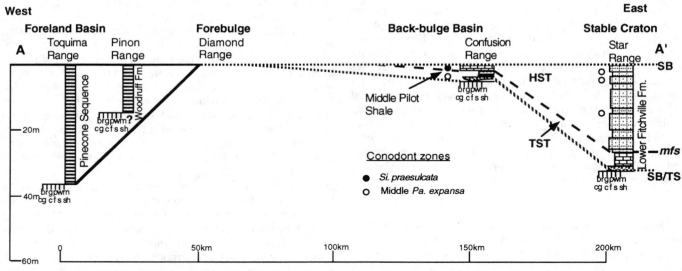

Fig. 9.—Cross section A to A' displaying Sequence 4 (Lower *Pa. expansa* conodont zone) foreland strata.

sequent erosion of the sequence. In the Confusion Range, a thin sequence of relatively deep-water siliciclastic material (middle Pilot Shale) was deposited above an erosional base (Gutschick and Rodriguez, 1979). Initial deposition consisted of planar cross-bedded, coarse-grained, quartz sandstone containing abundant abraded fish bones and teeth, conodonts, and phosphatic pellets. The basal sand grades abruptly into silty, black, bituminous marine shale containing sparse fauna of sponge spicules, radiolarians, and other microfossils. Pyrite and black chert are common in the black shale facies. The uppermost portion of Sequence 4 in the Confusion Range contains carbonate mudstone concretions.

Star Range deposition consists of a comparatively thick sequence, which initiates with a thin bed of coarse-grained, quartz sandstone grading into wavy-bedded, argillaceous lime mudstone and wackestone interbedded with calcareous siltstone (Fig. 9, upper Pinyon Peak Limestone). Wackestone interbeds contain an abundant normal marine megafauna consisting of brachiopods, echinoderms, and molluscs.

Interpretation of Sequence 4 Accommodation Trends.—

The distribution of Sequence 4 depositional facies indicates that the topographic profile and position of flexural features interpreted for sequences 1 through 3 remained fixed during Sequence 4 deposition.

Within the foreland basin (Toquima and Pinon range) coarse-grained, allochthonous carbonate debris (allodapic limestones and slump blocks) were derived from shallow-water carbonate complexes formed either to the west or east and transported into the foreland basin (Coles, 1988). Carbonate detrital influx and slumping are interpreted to represent instability of the shallow-water carbonate complex due either to tectonic motion or erosion during eustatic sea-level fall. The transition from allodapic limestone to calcareous argillite indicates termination of coarse-grained, carbonate detrital influx (LST) and a relative deepening (TST).

Back-bulge basin (Confusion Range) depositional facies are interpreted as representing a basal transgressive, lag quartz sandstone deepening into a highly restricted basinal setting where black shale was deposited (TST). The relatively thin sequence of dominantly black shale containing an abundant conodont fauna suggests sediment starvation (condensed interval) within the back-bulge basin. The uplifted forebulge area may have acted as a sill, restricting circulation within the back-bulge basin. This resulted in stratification of the water column and anaerobic bottom conditions. Carbonate concretions near the top of Sequence 4 in the Confusion Range may represent fine-grained, detrital carbonate material carried into the basin by density currents from the shallow, subtidal carbonate platform to the east during the late highstand systems tract (HST).

Sequence 4 sediment starvation in the back-bulge basin is in marked contrast to the high sedimentation rates documented for back-bulge basin strata in sequences 1 through 3. The erosional nature of the basal sequence boundary and thin stratigraphic thickness suggest the back-bulge basin was not an important depocenter during the lowstand systems tract but was more likely a zone of sediment bypass.

Within the Star Range, deposition initiated with a basal transgressive lag sandstone deepening into a mixed, subtidal siliciclastic and carbonate environment. The relatively thick sequence of subtidal, argillaceous carbonates in the Star Range and thin, black shale in the Confusion Range indicates termination of coarse detrital influx into the area and persistent, moderately deep-water conditions across the area throughout Sequence 4 deposition.

Sequence 5 (Middle Palmatolepis expansa to Siphonodella praesulcata)

Sequence 5 is Late Devonian (late Famennian) in age and spans portions of two conodont zones corresponding to approximately 2 my (Sandberg and others, 1988). The basal sequence boundary is a sub-regional, erosional unconformity surface that becomes conformable in the westernmost stratigraphic sections (Fig. 10, Toquima and Pinon ranges). The upper sequence boundary is also a sub-regional, erosional unconformity surface that becomes conformable in the westernmost stratigraphic sections.

Sequence 5 Depositional Facies Trends.—

Deposition of bedded chert and hemipelagic claystone continued in the westernmost stratigraphic sections (Toquima and Pinon ranges). No strata of Sequence 5 age are preserved in the Diamond Range due to either non-deposition or subsequent erosion. Within the Confusion Range, a comparatively thin sequence of dominantly calcareous strata was deposited (middle Pilot Shale). The base of the sequence contains a thin (less than .5 m), coarse-grained, quartz sandstone containing abundant, poorly sorted and abraded fish bones and plates, conodonts, and phosphatic pellets. Vertical burrows are common. The sandstone is interpreted to represent shallow-water siliciclastic deposition above an erosional base (Gutschick and Rodriguez, 1979). Gradationally overlying the basal sandstone is a thin, laminated, black, carbonaceous shale. The shale contains abundant plant material and crustacean shells (*Cyzicus*) interpreted to have lived in fresh to brackish water, lagoonal conditions (Gutschick and Rodriguez, 1979). Marine fauna (goniatites, brittle starfish, *Lingula*, and trilobites) are present ,suggesting conditions must at least brackish-water for this facies. The black shale grades upward into a thin, grayish-green shale containing normal marine fauna. The greenish-gray shale is gradationally overlain by a fossiliferous, oncolite-sponge-biostrome. The biostrome horizon consists of skeletal wackestone to packstone with abundant nodular oncolites intercalated with calcareous shale. Skeletal grains consist

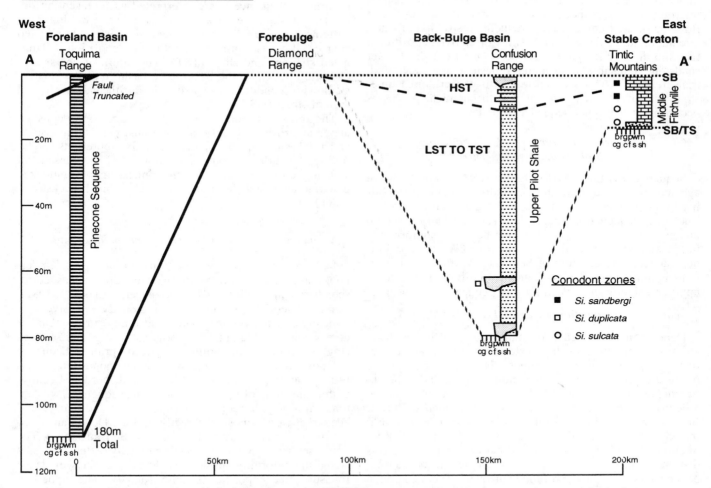

FIG. 10.—Cross section A to A' displaying Sequence 5 (Middle *Pa. expansa* to *Si. praesulcata* conodont zones) foreland strata.

of a diverse, normal marine fauna (Gutschick and Rodriguez, 1979).

Depositional facies within the Star Range consist of a very thin, coarse-grained, quartz sandstone horizon which grades abruptly into thin, wavy-bedded, argillaceous, crinoidal wackestone to packstone. Upward in the section, crinoidal packstone and grainstone form thick to massive beds that have been dolomitized near the top (lower Fitchville Formation).

Interpretation of Sequence 5 Accommodation Trends.—

Depositional facies patterns in Sequence 5 indicate that the topographic profile and position of flexural features remained in the same position as for previous sequences. The relatively thin sequence of deep-water, bedded chert and claystone in the foreland basin (Toquima and Pinon ranges) reflects continuation of deep-water, sediment-starved conditions. Depositional facies within the back-bulge basin indicate progressive deepening and aeration of bottom waters from highly restricted, anaerobic, shallow-water lagoonal conditions to dysaerobic, slightly deeper-water shale deposition (TST) and finally to aerobic normal marine conditions (oncolite-sponge-biostrome). The progressive change in aeration of the bottom waters in the back-bulge basin is interpreted to correspond to eustatic rise in sea level and flooding of the forebulge area, reducing restriction of water circulation in the back-bulge basin. Correlative depositional facies in the Star Range indicate deepening from shoal-water sandstone into subtidal, crinoidal limestone associated with transgression (TST). Dolomitization of the crinoidal limestone may have been associated with chemical processes occurring at the upper sequence boundary.

*Sequence 6 (Siphonodella sulcata to
Siphonodella sandbergi)*

Sequence 6 is Early Mississippian (early Kinderhook) in age and spans portions of three conodont zones corresponding to approximately 4.5 my (Sandberg and others, 1982). The basal sequence boundary is a sub-regional, erosional unconformity surface (Fig. 11). that becomes conformable in the westernmost stratigraphic sections (Toquima and Pinon ranges). Preservation of Sequence 6 stratigraphic sections is limited due to post-Sequence 6 differential tectonic uplift and significant erosion prior to Sequence 7 deposition.

Sequence 6 Depositional Facies Trends.—

Deposition of bedded phosphatic chert and hemipelagic claystone continued in the westernmost stratigraphic sections (Toquima and Pinon ranges). No strata of Sequence 6 age are preserved in the Diamond Range due either to non-deposition or, more likely, erosion prior to Sequence 7 deposition that removed strata down to Sequence 3 level. In the Confusion Range, a relatively thick sequence of laminated siltstone and fine-grained sandstone (upper Pilot Shale), containing sparse normal marine macrofauna, was deposited. Small-scale channels, containing siltstone, cross-bedded sandstone, and *Skolithos* burrows, are concentrated near the base of the sequence and appear again at the top of the sequence.

The Tintic Mountain section was substituted for the Star Range section (Fig. 11) because it contains similar depositional facies trends as the Star Range, but the conodont biostratigraphy of Sequence 6 age strata has been better documented. Depositional facies within the Tintic Mountains consist of a thin, coarse-grained, quartz sandstone horizon deposited above an erosional base. The basal quartz sandstone grades abruptly upward into a thick interval of argillaceous, peloidal, mudstone and wackestone (middle Fitchville Limestone). The argillaceous, peloidal mudstone beds are gradationally overlain by argillaceous, crinoidal-peloidal packstone which caps the sequence.

Interpretation of Sequence 6 Accommodation Trends.—

The distribution of Sequence 6 depositional facies indicates that the topographic profile and thus the interpreted position of flexural features remained the same as for sequences 1 through 5. Within the foreland basin, depositional facies indicate that conditions of low sedimentation rate or sediment starvation continued. Coarser-grained siliciclastic deposition was concentrated within the back-bulge basin (Confusion Range) similar to sequences 1 through 3. Coarse-grained, channelized siliciclastic intervals in the lower portion of Sequence 6 are interpreted as representing detrital influx into the back-bulge basin during the lowstand systems tract (LST). Subsequent deepening terminated coarse, siliciclastic influx, resulting in deposition of laminated siltstone and shale (TST). The return of coarse siliciclastic influx into the back-bulge basin indicates relative shallowing conditions occurring during the late highstand systems tract (HST).

The transition from coarse, quartz sandstone deposited above an erosional base to argillaceous, peloidal mudstone and wackestone in the Tintic Mountains represents a relative sea-level rise over the stable craton (TST). Subsequent shoaling conditions resulted in winnowing of mud-sized material and deposition of crinoidal-peloidal packstones (HST).

Sequence 7 (Lower to Upper Siphonodella crenulata)

Sequence 7 is Early Mississippian in age (Late Kinderhook) and spans portions of two conodont zones corresponding to approximately 4 my (Sandberg and others, 1982). The basal sequence boundary is a regionally extensive, tectonically enhanced, erosional unconformity surface that becomes conformable in the westernmost stratigraphic section (Fig. 12). Differential uplift and erosion prior to Sequence 7 deposition affected a zone from the Diamond Range eastward to approximately the Nevada/Utah border. Locally, within this zone of uplift the entire pre-Sequence 7 syntectonic stratigraphic sequence was removed, and Sequence 7 age strata were deposited directly on Late Devonian (Frasnian) age passive margin strata (Guilmette Limestone). The upper-sequence boundary is a regionally extensive, erosional unconformity surface that locally displays evidence for subaerial exposure (Needle Range). The upper-sequence boundary becomes conformable west of the Needle and Confusion ranges.

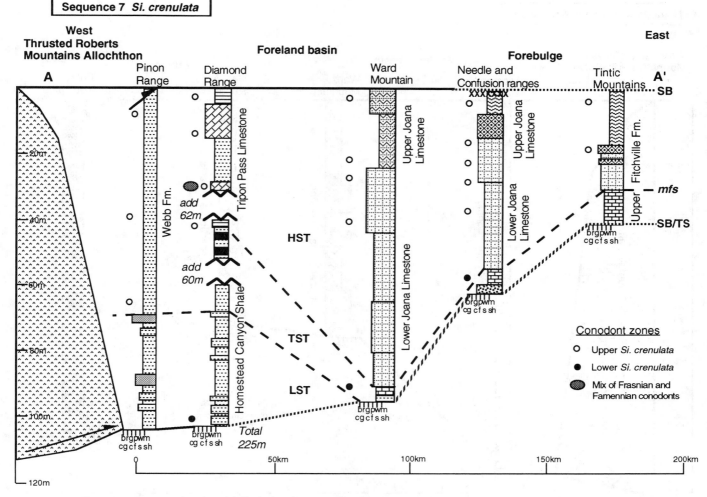

FIG. 11.—Cross section A to A' displaying Sequence 6 (*Si. sulcata* to *Si. sandbergi* conodont zones) foreland strata.

Sequence 7 depositional facies patterns.—

Prior to Sequence 7 deposition, the Pinecone sequence (containing strata of Sequence 1 through 6 age, Toquima Range) was incorporated into the Roberts Mountains allochthon thrust wedge and transported eastward (Coles, 1988). The general geometry of Sequence 7 is a westward thickening sedimentary wedge. This geometry is in marked contrast to the irregular geometries displayed by sequences 1 through 6 across the same area.

The westernmost stratigraphic sections containing Sequence 7 age strata (Pinon and Diamond ranges) consist of laminated claystone, shale, and siltstone with local, lithic sandstone horizons (Webb Formation and Homestead Canyon Shale). The fine-grained, lithic sandstone horizons present in the lower portion of the Webb Formation contain distinctive, green chert lithoclasts that have been interpreted as turbiditic sandstones derived from the Roberts Mountains allochthon (Murphy and others, 1984). These distinctive lithoclasts are the first definitive indication of allocthon-derived, sandy sediment in the syntectonic strata.

In the Pinon Range, the fine-grained lithic sandstone and siltstone are overlain by a thick monotonous shale sequence. This same shale sequence is also present in the Diamond Range where towards the top it contains black, fissile shale. In the Diamond Range, the monotonous shale sequence is gradationally overlain by a thick succession of carbonate turbidites intercalated with calcareous siltstone and shale (Tripon Pass Limestone). The turbidite horizons are characterized by channelized or planar, fining-upward, graded beds containing coarse quartz sand grains, abraded crinoidal debris, and granule- to pebble-sized, subangular clasts. Clast lithofacies are highly variable and include calcareous quartz sandstone, dolomitic and calcitic mudstone, peloidal packstone and grainstone, and crinoidal wackestone and packstone. Conodonts derived from lithoclasts are of mixed ages spanning Late Devonian (Frasnian and Famennian) through Early Mississippian (Kinderhook), indicating major local uplift and erosion of Late Devonian to Early Mississippian strata. Intercalated fine-grained, crinoidal wackestone beds, also interpreted as turbidites, do not yield mixed faunas but contain Upper *Si. crenulata* fauna

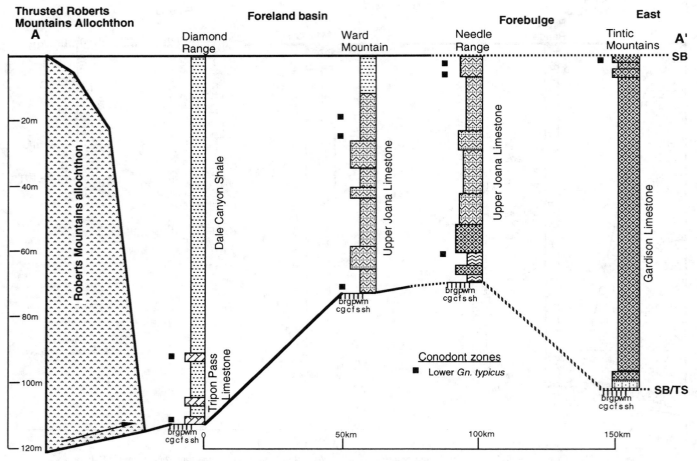

Fig. 12.—Cross section A to A' displaying Sequence 7 (Lower to Upper *Si. crenulata* conodont zones) foreland strata.

near the base of the Tripon Pass Limestone and Lower *Gn. typicus* fauna at the top. This indicates the Tripon Pass Limestone deposition spans across the Sequence 7 boundary. At the type Tripon Pass section at Tripon Pass (Fig. 2, locality TP), time equivalent strata are organized into a thick, carbonate turbidite fan geometry with current-rippled horizons indicating an eastern source of detritus.

Strata to the east at Ward Mountain are dominated by limestone (lower and upper Joana Limestone). Sequence 7 initiates here with a relatively thin horizon of planar-bedded, calcareous siltstone and very fine-grained, quartz sandstone. The thin siliciclastic horizon grades upward into argillaceous, wavy-bedded, fossiliferous wackestone to packstone. This facies contains a very diverse, normal marine, megafauna assemblage consisting of crinoids, brachiopods, solitary and colonial corals, and bryozoan. The fossiliferous wackestone grades abruptly upward into highly bioturbated, thick to massive beds of crinoidal wackestone to packstone. Disarticulated crinoidal debris dominates this depositional facies, but subordinate amounts of other normal marine fauna are present. Locally, discontinuous crinoidal grainstone lenses are present. Towards the top of this facies, the crinoidal material is coarser grained and forms packstone to grainstone beds.

The bioturbated crinoidal wackestone to packstone facies is overlain by thinly interbedded, deep subtidal, crinoidal mudstone and wackestone with crinoidal packstone (upper Joana Limestone). The packstone beds generally contain coarser crinoidal material than the mudstone and wackestone intervals and a more diverse normal marine fauna. The thinly interbedded crinoidal mudstone and packstone grade upward into thick-bedded, crinoidal packstone with crinoidal grainstone lenses. The packstone beds are often graded (fining upward) and locally hummocky cross-stratified with erosionally scoured bases and rip-up clasts. These graded beds are interpreted as representing current reworking by storm waves (tempestites) in a relatively deep subtidal ramp setting (Goebel, 1991b).

The Confusion Range and Needle Range sections were combined for Sequence 7 to derive the most biostratigraphical information with the best facies control. Deposition in this area was dominated by carbonate sedimentation (lower and upper Joana Limestone). The sequence initiated with siliciclastic strata deposited above an erosional base that truncates sedimentary structures within strata of the underlying Sequence 6. The basal siliciclastic deposits consist of thick, planar and cross-bedded quartz sandstone that is texturally and compositionally mature. The quartz sandstone grades upward into argillaceous, wavy-bedded, skeletal wackestone to packstone facies followed by thick- to massive-bedded, crinoidal wackestone and packstone. This is the same facies succession described for Ward Mountain. The crinoidal packstone to grainstone facies is overlain by thin beds that grade upward to thick beds of very fine-grained, crinoidal-peloidal packstone to grainstone. Internal cross-bedding is bi-directional with paleocurrent directions of south-southwestward and north-northeastward. This facies is interpreted as representing current-agitated, carbonate deposition in a shallow, subtidal setting.

Sequence 7 is capped by beds of irregularly laminated (stromatolitic), dolomitic, peloidal mudstone grading upward into limestone breccia. Breccia clasts are granule to cobble-sized and composed of irregularly laminated, peloidal mudstone identical to the underlying strata. The brecciated horizon is interpreted as representing subaerial exposure and karsting of the peloidal mudstone beds at the Sequence 7 boundary.

Sequence 7 strata have been pervasively dolomitized in the Tintic Mountains, but discernible depositional facies show a similar succession of facies to those described for the Confusion and Needle ranges. Notable differences from the Confusion and Needle ranges are the absence of a basal sandstone horizon and the presence of a much thicker, irregularly laminated, peloidal mudstone facies. The uppermost part of the irregularly laminated, peloidal mudstone facies contains laterally linked, stromatolite heads with birds-eye fabric (Proctor and Clark, 1956).

Interpretation of Sequence 7 Accommodation Trends.—

Accommodation trends for Sequence 7 reflect profound changes from trends documented for sequences 1 through 6. In general, a westward increase in accommodation space is interpreted for Sequence 7, based on a westward thickening sediment wedge and westward deepening of depositional facies. This westward increase in accommodation is interpreted to represent flexural downwarping of the lithosphere and migration of the flexural features in response to migration of the thrust load prior to Sequence 7 deposition. Differential uplift and erosion in the zone from the Diamond Range to the Utah/Nevada border is interpreted as produced by eastward migration of the forebulge. This period of flexural uplift coincided with third-order sea-level fall (Fig. 4) which resulted in erosional beveling of the uplifted area.

The former uplifted forebulge area for sequences 1 through 6 became a deep depocenter prior to Sequence 7 deposition, and the former topographically low area (back-bulge basin) during Sequences 1 through 6 was flexurally uplifted over the forebulge prior to Sequence 7 deposition, reflecting regional-scale inversion of topography.

Deposition in the Pinon Range indicates an increase in sedimentation rate and change in sedimentary facies from hemipelagic claystone and bedded chert (found at this locality for sequences 1 through 6) to coarser sandstone, siltstone, and shale. Depositional facies in the Pinon Range display a deepening upward trend. The basal coarse-grained siliciclastic sediments are interpreted to have been derived from erosion of the allochthon and deposited within the foreland basin during the lowstand systems tract (LST). Coarse-grained sediment influx was terminated by sea-level rise (TST) trapping sediments shoreward. Depositional facies trends in the Diamond Range show a similar deepening upward trend from dominantly siltstone to a thick monotonous shale sequence. The thick succession of relatively deep-water siltstone and shale indicates major post-Sequence 6 accommodation increase and has been interpreted as representing flexural downwarping of the former forebulge and incorporation into the foreland basin. The overlying Tripon Pass limestone contains clasts of Late Devonian through Early Mississippian age, indicating large scale uplift and erosion of strata of these ages somewhere to the east. Stratigraphic sections directly to the east contain relatively complete Upper Devonian through Early Mississippian stratigraphic sections indicating the Tripon Pass was deposited longitudinally into the foreland basin (Goebel, 1991b). Stratigraphic sections in the Goose Creek and Grouse Creek mountains in northeastern Utah are missing strata of this age and may represent erosion of locally uplifted areas, which supplied sediment for the Tripon Pass deposits.

Along the eastern margin of the foreland basin, laterally extensive, shallow subtidal facies indicate gradual deepening (TST) over an extensive, gently westward-dipping carbonate ramp with little topographic relief (lower Joana Limestone). A basal transgressive lag sandstone was deposited locally over the erosional surface in a shoal-water, possibly shoreface environment. The sandstone deepens upward into subtidal, argillaceous crinoidal wackestone and packstone (TST). The transgressive systems tract is abruptly overlain by thick, subtidal, bioturbated to massive, crinoidal wackestone and packstone. The crinoidal wackestone to packstone deepens upward into deep, subtidal ramp facies. In the Confusion Range and Tintic Mountains, the sequence follows Ward Mountain by an initial transgression (TST). Following transgression, Sequence 7 shoals upward and is ultimately subaerially exposed, forming a karst surface (Confusion Range). The shoaling upward carbonate sequence represents a decrease in accommodation space during highstand deposition. Local exposure and karsting is interpreted to represent accommodation decrease due to tectonic uplift and exposure over the migratory forebulge. Locally in areas to the north there was significant forebulge uplift and erosion of the entire Late Devonian and Early Mississippian sections (as much as 300 m of strata are estimated to have been removed). The material was transported into the foreland basin forming local carbonate submarine turbidite fans.

Sequence 8 (lower Gnathodus typicus)

Sequence 8 is Early Mississippian in age (early Osage) and spans one conodont zone, approximately 1.5 my in duration (Sandberg and others, 1982). The basal sequence boundary is a regionally extensive, erosional unconformity surface that locally displays evidence of subaerial exposure (Fig. 13, Needle Range). The upper sequence boundary is an unconformity surface in the easternmost stratigraphic sections (Needle Range and Tintic Mountains) overlain by a distinctive phosphatic facies (Delle Phosphatic member of the Deseret Limestone) that represents a regionally correlatable anoxic event (Silberling and Nichols, 1991). The boundary here is marked by corrosion and dissolution of Sequence 8 strata and represents an interruption in deposition (Silberling and Nichols, 1991). Westward the upper sequence boundary is apparently conformable.

Sequence 8 Depositional Facies.—

In the Diamond Range, deposition of intercalated carbonate turbidites with calcareous siltstone and shale (Tripon Pass Limestone) reflects apparently continuous sedimentation from Sequence 7 into Sequence 8 time. The carbonate turbidite lithofacies are the same as those described for the Tripon Pass Limestone deposited during Sequence 7 time. The carbonate turbidites grade upward into a monotonous sequence of laminated calcareous siltstone and shale (Chainman Shale).

At Ward Mountain, Sequence 8 strata are dominated by thinly interbedded, subtidal crinoidal packstone with crinoidal mudstone and wackestone (upper Joana Limestone). The packstone beds generally contain coarser crinoidal material than the mudstone and wackestone intervals and a more diverse, normal marine fauna. The thinly-interbedded, crinoidal mudstone and packstone grade into intervals of thick-bedded, crinoidal packstone with crinoidal grainstone lenses. The packstone beds are often graded (fining-upward) and locally hummocky cross-stratified, with erosionally scoured bases and rip-up clasts. These graded beds are interpreted as representing current reworking by storm waves (tempestites) in a relatively deep, subtidal ramp setting. The carbonate strata grade abruptly upward into laminated siltstone and shale (Chainman Shale).

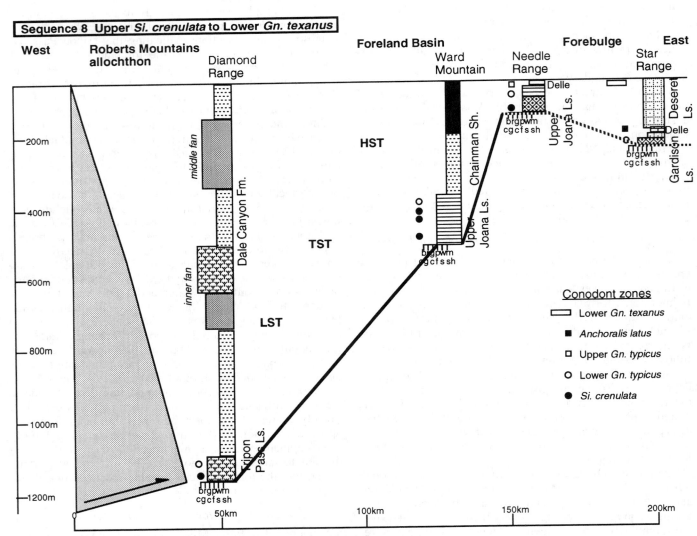

FIG. 13.—Cross section A to A' displaying Sequence 8 (Lower *Gn. typicus* conodont zones) foreland strata.

Deposited above the karsted sequence-boundary surface in the Needle Range was irregularly laminated (stromatolitic), peloidal mudstone containing peloidal mudstone rip-up clasts. Locally, the irregular laminae are organized into laterally-linked stromatolite heads. This facies is interpreted as having been deposited under highly restricted, shallow-water, intertidal conditions. This facies is interbedded with thick to massive beds of peloidal grainstone and oolitic grainstone with large cross-stratification. The peloidal and oolitic grainstones are interpreted as representing shallow subtidal, migrating dune forms. Abruptly overlying the subtidal peloidal and oolitic grainstones are a thick succession of interbedded subtidal crinoidal packstone with crinoidal mudstone and wackestone facies interpreted to represent deep subtidal deposits reworked by storm waves.

Sequence 8 strata in the Tintic Range contain a very thick, uniform succession of thick, irregular-bedded, interstratified crinoidal wackestone and packstone. An abundant, normal marine megafauna is present (corals, gastropods, bryozoa, and brachiopods) along with peloidal material. The basal unit is a cross-bedded crinoidal grainstone which channels into the underlying Sequence 7 strata. This facies is interpreted as representing shallow, current agitated, subtidal, open marine deposition (Sandberg and Gutschick, 1979; Silberling and Nichols, 1992).

Interpretation of Sequence 8 Accommodation Trends.—

The topographic configuration present for Sequence 8 is very similar to that of Sequence 7. Depositional facies suggest a relatively deep foreland basin forming a ramp on the distal margin which slopes upward to a topographic high (forebulge) in the area of the Needle Range. All stratigraphic sections display patterns of accommodation increase. The regionally uniform accommodation behavior suggests eustatic sea-level rise controlling stratigraphic patterns. Eustatic sea-level rise is interpreted to have reached a maximum for the Mississippian during early Osage time (Fig. 4). Strata from the Lower *Gn. typicus* conodont zone were deposited during this major rise in sea level.

Within the basinal portion of the foreland basin (Diamond Range), locally thick, polymict carbonate turbidite systems formed at or near the base of Sequence 8 (LST). Carbonate turbidite debris was most likely associated with continued local erosion of portions of the Lower Mississippian and Upper Devonian platform uplifted over the forebulge to the northeast and exposed during sea-level lowstand. The carbonate turbidites are overlain by a monotonous shale sequence interpreted as termination of coarse-grained detrital material due to sea-level rise that flooded forebulge uplifted areas. The strata at Ward Mountain and the Needle Range both display deepening-upward facies trends from deep-water, slopal carbonates to basinal shale at Ward Mountain and shallow-water, subtidal and intertidal carbonates to deep-water, slopal carbonates in the Needle Range. In the Tintic Range, carbonate sedimentation kept pace with accommodation increase, producing a thick, aggradational, shallow subtidal, carbonate system.

POST-TECTONIC STRATIGRAPHIC SUCCESSION

Flexural subsidence induced by thrust loading ended following cessation of thrust advance in Early Mississippian time and marks the climax of the syntectonic phase of foreland sedimentation. The transition from syntectonic to post-tectonic deposition corresponds to the time of maximum eustatic sea-level rise that resulted in flooding of the cratonic shelf (Fig. 4).

Sedimentation patterns during the post-tectonic phase reflect infilling of relict topography produced by flexural subsidence during Sequence 8 time. Allochthon-derived detrital influx into the foreland basin increased significantly at this time producing a thick siliciclastic turbidite system within a deep trough near the thrust front (Fig. 14, Diamond Range, Dale Canyon Formation). Foreland basin sedimentation consisted of combined basin floor, slope, inner fan, and middle-fan deposits. In the Diamond Range, these strata have been mapped as Chainman Shale (Larson and Riva, 1963) but are here assigned to the Dale Canyon Formation (after the convention of Johnson and Visconti, 1992). In exposures on the west flank of the Diamond Range (Harbaugh and Dickinson, 1981), the following facies occur in retrogradational succession: (a) massive shale and siltstone slope facies (350 m) containing lenses of olistostromal pebbly mudstone, (b) inner-fan facies (350 m) composed of amalgamated conglomerate-sandstone channel complexes encased in shale containing thin overbank turbidites, and (c) sandy midfan facies (450 m) composed of thinning-upward channel sandstone packets and intervening shale interchannel deposits. Paleocurrent data are inadequate to distinguish between turbidite bodies built into the trough as transverse fans from analogous facies diverted longitudinally along the trough axis.

On the distal margin of the foreland basin approximately coeval with submarine fan deposition in the foreland basin, the regionally extensive Delle Phosphatic Member of the Deseret Limestone and Woodman Formation was deposited. Depositional facies consist of a comparatively very thin sequence of phosphatic, calcareous mudstone and silicified mudstone. The mudstone represents dysaerobic to anaerobic, sediment-starved conditions during maximum transgression of sea level during Early Mississippian (Osage) time. The depositional setting of the Delle Phosphatic Member has been interpreted as a very deep-water (>300 m), basinal facies (Sandberg and Gutschick, 1980, 1989) and as a restricted, relatively shallow-shelf facies (Nichols and Silberling, 1990; Silberling and Nichols, 1991). Our observations of depositional setting and accommodation trends favor the latter interpretation. The Delle Phosphatic Member anoxic event in the easternmost stratigraphic sections (Tintic Mountains and Star Range) was followed by progradational, shallow subtidal carbonate deposition (Deseret Limestone). Progradation of the carbonate platform did not reach the edge of the former drowned-shelf margin and is reflected in the thin sequence of black, unfossiliferous mudstone in the Needle Range (Fig. 14, portion of the Delle Phosphatic Member).

Overlying the submarine fan deposits (Dale Canyon Formation) in the foreland basin are delta-slope deposits of the Chainman Shale. The delta-slope facies is composed of 450 m of massive siltstone containing lenses of ravine-fill grain-flow conglomerate (Harbaugh and Dickinson, 1981). The overlying delta-front and delta-plain deposits form the delta-

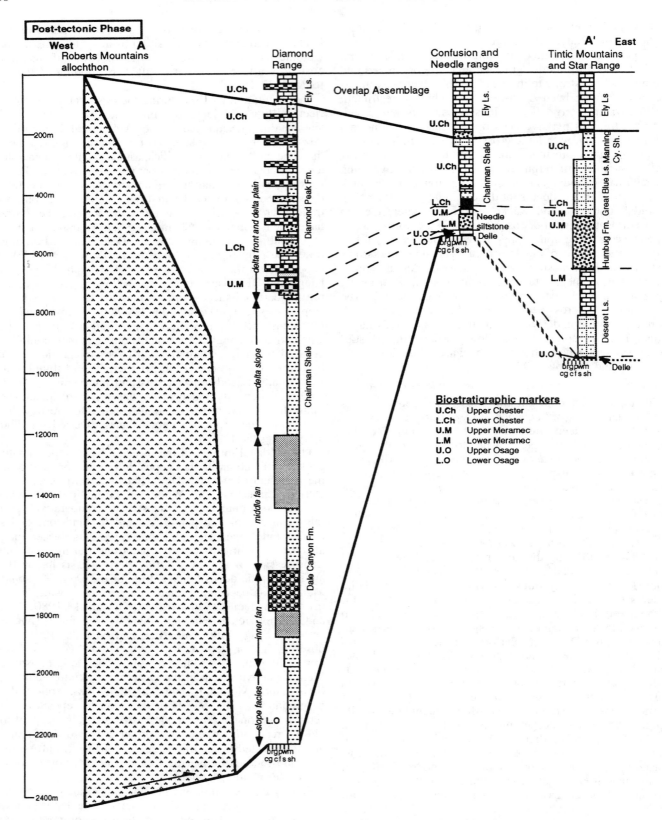

FIG. 14.—Cross section A to A' displaying post-tectonic strata (Lower to Upper Mississippian); lower correlation line is base of Upper *Gn. typicus* conodont zone and upper correlation line is near top of Chesterian stage, but basal Ely Limestone (part of the overlap assemblage) is also Chesterian in age (see text). Ages of biostratigraphic markers based on conodont, ammonoid, brachiopod, foram, and coral data derived from Poole and Sandberg (1991). Note change of scale from Figures 6–13.

the vertical stacking of facies follows the predicted Waltherian pattern for continuously decreasing accommodation space. We interpret the large-scale, shoaling-upward sequence to have resulted from a combination of second-order, eustatic sea-level lowering (Fig. 4), sediment infilling of the foreland basin, and isostatic rebound as the thrust load is eroded and the load redistributed. The third-order eustatic sea-level signature was apparently masked in the foreland basin by high sedimentation rates.

When available accommodation space in proximal portions of the foreland basin was filled, clastic sediment prograded cratonward to the distal margin of the foreland basin (Needle Range). The provenance of fine-grained, lithic sandstone in the Needle Siltstone Member of the Chainman Shale (Needle Range) is the first indication of a western detrital source so far east. In our view, the several named and carefully mapped Meramecian and Chesterian members of the Chainman Shale in the Needle Range along the Nevada-Utah border (Hintze, 1986; Hintze and Best, 1987) were deposited in a shallow marine strait between the Antler highlands and the craton. This area was in a distal position to receive detritus from the westward Antler highlands and from cratonal sources to the east. As a result, the area was relatively sediment-starved and formed a comparatively thin sequence of strata. Depositional facies in this area include a variety of sparingly fossiliferous limestones, siltstones, and shales correlative with both the prodeltaic Chainman Shale and the deltaic Diamond Peak Formation of the Diamond Range along the axis of the foreland basin. Westward progradation from the craton of an extensive carbonate platform (Great Blue Limestone) was also initiated during Late Mississippian time (Rose, 1976) coeval with eastward progradation of Diamond Peak delta front and delta plain facies.

The conformable and gradational contact with the overlying Ely Limestone (Fig. 14), which overlaps the Roberts Mountains allochthon, marks the end of the direct influence of Antler tectonism and associated orogenic relief on foreland sedimentation. The basal part of the Ely Limestone is latest Mississippian (Late Chesterian) age, but most of the unit is Pennsylvanian age. Its prominently cyclic internal stratigraphy reflects the pronounced influence of late Paleozoic glacio-eustasy on shallow-marine depositional systems.

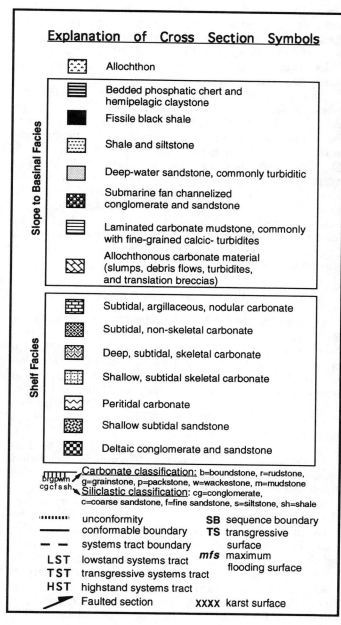

FIG. 14.—Continued.

platform facies (650 m) of the Diamond Peak Formation. We regard the local, angular discordance between delta-slope and delta-platform facies, interpreted previously to be a sequence boundary (Trexler and Nitchman, 1990), as a toplap contact developed as the platform of the Diamond Peak braid delta prograded across the delta foreslope.

This long duration shoaling-upward of depositional facies from submarine fan to deltaic deposits reflects continuously decreasing accommodation space in the foreland basin. The third-order, eustatic sea-level curve shows many major sea-level rises and falls over this time period, yet no regionally correlatable unconformity surfaces have been identified in the thick sequence of foreland basin strata, and

DISCUSSION

Discontinuous Migration of Flexural Features

The syntectonic stratigraphic succession indicates episodic, eastern migration of a structural high or forebulge. The position of the flexural forebulge did not progress continuously cratonward through time but appeared to stall at its initial position near the area of the Diamond Range during deposition of Sequences 1 through 6 and then moved rapidly cratonward into western Utah during Sequences 7 and 8. Two possible reasons for discontinuous migration of flexural features that may apply to the Antler foreland are episodic migration of the thrust load and inhomogeneities within the lithosphere.

If the thrust load is not migrating, the flexural features will not migrate. The buildup and migration of an accretionary prism or thrust load is commonly modeled involving "sled-runner" thrust imbrication, where progressive cratonward outstepping of the thrust front incorporates new material into the thrust load as it migrates. Thrust loads may also build up by almost vertical stacking of thrust sheets along ramps within the hinterland, producing the critical taper needed for the thrust system to migrate (Boyer and Elliot, 1982). In this case, flexural features would not continuously migrate cratonward because the thrust load itself is not continuously migrating cratonward.

Waschbusch and Royden (1992) suggest that discontinuous migration of flexural features may be caused by inhomogeneities within the lithosphere that fix the position of the forebulge to a weak segment of the foreland lithosphere. In this case, the forebulge remains in a fixed position until the subduction system and thrust belt advance to some critical distance near the weak segment. At that time, the forebulge may abandon the weak segment, and upwarping may progress rapidly across the foreland toward the next weak segment or to the proper flexural position, assuming a homogenous lithosphere.

Using the concepts outlined by Waschbusch and Royden (1992), the initial Antler forebulge may have been fixed to a weak zone corresponding to the former miogeoclinal shelf margin located in the Diamond Range. In their model, "a forebulge that remains fixed for long periods of time undergoes amplitude increase and reaches a maximum just prior to abandonment of the weak zone as the locus of the flexural bulge." Applying this concept to the Antler system, the Roberts Mountains allochthon and associated subduction zone progressively migrated toward the fixed forebulge until stresses reached a threshold during Sequence 6. At this time the forebulge migrated rapidly eastward causing differential block uplift and erosion of the former back-bulge basin strata. Following migration of the forebulge, the area flexurally subsided to form a gentle ramp with little topographic relief.

Impact of Late Precambrian Rifted Margin on Flexural Subsidence

Flexural modeling of the Antler system using acceptable parameters indicates that the flexural back-bulge basin should be several hundred kilometers wide (Goebel, 1991c). From the interpreted position of the forebulge and back-bulge basin in sequences 1 through 6, strata in the Star Range should have been deposited within the back-bulge basin and therefore shoud display similar depositional facies and accommodation trends as are found in the Confusion and Cherry Creek ranges. Sequences 1 through 6 display a major discordance in depositional facies between the Confusion Range and the Star Range. Thick, relatively deep-water siliciclastic strata were deposited in the Confusion Range synchronous with the deposition of relatively thin sequences of shoal-water, shelfal carbonates in the Star Range directly to the east. The fact that shelf deposits did not prograde eastward into the deep-water siliciclastic realm suggests there was a structural discontinuity between the two areas creating a very steep slope. Consistent with this observation, north along strike of the Confusion Range (Fig. 2, locality CK) is a thick (300 m) succession of deltaic deposits requiring significant local subsidence to provide sufficient accommodation space. Directly to the east only about 100 m of shoal-water carbonate strata were deposited.

Approximately coincident with the position of the topographic discontinuity (Fig. 2) is the trend of the Late Precambrian rifted margin (Picha and Gibson, 1985). East of the margin, the homogeneity and effective elastic thickness of the lithosphere would be greater than west of the margin. The flexural behavior of these two areas would therefore by markedly different. The rifted margin may have also served as a break in the lithosphere so that loading to the west would not have affected the eastern side of the rifted margin. We suggest that the western side of the Precambrian rifted margin underwent subsidence related to flexural downwarping during the Late Devonian and the eastern side was relatively unaffected.

Underfilled Foreland Basin

The initially formed Antler foreland basin was sediment-starved and can be considered to be "underfilled," that is, accommodation space produced by flexural loading and eustasy was not completely filled by sediment. Underfilled foreland basins are characteristic of the rapid tectonic loading phase of marine foreland systems (Heller and others, 1988). In fact, the interpreted foreland basin stratigraphic sections were occasionally significantly thinner than the back-bulge basin sections, yet flexural accommodation in the back-bulge basin is at least an order of magnitude less than the foreland basin. This illustrates the importance of variations in sediment supply in forming stratigraphic sequences in foreland systems.

With cessation of thrust migration (post-tectonic phase), the rate of flexural subsidence decreased significantly because the thrust load was no longer producing any new flexural subsidence. At this time, adjacent to the thrust front, relative uplift or flexural rebound may have occurred in response to erosion of the allochthon which effectively redistributed the load. The post-tectonic Antler foreland basin strata (Dale Canyon Formation, Chainman Shale, and Diamond Peak Formation) form a progradational sequence of submarine fan to delta-slope and delta-platform deposits that "filled" the foreland basin and prograded eastward (Harbaugh and Dickinson, 1981).

CONCLUSIONS

This study has examined depositional facies and interpreted relative accommodation trends for eight, unconformity bounded, third-order stratigraphic sequences that were deposited in the Antler foreland in Nevada and Utah during Antler orogenesis. Stratigraphic analysis of accommodation trends within the syntectonic sequences suggests that both lithospheric flexure and eustasy were important factors in controlling accommodation trends. The dominance of one factor versus another apparently varied throughout the syntectonic phase of deposition.

Sequence 1 strata reflect synchronous accommodation increase and decrease across the foreland system. The opposing accommodation trends have been interpreted as reflecting differential subsidence and uplift associated with flexure of the lithosphere during thrust loading. The position of flexurally produced, topographic highs and lows are interpreted to have remained static until Sequence 7 time. Deposition of Sequences 2 through 6 were superimposed on the initial flexural topography, which controlled the position and geometry of depocenters and sediment source areas. Sequences 2 through 6 display congruous accommodation trends in coeval stratigraphic sections, indicating that eustasy was the primary control on changes in accommodation space for these sequences. Sequence 7 displays opposing accommodation trends in coeval stratigraphic sections, reflecting the cratonward migration of flexural features and renewed flexural subsidence and uplift. Sequence 8 displays congruous, deepening upward accommodation trends across the entire foreland, reflecting major eustatic sea-level rise in Early Mississippian time controlling accommodation trends.

The vertical succession of depositional facies and accommodation trends in the Diamond Range stratigraphic section (Fig. 5) reflect change in depositional regime from a dominantly erosional, topographic high to a major subsiding trough. Coeval vertical successions in the Needle and Confusion ranges reflected uplift of former topographic lows. These documented "inversions of topography" are interpreted to reflect the cratonward migration of flexural features in response to cratonward migration of the Roberts Mountains allochthon.

The initially formed forebulge is interpreted to have been in the Diamond Range area, and the final resting position of the forebulge was east of the Confusion and Needle ranges. Thus the forebulge migration distance was approximately 120 km which is consistent with the distance the Roberts Mountains allochthon was estimated to have been overthrust.

The post-tectonic Lower to Upper Missisippian succession displays neither opposing accommodation trends in coeval strata nor topographic inversion through time. We interpret this succession to represent passive filling of the flexurally produced foreland basin (post-tectonic phase). Stratigraphic sequences in this phase were primarily controlled by eustatic sea-level changes and variations in sediment supply.

ACKNOWLEDGMENTS

The authors would like to thank the reviewers, Steve Dorobek, Steve Reid, and Mark Harris for their comments which significantly improved the manuscript.

REFERENCES

ALLEN, P. A., HOMEWOOD, P., AND WILLIAMS, G. D., 1986, Foreland basins: An introduction, *in* Allen, P. A. and Homewood, P., eds., Foreland Basins: International Association of Sedimentologists Special Publication 8, p. 3–12.

BEAUMONT, C., 1981, Foreland Basins: Geophysical Journal of the Royal Astronomical Society, v. 65, p. 291–329.

BOGEN, N. L. AND SCHWEIKERT, R. A., 1985, Magnitude of crustal extension across the northern Basin and Range province: constraints from paleomagnetism: Earth and Planetary Science Letters, v. 75, p. 93–100.

BOYER, S. E. AND ELLIOT, D., 1982, Thrust systems: American association of Petroleum Geologists Bulletin, v. 66, p. 1196–1230.

BURCHFIEL, B. C., AND ROYDEN, L. H., 1991, Antler orogeny: A Mediterranean-type orogeny: Geology, v. 19, p. 66–69.

CALDWELL, J. G., HAXBY, W. F., KARIG, D. E., AND TURCOTTE, D. L., 1976, On the applicability of a universal elastic trench profile: Earth and Planetary Science Letters, v. 31, p. 239–246.

CARPENTER, J. A., CARPENTER, D. G., AND DOBBS, S. W., 1993a, Structural analysis of the Pine Valley area, Nevada: *in* Gillespie, C. W., ed., Structural and Stratigraphic Relationships of Devonian Reservoir Rocks, East-central Nevada: Reno, Nevada Petroleum Society 1993 Field Conference Guidebook, p. 9–49.

CARPENTER, J. A., CARPENTER, D. G., AND DOBBS, S. W., 1993b, Fault reactivation and deactivation in the Basin-Range, western United States: *in* Gillespie, C. W., ed., Structural and Stratigraphic Relationships of Devonian Reservoir Rocks, East-central Nevada: Nevada Petroleum Society 1993 Field Conference Guidebook, p. 73–87.

COLES, K. S., 1988, Stratigraphy and structure of the Pinecone sequence, Roberts Mountains allochthon, Nevada, and aspects of Mid-Paleozoic sedimentation and tectonics in the Cordilleran geosyncline: Unpublished Ph.D. Dissertation, Columbia University, New York, 246 p.

COOK, H. E, HINE, A. C., AND MULLINS, H. T., 1983, Platform margin and deep water carbonates: Tulsa, Society of Economic Paleontologists and Mineralogists Short Course 12, 189 p.

DICKINSON, W. R., HARBAUGH, D. W., SALLER, A. H., HELLER, P. L., AND SNYDER, W. S., 1983, Detrital modes of Upper Paleozoic sandstones derived from Antler orogen in Nevada: Implications for nature of Antler Orogeny: American Journal of Science, v. 283, p. 481–509.

DOBBS, S. W., CARPENTER, J. A., AND CARPENTER, D. G., 1993, Structural analysis from the Roberts Mountains to the Diamond Mountains, Nevada: Estimates on the magnitude of contraction and extension, *in* Gillespie, C. W., ed., Structural and Stratigraphic Relationships of Devonian Reservoir Rocks, East-central Nevada: Reno, Nevada Petroleum Society 1993 Field Conference Guidebook, p. 51–57.

FLEMINGS, P. B. AND JORDAN, T. E., 1989, A synthetic stratigraphic model of foreland basin development: Journal of Geophysical Research, v. 94, p. 3851–3866.

GANS, P. B., 1987, An open-system, two-layer crustal stretching model for the eastern Great Basin: Tectonics, v. 6, p. 1–12.

GOEBEL, K. A., 1991a, Paleogeographic setting of Late Devonian to Early Mississippian transition from passive to collisional margin, Antler foreland, eastern Nevada and western Utah, *In* Cooper, J. D., and Stevens, C. H., eds., Paleozoic Paleogeography of the Western United States-II: Los Angeles, Society of Economic Paleontologists and Mineralogists, Pacific Section, v. 67, p. 401–418.

GOEBEL, K. A., 1991b, Interpretation of the Lower Mississippian Joana Limestone and Implications for the Antler orogenic system: Unpublished Ph.D. Dissertation, University of Arizona, Tucson, 222 p.

GOEBEL, K. A., 1991c, Late Devonian to Early Mississippian shelf margin retreat related to lithospheric flexure, Antler foreland, eastern Nevada and western Utah (abs.): Geological Society of America Abstracts with Programs, v. 23, p. 347.

GOLDHAMMER, R. K., OSWALD, E. J., AND DUNN, P. A., 1991, The hierarchy of stratigraphic forcing: an example from the Middle Pennsylvanian shelf carbonates of the Paradox basin, *in* Franseen, E. K., Watney, W. L., Kendall, G. C. St. C., and Ross, W., eds., Sedimentary Modeling: Computer Simulations and Methods for Improved Parameter Definition: Lawrence, Kansas Geological Survey Bulletin 233, p. 361–413.

GUTSCHICK, R. C. AND RODRIGUEZ, J., 1979, Biostratigraphy of the Pilot Shale (Devonian-Mississippian) and contemporaneous strata in Utah, Nevada, and Montana: Brigham Young University Geological Studies, v. 26, p. 37–63.

HARBAUGH, D. W. AND DICKINSON, W. R., 1981, Depositional facies of Mississippian clastics, Antler foreland basin, central Diamond Range, Nevada: Journal of Sedimentary Petrology, v. 51, p. 1223–1234.

HELLER, P. L., ANGEVINE, C. L., WINSLOW, N. S., AND PAOLA, C., 1988, Two-phase stratigraphic model of foreland-basin sequences: Geology, v. 16, p. 501–504.

HINTZE, L. F., 1986, Geologic map of the Morman Gap and Tweedy Wash quadrangles, Millard County, Utah, and Lincoln and White Pine

Counties, Nevada: Washington, D. C., United States Geological Survey Miscellaneous Field Studies Map MF-1872, 1:24,000.

HINTZE, L. F. AND BEST, M. G., 1987, Geologic map of the Mountain Home Pass and Miller Wash quadrangles, Millard and Beaver Wash counties, Utah, and Lincoln County, Nevada: Washington, D. C., United States Geological Survey Miscellaneous Field Studies Map MF-1950, 1:24,000.

JACOBI, R. D., 1981, Peripheral bulge- a causal mechanism for the Lower/Middle Ordovician unconformity along the western margin of the northern Appalachians: Earth and Planetary Science Letters, v. 56, p. 245-251.

JOHNSON, J. G., 1970, Taghanic onlap and the end of North American Devonian provinciality: Geological Society of America Bulletin, v. 81, p. 2077-2105.

JOHNSON, J. G. AND PENDERGAST, A., 1981, Timing and mode of emplacement of the Roberts Mountains allochthon, Antler orogeny: Geological Society of America Bulletin, v. 92, p. 648-658.

JOHNSON, J. G., SANDBERG, C. A., AND POOLE, F. G., 1991, Devonian Lithofacies of western United States, in Cooper, J. D., and Stevens, C. H., eds., Paleozoic Paleogeography of the Western United States-II: Tulsa, Society of Economic Paleontologists and Mineralogists, Pacific Section, v. 67, p. 83-105.

JOHNSON, J. G. AND VISCONTI, R., 1992, Roberts Mountains thrust relationships in a critical area, northern Sulfur Spring Range, Nevada: Geological Society of America Bulletin, v. 104, p. 1208-1220.

JONES, M. E., 1990, Stratigraphy and sedimentology of the Lower Member of the Pilot Shale in White Pine County, Nevada: Unpublished Masters Thesis, Eastern Washington University, Cheney, 113 p.

JORDAN, T. E., 1981, Thrust loads and foreland basin evolution, Cretaceous, western United States: American Association Petroleum Geologists Bulletin, v. 6, p. 2506-2520.

LARSON, B. R., CHAN, M. A., AND BERESKIN, S. R., 1989, Cyclic stratigraphy of the upper member of the Guilmette Formation (uppermost Givetian, Frasnian), west-central Utah, in McMillan, N. J., Embry, A. F., and Glass, D. J., eds., Devonian of the World: Calgary, Canadian Society of Petroleum Geologists Memoir 14, p. 183-220.

LARSON, E. R. AND RIVA, J. F., 1963, Preliminary geologic map of the Diamond Springs quadrangle, Nevada: Washington, D. C., Nevada Bureau of Mines Map 20, 1:62,500.

LEVY, M. AND CHRISTIE-BLICK, N., 1989, Pre-Mesozoic palinspastic reconstruction of the eastern Great Basin (western United States): Science, v. 245, p. 1454-1462.

MADRID, R. J., 1987, Stratigraphy of the Roberts Mountains allochthon in north central Nevada: Unpublished Ph.D. Dissertation, Stanford University, Stanford, 341 p.

MURPHY, M. A., POWERS, J. D., AND JOHNSON, J. G., 1984, Evidence for Late Devonian movement within the Roberts Mountains allochthon, Roberts Mountains, Nevada: Geology, v. 12, p. 20-23.

NICHOLS, K. M. AND SILBERLING, N. J., 1990, Delle Phosphatic: An anomalous phosphatic interval in the Mississippian (Osage-Meramecian) shelf sequence of central Utah: Geology, v. 18, p. 46-49.

NILSEN, T. H. AND STEWART, J. H., 1980, The Antler orogeny- Mid-Paleozoic tectonism in western North America: Penrose conference report: Geology, v. 8, p. 298-302.

PICHA, F. AND GIBSON, R. I., 1985, Cordilleran hingeline: Late Precambrian rifted margin of the North American craton and its impact on the depositional and structural history, Utah and Nevada: Geology, v. 13, p. 465-468.

POOLE, F. G., 1974, Flysch deposits of the Antler foreland basin, western United States, in Dickinson, W. R., ed., Tectonics and Sedimentation: Tulsa, Society of Econonmic Paleontologists and Mineralogists Special Publication 22, p. 58-82.

POOLE, F. G. AND SANDBERG, C. A., 1991, Mississippian Paleogeography and conodont biostratigraphy of the western United States, in Cooper, J. D., and Stevens, C. H., eds., Paleozoic Paleogeography of the Western United States-II: Society of Economic Paleontologists and Mineralogists, Pacific Section, v. 67, p. 107-136.

PROCTOR, P. D. AND CLARK, D. L., 1956, The Curley Limestone: An unusual biostrome in central Utah: Journal of Sedimentary Petrology, v. 26, p. 313-321.

QUINLAN, G. M. AND BEAUMONT, C., 1984, Appalachian thrusting, lithospheric flexure, and the Paleozoic stratigraphy of the eastern interior of North America: Canadian Journal of Earth Sciences, v. 21, p. 973-996.

ROBERTS, R. J., HOTZ, P. E., GILLULY, J., AND FERGUSAN, H. G., 1958, Paleozoic rocks of north central Nevada: American Association of Petroleum Geologists Bulletin, v. 42, p. 2813-2857.

ROSE, P. R., 1976, Mississippian carbonate shelf margins, western United States: Journal of Research of United States Geological Survey, v. 4, p. 449-466.

ROSS, C. A. AND ROSS, J. R. P., 1987, Late Paleozoic sea-levels and depositional sequences: Cushman Foundation for Foraminiferal Research Special Publication No. 24, Plate 2.

SANDBERG, C. A. AND GUTSCHICK, R. C., 1979, Guide to conodont biostratigraphy of Upper Devonian and Mississippian rocks along the Wasatch front and Cordilleran hingeline, Utah: Bringham Young University Geology Studies, v. 26, p. 107-133.

SANDBERG, C. A. AND GUTSCHICK, R. C., 1980, Sedimentation and biostratigraphy of Osagean and Meramecian starved basin and foreslope, western United States, in Fouch, T. D., and Magathan, E., eds., Paleozoic Paleogeography of west-central United States: Rocky Mountain Section, Society of Economic Paleontologists and Mineralogists, West-central United States Paleogeography Symposium 1, p. 129-147.

SANDBERG, C. A. AND GUTSCHICK, R. C., 1989, Deep-water phosphorite in the early Carboniferous Deseret starved basin, Utah, in Notholt, A. J. G., Sheldon, R. P., and Davidson, D. F., eds., Phosphate deposits of the world, v. 2, Phosphate rock resources: Cambridge Earth Science Series, p. 18-23.

SANDBERG, C. A. AND POOLE, F. G., 1977, Conodont biostratigraphy and depositional complexes of the Upper Devonian cratonic platform and continental-shelf rocks in the western United States, in Murphy, M. A., Berry, W. B. N., and Sandberg, C. A., eds., Western North America: Devonian: Riverside, Californian University, Campus Museum Contributions 4, p. 144-182.

SANDBERG, C. A., GUTSCHICK, R. C., JOHNSON, J. G., POOLE, F. G., AND SANDO, W. J., 1982 Middle Devonian to Late Mississippian geologic history of the overthrust belt region, western United States: Denver, Rocky Mountain Association of Geologists, Geologic Studies of the Cordilleran Thrust Belt, v. 2, p. 691- 719.

SANDBERG, C. A., POOLE, F. G., AND JOHNSON, J. G., 1988, Upper Devonian of western United States, in McMillan, N. J., Embry, A. F., and Glass, D. J., eds., Devonian of the World: Calgary, Canadian Society of Petroleum Geologists Memoir 14, p. 183-220.

SILBERLING, N. J. AND NICHOLS, K. M., 1991, Petrology and regional significance of the Mississippian Delle Phosphatic Member, Lakeside Mountains, northwestern Utah, in Cooper, J. D. and Stevens, C. H., eds., Paleozoic Paleogeography of the Western United States-II: Los Angeles, Society of Economic Paleontologists and Mineralogists, Pacific Section, v. 67, p. 425-438.

STEWART, J. H. AND POOLE, F. G., 1974, Lower Paleozoic and uppermost Precambrian Cordilleran miogeocline, Great Basin, western United States, in Dickinson, W. R., ed., Tectonics and Sedimentation: Tulsa, Society of Econonmic Paleontologists and Mineralogists Special Publication 22, p. 28-57.

TREXLER, J. H., JR. AND NITCHMAN, S. P., 1990, Sequence stratigraphy and evolution of the Antler foreland basin, east-central Nevada: Geology, v.18, p. 422-425.

TURCOTTE, D. L. AND SCHUBERT, G., 1982, Geodynamics: Applications of Continuum Physics to Geological Problems: New York, John Wiley and Sons, 450 p.

VAIL, P. R., MITCHUM, R. M., TODD, R. G., WIDMIER, J. M., THOMPSON, S., III, SANGREE, J. B., BUBB, J. N., AND HATLELID, W. G., 1977, Seismic stratigraphy and global changes of sea-level, in Payton, C. E., ed., Seismic Stratigraphy- Applications to Hydrocarbon Exploration: Tulsa, American Association of Petroleum Geologists Memoir 26, p. 49-212.

VAN WAGONER, J. C., MITCHUM, R. M., CAMPION, K. M., AND RAHMANIAN, V. D., 1990, Siliciclastic sequence stratigraphy in well logs, cores, and outcrops: Tulsa, American Association of Petroleum Geologists Methods in Exploration Series, no. 7, 55 p.

WALCOTT, R. I, 1970, Isostatic response to loading of the crust in Canada: Canadian Journal of Earth Sciences, v. 7, p. 2-13.

WASCHBUSCH, P. J. AND ROYDEN, L. H., 1992, Episodicity in foredeep basins: Geology, v. 20, p. 915-918.

APPENDIX—SOURCES OF LITHOSTRATIGRAPHIC AND
BIOSTRATIGRAPHIC DATA USED FOR STRATIGRAPHIC SECTIONS

Cherry Creek Range
    Goebel (1991b)
    Harmala (1982)
    Johnson and others (1991)
    Jones (1990)
    Sandberg, Poole, and Johnson (1988)
Confusion Range
    Goebel (1991b)
    Gutschick and Rodriguez (1979)
    Poole and Sandberg (1991)
    Sandberg, Poole, and Johnson (1988)
Diamond Range
    Goebel (1991b)
    Poole and Sandberg (1991)
    Sandberg and Poole (1977)
    Sandberg, Poole, and Johnson (1988)
Needle Range
    Goebel (1991b)
Pinon Range
    Poole and Sandberg (1991)
    Smith and Ketner (1975)
Star Range
    Poole and Sandberg (1991)
    Sandberg and Poole (1977)
Tintic Range
    Sandberg and Gutschick (1979)
    Sandberg and Poole (1977)
    Poole and Sandberg (1991)
Toquima Range
    Coles (1988)

CRATONIC-MARGIN AND ANTLER-AGE FORELAND BASIN STRATA (MIDDLE DEVONIAN TO LOWER CARBONIFEROUS) OF THE SOUTHERN CANADIAN ROCKY MOUNTAINS AND ADJACENT PLAINS

LAURET E. SAVOY

Department of Geography and Geology, Mount Holyoke College, South Hadley, MA 01075

AND

ERIC W. MOUNTJOY

Department of Earth and Planetary Sciences, McGill University, 3450 University Street, Montreal, H3A 2A7, Quebec, Canada

ABSTRACT: Upper Devonian and Lower Carboniferous strata in the Rocky Mountains and adjacent subsurface of southern Canada and adjacent Montana record a series of widely correlated, transgressive-regressive sequences that accumulated along the westward deepening cratonic margin of North America. Several Frasnian to Tournaisian megasequences are recognized: (1) upper Givetian Swan Hills platform and reefs (Beaverhill Lake megacycle); (2) extensive Frasnian Cairn-Peechee and Leduc reefs (Woodbend megacycle); (3) Nisku-Arcs carbonate bank and reefs (Winterburn megacycle); (4) upper Frasnian Ronde-Simla-Winterburn carbonate banks; (5) Famennian Sassenach basin filling; (6) Palliser-Wabamun carbonate ramp; and (7) Famennian-Tournaisian Exshaw and Banff deep-water, partly anoxic deposits. The Famennian and lower Tournaisian strata record a general upward deepening and progressive flooding of the continental margin, and may reflect the development of marginal tectonism related to Antler-age orogenesis. Quartzofeldspathic clastic deposits in the Sassenach(?), Exshaw, and lower Banff Formations reflect the periodic influx of westerly-derived, Antler(?) orogenic-sourced detritus. Depositional patterns, including the westward thickening of most units in the southern Canadian Rocky Mountains, support marked tectonic subsidence in the region from Frasnian to Tournaisian time.

The Givetian to uppermost Frasnian sequence represents a major transgression across Alberta that is divisible into three transgressive cycles, each consisting of a basal carbonate platform overlain by reefs and ending with basin filling (Swan Hills, Leduc, and Nisku reef-building cycles). Dominantly anoxic conditions (overlying Duvernay-Perdrix source beds) occurred immediately adjacent to the reefs when they were of low relief and throughout their development. The reefs kept pace with sea-level changes, resulting in a series of shallowing upward cycles broken by periods of non-deposition. The progressive southwestward thickening of the reefs, reaching three times their thickness in the subsurface, records regional differential subsidence that appears to be related to both tectonic and sedimentary loading of the continental margin and more regional crustal subsidence.

The western part of the Western Canada Sedimentary Basin was filled during the initial Famennian sea-level rise by Sassenach siliciclastic deposits, which may have been derived from a landmass to the southwest. A westward deepening and thickening carbonate ramp (Palliser Formation and Wabamun Group) was bordered to the west by a deep basin (Lussier region) in Famennian time. Carbonate ramp sedimentation ended by late Famennian time with the initial deposition of deep-water Exshaw black shale, roughly coincident with a major sea-level rise and the widespread deposition of similar low-oxygen facies in western North America. The lower Banff Formation consists of starved-basin to deep-ramp lithofacies, which are overlain by shallower-water carbonates of the middle and upper Banff; this sequence records basinward (westward) progradation of the Banff ramp.

INTRODUCTION

The middle Paleozoic sedimentary record reflects a change from a passive to a tectonically active margin along much of western North America (e.g., Richards, 1989; Bond and Kominz, 1991; Kominz and Bond, 1991). Evidence of convergent, Antler orogenic activity is well documented in parts of the western United States (e.g., Poole and Sandberg, 1977, 1991; Poole and others, 1977; Dickinson, 1977; Nilsen and Stewart, 1980; Speed and Sleep, 1982; Burchfiel and Royden, 1991; Fig. 1). Although the type of deformation has been difficult to identify conclusively, contemporaneous(?) tectonic activity occurred along the Canadian Cordillera and resulted in the deposition of clastic rocks such as the Earn Group in the central and northern parts (Fig. 1). Different tectonic settings have been proposed to account for this activity and include an extensional setting related to rifting or transtension (e.g., Tempelman-Kluit, 1979; Struik, 1986; and others, 1987; Gordey, 1988; Turner and others 1989), strike-slip or transcurrent faulting (e.g., Eisbacher, 1983), and flexural extension behind a convergent margin (e.g., Gordey, 1988; Richards, 1989; Smith and others, 1993).

The paleogeography and tectonic history of the margin near what is now the United States-Canada international border (49°N) north to 53°N, however, are incompletely understood. Results of recent quantitative subsidence analyses of miogeoclinal strata of the cratonic margin in Alberta (e.g., Bond and Kominz, 1991) and Antler foreland in Idaho and Montana (e.g., Dorobek and others, 1991a, b) suggest that significant tectonic subsidence may have occurred in these areas during Late Devonian-Mississippian time. Quartzofeldspathic clastic units and volcanic detritus in strata of this age (e.g., Exshaw Formation) in the southern Canadian Rocky Mountains provide a record of tectonism in the region. In addition, there is a growing body of evidence for the occurrence of Middle Devonian to Mississippian plutonism and deformation from the Purcell Mountains, Kootenay arc, and adjacent areas in southeastern British Columbia (Fig. 1; see below for references). Evidence of the causes of this subsidence and tectonism, however, is largely obscured by post middle-Paleozoic metamorphism, terrane accretion, tectonic deformation, and erosion.

This paper evaluates the Devonian-Mississippian (Lower Carboniferous) depositional framework, subsidence history, and tectonic setting for the region by relating the litho- and biostratigraphy of sections in the southwestern part of the Western Canada Sedimentary Basin and comparing trends to global sea-level curves of Johnson and others (1985), Ross and Ross (1985, 1987a, b) and Johnson and Sandberg (1988) in order to develop a sea-level curve for the region. Upper Devonian-Lower Mississippian strata that are exposed in the Front and eastern Main ranges of the foreland thrust and fold belt in the southeastern Canadian Cordillera

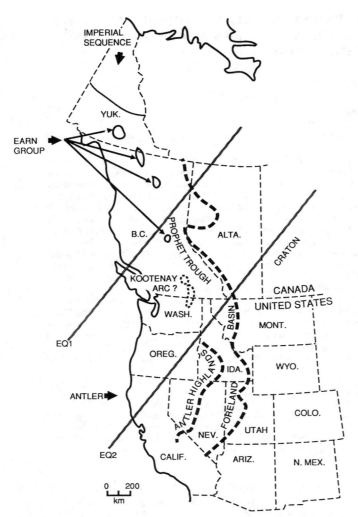

FIG. 1.—Major tectonic features of western North America in Late Devonian and Early Mississippian time (from Savoy, 1992). The Kootenay arc is an arcuate belt of deformed and metamorphosed rocks that contains evidence of middle Paleozoic tectonism. The Earn Group and Imperial sequence are Devonian-Mississippian clastic assemblages derived from orogenic or uplifted regions. EQ1 is the approximate Tournaisian position of the paleoequator based on lithic paleoclimatic data of Witzke (1990). Scotese and McKerrow (1990) placed the paleoequator in a southern position (western U. S.) at this time; EQ2 is the approximate Famennian position of the paleoequator based on their data.

and adjacent Montana (Fig. 2) accumulated on the outer craton and continental margin of western North America. This interval was marked by a general expansion of widespread epicontinental seas, and middle to upper Paleozoic strata in this region record parts of several transgressive-regressive (T-R) sequences (e.g., Sandberg and others, 1983; Workum, 1993; Johnson and others, 1985; Johnson and Sandberg, 1988; Morrow and Geldsetzer, 1988; Richards, 1989; and information in this paper).

During the deposition of three reef stages in Late Givetian and Frasnian time (Swan Hills, Leduc, and Nisku), the Western Canada Sedimentary Basin subsided differentially so that more than twice the thickness of sediments ac-

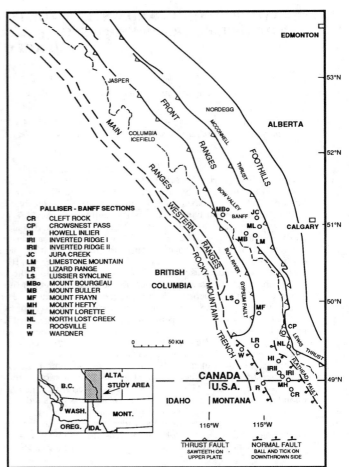

FIG. 2.—Index map showing major structural regions of the Cordilleran thrust and fold belt in the southern Canadian Rocky Mountains and the location of Palliser-Banff sections referred to in the text.

cumulated during Frasnian time in the Main Ranges (about 800 m) than to the east in the Front Ranges (400 m; Mountjoy, 1978; Mallamo and Geldsetzer, 1991; McLean and Mountjoy, 1993; Fig. 3). Fine-grained siliciclastic sediments were transported southward along the eastern margin of the basin and diluted by locally derived carbonate muds. Deposition was cyclic consisting of a series of shallowing upward, alternating shales and carbonates, the carbonate portions being deposited mainly during times of rising sea level. Following deposition of clastic sediments (Sassenach Formation) in early Famennian time, a carbonate ramp and bank (Palliser Formation and Wabamun Group) was established in Alberta and adjacent British Columbia. The Palliser Formation also thickens westward, and in the eastern Main Ranges south of Jasper (Sunwapta Pass, Columbia Icefield area) the unit is approximately three times thicker than it is in the eastern Front Ranges and Foothills (Fig. 3). In the southern Canadian Rocky Mountains, ramp-carbonate deposition was ultimately followed by the accumulation of Famennian-Tournaisian organic-rich, deeper water facies (Exshaw and lower Banff Formations) under low-oxygen conditions. The return to carbonate-dominated

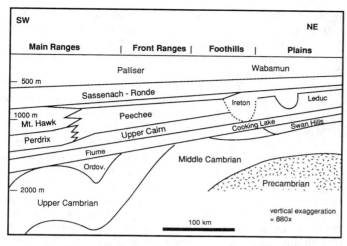

FIG. 3—Schematic SW-NE cross section from the Main Ranges to Calgary near latitude 51°N showing the westward thickening in the Frasnian (Flume, Cairn, Peechee) buildup units and the Famennian Palliser Formation.

deposition (middle to upper Banff Formation and correlative units) occurred in middle to late Tournaisian time in this southern area.

On the basis of stratigraphic, paleoenvironmental, and paleontological data, we interpret that the continental margin near and north of the international border was subjected to considerable subsidence that in part may have resulted from tectonic loading caused by the emplacement of thrust sheets to the southwest together with sedimentary loading by a clastic wedge derived from Antler or related highlands located somewhere in what is now Idaho, Oregon, and southern British Columbia (e.g., Dorobek and others, 1991a, b). Evidence for westerly derived siliciclastic rocks in the southern Canadian Rocky Mountains first occurs in Sassenach strata of earliest Famennian age (e.g., Geldsetzer and Upitis, 1993), and also occurs in Exshaw and lower Banff strata of latest Famennian and Tournaisian age.

This work builds upon previous stratigraphic studies of Devonian-Mississippian units in the Canadian Rocky Mountains, which have focused on the Bow Valley area of the Front Ranges (51°N) and regions to the north (e.g., Harker and McLaren, 1958; Macqueen and Sandberg, 1970; Richards, 1989; Richards and others, 1991; McLean and Mountjoy, 1993, 1994); it also builds on recent analyses of the paleotectonic setting (e.g., Richards, 1989; Smith and others, 1993). The Famennian and Tournaisian strata were evaluated in this study primarily in the region south of 51°N (southern British Columbia and Alberta, and Montana), whereas the Givetian to lower Famennian record is based on studies in the Rocky Mountains between Banff and Jasper in Alberta (50°N to 53°N) and on published data from the adjacent Alberta subsurface to the east (Figs. 2, 4).

LITHOFACIES AND DEPOSITIONAL SETTING

Space does not permit us to cover many specific aspects of the upper Givetian through Tournaisian stratigraphy of this region. Instead, this paper outlines the overall cyclic

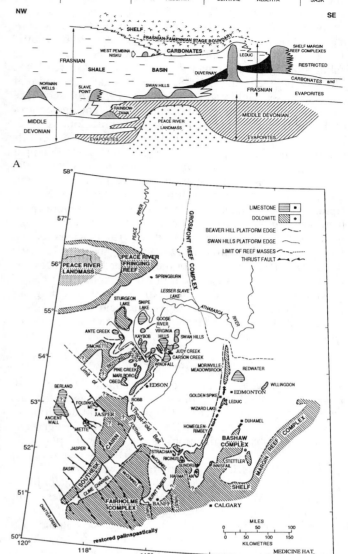

FIG. 4—(A) Schematic northwest-southeast cross section of the Middle and Upper Devonian succession from the Northwest Territories to southern Alberta and adjacent Saskatchewan, illustrating five levels of reef development (stippled): Keg River (Rainbow-Zama), Slave Point, Swan Hills, Leduc and Nisku (after Bassett and Stout, 1968). Right part of section crosses central part of Figure 4b. Hachured pattern—evaporites, solid black represents Duvernay source beds. (B) Distribution of Upper Devonian Leduc and Swan Hills reefs and carbonate platforms, Alberta with Rocky Mountain portion palinspastically restored (updated from Mountjoy, 1980). Most reefs and platforms are dolomitized.

nature of these strata from which a revised sea-level curve can be constructed, based on the curve of Johnson and others (1985), Ross and Ross (1985, 1987a, b) and Johnson and Sandberg (1988). The revised sea-level curve is based on successions that have undergone greater subsidence than those of the adjacent continental interior. The upper Givetian-Frasnian stratigraphy of the Western Canada Sedimentary Basin (between 49°N and 56°N, Fig. 4B) is divisible into three megacycles: Beaverhill Lake, Woodbend, and Winterburn. The upper Famennian and Tournaisian stratigraphy also forms at least two major megacycles: Wabamun/Palliser, and uppermost Palliser/Exshaw and Banff. The Devonian stratigraphy has been outlined in numerous publications (e.g., for regional summaries see Moore, 1988; Morrow and Geldsetzer, 1988; for individual reefs and reef complexes see Geldsetzer and others, 1989; for sequence stratigraphy see Wendte and others, 1992). The age control of Frasnian strata is based mainly on brachiopods (McLaren, 1962; Sartenaer, 1969; Raasch, 1988), corals (McLean and Pedder, 1984, 1987) and conodonts (Klapper and Lane, 1988; Orchard, 1988; Weissenberger, 1988, 1994; Uyeno, 1991; Mallamo and Geldsetzer, 1991, pers. commun., 1992). The facies relationships of the lower Mississippian sequence have also been discussed in several publications (e.g., Harker and McLaren, 1958; Macqueen and Sandberg, 1970; Bamber and others, 1984; Richards and Higgins, 1988; Richards, 1989; Richards and others, 1991; Savoy, 1992). Age control for the Palliser-Banff succession is based primarily on conodont biostratigraphy documented in Macqueen and Sandberg (1970), Richards and Higgins (1988), Savoy (1990, 1992), Higgins and others (1991), Johnston and Chatterton (1991), Richards and others (1991), and Savoy and Harris (1993a, b).

Beaverhill Lake Megacycle

This megacycle began with a carbonate platform followed by the buildup of platform reefs (Swan Hills) and ended with a phase of basin filling by lateral accretion of mixed carbonate and siliciclastic sediments (Waterways) accompanied by progradation of the carbonate platforms (Moore, 1989; Wendte, 1992; Fig. 5). This megacycle is divisible into a series of shallowing upward cycles.

Following the early Givetian Watt Mountain unconformity above the Muskeg and Fort Vermilion evaporites of the Elk Point Basin and progressively older strata to the west, the widespread Slave Point Platform developed on the eastern flanks of the Peace River and West Alberta Arch, upon which the Swan Hills Formation built up locally into reefs following a rapid sea-level rise (Fig. 5). In the latter part of the transgressive phase, these reefs built up in a series of stages tracking specific rises in sea level, creating significant reef to basin relief (Wendte and Stoakes, 1982; Wendte, 1992). The upper regressive or progradational phase is marked by pronounced asymmetric filling of the basin by the Waterways Formation, which resulted in increasingly turbid-water conditions that progressively stopped reef growth from east to west across the basin.

Woodbend Megacycle

The Woodbend megacycle is essentially a repetition of the Beaverhill Lake megacycle (Fig. 5). The sediments of early Frasnian age gradually onlapped and buried the West Alberta Ridge forming the extensive Flume carbonate platform upon which the Leduc and Fairholme reefs were built. Following a rapid rise in sea level, most of this platform was submerged and covered by argillaceous subtidal carbonate deposits (Maligne) and in turn by anoxic black shales of the Perdrix (Duvernay in the subsurface). Carbonate sedimentation continued only above the higher portions of the platforms forming the Leduc reef chains and reef clusters of the subsurface and several reef complexes in the Rocky Mountains (Fig. 4B).

Platform reefs developed during the main part of this sequence, and in the Rocky Mountains they comprise two vast complexes (the Fairholme and Southesk-Cairn separated from each other by the narrow Cline Channel, 116° to 118°W and 51° to 52°N, Fig. 4B), with a number of isolated platform reefs occurring to the north (Miette, Ancient Wall, etc.). In the adjacent subsurface, a series of platform reefs grew and to the north fringe the Peace River Arch. On the eastern side of the Alberta Basin, the reefs occur in two, north-northeast trending reef chains, the Rimbey-Meadowbrook and Bashaw-Duhamel (Fig. 4B). These reef chains appear to have been controlled by regional syndepositional tectonic flexures, likely related to zones of weakness that follow tectonic boundaries in the underlying Precambrian Shield (Mountjoy, 1978, 1980, 1987; Ross and Stephenson, 1989; Ross, 1990; Ross and others, 1991). Some platform reefs are not aligned, Redwater (Klovan, 1964) being the largest and most important commercially. Instead it is controlled by highs (thicks) in the underlying Cooking Lake platform (Fig. 5). Farther southeast in southern Alberta shallow-water, restricted and evaporitic lithofacies developed behind the Leduc shelf margin reef complex (Fig. 4B).

Buildups generally tracked sea level. Stillstands or slight falls of relative sea level (and possibly decreasing rates of subsidence) caused interruptions in reef growth and, occasionally, emergence of the reef tops. There is little evidence of lowstand systems sequence wedges (Wendte, 1992; McLean and Mountjoy, 1993).

During the regressive phase, fine-grained siliciclastic deposits were transported southward along the eastern margin of the basin east of the Rimbey-Meadowbrook reef trend and were diluted by locally derived carbonate muds (Fig. 6; Oliver and Cowper, 1963; Mountjoy, 1980; Stoakes, 1979, 1980, 1992). These sediments began to fill in the basin towards the south and southwest (Stoakes, 1980) and eventually most of the basin (Fig. 6). Turbid-water conditions limited reef growth except in the western part of the basin (Mountjoy, 1980). Deposition of siliciclastic units was episodic, and the strata consist of a series of shallowing upward, alternating shale and carbonate cycles, the carbonate portions being deposited mainly during times of rising sea level. Following Stoakes (1979, 1992), we infer that the most reasonable source for these siliciclastic units was the Ellesmerian fold belt in the Canadian Arctic because no other sources were available to the east and northeast; the Canadian Shield was covered by Paleozoic strata at this time.

Winterburn Megacycle

Following a major sea-level rise, Nisku reefs of late Frasnian age developed in the West Pembina shale basin southwest of Edmonton (53°N) above the center of the previous shale basin. The Nisku carbonate platform bordered the shale basin and locally developed a reef margin and numerous

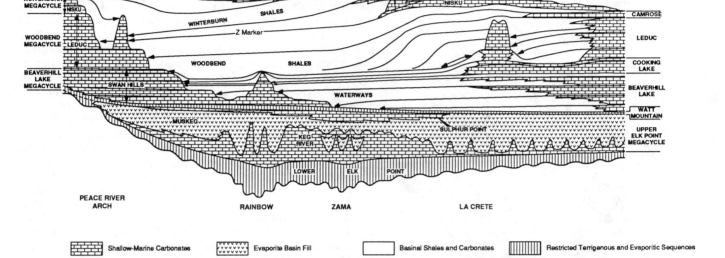

FIG. 5.—Central Alberta composite schematic, west-east cross section illustrating megacycles in Middle and Upper Devonian sequences and major facies (from Wendte, 1992).

Zeta Lake Member coral-stromatoporoid reefs (Stoakes, 1992). In contrast, an extensive Arcs carbonate platform developed instead of reefs in the Rocky Mountain region south of 54°N, except for some very small patch reefs around the Jasper basin.

During deposition of the above three reef stages (Swan Hills, Leduc, and Nisku) in late Givetian and Frasnian time, the basin subsided differentially so that more than twice the thickness of sediments accumulated in the Leduc and Arcs reefs of the Main Ranges (about 800 m) than in the Front Ranges (400 m) of the Rocky Mountains (McLean and Mountjoy, 1993; Mountjoy, unpub. data).

Wabamun/Palliser Megacycle

Sassenach Formation.—

The terminal Frasnian extinction event had a profound effect on subsequent reef building organisms (e.g., McLaren, 1982), and the only Famennian reefs known are small labechiid mounds. A major paraconformity occurs nearly everywhere between Frasnian and Famennian strata—the Simla (Ronde)-Sassenach in the Rocky Mountains and the Blue Ridge-Graminia contact in the subsurface. Missing are the *linguiformis* Zone (former Uppermost *gigas* Zone) and Lower *triangularis* Zone. In the Jasper area, this unconformity records a major sea-level fall followed by a major rise as the entire 200-m-thick Sassenach Formation onlapped the margins of the Ancient Wall reef complex but did not cover the top of this complex (McLaren and Mountjoy, 1962; Mountjoy, 1978, 1980). Similar relationships occur along the southwestern margin of the Fairholme reef complex (Mallamo and Geldsetzer, 1991, pers. commun., 1992).

Along the westernmost part of the Western Canada Sedimentary Basin from Jasper south to Fernie (southeastern B. C.), the Sassenach Formation filled a sedimentary depression (Jasper Basin and extensions to the south) west of the reef domains that was not completely filled by Mount Hawk siliciclastics and carbonate muds and overlying Ronde and equivalent carbonate sediments (Fig. 6). The incomplete filling of the western part of the basin is shown by the westward thinning of Mount Hawk basin strata around the Ancient Wall reef complex and along the western margin of the Fairholme reef complex (Mountjoy, 1980, 1987; Mallamo, pers. commun., 1992).

The Sassenach Formation has only been studied in reconnaissance fashion; the basal few meters are undated. At the type section on the southeast flank of the Ancient Wall buildup, the main lower 160 m of the formation consist of dark grey, silty, calcareous mudstones. Silt beds are planar laminated and commonly cross-laminated and represent deposition below wave base (McLaren and Mountjoy, 1962; Coppold, 1976). The upper sandy member is about 35 to 50 m thick adjacent to the Ancient Wall buildup and consists of unfossiliferous, silty and sandy medium to coarse-grained limestones and strongly cross-bedded, calcareous, fine- to medium-grained, quartzose sandstones representing deposition above wave base. Geldsetzer and Upitis (pers. commun., 1993) report that Sassenach sands southwest of Banff contain feldspar as do our preliminary unpublished data on the type Sassenach adjacent to the Ancient Wall reef complex (see also McLaren and Mountjoy, 1962). Both the westward thickening of the unit and the presence of

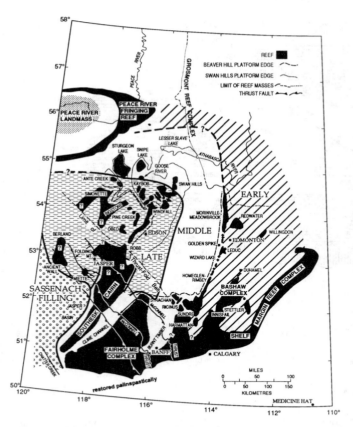

FIG. 6—Suggested stages of filling of the Alberta Basin by Ireton and Mount Hawk siliciclastics during late Frasnian and early Famennian (Sassenach) time (from Mountjoy, 1980, Fig. 7). Areas labelled "Early" to "Late" represent progressive filling of the basin around the reefs during Frasnian time. The area labelled "Late" represents what is termed the "Deep Basin" northeast of the Thrust-Fold belt. The Jasper basin now exposed in the Rocky Mountain Front Ranges received a condensed sequence of Frasnian siliciclastic deposits and was not filled until early Famennian time with siliciclastic sediments derived from the west and southwest(?).

feldspars support an inferred western source for at least some of these clastic deposits.

Palliser Formation.—

In the Rocky Mountains, the Famennian Palliser Formation (Fig. 7), which consists of the lower Morro Member and overlying Costigan Member, is laterally contiguous with the Wabamun Group to the east in the subsurface of Alberta. A carbonate ramp profile for the Palliser Formation in the study region is based on broad and simple facies tracts and the lack of any observed sedimentological evidence for a steep shelf-slope break.

Near the international border, the lower to middle Famennian Morro Member formed in an extensive, open-marine to partly restricted subtidal setting on the Palliser ramp (see also Johnston and Chatterton, 1991; Richards and others, 1991; Savoy, 1992; Savoy and Harris, 1993a, b). Morro lithofacies are similar to the north in the Bow Valley near Banff (see Fig. 2; e.g., Johnston and Chatterton, 1991; Richards and others, 1991).

Deposits assigned to the Costigan Member (see also Fig. 8) near the international border accumulated on a more open-marine position on the carbonate ramp in middle to late Famennian time (Savoy, 1990, 1992; Savoy and Harris, 1993a, b). In this southern area, the member contains a greater abundance of more open-marine, skeletal benthos than the Morro Member. In contrast, many workers have noted that much of the Costigan Member to the north in the Bow Valley consists of laminated and brecciated limestone and apparently was deposited in a shallow, restricted-marine setting (Geldsetzer, 1982; Richards and Higgins, 1988; Richards and others, 1991; Johnston and Chatterton, 1991). The uppermost beds of the Costigan Member in this general area, however, are burrowed, skeletal carbonate lithofacies deposited in an open-marine setting. Latest Famennian conodonts (Lower to middle *expansa* Zones) have been recovered from the uppermost Costigan Member at Mount Lorette (Savoy, 1992; Savoy and Harris, 1993a, b) and elsewhere in the Bow Valley area (e.g., Jura Creek; Richards and Higgins, 1988; Higgins and others, 1991; Johnston and Chatterton, 1991; Richards and others, 1991). The upper part of the member is interpreted to record a deepening in this region by late Famennian time (e.g., Moore, 1989; Richards and others, 1991) that is representative of the base of the Exshaw megacycle. A regional unconformity also is present below the correlative Big Valley Formation (Wabamun Group) to the east (e.g., Richards and others, 1991).

Exshaw—Banff Megacycle(s)

Exshaw Formation.—

Carbonate-ramp sedimentation ended by latest Famennian time with the influx of deep, oxygen-depleted water that likely accompanied the T-R IIf transgression of Johnson and others (1985) (Savoy, 1990, 1992). Although Exshaw lithofacies and thickness trends vary significantly in the study region (south of 51°N; Fig. 9), the organic-rich, grayish-black to black shale, radiolarian chert, and/or radiolarian-spicular chert/siliceous mudstone formed in a relatively deep-water, anaerobic to dysaerobic environment (>50?—>200-m depth) during latest Famennian (*expansa* Zone) to middle(?) Tournaisian time (Fig. 7; Savoy, 1990, 1992; see also Macqueen and Sandberg, 1970; Richards and Higgins, 1988; Sandberg and others, 1988). Thick "Exshaw" chert sequences in the southwest (Mount Frayn; Fig. 2) could have formed in an area of nutrient recharge (upwelling) and high silica productivity. The Exshaw Formation contains an upper siltstone member in the Bow Valley area and locally near the international border, which lacks obvious shallow-water sedimentary structures, and may represent clastic influx deposited below wave base in a dysaerobic to aerobic setting (see also Richards and Higgins, 1988; Richards and others, 1991). The lower siltstone member in the Bow Valley is of an Early Mississippian age (e.g., Macqueen and Sandberg, 1970); there is limited biostratigraphic control for the upper part of the Exshaw siltstone member. Bamber and others (in Gordey and others, 1991) suggest that this siltstone was deposited under shallowing upward conditions.

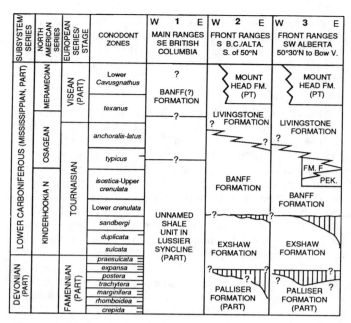

FIG. 7—Correlation chart of Upper Devonian and Lower Mississippian stratigraphic units in western Canada and northwestern United States. Ticks on the right side of the conodont zones indicate subzonal boundaries. Chart is based on the following sources: (1) Savoy (1990, 1992); (2, 3) Higgins and others (1991), Richards and others (1991), Savoy (1990, 1992), and Savoy and Harris (1993a).

Interlaminae of subangular quartzofeldspathic siltstone and sandstone occur within the Exshaw black shale near the international border. A thin bed (<6 cm thick) of medium- to coarse-grained, phosphatic, feldspathic arenite occurs in the basal Exshaw Formation at its type section (Jura Creek) and locally elsewhere in the Bow Valley (e.g., Mountjoy, 1956; Harker and McLaren, 1958; Macqueen and Sandberg, 1970). A thin bed of volcanic tuff has also been recognized in the organic-rich shale (see Richards, 1989; Richards and others, 1991).

The environmental significance of the Palliser/Exshaw contact is ambiguous. Although a subaerial exposure surface has been proposed for the top of the Palliser Formation at some areas in the Canadian Rocky Mountains, the sharp contact between Palliser and Exshaw strata in many outcrops exhibits little evidence of an emersion surface. Instead, the contact in the study area probably represents an interval of non-deposition associated with submergence of the Palliser ramp, and thus supports a submarine discontinuity interpretation, particularly north of the international border region (see also Richards and Higgins, 1988; Richards and others, 1991; Savoy, 1990, 1992).

Banff Formation.—

Although lithofacies vary vertically and laterally across thrust sheets, the Tournaisian Banff succession in the southern Canadian Rocky Mountains (south of 51°N) forms an overall shallowing upward, progradational sequence representing an oxygen-depleted, deep ramp to basin that was succeeded by a shallow marine, carbonate ramp (Fig. 9;

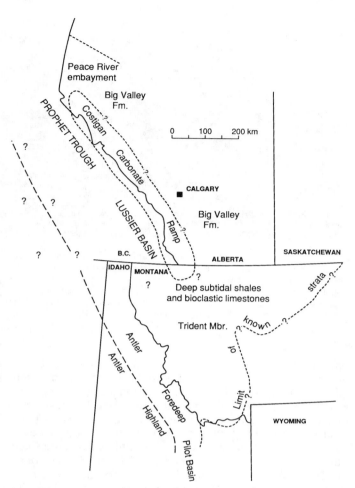

FIG. 8—Schematic, upper Famennian paleogeographic map of the Rocky Mountain region in southwestern Canada and northwestern United States showing the Palliser ramp (Costigan Member) bordered to the west by the Lussier basin and to the southwest by the Antler foredeep and highlands.

Richards, 1989; Savoy, 1990, 1992; Stoakes, 1992). A low-gradient, carbonate ramp profile is suggested by the broad and generally simple facies geometry; finer grained, lower energy sediments downslope; and the lack of evidence for a significant paleoslope, although slump deposits are locally common in the lower Banff Formation of the Bow Valley (see also Chatellier, 1988; Richards, 1989; Richards and others, 1991). The Exshaw-Banff contact has been interpreted as unconformable in parts of the southern Canadian Rocky Mountains (e.g., Macqueen and Sandberg, 1970; Richards and Higgins, 1988; Bamber and others in Gordey and others, 1991), yet it may be at least locally conformable near the international border (Savoy, 1992). Lower Banff lithofacies observed in the study area include organic-rich, black to grayish-black laminated, shale; phosphatic and/or siliceous (spicular or radiolarian-bearing) mudstone; glauconitic and phosphatic mudstone and siltstone (local); and/or thin interbeds of quartzofeldspathic siltstone and sandstone. In the northern area of this study, these units accumulated in a moderately deep-ramp setting until at least

FIG. 9—Uppermost Palliser, Exshaw, and Banff sequence at North Lost Creek (Lewis thrust sheet), southwestern Alberta. The top of the cliff-forming Palliser Formation is at the base of the section (far left of photograph). The Exshaw Formation is very thin at this section, corresponding to less than a meter of the recessive black shale zone above the Palliser Formation. The overlying strata are part of the Banff Formation.

middle Tournaisian(?) time, and in the international border area, a deep-ramp to basin environment existed until middle to late Tournaisian (*anchoralis-latus* Zone) time (Fig. 7). Richards (1989) interpreted Lower Banff deposition to have occurred in the early phase of a regional Mississippian transgression that followed Exshaw deposition. However, at least in the southern area of this study, it is more likely that deep-water deposition continued from Late Devonian to Early Mississippian (middle to late Tournaisian) time.

Overlying, bioclastic spicular mudstone, peloidal/bioclastic limestone, and fine-grained, quartzo-feldspathic sandstone are interpreted as primarily reworked debris derived from shallower positions on the Banff ramp. Grain-supported pelmatozoan limestone in the upper part of the Banff may, in part, represent redeposited debris as well as skeletal sand banks. This unit is of middle to late Tournaisian age in the Bow Valley (Richards, 1989; Savoy, 1990, 1992; Higgins and others, 1991; Richards and others, 1991; Savoy and Harris, 1993a, b) and late Tournaisian age near the international border (Savoy, 1990, 1992; Savoy and Harris, 1993a, b).

Lussier Syncline Units (Famennian—Visean)

The Palliser-Banff lithofacies succession is absent in the Lussier syncline (Main Ranges), the southwesternmost (basinward) section (Figs. 2, 8). Instead, a sequence (>120 m) of dark-gray to black shale and mudstone formed in a basin to deep-ramp setting west of the Palliser carbonate ramp from early Famennian (by Upper *crepida* Zone—Lower *rhomboidea* Zone) to at least middle Tournaisian time (Fig. 7; Savoy, 1990, 1992; Savoy and Harris, 1993a, b). This unit overlies undated and poorly exposed quartzofeldspathic siltstone (of unknown thickness at this section) that is interpreted to be part of the Upper Devonian succession of Leech (1958, 1979). Part of this succession may be correlative to the Sassenach Formation. Bioclastic limestone, chert, and siltstone of early Visean age overlie the shale-mudstone sequence and are interpreted as allochthonous skeletal sands derived from the westward prograding Banff(?) ramp.

SEA-LEVEL CHANGES AND BASIN EVOLUTION

At least two major, large-scale cycles can be recognized on the basis of major deepening and shallowing trends in our interpretive sea-level curve (Fig. 10): (1) the late Givetian to latest Frasnian T-R IIa—IId of Johnson and others (1985) and Johnson and Sandberg (1988), representing the main, reef-building phases and (2) Early Famennian to latest Tournaisian, T-R IIe—IIf of Johnson and others (1985) and Johnson and Sandberg (1988) and curve of Ross and Ross (1987a, b), represented by the Sassenach, Palliser, Exshaw, and Banff Formations (Fig. 10). These major cycles can be divided into smaller-scale subcycles.

The Devonian portion of the proposed regional sea-level curve follows the curve and T-R cycles of Johnson and others (1985) except in the latest Frasnian and Famennian intervals. The revised Frasnian segment of the curve is based on long-term research in Alberta and adjacent areas by Mountjoy and recent research of McLean (1992). The early Famennian part of the curve is based on stratigraphic data in the Rocky Mountains, especially in the Jasper areas (McLaren and Mountjoy, 1962; Coppold, 1976; Mountjoy, 1987). The remainder of the Famennian segment is based on Savoy (1990, 1992) and our joint research in the Rocky Mountains in conjunction with other studies and recent subsurface data (e.g., Halbertsma and Meijer Drees, 1987; Stoakes, 1992). The Tournaisian segment of the curve is from Ross and Ross (1987a, b).

T-R IIa to IId (Late Givetian and Frasnian)

The major transgression represented by T-R IIa to IId can be divided in western Canada into three subcycles or transgressive phases or floodings (Fig. 10). These three subcycles are: (1) the Swan Hills platform and reefs (Beaverhill Lake Group), (2) the Flume platform and Cairn-Peechee reefs (and equivalent subsurface Cooking Lake platform and Leduc reefs), and (3) the Arcs-Nisku (Zeta Lake) and Meekwap reefs. Each subcycle consists of a basal transgressive carbonate platform overlain by reefs and ending with basin filling.

Each subcycle represents a renewed transgression and deepening, and each took place in a series of relatively rapid sea-level rises, shown as steps on the sea-level curve (Fig. 10). Basin filling stages consist of a mixture of carbonate muds derived from the carbonate shelves that prograded into the basin and siliciclastic sediments transported southward from the Arctic. These reef building phases have been outlined by a number of authors (see Moore, 1988, 1989; Morrow and Geldsetzer, 1988). All limestone and most dolomitized platforms and reefs exhibit meter-scale, shallowing upward cycles.

First flooding—Swan Hills Beaverhill Megacycle (T-R cycle IIa)

The Swan Hills platform and reefs were deposited in a series of shallowing-upward cycles (Fischbuch, 1968; Wong

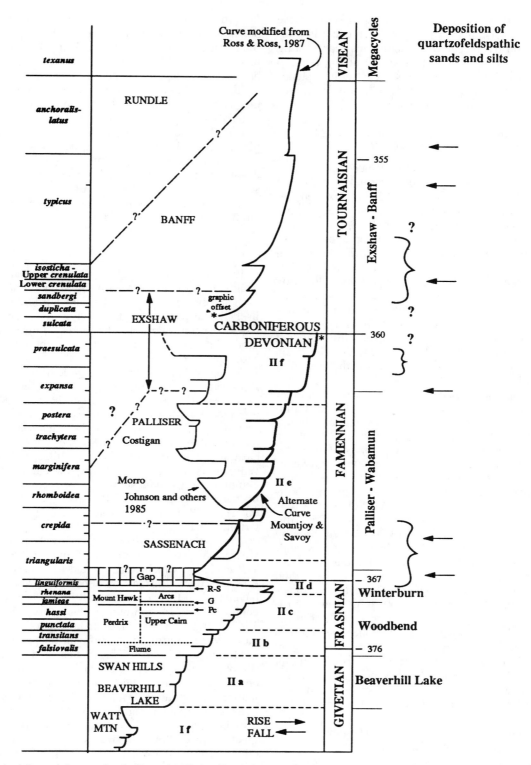

Fig. 10—Revised "eustatic" curve for the Devonian-Carboniferous succession of the southern Canadian Rocky Mountains. The Devonian part is modified from Johnson and others (1985) and Johnson and Sandberg (1988). The Carboniferous portion of the curve is taken directly from Ross and Ross (1987a). Devonian and Lower Carboniferous stratigraphic units are plotted by age. Arrows and brackets indicate intervals when siliciclastic and feldspathic sands were transported into the basin from the west, presumably derived from Antler-age highlands to the west. The data from southeastern British Columbia and southwestern Alberta support continued transgression from late Famennian into Tournaisian time (see text). Both the Devonian Johnson and others' (1985) curve and our curve are shown for comparison. Note that the curve should be continuous across the Devonian-Carboniferous boundary but because of space limits has been offset graphically. (Pc = Peechee, G = Grotto, and R-S = Ronde-Simla.)

and Oldershaw, 1980; Wendte and others, 1992). These cycles are clearly episodic as the top of each cycle is truncated and overlain by deeper-water sediments representing the base of the next cycle (best observed at the buildup margins). Basin sediments are represented by dark shales at the base overlain by lighter colored and more calcareous shales with limestones in the upper part.

Second flooding—Leduc Woodbend Megacycle (T-R cycle IIb and lower IIc)

The Flume, Cairn, Peechee (= Leduc) succession is similar to the Swan Hills cycle but formed thicker and more widespread reefs. Initially the basin became oxygen depleted (Duvernay and Perdrix strata), but later became oxygenated when the relief between the reefs and basin was near a maximum. The causes of oxygen depletion are uncertain but could be related to density stratification within the shallow basin (Mountjoy, 1980; McLean and Mountjoy, 1993) and not directly related to upwelling (Mountjoy in Geldsetzer and Mountjoy, 1992). Oxygen-deficient water, as reflected by black shales in equivalent strata to the northwest along the cratonic margin, may have, in part, controlled anoxia in the Alberta Basin. Basin filling did not take place until after most of the reef growth. The basin was filled towards the west and southwest with a series of siliciclastic cycles (Stoakes, 1980, 1992).

Cyclic stacking patterns in the Fairholme Group reef complexes (Miette, Southesk-Cairn, and Fairholme buildups) of the southern Canadian Rocky Mountains suggest that superimposed short-term and long-term fluctuations in relative sea level controlled buildup stratigraphy. The Flume platform and overlying Upper Cairn biostrome consist of meter-scale, shallowing-upward hemicycles. These hemicycles are interpreted to have been deposited during short-term, high-frequency (fifth-order) sea-level oscillations (McLean, 1992; McLean and Mountjoy, 1994). Superimposed on this high-frequency cyclicity are larger, broadly shallowing-upward trends in which dominantly subtidal, meter-scale hemicycles or subtidal, non-cyclic intervals gradually (sometimes abruptly) pass upward into peritidal hemicycles of comparable thickness. These intermediate-scale sequences are the product of both fourth- and third-order driving mechanisms of uncertain origin. The Flume, Upper Cairn, and overlying Peechee members represent a low-frequency, third-order depositional sequence (e.g., Van Wagoner and others, 1988). The various orders of cyclicity are best developed in the eastern Main Ranges where greater syndepositional subsidence allowed for increased sediment accumulation (McLean, 1992; McLean and Mountjoy, 1993).

Third flooding—Nisku Winterburn Megacycle (T-R cycle IIc upper)

During Nisku time an extensive carbonate shelf formed around the eastern, southern, and western sides of the basin, except in the west-central Alberta subsurface (118°W and 53°N). These reefs developed on the platform margins and adjacent slopes (above the center of the underlying Ireton basin). Although this event was related to a sea-level rise, it was not as large nor as long as the first and second floodings. The Nisku represents a shorter interlude of carbonate sedimentation and local reef development during the final stages of Frasnian basin filling by siliciclastic sediments. In the Main Ranges of the Rocky Mountains west of the deep West Pembina basin, Fejer and Narbonne (1992) documented buildup interior cycles in the Grotto and Arcs members in the Fairholme reef complex south of the Columbia Icefield.

The inter-reef areas in the Jasper Basin were never completely filled during Frasnian time but were filled later by Sassenach siliciclastic sediments (Fig. 6). This occurred during a sea-level rise that followed a major and marked sea-level lowering at the end of the Frasnian epoch (Fig. 10). These sediments may represent the distal deposits of an orogenic (Antler?) clastic wedge.

Famennian—Tournaisian Cycles

The second major cycle is divided into at least two subcycles in most of the region: Sassenach-Palliser (IIe) and, particularly near the international border, uppermost Palliser/Exshaw (IIf) and Banff (Fig. 10). The lower Banff has also been interpreted as representing a subsequent regional transgression in part of the southern Canadian Rocky Mountains (Richards, 1989).

An early Famennian (T-R cycle IIe) deepening event, followed by a prolonged regression interrupted by transgressive pulses in middle and late Famennian time was interpreted by Johnson and others (1985) and Johnson and Sandberg (1988). The Sassenach siliciclastic sediments accumulated in basins outboard of the reef complexes and the eastern part of the basin that was filled earlier by Frasnian sediments (Figs. 6, 10), whereas the Palliser Formation accumulated on a widespread, west-facing, carbonate ramp primarily during this T-R IIe cycle (see also Sandberg and others, 1988; Johnson and others, 1991; Johnston and Chatterton, 1991).

The uppermost Palliser in the Bow Valley area and elsewhere may have been deposited during the early phase of T-R cycle IIf, which began in latest Famennian time (Lower *expansa* Zone). Termination of carbonate sedimentation could have resulted from drowning of the Palliser ramp below the euphotic zone of optimum growth (e.g., Schlager, 1981, 1989; Sarg, 1988), and/or from incursion of oxygen-depleted water during the significant deepening at this time.

The anaerobic-dysaerobic facies of the Exshaw Formation were initially deposited during this T-R cycle IIf deepening. Similar black shales, such as part of the Bakken Formation, Sappington Member (Three Forks Formation), Leatham Formation, and Leatham Member (Pilot Shale) were also initially deposited in western North America during this time (e.g., Johnson and Sandberg, 1988; Sandberg and others, 1988). Anoxic conditions in this region may have developed with the inundation of low-latitude, low-gradient, west-facing, epi- and pericontinental seas by oxygen-depleted water derived from the oceanic oxygen-minimum zone (OMZ) (Savoy, 1990, 1992; see also Witzke and Heckel, 1988). With a Famennian flooding event, shelf/epeiric sea depths may have been sufficient to maintain density-stratified conditions with oxygen-depleted water below the surface, wind-mixed layer. Increased area of the epicontinental and shelf seas could have promoted biologic productivity in this low-latitude setting and, possibly with upwelling, resulted in vertical and lateral expansion of the OMZ.

Deeper-water, black-shale deposition with coarser siliciclastic input (Fig. 10) occurred at least locally in the study region until early-middle? Tournaisian time. Although a major eustatic fall has been documented elsewhere across the Devonian-Mississippian boundary (e.g., Sandberg and others, 1983, 1988; Johnson and others, 1985; Ross and Ross, 1987a), no conclusive evidence of subaerial exposure at this time was observed in the area of this study (i.e., Bow Valley to northwestern Montana). Transgressive-regressive cycles in Europe and northwestern Africa in Middle to Late Devonian time are shown in Figure 11.

The lower Banff accumulated in an anaerobic to marginally aerobic, deep-water setting during much of Tournaisian time. Deposition of this unit has been interpreted as an early phase of a subsequent regional transgression and deepening in the Rocky Mountains south of 56°N (e.g., Richards, 1989) following deposition of the inferred, shallowing upward Exshaw siltstone (Bamber and others in Gordey and others, 1991). Near the international border, however, lithofacies relationships support a model of continued deep-water deposition beginning with the late Famennian (IIf) transgression and continuing to late Tournaisian time (see also Wendte and others, 1992). The mixed carbonate-siliciclastic lithofacies of the middle and upper Banff Formation record the return to aerobic, carbonate-dominated, shallower-water, ramp deposition by middle-late Tournaisian time. Thus, the occurrence of westward-thickening, deep-water (including anaerobic) lithofacies and quartzofeldspathic clastic units in the Exshaw and lower Banff, and the lack of evidence for subaerial exposure across the Devonian-Mississippian boundary, support an inferred period of significant subsidence in the region associated with Antler-age tectonism.

REGIONAL TECTONIC SUBSIDENCE

The Frasnian reef complexes and the Famennian Palliser Formation all thicken westward from the subsurface across the Foothills and Front Ranges to the Main Ranges (Fig. 3) and indicate that the Western Canada Sedimentary Basin subsided differentially. As noted, the Frasnian reef complexes and Palliser Formation in the Main Ranges are nearly three times as thick as correlative strata in the eastern Front Ranges and the adjacent subsurface. The Banff Formation also generally thickens westward in the study area. These relationships indicate that the western part of the Western Canada Sedimentary Basin subsided differentially more in the west than beneath the Plains. In contrast, during early

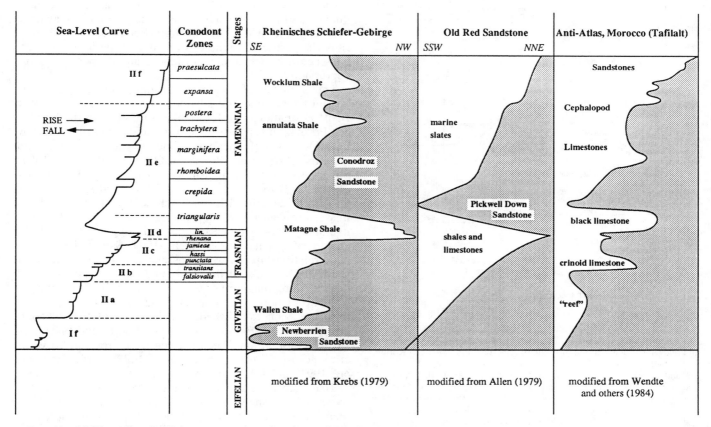

Fig. 11—Middle to Late Devonian transgressive-regressive cycles in Europe and Morocco compared to our revised curve for western Canada (modified from Buggisch, 1991).

Famennian (Sassenach) time apparently little subsidence took place since the Frasnian reef complexes are either not overlapped by (as at Ancient Wall) or are only covered by a few meters of Sassenach strata (e.g., Miette; Mountjoy, 1980, 1987).

A marked increase in the subsidence rate from late Famennian to early Visean time has been reported in the Antler foreland basin in western Montana and east-central Idaho (Dorobek and others, 1991a, b), as well as to the north in the southeastern Canadian Cordillera (Prophet Trough of Richards, 1989; Bond and Kominz, 1991; Kominz and Bond, 1991; Fig. 12). Quantitative reductions of cumulative stratigraphic thickness curves for the southern Canadian Cordillera (and for western Montana and east-central Idaho) suggest that even though the latest Devonian sea-level rise was large, a significant tectonic component of subsidence was superimposed over any eustatic sea-level rise (Fig. 12; Bond and others, 1989; Figs. 6, 7 of Bond and Kominz, 1991; Dorobek and others, 1991a, b). This deepening was interpreted to be due to convergence by Richards (1989) and Dorobek and others (1991a, b). Thus, some variations in crustal loading, along with other poorly understood crustal processes, may have been responsible for some of the larger-scale cycles and depositional sequences present in the Late Devonian to Early Mississippian succession of the southern Canadian Rocky Mountains.

PROVENANCE OF CLASTIC ROCKS

In addition to significant regional subsidence, evidence of the Antler-age orogenic activity comes from inferred, westerly derived, clastic sediments in the Exshaw and lower Banff Formations (e.g., Richards, 1989; Savoy, 1992), and possibly the older Sassenach Formation (Mountjoy, 1980; Geldsetzer and Upitis, 1993; and this paper). Although the sources are uncertain, the westward thickening, general southwestward coarsening of the Sassenach Formation and the presence of feldspar suggest at least a partial western or southwestern source area. This unit may have been deposited in the distal part of the foreland basin reported to the south in Montana (Fig. 1), and according to H. Geldsetzer (pers. commun., 1992), grain size appears to increase southward from Banff toward the international border area.

Following deposition of the Sassenach Formation, the Palliser-Wabamun carbonate ramp was established in the Western Canada Sedimentary Basin. The lack of extensive siliciclastic sediments in the Palliser carbonates and the great thickness of the carbonates suggest that tectonism may have been episodic or that any western-derived siliciclastic detritus was restricted to the western side of the foredeep, perhaps because of tectonic subsidence and deepening of this depression.

The presence of quartzofeldspathic (including sanidine bearing) sandstone and siltstone in the Exshaw and lower Banff Formations and feldspathic sandstone in the Sassenach Formation supports a western, active-margin setting by Late Devonian (Famennian) time (see also Richards and Higgins, 1988; Richards, 1989; Gordey and others, 1991; Savoy, 1992). Volcanic ash beds (dated as ca. 365 Ma (U-Pb) by G. Ross, 1992, pers. commun.) and sanidine have also been recognized in the Exshaw at different localities in western Canada (Macqueen and Sandberg, 1970; Lethiers and others, 1986; Richards and Higgins, 1988; Richards, 1989; Richards and others, 1991; Savoy, 1992). Genetic relationships between Sassenach, Exshaw, and lower Banff clastic rocks of this study and Paleozoic plutons and detritus in regions farther west such as in the Kootenay arc area (e.g., Gabrielse and others, 1982; Okulitch, 1985, 1993) have not been established. On the basis of limited compositional data, Exshaw and lower Banff sandstones and siltstones might have been derived from erosion of a western volcanic arc (see also Richards, 1989; Gordey and others, 1991) and deposited in the distal parts of a clastic wedge.

There are three potential sources for siliciclastics in the Sassenach, Exshaw, and Banff Formations: (1) the Canadian Shield, (2) Ellesmerian fold belt in the Arctic (same source as for the Frasnian basin fill), and (3) western sourced Antler clastics. A Canadian Shield source essentially can

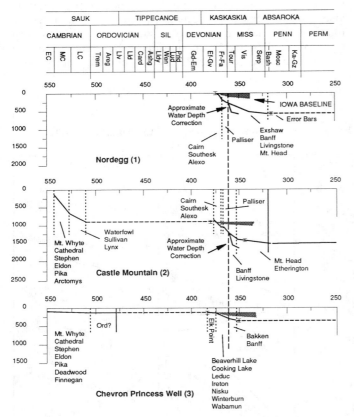

FIG. 12—First reduction subsidence curves for two Devonian-Mississippian outcrop sections in the southern Canadian Rocky Mountains and one subsurface location in southern Alberta (from Bond and Kominz, 1991). Curves were constructed by removing the effects of sediment loading and compaction, producing curves that approximate the subsidence that would occur if the basin had subsided in water without deposition of sediment. Broken horizontal lines represent unconformities (from Bond and Kominz, 1991). Nordegg (northernmost section, Foothills; Fig. 2) and Castle Mountain (Rocky Mountains) are in western to southwestern Alberta. The Chevron Princess well is in southeastern to south-central Alberta.

be ruled out because most of the shield was covered in Devonian time by Paleozoic strata and because there is little evidence of eastward thickening of units or of siliciclastic detritus of definite shield origin in channels in the Graminia, Wabamun (Palliser equivalent), Bakken, and Lodgepole Formations. An Arctic source also seems unlikely because the Frasnian basin fill lacks fine-grained and feldspathic sands, suggesting that such sands were not transported this far south. A small part of the Peace River Arch was exposed at this time, but during Sassenach time it would not have supplied sufficient sediment to account for the 200 m of accumulated sediment. The Peace River Arch was covered by uppermost Wabamun strata and was not a source for the Exshaw and Banff clastics. Given the data available, a western provenance region may be the most likely candidate for the quartzo-feldspathic clastic units in these formations.

DISCUSSION—EVIDENCE FOR ANTLER-AGE OROGENESIS

There is little direct evidence of middle to late Paleozoic orogenesis, its causes, and tectonic control of subsidence in the southern Canadian Rocky Mountains, although episodes of tectonism have been recorded from southeastern British Columbia to the northern Yukon Territory in Canada and from Idaho to Nevada in the United States (see below for references). Rubin and others (1991) proposed that a magmatic arc system (volcano-plutonic complexes) existed along much of the length of the Cordilleran continental margin in middle Paleozoic time. In addition, Smith and others (1993) suggested, on the basis of regional tectonic analyses, that the margin from Nevada to the Yukon Territory may have undergone contractional, Antler-age orogenesis that involved emplacement of deep-water, sedimentary, and mafic volcanic rocks over the continental margin during Late Devonian to Mississippian time.

The convergent Antler orogeny was originally defined in Nevada, extended to central Idaho, and involved the termination of passive continental-margin sedimentation in Late Devonian—Early Mississippian time (e.g., Roberts and others, 1958; Roberts and Thomasson, 1964; Nilsen, 1977; Nilsen and Stewart; 1980; Johnson and Pendergast, 1981; Speed and Sleep, 1982). The Antler orogenic belt consists of imbricated, deep-water Cambrian to Upper Devonian (and locally, Mississippian) sedimentary rocks (Roberts Mountains allochthon and northern equivalents) that were emplaced onto the edge of the continental shelf along the Roberts Mountains thrust (Roberts and others, 1958; Kay and Crawford, 1964; Turner and others, 1989; Burchfiel and Royden, 1991). The Antler orogeny is generally interpreted as an episode of deformation involving arc-continent collision. In addition to displacement, principal events include marked increase in the rate of subsidence, the development of the Antler foreland basin, and the migration of a flexural forebulge (e.g., Poole, 1974; Dickinson, 1977; Poole and Sandberg, 1977; Johnson and Pendergast, 1981; Speed and Sleep, 1982; Dickinson and others, 1983; Sandberg and others, 1983; Burchfiel and Royden, 1991; Goebel, 1991). On the basis of structural and stratigraphic evidence, Antler deformation and motion of the allochthon may have occurred between Late Devonian (Frasnian) and Kinderhookian or Osagean (Early Mississippian) time (e.g., Johnson and Pendergast, 1981; Murphy and others, 1984; Goebel, 1991).

The nature and timing of tectonism in the southern Canadian Cordillera, however, have yet to be clearly identified. Pre-Mississippian shortening and deformation may have occurred in the Kootenay arc to the west and southwest of the study region in southeastern British Columbia and northeasternmost Washington (e.g., Read and Wheeler, 1976; Klepacki, 1985; Klepacki and Wheeler, 1985; Gehrels and Smith, 1987; Smith and Gehrels, 1992) although the timing and nature of deformation are not well constrained. In southeastern British Columbia, Middle Paleozoic igneous activity has been noted to the west and north of the Kootenay arc (e.g., Okulitch, 1985), and local outcrops of Middle to Upper Devonian strata in the Purcell Mountains to the east of the arc have been interpreted as foreland basin deposits (Root, 1987, 1992, pers. commun., 1992). Contractional deformation was also proposed by White (1959) in the Cariboo Mountains of central British Columbia and termed the Cariboo orogeny, whereas Struik (1981, 1986) interpreted this region as an area of extensional tectonism. Struik (1986) further interpreted the Exshaw Formation of the adjacent Rocky Mountains to the east as a distal part of a graben-edge uplift sequence. Smith and others (1993), however, have suggested that the Guyet conglomerate in the Cariboo Mountains could represent a proximal foreland assemblage as proposed by Campbell and others (1973), with the Exshaw Formation representing distal foreland deposits. Richards (pers. commun., 1993) also interprets the Cariboo orogeny as a contractional event.

Episodes of Devono-Mississippian tectonism have also been reported in the northern Cordillera of northern British Columbia and the Yukon Territory. Gordey and others (1987) proposed that Devonian-Mississippian coarse clastic strata (Earn Group) in the Selwyn Basin and adjacent areas were deposited in graben structures associated with regional extension or strike-slip faulting. Smith and others (1993) suggested that extensional features could have formed due to flexural extension during foreland deformation rather than by rifting or transtension.

On the basis of indirect evidence, including sedimentation and thickness patterns, tectonic loading of the continental lithosphere in conjunction with other tectonic processes (see below) may be inferred in the southern Canadian Cordillera, thus supporting a model of contractional Antler-age orogenic activity in this area (see also Richards, 1989). High rates of Late Devonian subsidence in the southern Canadian Rocky Mountains calculated by Bond and Kominz (1991) were coincident with the formation of the Antler basin in the western United States beginning in Frasnian time and continuing into Tournaisian time (Fig. 12; see also Richards, 1989).

In addition, the presence of sandstones of Famennian and Tournaisian age in the southern Canadian Rocky Mountains that are interpreted as westerly derived in this and other studies suggests that a land area was uplifted west and outboard of the cratonic margin (e.g., Mountjoy, 1980; Morrow and Geldsetzer, 1988; Richards, 1989; Savoy, 1992), and that this land area began to be eroded by late Frasnian(?) to early Famennian time. Sedimentary sequences from foreland basins commonly consist of siliciclastic units, such as, for example, the Cretaceous-Lower Tertiary succession of western Canada (e.g., Cant and Stockmal, 1989; Leckie and Smith, 1992) or the Tertiary Eurekan orogeny (De Paor and others, 1989). The stratigraphic succession of the southern Canadian Rocky Mountains and adjacent Montana bears some similarities with the sedimentary record of the Devonian Acadian orogenic belt in the eastern United States and adjacent Canada. There the lateral facies relationships are somewhat better displayed between carbonate shelves and ramps of the continental margin and the associated siliciclastic sediments of the foreland clastic wedge (e.g., Ettensohn, 1985; Faill, 1985). The Devonian, dark-gray to black basinal shales of the Catskill Delta complex locally grade westward into a carbonate ramp or platform succession that may be similar to the distal part of the Palliser ramp in southwestern Canada. The Famennian and Tournaisian quartzofeldspathic siltstones and sandstones of the southern Canadian Rocky Mountains at least superficially resemble portions of the Catskill clastic-wedge lithofacies successions.

It has been suggested that in widely disparate geographic and tectonic settings throughout North America the onset of increased tectonic subsidence in Frasnian time is roughly synchronous irrespective of the local tectonic setting (Bond and Kominz, 1991). Moreover, this synchronous event corresponds in time with the onset of an eustatic sea-level rise (Lenz, 1982; Johnson and others, 1985).

Foreland basins are controlled by the physical properties of the crust, and thus the subsidence of foreland basins is restricted to a 500- to 700-km width governed by the flexural half-wavelengths of continuous and broken continental plates (Stockmal and others, 1992). The restored width of the Devonian sequence in the Canadian Rocky and Purcell mountains would be more than 500 km (Price and Mountjoy, 1970; Price, 1981, 1986). Clearly subsidence has extended well into the interior of the continent at various times in the geologic past (Bond and Kominz, 1984, 1991; Mitrovica and others, 1989; Kominz and Bond, 1991; Gurnis, 1993a). Unfortunately, the mechanisms and processes that control regional subsidence and uplift of large areas of the crust are not well understood. Mitrovica and others (1989) showed that chronostratigraphic sequences of Late Cretaceous and Early Tertiary age in western North America are regionally tilted and that this tilting probably led to sedimentation in the interior of the craton. This large-scale tilting overlaps in time and space and coincides with the formation of foreland basins, making it difficult to separate these two mechanisms in some areas.

The regional differential tilt observed in the Upper Devonian and Lower Carboniferous sequences of western Canada may be due to a combination of foreland basin development and more regional-scale epeirogenic tilting based on evidence that subsidence has affected large parts of North America and that this subsidence was greater in most regions relative to the Iowa baseline reported by Bond and Kominz (1991) and Kominz and Bond (1991). Until we analyze the tilt of the different Devonian and Carboniferous sequences, it is not possible to distinguish these processes.

Crustal researchers have attempted to explain and model crustal subsidence in a number of different ways. Mitrovica and others (1989) showed, using numerical analysis models, that large-scale subsidence over widths of 1000 km or more will occur when subduction angles are less than 45°. Gurnis (1993a) observed that the continental crust behind oceanic trenches and above subduction slabs is 400 to 500 m lower than the global average, suggesting that the stress of the mantle flow pulls the surface of the Earth downwards above heavy slabs. Other mantle convection controls include changes in ridge spreading rates, the motion of continents, and changes in dynamic topography by supercontinent insulation (Gurnis, 1990, 1992, 1993a, b). A strong pulse of compressive, intraplate stress may have initiated tectonic subsidence in Frasnian time (Bond and Kominz, 1991), which would explain the plate-wide synchroneity of relatively rapid Late Devonian subsidence. Kominz and Bond (1991) and Bond and Kominz (1991) proposed that compressive stresses may have developed during the "capture" of the Laurentian plate in a cool region of mantle downwelling. Alternatively, these compressive stresses may have developed during Devono-Mississippian plate convergence, initiated by middle Paleozoic orogenesis around the perimeter of Laurentia (Howell, 1989; Howell and van der Pluijm, 1990; Kolata and Nelson, 1990; Bond and Kominz, 1991).

Cathles and Hallam (1991) hypothesized that many rapid changes in sea level can be caused by stress-induced changes in plate density that can propagate across large lithospheric plates in less than 30,000 years. They suggested that plate elevation changes of between 50 and 200 m occurring in less that one million years could result from the creation of new rifts or from increased plate compression during continental collisions. During Late Devonian time, the North American plate was undergoing collisions on three sides (Catskill, Antler, and Ellesmerian orogenies). These collisions would have created stress-induced changes in plate density that might have been at least partly responsible for the overall transgressive nature of the Devonian to Lower Carboniferous sequences. Also the dip of subduction slabs around the North American plate may have controlled to some extent the subsidence of the continental margins.

We propose that marked acceleration of subsidence in southwestern Canada in Late Devonian time may be attributable, at least in part, to crustal loading during the incipient stages of the Antler-age orogenesis. Our subsidence data in conjunction with the presence of Famennian to Tournaisian, westerly derived clastic deposits support the presence of a tectonically uplifted source region to the west-southwest(?). Clearly our assessments are based on limited data. More extensive analyses of the relationships between middle Paleozoic sedimentation and tectonic setting are required in the Canadian Rocky Mountains, particularly where only regional scale studies have been conducted on many of the stratigraphic units. In addition, crustal models must be developed to explain the subsidence over the large width of continental crust noted in this study. We hope that this paper will encourage others to pursue this complex topic.

SUMMARY AND CONCLUSIONS

This paper outlines the Devonian-Early Mississippian (Early Carboniferous) history and evolution of the continental margin in the southern Canadian Rocky Mountains of southern British Columbia and Alberta, and adjacent Montana.

1. The westward thickening of most Upper Devonian-Lower Mississippian units in this region supports a model of regional differential subsidence interpreted to be related to Antler-age tectonic and sedimentary loading and unknown crustal processes along the foreland basin in this part of western North America. The subsidence began in this region by Frasnian(?) time and continued into Tournaisian time.
2. Quartzofeldspathic clastic rocks in the Exshaw, lower Banff, and Sassenach(?) Formations reflect periodic influx of westerly-derived, orogenic-sourced detritus and support the presence of a foreland basin and a model of convergent tectonism along adjacent parts of the margin. These siliciclastic deposits are interpreted to represent the distal parts of a clastic wedge built out from the western side of a foreland basin.
3. The megacycles described in this paper are part of two and perhaps three major cycles: (a) Givetian to late Frasnian T-R IIa to IIc (lower) of Johnson and others (1985) and Johnson and Sandberg (1988) representing the main reef-building phases, followed by two carbonate banks (Arcs-Nisku, Simla-Ronde; IIc (upper) and IId). (b) Early Famennian to end of Tournaisian—Sassenach, Palliser, Exshaw, and Banff Formations.
4. The second of these major cycles is divisible into at least two subcycles: a lower Sassenach-Palliser and an upper uppermost Palliser/Exshaw-Banff. The uppermost part of the Palliser Formation (upper Costigan) in the Bow Valley-Jasper region records the early stage of the major regional transgression of the Exshaw-Banff cycle. North of the international border area (e.g., Bow Valley and other regions to the north of this study), the basal Banff may represent the transgressive phase of a subsequent transgressive-regressive cycle.
5. The transgression and deepening phases of the Exshaw-Banff subcycle are, in part, related to the influence of accelerated, Antler-age tectonic subsidence and more regional crustal subsidence that submerged the Palliser ramp.

ACKNOWLEDGMENTS

Lauret Savoy gives special thanks to Anita Harris (USGS) and Cathryn Newton. She also thanks the following people for critical advice and assistance: Barry Richards, Kevin Root, John Brady, Helmut Geldsetzer, and Andrew Okulitch. Eric Mountjoy thanks the Geological Survey of Canada (since 1957), the Natural Science and Engineering Research Council of Canada (NSERC), David McLean, H. Geldsetzer, M. Mallamo, R. A. Price, Kevin Root, Paul Hoffman, and numerous other geologists who have freely shared their ideas and comments. Mountjoy especially thanks David McLean for discussions about sequence stratigraphy and constructing and interpreting sea-level curves, as well as for his part in revising the Frasnian portion of the curve (McLean and Mountjoy, 1993). Many other colleagues generously provided assistance and advice in various phases of this study. We are grateful to Kevin Root, Barry Richards, Moira Smith, and Gerry Ross, whose critical and constructive reviews helped us improve this paper. Savoy acknowledges the Donors of The Petroleum Research Fund, administered by the American Chemical Society, for partial support of this research, as well as the U. S. Geological Survey Graduate Internship program, Geological Society of America research grant, and Shell Canada.

REFERENCES

ALLEN, J. R. L., 1979, Old Red Sandstone facies in external basins, with particular reference to southern Britain: London, Special Papers in Palaeontology No. 23, p. 65–80.

BAMBER, E. W., MACQUEEN, R. W., AND RICHARDS, B. C., 1984, Facies relationships at the Mississippian carbonate platform margin, western Canada: Ninth International Congress of Carboniferous Stratigraphy and Geology, 1979, Compte Rendu, v. 3, p. 461–478.

BASSETT, H. G. AND STOUT, J. G., 1967, Devonian of western Canada, in Oswald, D. H., ed., International Symposium on the Devonian System: Calgary, Alberta Society of Petroleum Geologists, v. 1, p. 717–752.

BOND, G. C. AND KOMINZ, M. A., 1984, Construction of tectonic subsidence curves for the early Paleozoic miogeocline, southern Canadian Rocky Mountains: Implications for subsidence mechanisms, age of breakup, and crustal thinning: Geological Society of America Bulletin, v. 95, p. 155–173.

BOND, G. C. AND KOMINZ, M. A., 1991, Disentangling middle Paleozoic sea level and tectonic events in cratonic margins and cratonic basins of North America: Journal of Geophysical Research, v. 96, p. 6619–6639.

BOND, G. C., KOMINZ, M. A., STECKLER, M. S., AND GROTZINGER, J. P., 1989, Role of thermal subsidence, flexure, and eustasy in the evolution of early Paleozoic passive-margin carbonate platforms, in Crevello, P. D., Wilson, J. L., Sarg, J. F., and Read, J. F., eds., Controls on Carbonate Platform and Basin Development: Tulsa, Society of Economic Paleontologists and Mineralogists Special Publication 44, p. 39–61.

BUGGISCH, W., 1991, The global Frasnian-Famennian "Kellwasser Event": Geologische Rundschau, v. 80 p. 49–72.

BURCHFIEL, B. C. AND ROYDEN, L. H., 1991, Antler orogeny: A Mediterranean-type orogeny: Geology, v. 19, p. 66–69.

CAMPBELL, R. B., MOUNTJOY, E. W., AND YOUNG, F. G., 1973, Geology of the McBride map area, British Columbia: Ottawa, Geological Survey of Canada Paper 72-35, 104 p.

CANT, D. J. AND STOCKMAL, G. S., 1989, The Alberta foreland basin: Relationship between stratigraphy and Cordilleran terrane-accretion events: Canadian Journal of Earth Sciences, v. 26, 1964–1975.

CATHLES, L. M. AND HALLAM, A., 1991, Stress-induced changes in plate density, Vail sequences, epeirogeny, and short-lived global sea-level fluctuations: Tectonics, v. 10, p. 659–671.

CHATELLIER, J.-Y., 1988, Carboniferous carbonate ramp, the Banff Formation, Alberta, Canada: Bulletin des Centres Recherches Exploration-Production Elf-Aquitaine, v. 12, p. 569–599.

COPPOLD, M. P., 1976, Buildup to basin transition at the Ancient Wall complex (Upper Devonian), Alberta: Bulletin of Canadian Petroleum Geology, v. 24, p. 154–192.

DE PAOR, D. G., BRADLEY, D. C., EISENSTADT, G., AND PHILLIPS, S. M., 1989, The Arctic Eurekan orogen: A most unusual fold-and-thrust belt: Geological Society of America Bulletin, v. 101, 952–967.

DICKINSON, W. R., 1977, Paleozoic plate tectonics and the evolution of the Cordilleran continental margin, in Stewart, J. H., Stevens, C. H., and Fritsche, A. E., eds., Paleozoic Paleogeography of the Western United States: Los Angeles, Society of Economic Paleontologists and Mineralogists Pacific Coast Paleogeography Symposium 1, p. 137–155.

DICKINSON, W. R., HARBAUGH, D. W., SALLER, A. H., HELLER, P. L., AND SNYDER, W. S., 1983, Detrital modes of upper Paleozoic sandstones derived from the Antler orogen in Nevada: Implications for the nature of the Antler orogeny: American Journal of Science, v. 283, p. 481–509.

DOROBEK, S. L., REID, S. K., AND ELRICK, M., 1991a, Antler foreland stratigraphy of Montana and Idaho: The stratigraphic record of eustatic fluctuations and episodic tectonic events, in Cooper, J. D., and Stevens, C. H., eds., Paleozoic Paleogeography of the Western United States-II: Los Angeles, Pacific Section, SEPM (Society for Sedimentary Geology), v. 67, p. 487–508.

DOROBEK, S. L., REID, S. K., ELRICK, M., BOND, G. C., AND KOMINZ, M. A., 1991b, Subsidence across the Antler foreland of Montana and Idaho: Tectonic versus eustatic effects, in Franseen, E. K., Watney, W. L., Kendall, C. G. St. C., and Ross, W., eds., Sedimentary Modeling: Computer Simulation and Methods for Improved Parameter Definition: Kansas Geological Society Bulletin 233, p. 231–251.

EISBACHER, G. H., 1983, Devonian-Mississippian sinistral transcurrent faulting along the cratonic margin of western North America: A hypothesis: Geology, v. 11, p. 7–10.

ETTENSOHN, F. R., 1985, The Catskill Delta complex and the Acadian orogeny: A model, in Woodrow, D. L., and Sevon, W. D., eds., The Catskill Delta: Boulder, Geological Society of America Special Paper 201, p. 39–49.

FAILL, R. T., 1985, The Acadian orogeny and the Catskill Delta, in Woodrow, D. L., and Sevon, W. D., eds., The Catskill Delta: Boulder, Geological Society of America Special Paper 201, p. 15–37.

FEJER, P. E. AND NARBONNE, G. M., 1992, Controls on Upper Devonian metre-scale carbonate cyclicity, Icefall Brook, southeast British Columbia, Canada: Bulletin of Canadian Petroleum Geology, v. 40, p. 363–380.

FISCHBUCH, N. R., 1968, Stratigraphy, Devonian Swan Hills reef complexes of central Alberta: Bulletin of Canadian Petroleum Geology, v. 16, p. 444–556.

GABRIELSE, H., LOVERIDGE, W. D., SULLIVAN, R. W., AND STEVENS, R. D., 1982, U-Pb measurements on zircon indicate Paleozoic plutonism in the Omineca crystalline belt, north-central British Columbia: Ottawa, Geological Survey of Canada Paper 82-1C, p. 139–146.

GEHRELS, G. E. AND SMITH, M. T., 1987, "Antler" allochthon in the Kootenay arc?: Geology, v. 15, p. 769–770.

GELDSETZER, H. H. J., 1982, Depositional history of the Devonian succession in the Rocky Mountains southwest of the Peace River Arch: Ottawa, Geological Survey of Canada Paper 82-C, p. 55–64.

GELDSETZER, H. H. J., JAMES, N. P., AND TEBBUTT, G. E., 1989, Reefs—Canada and Adjacent Areas: Calgary, Canadian Society of Petroleum Geologists Memoir 13, 775 p.

GELDSETZER, H. H. J. AND MOUNTJOY, E. W., 1992, Upper Devonian platform reefs and inter-platform basins, Canmore to Jasper, Alberta: Tulsa, Society for Sedimentary Geology (SEPM) Fieldtrip Guidebook No. 23, 61 p.

GELDSETZER, H. H. J. AND UPITIS, G. W., 1993, The Sassenach Formation, indirect evidence of Late Devonian tectonism: Geological Association of Canada/Mineralogical Association of Canada Joint Annual Meeting Programs and Abstracts, p. A34.

GOEBEL, K. A., 1991, Paleogeographic setting of Late Devonian to Early Mississippian transition from passive to collisional margin, Antler foreland, eastern Nevada and western Utah, in Cooper, J. D., and Stevens, C. H., eds., Paleozoic Paleogeography of the Western United States-II: Los Angeles, Pacific Section, SEPM (Society for Sedimentary Geology), v. 67, p. 401–418.

GORDEY, S. P., 1988, Devono-Mississippian clastic sedimentation and tectonism in the Canadian Cordilleran miogeocline, in McMillan, N. J., Embry, A. F., and Glass, D. J., eds., Devonian of the World, v. II: Calgary, Canadian Society of Petroleum Geologists Memoir 14, p. 1–14.

GORDEY, S. P., ABBOTT, J. G., TEMPELMAN-KLUIT, D. J., AND GABRIELSE, H., 1987, "Antler" clastics in the Canadian Cordillera: Geology, v. 15, p. 103–107.

GORDEY, S. P., GELDSETZER, H. H. J., MORROW, D. W., BAMBER, E. W., HENDERSON, C. M., RICHARDS, B. C., MCGUGAN A., GIBSON, D. W., AND POULTON, T. P., 1991, Upper Devonian to Middle Jurassic assemblages, in Gabrielse, H. and Yorath, C. J., eds., Geology of the Cordilleran Orogen in Canada: Ottawa, Geological Survey of Canada, Geology of Canada 4, p. 219–327.

GURNIS, M., 1990, Ridge spreading, subduction, and sea-level fluctuations: Science, v. 250, p. 970–972.

GURNIS, M., 1992, Rapid continental subsidence following the initiation and evolution of subduction: Science, v. 255, p. 1556–1558.

GURNIS, M., 1993a, Depressed continental hypsometry behind oceanic trenches: A clue to subduction controls on sea-level change: Geology, v. 21, p. 29–32.

GURNIS, M., 1993b, Phanerozoic marine inundation of continents driven by dynamic topography above subducting slabs: Nature, v. 364, p. 589–593.

HALBERTSMA, H. L. AND MEIJER DREES, N. C., 1987, Wabamun limestone sequences in north-central Alberta: Calgary, Canadian Society of Petroleum Geologists, 13th Core Conference and Display, 2nd International Symposium on the Devonian System, p. 21–37.

HARKER, P. AND MCLAREN, D. J., 1958, The Devonian-Mississippian boundary in the Alberta Rocky Mountains, in Goodman, A. J., ed., Jurassic and Carboniferous of Western Canada: Tulsa, American Association of Petroleum Geologists John Andrew Allan Memorial Volume, p. 244–259.

HIGGINS, A. C., RICHARDS, B. C., AND HENDERSON, C. M., 1991, Conodont biostratigraphy and paleoecology of the uppermost Devonian and Carboniferous of the Western Canada Sedimentary Basin, in Orchard, M. J., and McCracken, A. D., eds., Ordovician to Triassic Conodont Paleontology of the Canadian Cordillera: Geological Survey of Canada Bulletin 417, p. 215–251.

HOWELL, P. D., 1989, Epeirogeny, cratonic basin reactivation and stress-induced weakening of the lower crust (abs.): Geological Society of America, Abstracts and Programs, v. 21, p. A81.

HOWELL, P. D. AND VAN DER PLUIJM, B. A., 1990, Early history of the Michigan basin: Subsidence and Appalachian tectonics: Geology, v. 18, p. 1195–1198.

JOHNSON, J. G., KLAPPER, G., AND SANDBERG, C. A., 1985, Devonian eustatic fluctuations in Euramerica: Geological Society of America Bulletin, v. 96, p. 567–587.

JOHNSON, J. G. AND PENDERGAST, A., 1981, Timing and mode of emplacement of the Roberts Mountain allochthon, Antler orogeny: Geological Society of America Bulletin, v. 92, p. 648–658.

JOHNSON, J. G. AND SANDBERG, C. A., 1988, Devonian eustatic events in the western United States and their biostratigraphic responses, in McMillan, N. J., Embry, A. F., and Glass, D. J., eds., Devonian of the World, v. III: Calgary, Canadian Society of Petroleum Geologists Memoir 14, p. 9–22.

JOHNSON, J. G., SANDBERG, C. A., AND POOLE, F. G., 1991 Devonian lithofacies of western United States, in Cooper, J. D., and Stevens, C. H., eds., Paleozoic Paleogeography of the Western United States-II: Los Angeles, Pacific Section, SEPM (Society for Sedimentary Geology), v. 67, p. 83–105.

JOHNSTON, D. I. AND CHATTERTON, B. D. E., 1991, Famennian conodont biostratigraphy of the Palliser Formation, Rocky Mountains, Alberta and British Columbia, Canada in Orchard, M. J., and McCracken, A. D., eds. Ordovician to Triassic Conodont Paleontology of the Canadian Cordillera: Geological Survey of Canada Bulletin 417, p. 163–183.

KAY, M. AND CRAWFORD, J. P., 1964, Paleozoic facies from the miogeosynclinal to the eugeosynclinal belt in thrust slices, central Nevada: Geological Society of America Bulletin, v. 75, p. 425–454.

KLAPPER, G. AND LANE, H. R., 1988, Frasnian (Upper Devonian) conodont sequence at Luscar Mountain and Mount Haultain, Alberta Rocky Mountains, in McMillan, N. J., Embry, A. F., and Glass, D. J., eds., Devonian of the World, v. III: Calgary, Canadian Society of Petroleum Geologists Memoir 14, p. 469–478.

KLEPACKI, D. W., 1985, Stratigraphy and structural geology of the Goat Range area, southeastern British Columbia: Unpublished Ph.D. Dissertation, Massachusetts Institute of Technology, Cambridge, 268 p.

KLEPACKI, D. W. AND WHEELER, J. O., 1985, Stratigraphic and structural relations of the Milford, Kaslo, and Slocan Groups, Goat Range, Lardeau and Nelson map areas, British Columbia: Ottawa, Geological Survey of Canada Paper 85-1A, p. 277–286.

KLOVAN, J. E., 1964, Facies analysis of the Redwater reef complex, Alberta, Canada: Bulletin of Canadian Petroleum Geology, v. 12, p. 1–100.

KOLATA, D. R. AND NELSON, J. W., 1990, Tectonic history of the Illinois Basin (abs.): American Association of Petroleum Geologists Bulletin, v. 74, p. 696.

KOMINZ, M. A. AND BOND, G. C., 1991, Unusually large subsidence and sea-level events during middle Paleozoic time: New evidence supporting mantle convection models for supercontinent assembly: Geology, v. 19, p. 56–60.

KREBS, W., 1979, Devonian basinal facies: London, Special Papers in Palaeontology 23, p. 125–139.

LECKIE, D. A. AND SMITH, D. G., 1992, Regional setting, evolution, and depositional cycles of the Western Canada Foreland basin, in Macqueen, R. W., and Leckie, D. A., eds., Western Canada Foreland Basin: Tulsa, American Association of Petroleum Geologists Memoir 55, p. 9–30.

LEECH, G. B., 1958, Fernie map-area, west half, British Columbia: Ottawa, Geological Survey of Canada Paper 58-10, 40 p.

LEECH, G. B., 1979, Kananaskis Lakes, west half: Ottawa, Geological Survey of Canada Open File Map 634.

LENZ, A. C., 1982, Ordovician to Devonian sea-level changes in western and northern Canada: Canadian Journal of Earth Sciences, v. 19, p. 1919–1932.

LETHIERS, F., BRAUN, W. K., CRASQUIN, S., AND MANSY, J.-L., 1986, The Strunian event in western Canada with reference to ostracode assemblages: Annales de la Societe Geologique de Belgique, v. 109, p. 149–157.

MACQUEEN, R. W. AND SANDBERG, C. A., 1970, Stratigraphy, age, and interregional correlation of the Exshaw Formation, Alberta Rocky Mountains: Bulletin of Canadian Petroleum Geologists, v. 18, p. 32–66.

MALLAMO, M. P. AND GELDSETZER, H. H. J., 1991, The western margin of the Upper Devonian Fairholme reef complex, Banff-Kananaskis area, southwestern Alberta: Ottawa, Geological Survey of Canada Paper 91-1B, p. 59–69.

MCLAREN, D. J., 1962, Middle and Early Upper Devonian rhynchonellid brachiopods from western Canada: Geological Survey of Canada Bulletin 86, 122 p.

MCLAREN, D. J., 1982, Frasnian-Famennian extinctions, in Silver, L. T., and Schultz, P. H., eds., Geological Implications of Large Asteroids and Comets on the Earth: Boulder, Geological Society of America Special Paper 190, p. 477–484.

MCLAREN, D. J. AND MOUNTJOY, E. W., 1962, Alexo equivalents in the Jasper region, Alberta: Ottawa, Geological Survey of Canada Paper 62-23, 36 p.

MCLEAN, D. J., 1992, Upper Devonian buildup development in the southern Canadian Rocky Mountains: a sequence stratigraphic approach: Unpublished Ph.D. Dissertation, McGill University, Montreal, 240 p.

MCLEAN, D. J. AND MOUNTJOY, E. W., 1993, Upper Devonian buildup, margin and slope development in the southern Canadian Rocky Mountains: Geological Society of America Bulletin, v. 105, p. 1263–1283.

MCLEAN, D. J. AND MOUNTJOY, E. W., 1994, Allocyclic control on Late Devonian buildup development, southern Canadian Rocky Mountains: Journal of Sedimentary Research, Stratigraphy and Global Studies, v. B64, p. 326–340.

MCLEAN, R. A. AND PEDDER, A. E. H., 1984, Frasnian rugose corals of western Canada, part 1: Smithiphylum: Palaeontographica Abteilung A, Bd. 195, p. 133–173.

MCLEAN, R. A. AND PEDDER, A. E. H., 1987, Frasnian rugose corals of western Canada, part 2: Chonophyllidae and Kyphophyllidae: Palaeontographica Abteilung A, Bd. 185, p. 1–38.

MITROVICA, J. X., BEAUMONT, C., AND JARVIS, G. T., 1989, Tilting of continental interiors by the dynamical effects of subduction: Tectonics, v. 8, p. 1079–1094.

MOORE, P. F., 1988, Devonian reefs in Canada and some adjacent areas, in Geldsetzer, H. J. J., James, N. P., and Tebbutt, G. E., eds., Reefs—Canada and Adjacent Areas: Calgary, Canadian Society of Petroleum Geologists Memoir 13, p. 367–390.

MOORE, P. F., 1989, The Kaskaskia sequence: Reefs, platforms and foredeeps, the lower Kaskaskia sequence—Devonian, in Ricketts, B. D., ed., Western Canada Sedimentary Basin, A Case History: Calgary, Canadian Society of Petroleum Geologists, p. 139–164.

MORROW, D. W. AND GELDSETZER, H. H. J., 1988, Devonian of the eastern Canadian Cordillera, in McMillan, N. J., Embry, A. F., and Glass, D. J., eds., Devonian of the World, v. I: Calgary, Canadian Society of Petroleum Geologists Memoir 14, p. 85–121.

MOUNTJOY, E. W., 1956, The Exshaw Formation, Alberta: Transactions of Canadian Institute of Mining and Metallurgy, v. 59, p. 376–380.

MOUNTJOY, E. W., 1978, Upper Devonian reef trends and configuration of the western portion of the Alberta Basin, in McIlreath, I. A. and Jackson, P. C., eds., The Fairholme Carbonate Complex: Calgary, Canadian Society of Petroleum Geologists, Guidebook, p. 1–30.

MOUNTJOY, E. W., 1980, Some questions about the development of Upper Devonian carbonate buildups (reefs), Western Canada: Bulletin of Canadian Petroleum Geology, v. 28, p. 315–344.

MOUNTJOY, E. W., 1987, The Upper Devonian Ancient Wall reef complex, Jasper National Park, Alberta: Calgary, Field Excursion A5 Guidebook, Second International Symposium on the Devonian System, Canadian Society of Petroleum Geologists, 50 p.

MURPHY, M. A., POWER, J. D., AND JOHNSON, J. G., 1984, Evidence for Late Devonian movement within the Roberts Mountains allochthon, Roberts Mountains, Nevada: Geology, v. 12, p. 20–23.

NILSEN, T. H., 1977, Paleogeography of Mississippian turbidites in south-central Idaho, in Stewart, J. H., Stevens, C. H., and Fritsche, A. E., eds., Paleozoic Paleogeography of the Western United States: Los Angeles, Society of Economic Paleontologists and Mineralogists Pacific Coast Paleogeography Symposium 1, p. 275–300.

NILSEN, T. H. AND STEWART, J. H., 1980, The Antler Orogeny: Mid-Paleozoic tectonism western North America: Geology, v. 8, p. 298–302.

OKULITCH, A. V., 1985, Paleozoic plutonism in southeastern British Columbia: Canadian Journal of Earth Science, v. 22, p. 1409–1424.

OKULITCH, A. V., 1993, Paleozoic orogenesis in the southern Canadian Cordillera, British Columbia (abs.): An historical perspective: Geological Association of Canada/Mineralogical Association of Canada Joint Annual Meeting Programs and Abstracts, p. A79.

OLIVER, T. A. AND COWPER, N. W., 1963, Depositional environments of the Ireton Formation, central Alberta: Bulletin of Canadian Petroleum Geology, v. 11, p. 183–202.

ORCHARD, M. J., 1988, Conodonts from the Frasnian-Famennian boundary interval in western Canada, in McMillan, N. J., Embry, A. F., and Glass, D. J., eds., Devonian of the World, v. III: Calgary, Canadian Society of Petroleum Geologists Memoir 14, p. 35–52.

POOLE, F. G., 1974, Flysch deposits of Antler Foreland Basin, western United States, in Dickinson, W. R., ed., Tectonics and Sedimentation: Tulsa, Society of Economic Paleontologists and Mineralogists Special Publication 22, p. 58–82.

POOLE, F. G. AND SANDBERG, C. A., 1977, Mississippian paleogeography and tectonics of the western United States, in Stewart, J. H., Stevens, C. H., and Fritsche, A. E., eds., Paleozoic Paleogeography of the Western United States: Los Angeles, Society of Economic Paleontologists and Mineralogists Pacific Coast Paleogeography Symposium 1, p. 67–85.

POOLE, F. G. AND SANDBERG, C. A., 1991, Mississippian paleogeography and conodont biostratigraphy of the western United States, in Cooper, J. D., and Stevens, C. H., eds., Paleozoic Paleogeography of the Western United States-II: Los Angeles, SEPM (Pacific Section, Society for Sedimentary Geology), v. 67, p. 107–136.

POOLE, F. G., SANDBERG, C. A., AND BOUCOT, A. J., 1977, Silurian and Devonian paleogeography and tectonics of the western United States, in Stewart, J. H., Stevens, C. H., and Fritsche, A. E., eds., Paleozoic Paleogeography of the Western United States: Los Angeles, Society of Economic Paleontologists and Mineralogists Pacific Coast Paleogeography Symposium 1, p. 39–65.

PRICE, R. A., 1981, The Cordilleran foreland thrust and fold belt in the southern Canadian Rocky Mountains, in McClay, K. R., and Price, N. J., eds., Thrust and Nappe Tectonics: London, Geological Society of London Special Publication 9, p. 427–488.

PRICE, R. A., 1986, The southeastern Canadian Cordillera: Thrust faulting, tectonic wedging and delamination of the lithosphere: Journal of Structural Geology, v. 8, p. 239–254.

PRICE, R. A. AND MOUNTJOY, E. W., 1970, Geological structure of the Canadian Rockies between Bow and Athabasca rivers—A progress report, in Wheeler, J. O., ed., Structure of the Southern Canadian Cordillera: Ottawa, Geologic Association of Canada Special Publication 6, p. 7–25.

RAASCH, G. O., 1988, Famennian faunal zones in Western Canada: in McMillan, N. J., Embry, A. F., and Glass, D. J., eds., Devonian of the World, v. III: Calgary, Canadian Society of Petroleum Geologists Memoir 14, p. 619–632.

READ, P. B. AND WHEELER, J. O., 1976, Geology, Lardeau west half, British Columbia: Ottawa, Geological Survey of Canada Open File Map 432.

RICHARDS, B. C., 1989, Upper Kaskaskia Sequence: Uppermost Devonian and Lower Carboniferous, in Ricketts, B. D., ed., Western Canada Sedimentary Basin, A Case History: Calgary, Canadian Society of Petroleum Geologists, p. 165–201.

RICHARDS, B. C., HENDERSON, C. M., HIGGINS, A. C., JOHNSTON, D. I., MAMET, B. L., AND MEIJER DREES, N. C., 1991, The Upper Devonian (Famennian) and Lower Carboniferous (Tournaisian) at Jura Creek, southwestern Alberta, in Smith P. L., ed., A Field Guide to the Paleontology of Southwestern Canada: Geological Association of Canada, p. 34–81.

RICHARDS, B. C. AND HIGGINS, A. C., 1988, Devonian-Carboniferous boundary beds of the Palliser and Exshaw formations at Jura Creek, Rocky Mountains, southwestern Alberta, in McMillan, N. J., Embry, A. F., and Glass, D. J., eds., Devonian of the World, v. II: Calgary, Canadian Society of Petroleum Geologists Memoir 14, p. 399–412.

ROBERTS, R. J., HOTZ, P. E., GILLULY, J., AND FERGUSON, H. G., 1958, Paleozoic rocks of north central Nevada: American Association of Petroleum Geologists Bulletin, v. 42, p. 2813–2857.

ROBERTS, R. J. AND THOMASSON, M. R., 1964, Comparison of late Paleozoic depositional history of northern Nevada and central Idaho: Washington, D. C., United States Geological Survey Professional Paper 475-D, p. D1-D6.

ROOT, K. G., 1987, Geology of the Delphine Creek area, southeastern British Columbia: Implications for the Proterozoic and Paleozoic development of the Cordilleran divergent margin: Unpublished Ph.D. Dissertation, University of Calgary, Calgary, 446 p.

ROOT, K. G., 1992, Middle Devonian thrust belt and foreland basin development, southeastern British Columbia (abs.): 1992 American Association for Petroleum Geologists Annual Convention, p. 111.

ROSS, C. A. AND ROSS, J. R. P., 1985, Late Paleozoic depositional sequences are synchronous and worldwide: Geology, v. 13, p. 194–197.

ROSS, C. A. AND ROSS, J. R. P., 1987a, Late Paleozoic sea levels and depositional sequences, in Ross, C. A., and Haman, D., eds., Timing and Depositional History of Eustatic Sequences: Constraints on Seismic Stratigraphy: Cushman Foundation for Foraminiferal Research Special Publication 24, p. 137–149.

ROSS, C. A. AND ROSS, J. R. P., 1987b, Biostratigraphic zonation of late Paleozoic depositional sequences, in Ross, C. A., and Haman, D., eds., Timing and Depositional History of Eustatic Sequences: Constraints on Seismic Stratigraphy: Cushman Foundation for Foraminiferal Research Special Publication 24, p. 151–168.

ROSS, G. M., 1990, Basement structure, in-plane stress and the rise and fall of the Rimbey Trend: Bulletin of Canadian Petroleum Geologists, v. 38, p. 179.

ROSS, G. M. AND STEPHENSON, R. A., 1989, Crystalline basement: The foundations of Western Canada Sedimentary Basin, Chapter 3, in Ricketts, B. D., ed., Western Canada Sedimentary Basin, A Case History: Calgary, Canadian Society of Petroleum Geologists, p. 33–45.

ROSS, G. M., PARRISH, R. R., VILLENEUVE, M. E., AND BOWRING, S. A., 1991, Geophysics and geochronology of the crystalline basement of the Alberta Basin, western Canada: Canadian Journal of Earth Sciences, v. 28, p. 512–522.

RUBIN, C. M., MILLER, M. M., AND SMITH, G. M., 1991, Tectonic development of Cordilleran middle Paleozoic volcano-plutonic provinces: Evidence for convergent margin tectonism, in Harwood, D. S., and Miller, M. M., eds., Paleozoic and Early Mesozoic Paleogeographic Relations of the Sierra Nevada, Klamath Mountains, and Related Terranes: Boulder, Geological Society of America Special Paper 255, p. 1–16.

SANDBERG, C. A., GUTSCHICK, R. C., JOHNSON, J. G., POOLE, F. G., AND SANDO, W. J., 1983, Middle Devonian to Late Mississippian geologic history of the Overthrust Belt region, western United States: Denver, Rocky Mountain Association of Geologists, Geologic Studies of the Cordilleran Thrust Belt, v. 2, pp. 691–719.

SANDBERG, C. A., POOLE, F. G., AND JOHNSON, J. G., 1988, Upper Devonian of western United States, in McMillan, N. J., Embry, A. F., and Glass, D. J., eds., Devonian of the World, v. I: Calgary, Canadian Society of Petroleum Geologists Memoir 14, p. 183–220.

SARG, J. F., 1988, Carbonate sequence stratigraphy, in Wilgus, C. K., Hastings, B. S., Posamentier, H., Van Wagoner, J., Ross, C. A., and Kendall, C. G. St. C., eds., Sea-Level Changes: An Integrated Approach: Tulsa, Society of Economic Paleontologists and Mineralogists Special Publication 42, p. 155–181.

SARTENAER, P., 1969. Late Upper Devonian (Famennian) Rhynchonellid brachiopods from western Canada: Geological Survey of Canada Bulletin 169, 269 p.

SAVOY, L. E., 1990, Sedimentary record of Devonian-Mississippian carbonate and black shale systems, southernmost Canadian Rockies and adjacent Montana: Facies and processes: Unpublished Ph.D. Dissertation, Syracuse University, Syracuse, 226 p.

SAVOY, L. E., 1992, Environmental record of Devonian-Mississippian carbonate and low-oxygen facies transitions, southernmost Canadian Rocky Mountains and northwesternmost Montana: Geological Society of America Bulletin, v. 104, p. 1412–1432.

SAVOY, L. E. AND HARRIS, A. G., 1993a, Conodont biofacies in a ramp to basin setting (latest Devonian and earliest Carboniferous) in the Rocky Mountains of southernmost Canada and northern Montana: Washington, D. C., United States Geological Survey Open File Report 93–184, 38 p.

SAVOY, L. E. AND HARRIS, A. G., 1993b, Conodont biofacies and taphonomy along a carbonate ramp to black shale basin (latest Devonian and earliest Carboniferous), southernmost Canadian Cordillera and adjacent Montana: Canadian Journal of Earth Sciences, v. 30, p. 2404–2422.

SCHLAGER, W., 1981, The paradox of drowned reefs and carbonate platforms: Geological Society of America Bulletin, v. 92, p. 197–211.

SCHLAGER, W., 1989, Drowning unconformities on carbonate platforms, in Crevello, P. D., Wilson, J. L., Sarg, J. F., and Read, J. F., eds., Controls on Carbonate Platform and Basin Development: Tulsa, Society of Economic Paleontologists and Mineralogists Special Publication 44, p. 15–25.

SCOTESE, C. R. AND MCKERROW, W. S., 1990, Revised world maps and introduction, in McKerrow, W. S., and Scotese, C. R., eds., Palaeozoic Palaeogeography and Biogeography: London, Geological Society Memoir No. 12, p. 1–21.

SMITH, M. T., DICKINSON, W. R., AND GEHRELS, G. E., 1993, Contractional nature of Devonian-Mississippian Antler tectonism along the North American continental margin: Geology, v. 21, p. 21–24.

SMITH, M. T. AND GEHRELS, G. E., 1992, Structural geology of the Lardeau Group near Trout Lake, British Columbia: Implications for the structural evolution of the Kootenay Arc: Canadian Journal of Earth Sciences, v. 29, p. 1305–1319.

SPEED, R. C. AND SLEEP, N. H., 1982, Antler orogeny and foreland basin: A model: Geological Society of America Bulletin, v. 93, p. 815–828.

STOAKES, F. A., 1979, Sea-level control of carbonate deposition during progradational basin filling: The Upper Devonian Duvernay and Ireton formations of Alberta, Canada: Unpublished Ph.D. Dissertation, University of Calgary, Calgary, 346 p.

STOAKES, F. A., 1980, Nature and control of shale basin fill and its effect on reef growth and termination: Upper Devonian Duvernay and Ireton Formations of Alberta, Canada: Bulletin of Canadian Petroleum Geology, v. 28, p. 345–410.

STOAKES, F. A., 1992, Nature and Succession of Basin Fill Strata, in Wendte, J., Stoakes, F. A., and Campbell, C. V., eds., Devonian-Early Mississippian Carbonates of the Western Canada Sedimentary Basin: A Sequence Stratigraphic Framework: Tulsa, SEPM, Short Course No. 28, p. 127–144.

STOCKMAL, G. C., CANT, D. J., AND BELL, J. S., 1992, Relationship of the stratigraphy of the western Canada foreland basin to Cordilleran tectonics: Insights from geodynamic models, in Macqueen, R. W., and Leckie, D. L., eds., Foreland Basins and Fold Belts: Tulsa, American Association of Petroleum Geologists Memoir 55, p. 107–120.

STRUIK, L. C., 1981, A reexamination of the type area of the Devonian-Mississippian Cariboo orogeny, central British Columbia: Canadian Journal of Earth Sciences, v. 18, p. 1767–1775.

STRUIK, L. C., 1986, Imbricated terranes of the Cariboo gold belt with correlations and implications for tectonics in southeastern British Columbia: Canadian Journal of Earth Sciences, v. 23, p. 1047–1061.

TEMPELMAN-KLUIT, D. J., 1979, Transported cataclasite, ophiolite, and granodiorite in Yukon: Evidence of arc-continent collision: Ottawa, Geological Survey of Canada Paper 79-14, 27 p.

TURNER, R., MADRID, R., AND MILLER, E., 1989, Roberts Mountains Allochthon: Stratigraphic comparison with lower Paleozoic outer continental margin strata of the northern Canadian Cordillera: Geology, v. 17, p. 341-344.

UYENO, T. T., 1991, Pre-Famennian Devonian conodont biostratigraphy of selected intervals in the eastern Canadian Cordillera, in Orchard, M. J., and McCracken, A. D., eds., Ordovician to Triassic Conodont Paleontology of the Canadian Cordillera: Geological Survey of Canada Bulletin 417, p. 129-162.

VAN WAGONER, J. C., POSAMENTIER, H.W., MITCHUM JR., R. M., VAIL, P. R., SARG, J. F., LOUTIT, T. S., AND HARDENBOL, J., 1988, An overview of the fundamentals of sequence stratigraphy and key definitions, in Wilgus, C.K., Hastings, B.S., Kendall C.G.St.C., Posamentier, H.W., Ross, C.A., and Van Wagoner, J.C., eds., Sea-Level Changes: An Integrated Approach: Tulsa, Society of Economic Paleontologists and Mineralogists Special Publication 42, p. 39-46.

WEISSENBERGER, J. A. W., 1988, Sedimentology and preliminary conodont biostratigraphy of the Upper Devonian Fairholme Group, Nordegg area, west-central Alberta, Canada, in McMillan, N. J., Embry, A. F., and Glass, D. J., eds., Devonian of the World, v. II: Calgary, Canadian Society of Petroleum Geologists Memoir 14, p. 451-462.

WEISSENBERGER, J. A. W., 1994, Frasnian reef and basinal strata of West Central Alberta: A combined sedimentological and biostratigraphic analysis: Bulletin of Canadian Petroleum Geology, v. 42, p. 1-25.

WENDTE, J., 1992, Cyclicity of Devonian strata in the Western Canada Sedimentary Basin, in Wendte, J., Stoakes, F. A., and Campbell, C. V., eds., Devonian-Early Mississippian Carbonates of the Western Canada Sedimentary Basin: A Sequence Stratigraphic Framework: Tulsa, SEPM, Short Course No. 28, p. 25-40.

WENDTE, J., AIGNER, T., AND NEUGEBAUER, J., 1984, Cephalopod limestone deposition on a shallow pelagic ridge: The Tafilalt Platform (Upper Devonian, eastern Anti-Atlas, Morocco): Sedimentology, v. 31, p. 601-625.

WENDTE, J. AND STOAKES, F. W., 1982, Evolution and corresponding porosity distribution of the Judy Creek reef complex, Upper Devonian, Central Alberta, in Cutler, W. G. ed., Canada's Giant Hydrocarbon Reserves: Calgary, Canadian Society of Petroleum Geologists Core Reference Manual, p. 63-81.

WENDTE, J., STOAKES, F. W., AND CAMPBELL, C. V., 1992, Devonian-Early Mississippian Carbonates of the Western Canada Sedimentary Basin: A Sequence Stratigraphic Framework: Tulsa, SEPM, Short Course No. 28, 255 p.

WHITE, W. H., 1959, Cordilleran tectonics in British Columbia: American Association of Petroleum Geologists Bulletin, v. 53, p. 60-100.

WITZKE, B. J., 1990, Palaeoclimatic constraints for Palaeozoic palaeolatitudes of Laurentia and Euramerica, in McKerrow, W. S., and Scotese, C. R., eds., Palaeozoic Palaeogeography and Biogeography: London, Geological Society Memoir No. 12, p. 57-73.

WITZKE, B. J. AND HECKEL, P. H., 1988, Paleoclimatic indicators and inferred Devonian paleolatitudes of Euramerica, in McMillan, N. J., Embry, A. F., and Glass, D. J., eds., Devonian of the World, v. I: Calgary, Canadian Society of Petroleum Geologists Memoir 14, p. 49-63.

WONG, P. K. AND OLDERSHAW, A. E., 1980, Causes of cyclicity in reef interior sediments, Kaybob reef, Alberta: Bulletin of Canadian Petroleum Geology, v. 28, p. 411-424.

WORKUM, R. H., 1983, Patterns within the Devonian of the Alberta Rocky Mountains as analogs to the subsurface: Calgary, Canadian Society of Petroleum Geologists Short Course, 24 p.

ZIEGLER, P. A., 1988, Laurussia—The Old Red Continent, in McMillan, N. J., Embry, A. F., and Glass, D. J., eds., Devonian of the World, v. I: Calgary, Canadian Society of Petroleum Geologists Memoir 14, p. 15-48.

UPPER TRIASSIC-JURASSIC FORELAND SEQUENCES OF THE ORDOS BASIN IN CHINA

LI SITIAN AND YANG SHIGONG
China University of Geosciences, Beijing-Wuhan, People's Republic of China
AND
TOM JERZYKIEWICZ
Geological Survey of Canada, Institute of Sedimentary and Petroleum Geology, Calgary, Canada

ABSTRACT: The Ordos basin is one of the most important coal and hydrocarbon-bearing sedimentary basins in China. The sedimentary cover, which overlies crystalline basement, is subdivided into seven unconformity-bounded stratigraphic sequences. The Upper Triassic to Upper Jurassic sequences are interpreted to be of foreland basin deposits. Development of a rapidly subsiding foredeep, adjacent to the rising western and southern flanks of the Ordos basin, commenced in late Triassic time during the late phases of the Indosinian Orogeny. A lake which developed in the axial part of the foredeep was a site of sedimentation of organic-rich mudstones which are the major hydrocarbon source in the basin. The main, late Triassic phase of foreland deposition is represented by over 3,000 m of proximal, coarse-grained sediments interfingering with the lacustrine facies. Lacustrine fan deltas and steep-sloped deltas were the main sites of foreland deposition. The depositional history of the Ordos foreland basin includes an overfilled phase in early Jurassic time and changes of source area in late Jurassic time. Unlike the Mesozoic retroarc basins of North America, which developed under a generally unidirectional, west-east compressional regime, the Ordos foreland basin was developed under the influence of transpression related to collision in the Tethys tectonic domain of southwest China and the collision of North China and Yangtze blocks.

Foreland sedimentation was terminated in early Cretaceous time due to a continent-wide shift to an extensional tectonic regime.

INTRODUCTION

The Ordos basin is one of the largest (>250,000 km^2) and most important hydrocarbon-bearing basins in China. More than 500 billion metric tons of high-quality coal reserves have been found since the 1970's. Results of exploration for oil and natural gas carried out in the 1980's have also been very encouraging. Data on the stratigraphy and structure of the Ordos basin have been collected over the last decade due to exploration drilling, high quality seismic reflection profiles and field work. Some of this data have been published (Sun and others, 1985; Chen and others, 1987; Zhang, E., 1989; Zhang, K., 1989; Yang, 1990), but a concise account on the depositional history of the basin in Triassic-Jurassic has not yet been attempted. The Ordos basin provides an example of an inland basin isolated from direct marine influence throughout almost the entire Mesozoic history. It is therefore reasonable to assume that the clastic wedges and the boundaries of the stratigraphic sequences of the Ordos basin are controlled primarily by tectonics.

The Ordos basin has been described in terms of "an unstable cratonic interior superimposed basin" (Sun and others, 1989). Hsu (1989) pointed out that Ordos is in fact not located in a "cratonic interior"; it is surrounded by mountain chains and the basin subsidence at various times was largely related to compressional deformation. The subsidence of the Ordos basin in late Triassic time is related to the collision of the North China and Yangtze blocks. According to Li (1992b) the genesis of the Ordos basin is comparable to that of a foreland basin because the subsidence is related to the process of thrusting one segment of continental crust under another.

The purpose of this paper is to provide a stratigraphic, sedimentologic, and geotectonic framework for the late Triassic-Jurassic, Ordos foreland sequences. We hope that the brief descriptions of stratigraphic sequences and sedimentary geology which follow may be useful not only for those studying the sedimentary response to tectonism in sedimentary basins worldwide but also for sequence stratigraphers.

REGIONAL SETTING

The Ordos basin is situated in central China within the Yellow River drainage basin. It is limited by latitudes 34°00′N to 40°35′N and by longitudes 106°50′E to 111°10′E (Fig. 1).

Geologically, the Ordos basin is situated in the western part of the Sino-Korean continental massif. The basin is bordered on the north, west, and south by the Yinshan-Tianshan, Helan, and Qinling-Qilian Paleozoic orogenic belts respectively (Fig. 1). The crystalline basement of the Ordos basin, dated at 1900–3400 Ma (Wang, 1982; Zhang, E., 1989), is one of the oldest and most stable structural nuclei in east Asia.

The middle-to-late Proterozoic marine strata distributed in fault-bounded troughs of the Ordos basin are interpreted as deposited in aulacogens. Cambrian and Ordovician shallow-marine carbonates are part of large sedimentary cover of the Sino-Korean continental block. Near the southern margin of the Ordos basin, the early Paleozoic platform carbonate rocks grade into continental slope, deep-water deposits which represent the northern margin of Qilian-Qinling seaway. Situated on the western flank of the Ordos basin, the Helan aulacogen is related to early Paleozoic collisional events recorded in the Qilian-Qinling orogenic belt (Lin and others, 1991). Carboniferous, Permian, and early-to-middle Triassic paralic deposits are distributed in both the Ordos basin and the Helan aulacogen.

Base-level drop and deformation events which took place in the Ordos basin in mid-to-late Traissic, referred to as the Indosinian Orogenesis, coincide with the collision event at the Jinsha-Menglian suture of the Tethys tectonic domain in southwest China (Huang and Chen, 1987). The collision event led to transpression in the intercontinental area and produced thrust belts that formed the southwestern margin of the Ordos basin, the northwestern margin of Sichuan basin, and the late Triassic foreland basin which is the subject of this report.

Paleozoic seaways into the Ordos basin were closed by the end of the era in the north and by Triassic time in the

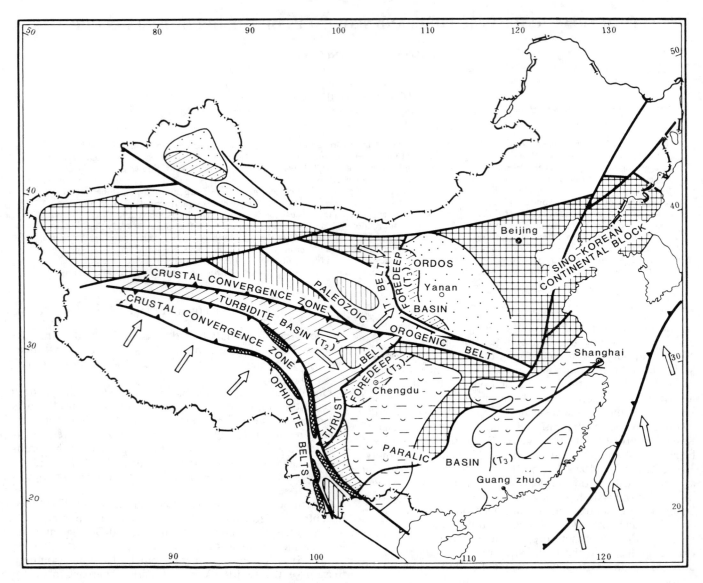

Fig. 1.—Late Triassic, geotectonic framework of the Ordos basin in China.

south; therefore, the Mesozoic stratigraphic record of the basin is exclusively nonmarine. The youngest marine fossils are of early Triassic age and occur in the southernmost Ordos basin (Yin and others, 1988).

STRATIGRAPHIC FRAMEWORK

The stratigraphic setting of the Upper Triassic and Jurassic foreland sequences, which exceed 5,000 m in thickness along the western margin of the Ordos basin, is shown in Figure 2. The sedimentary cover of the crystalline basement in the Ordos basin is subdivided into seven, unconformity bounded, stratigraphic megasequences. Proterozoic sedimentary and volcanic rocks are separated from the high grade metamorphic rocks of the Sino-Korean continental block by an unconformity that extends far beyond the Ordos basin (Yang and others, 1986). In the Ordos basin, this unconformity is clearly visible on the seismic profiles and in some surface sections exposed in the bordering mountain belts.

The middle-late Proterozoic megasequence (MS-I in Fig. 2) consists of silicious limestone, sandstone, and volcanic rocks which are distributed in the deep-buried, fault-bounded troughs underlying the Paleozoic strata. The upper Proterozoic silicious limestone is separated from the lower Cambrian trilobite-bearing limestone by a widespread unconformity (Zhang, E., 1989).

The lower Paleozoic megasequence (MS-II in Fig. 2) consists of up to 800 m shallow marine carbonates consisting of limestone, dolomite, and evaporite (Feng and others, 1991). These carbonate strata are the reservoir rocks for natural gas. The early Paleozoic platform carbonates are covered by Carboniferous continental, coal-bearing clastics.

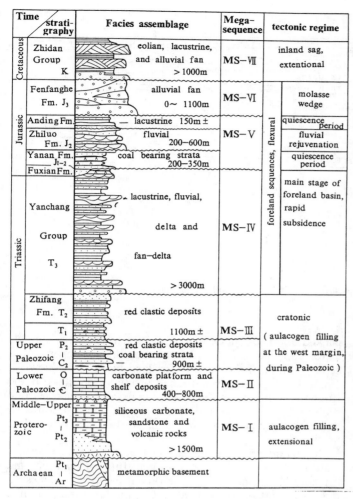

Fig. 2.—Stratigraphic sequences of the Ordos basin.

The upper Paleozoic to middle Triassic megasequence (MS-III in Fig. 2) in the Ordos basin is about 2,000 m thick and consists of clastics which were laid down largely in nonmarine environments extending north of the present boundaries of the basin. Sedimentation of the lower Carboniferous, coal-bearing strata was limited to the western margin of the Ordos basin. Gradually, the coal-bearing, Carboniferous sedimentation embraced eastern parts of the basin where the upper Carboniferous strata were documented immediately above the sub-upper Paleozoic disconformity (Sun and others, 1989, 1990). The lower part of the megasequence III is dominated by the coal-bearing strata of the upper Carboniferous and lower Permian. The upper part of the megasequence III consists of late Permian to middle Triassic red beds which are known to extend far beyond the present boundaries of the Ordos basin. The lower and middle Triassic red beds (Zhifang Formation) are separated by a second order unconformity in the western part of the Ordos basin (Zhang, E., 1989; Zhang, K., 1989; Sun and others, 1990).

The sub-upper Triassic unconformity occurs throughout the entire Ordos basin and has been observed in seismic reflection profiles, boreholes, and surface sections in the marginal area of the basin (Zhang, K., 1989; Sun and others, 1985). This unconformity is marked by a substantial change in sedimentation pattern and in the nature of erosional surfaces. The late Triassic sedimentation in the western and southwestern part of the Ordos Basin commenced with development of conglomerates. Proximal breccia and conglomerate were deposited near the western and southwestern flanks of the basin. Farther east, the conglomerates interfinger with channelized sandstone, known largely from borehole information. Distal facies in the central part of the Ordos basin are represented by fine-grained sediments with lacustrine fauna. The upper Triassic megasequence (MS-IV in Fig. 2), which corresponds to the Yanchang Group, forms more than a 3,000-m thick clastic wedge in the western and southwestern parts of the Ordos basin, tapering to the east to less than 1,000 m (Figs. 2, 3).

The sub-Jurassic unconformity is marked by a basin-wide erosional surface. The lithologic, faunistic, and floristic differences between the uppermost sediments of the Triassic Yanchang Group and Jurassic strata in the Ordos basin are distinct. The lowermost Jurassic Fuxian Formation, which shows rapid lateral facies changes from conglomerates to shale, infills local depressions on the pre-Jurassic erosion surface. The bulk of the lower to middle Jurassic megasequence (MS-V, Fig. 2) consists of fluvio-lacustrine strata with an economic coal-bearing unit in the lower part (the Yanan Formation).

The upper Jurassic sediments form an unconformity-bounded, coarse-grained clastic wedge along the western flank of the Ordos basin (MS-VI in Fig. 2). The wedge consists of breccia and conglomerate of alluvial fan origin and extends 100 km east from the western flank of the basin. Maximum thickness of the alluvial fan deposits near the western flank of the basin exceeds 1,000 m. Farther eastward, this upper Jurassic megasequence, known as the Fenfanghe Formation, pinches out between the middle Jurassic Anding Formation and Cretaceous Zhidan Group sediments.

Cretaceous sediments in the Ordos basin are represented by a thick succession of red beds of alluvial fan, fluvio-lacustrine, and eolian origin (Zhidan Group, MS-VII, Fig. 2). The Cretaceous red beds are underlain by an unconformity documented far beyond the Ordos basin (Yang and others, 1986). Their thickness across most of the Ordos basin is about 1,000 m. The thickest sections of the Cretaceous red beds, referred to as the Liaupanshan Group are preserved in tectonic grabens on the southwestern edge of the Ordos basin (west of Huating).

LATE TRIASSIC AND JURASSIC FORELAND DEPOSITION

Since early Triassic time, when the sea retreated from the southern part of the Ordos basin and the connection with the seaway located to the south was broken, the basin maintained a nonmarine character. The drainage systems developed in the Ordos basin in Mesozoic time were of internal types. A large inland lake, which is referred to here as the Ordos Lake, developed in the Ordos basin by late Triassic time. A rapidly subsiding foredeep, in front of the

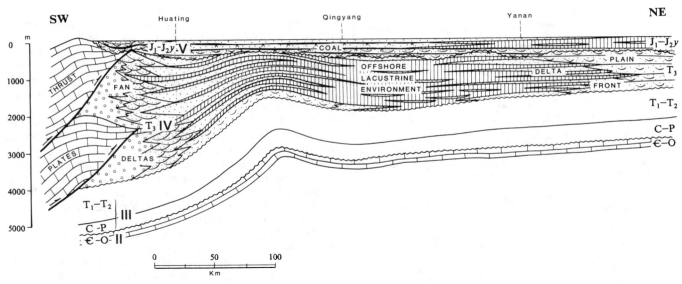

FIG. 3.—Section across the Ordos basin based on seismic data and exploration wells.

rising western and southwestern flanks of the Ordos basin, developed (or formed) a foreland lake (Ordos Lake) during the late phases of the Indosinian Orogeny (Sun and others, 1985, 1990). The western and southwestern Ordos foredeep was infilled by a succession of clastics more than 3,000 m thick derived from the emerging thrust plates to the west and southwest (Fig. 3).

The Ordos Lake, which developed in response to tectonically induced changes in the deposition pattern between middle and upper Triassic times, extended from NNW to SSE in the axial part of the Ordos basin. During the main phase of late Triassic foreland deposition, the center of the Ordos Lake near Huachi and Qingyang received up to 400 m of organic-rich, dark mudstones, covering an area of approximately 100,000 km². These mudstones interbedded with turbidites are the major source rock of the hydrocarbons in the Ordos basin.

These late Triassic Yanchang Group in the western Ordos (MS-IV) are interpreted to have formed as part of a fan delta system. Over 3,000 m of proximal, coarse-grained sediments which interfinger with fine-grained, offshore lacustrine mudstone are well exposed in several sections through the Helan Mountains west of Yinchuan and in the south-western margin of the Ordos basin near Huating and Kongtongshan (Figs. 4A-B, 5).

The base of the megasequence consists of conglomerates or breccias derived from the southwestern, thrusted margin of the basin. The coarsest and most proximal known facies are located in the western part of Helan Mountains and the southwestern part of the Ordos basin near Kongtongshan. There, the basal sediments of the Yanchang Group consist of poorly sorted, unstratified boulders, cobbles, and pebbles supported by arkosic matrix. Large angular clasts (some >1 m in diameter), poor sorting, and general lack of stratification clearly indicate an extremely proximal fanglomerate setting. Pebbles derived largely from Paleozoic and pre-Paleozoic rocks of the thrust belt are often very closely packed and show pock marks and fractures. Very similar, tectonically deformed conglomerates in the Upper Cretaceous portion of the Alberta foreland basin (Jerzykiewicz, 1985) developed adjacent to the Canadian Cordillera. In the Rugigou section near Yinchuan, the basal Yanchang Group sediments are represented by clast-supported conglomerate and coarse-grained sandstone of alluvial fan and braid-plain origin (Fig. 6).

Basal fanglomerate and alluvial fan deposits grade upward into several intervals of fluvial, fluvio-lacustrine, and offshore lacustrine facies. The fluvial facies are thickest in the lower part of the Yanchang Group above the proximal alluvial fan facies. Braid-plain sandstone facies in the Rugigou section, up to 500 m thick, may suggest steep paleoslope and rapid subsidence of the foredeep in front of an active basin margin.

The middle and the upper parts of the Yanchang Group in the western part of the Ordos basin (more than 2,000 m thick in the Rugigou section) consist of several upwards-coarsening parasequences of offshore lacustrine to fluvio-lacustrine origin (Fig. 4A). The coarsening upwards successions are of two types: (1) gradual transition from fine-grained mudrock with lacustrine fauna interstratified with thin and distal turbidites capped by medium grained sandstone layers of distributary channel and/or distributary mouth bar origin (Fig. 7A) and (2) several, vertically stacked, distributary channel sandstone layers interbedded with thin, offshore lacustrine turbidites of overall coarsening upward appearance (Figs. 7B, 8).

The distribution of facies within the Yanchang Group across the Ordos basin is strongly asymmetric (Figs. 3, 4A-B). Relatively rapid facies change from the fanglomerates to offshore lacustrine facies suggests a steep-sloped fan delta along the western and southwestern margin of the basin. In some areas, the alluvial fans prograded directly into the lake (Fig. 4A, west of Yinchuan), in others across a sandy delta plain.

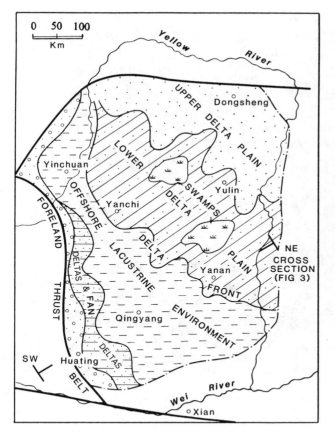
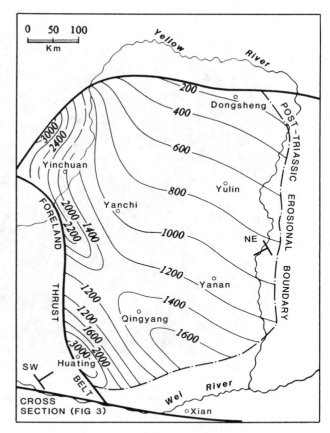

FIG. 4.—Distribution of facies (A) and coresponding isopach map (B) of the upper Triassic foreland sequence of the Ordos basin (modified from Guo, 1989).

FIG. 5.—Lacustrine mudstone interbedded with turbidite sandstone layers of Yanchang Group, west Ordos. Book is 20 cm high.

Another, much larger and gently-sloped delta complex was developed in late Triassic time in the northeastern part of the Ordos basin. This delta prograded from the northeast and north and covered most of the Ordos basin by the end of Triassic depostion. It has been the object of extensive hydrocarbon exploration. Relatively good reservoir sandbodies, such as those in the Ansai area (Mei and Lin, 1991), have been found within the system of lobes of this gently-sloped delta complex. Top set strata of the delta complex include economic coal deposits.

The upper Triassic, offshore lacustrine facies were limited to the northwest-southeast trending and southeasterly widening axial part of the Ordos Lake (Fig. 4B). The overfilled stage of the Ordos foredeep occurred at the end of Triassic time and is represented by a sub-Jurassic unconformity. The overlying Fuxian Formation represents a period of deposition in local depressions developed on the basin-wide, pre-Jurassic erosional surface. The paleovalley-fill deposits of the Fuxian Formation are one of the best hydrocarbon reservoirs in the Ordos basin (Song, 1989; Wu and Xue, 1992).

Rejuvenation of tectonic activity affecting the entire Ordos basin commenced with the deposition of coarse-grained,

Fig. 6.—Braided-plain deposits in the Ruqigou section (southwestern margin of the Ordos basin).

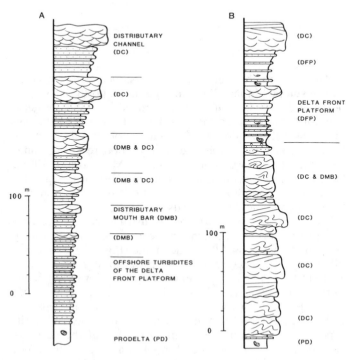

Fig. 7.—A typical coarsening-upward succession of the Yanchang Group. (A) Type 1 parasequences of fan-delta origin (cf. Fig. 8). (B) Parasequences of steep-sloped lacustrine delta origin.

fluvial sediments at the base of the Yanan Formation. These mainly braided, alluvial plain deposits are superceded by several parasequences consisting of fluvial and lacustrine deltaic coal-bearing deposits. The Yanan delta plain coal deposits are the most economically important in China (Li and others, 1990a, 1992a).

Paleotectonic and paleogeographic patterns in the Ordos basin have changed since Jurassic time. The reemergent Ordos Lake in the southeastern part of the Ordos basin (Yanan district) was gradually infilled by sediments of a large, gentle-sloped delta sourced from the uplifted flanks of the lake (Fig. 9; Li and others, 1992a).

The middle part of Jurassic megasequence V is represented by coarse-grained, fluvial and fluvio-lacustrine deposits of the Zhiluo Formation. The sub-Zhiluo erosional surface was likely produced by very large fluvial channels that formed paleovalleys in underlying Yanan, coal-bearing strata of the Ordos basin. The Zhiluo Formation forms an eastward-thinning clastic wedge developed adjacent to the western and northwestern tectonically active margin of the basin. The upper part of megasequence V, (i.e., Anding Formation) consists of purplish-red and variegate-colored strata deposited in a lacustrine environment in a semiarid climate. Megasequence V, as a whole, consists of two relatively fine-grained intervals the Yanan and Anding Formations interbedded with the coarse-grained, fluvial and fluvio-lacustrine Zhiluo Formation deposited during tectonically active periods.

FIG. 8.—Coarsening-upward parasequences of fan delta origin, Yanchang Group. Rigigou section near Yinchuan.

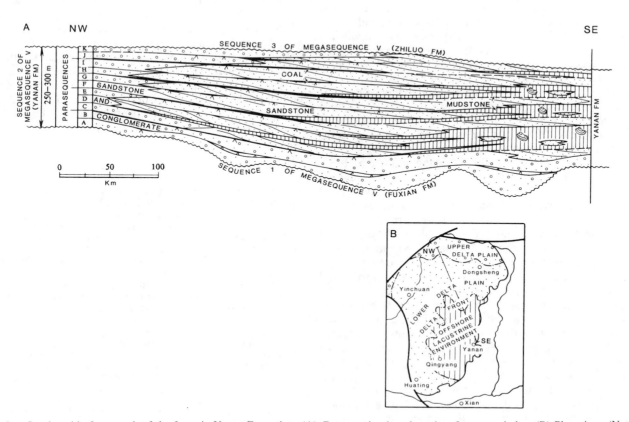

FIG. 9.—Stratigraphic framework of the Jurassic Yanan Formation. (A) Cross-section based on data from core holes. (B) Plan view. (Note the cross-section line).

The upper Jurassic Fenfanghe Formation (MS-VI in Fig. 2) was deposited in the last stage of foreland basin development and is interpreted to indicate rejuvenation of thrusting in the overthrust belt along the western margin of Ordos basin. An easterly tapering clastic wedge of the Fenfanghe Formation is marked very distinctly on seismic reflection data across the Ordos basin. The Fenfanghe clastic wedge consists predominantly of red beds of fanglomerate origin. It reaches about 1,300 m in thickness in the western part of the foredeep and exibits an angular, unconformable relationship with underlying strata.

GEOTECTONIC FRAMEWORK OF THE ORDOS FORELAND BASIN: DISCUSSION AND CONCLUSIONS

Upper Triassic and Jurassic, nonmarine clastics that infilled an asymmetric trough along the western and southwestern margins of the Ordos basin are interpreted as foreland sequences. Infilling of the foredeep in front of the evolving thrust belt adjacent to the Ordos basin occurred in several phases of pronounced subsidence separated by periods of relative quiescence. Nevertheless, the external geometries of the clastic wedges and paleocurrent patterns clearly suggest genetic links between the rising mountains and infilling of the foredeep.

The upper Triassic, Yanchang Group clastic wedge was deposited during the period of maximum total subsidence of the foredeep. The end of middle Triassic time was a period of great change in the tectonic and paleogeographic framework of mainland China (Indosinian Orogeny). This period was characterized by marine regression over a vast area of southern China and strong intra-continental deformation. Very thick foreland-type sequences were deposited in the western margins of both the Sino-Korea and the Yangtze massifs. Deposits similar to those of the Ordos foreland were laid down in the western part of the Sichuan basin near Chengdu (Fig. 1). The original extent of the area covered by foreland sedimentation can be inferred to have been much larger than the present size of the Ordos basin (Sun and others, 1985). There are also contrasting sedimentation patterns between middle and late Triassic times in many basins of southeast China (Huang and Chen, 1987; Yin, 1988). The shallow marine sedimentation typical for early Triassic time gave way to the paralic and continental sedimentation typical of the late Triassic in southeast China. Continental basins also developed in northern China. Collision of the North China and Yangtze blocs and closing of the north branch of the Tethys Ocean are considered to be the main reasons for the predominance of compressional regimes in southeastern Asia in late Triassic time (Wang, 1982; Sun and others, 1985; Hsu, 1989).

Unlike Mesozoic retroarc foreland basins of North America (Beaumont, 1981) which developed under a generally unidirectional, west-east compressional regime, the Ordos and Sichuan foredeeps were developed under the influence of a transpression. The transpressional regime may have developed as a result of collision in western Ordos and Sichuan (Huang and Chen, 1987; Xu and others, 1992; Sengor, 1990). According to this hypothesis, the Songpan-Garze block has moved to the west and produced the Longmen thrust belt and west Sichuan foredeep. The Qilian-Qinling, middle Triassic sea was closed by the effect of collision, and the transpressional regime that developed in the region led to the formation of the thrust belt and the Ordos foredeep.

By the early mid-Jurassic, the Ordos basin was infilled with sediments, the pace of the subsidence was lower than the rate of sediment supply, and the foreland basin reached an overfilled stage. The sedimentary pattern changed by the end of the Jurassic when easterly tapering, coarse-grained Fenfanghe clastic wedge developed as a response to thrusting which affected both the western and northern margins of the Ordos basin. By the beginning of Cretaceous time, the western margin of the Ordos basin had been overthrusted to the northeast forming the thrust belt.

Termination of foreland sedimentation in the Ordos basin is related to continental-wide change in the tectonic regime. The angular unconformity at the base of the Cretaceous succession in the Ordos basin is interpreted in terms of a major change in the tectonic pattern, namely as a shift from predominantly transpressional to extensional regime. The late Mezozoic rifting known from other regions of China (Li and others, 1990b; Glider and others, 1991) affected the Ordos basin beginning in early Cretaceous time. The Cretaceous, vertebrate-bearing red beds infilled numerous grabens and halfgrabens that developed along the marigins of the Ordos basin and within the adjacent areas of northern China and Mongolia (Jerzykiewicz and Russell, 1991).

ACKNOWLEDGMENTS

We would like to thank the National Science Foundation of China, Ministry of Geology and Mineral Resources of China, Geological Survey of Canada, Paleobiology Division of the National Museums of Canada, Ex Terra Foundation, Geological Bureaus of Ningxia, Shaanxi and Inner Mongolia, Shaanxi Company of Coal Geology, and many colleagues from China University of Geosciences for their contributions to our fieldwork in China. G. Stockmal and T. DeFreitas provided very useful reviews. Their detailed written comments resulted in many improvements in the text.

REFERENCES

Beaumont, C., 1981, Foreland Basin: Geophysical Journal of the Royal Astronomical Society, v. 65, p. 291–329.

Chen, F., Sun, J., Wang, P., Sun, G., and Liu, J., 1987, Structural features and prospects of the fold and thrust belt in the western margin of the Ordos basin: Geosciences, v. 1, p. 103–113 (in Chinese).

Feng, Z., Chen J., and Zhang, J., 1991, Lithofacies and paleogeography of early Paleozoic of the Ordos basin: Beijing, Geological Publishing House (in Chinese).

Glider, S. A., Keller, G. R., Luo, M., and Goodell, P. C., 1991, Timing and spatial distribution of rifting in China, in Gangi, A. F., ed., World Rift Systems: Special Issue of Tectonophysics, v. 197, p. 225–243.

Guo, Z., 1989, Structural Characteristics and oil/gas exploration in Shanxi-Gansu-Ningxia basin, in Advances in Evaluation and Research of Oil and Gas Resources: Beijing, Petroleum Industry Press, v. 1, p. 200–210 (in Chinese).

Hao, Y., 1986, The Cretaceous System of China, in Stratigraphy of China: Beijing, Geological Publishing House, No. 12, 301 p. (in Chinese).

Hsu, K. J., 1989, Origin of sedimentary basins of China, in Zhu, X., ed., Chinese Sedimentary Basins: Amsterdam, Elsevier, p. 207–227.

Huang, J. and Chen, B., 1987, The evolution of the Tethys in China and adjacent regions: Beijing, Geological Publishing House (in Chinese).

Hu, J. AND Huang, D., 1991, Fundamentals of Nonmarine Petroleum Geology in China: Beijing, The Petroleum Industry Press, 322 p. (in Chinese).

Jerzykiewicz, T., 1985, Tectonically deformed pebbles in the Brazeau and Paskapoo Formations, central Alberta Foothills, Canada: Sedimentary Geology, v. 42, p. 159–180.

Jerzykiewicz, T. AND Russell, D. A., 1991, Late Mesozoic stratigraphy and vertebrates of the Gobi Basin: Cretaceous Research, v. 12, p. 345–377.

Li, S., Cheng, S., AND Yang, S., 1992b, Sequence Stratigraphy and Depositional System Analysis of the Northeastern Ordos Basin: Beijing, Geological Publishing House, 194 p. (in Chinese).

Li, S., Yang, S., Hu, Y., AND Xu, L., 1990a, Analysis of depositional processes and architecture of the lacustrine delta, Jurassic Yanan Formation, Ordos basin: China Earth Sciences, v. 1, p. 217–231.

Li, S., Yang, S., AND Lin, C., 1992a, On the chronostratigraphic framework and basic building blocks of sedimentary basin: Acta Sedimentologica Sinica.

Li, S., Yang, S., Wu, C., AND Cheng, S., 1990b, Geotectonic background of the Mesozoic and Cenozoic rifting in east China and adjacent areas, in Wang, H., ed., Tectonopalaeography and Palaeobiologeography of China and Adjacent Regions: Beijing, China University of Geosciences Press, p. 109–126 (in Chinese).

Lin, C., Yang, Q., Li, S., AND Li, Z., 1991, Sedimentary characters of the early Paleozoic deep water gravity flow systems and basin filling style in the Helan aulacogen, northwest China: Geosciences, p. 252–262 (in Chinese).

Mei, Z. AND Lin, J., 1991, Stratigraphic pattern and character of skeletal sand bodies in lacustrine deltas: Acta Sedimentologica Sinica, v. 9, p. 1–11 (in Chinese).

Sengor, A. M. C., 1990, Plate tectonics and orogenic research after 25 years: Tethyan perspective: Earth Sciences Review, v. 27, p. 1–201.

Song, G., 1989, Jurassic channel deposits and the formation of oil pool, southern Ordos basin, in Sedimentary Facies and Oil, Gas Distribution of Oil-bearing Basins in China: Beijing, Petroleumm Industry Press, p. 217–229 (in Chinese).

Sun, G., Liu, J., Liu, K., AND Yuan, W., 1985, Evolution of a major Mesozoic continental basin within Huabei plate and its geodynamic setting: Oil and Gas Geology, v. 6, p. 278–287 (in Chinese).

Sun, Z., Xie, Q., AND Yang, J., 1989, Ordos Basin—a typical example of an unstable cratonic interior superimposed basin, in Zhu., ed., Chinese Sedimentary Basins: Amsterdam, Elsevier, p.63–75.

Sun, Z., Xie, Q., AND Yang, J., 1990, Ordos basin—a typical case of unstable intracratonic superimposed basin, in Zhu, X., and Shu, W., eds., The Mesozoic and Cenozoic Basins of China: Beijing, Petroleum Industry Press (in Chinese).

Tian, H., 1990, The origin of Leshan Great Buddha sandstone in Sichuan province: Acta Sedimentologica Sinica, v. 8, p. 41–47 (in Chinese).

Wang, H., 1982, The main stages of crustal development of China: Earth Science Journal of China University of Geosciences, v. 3, p. 155–177 (in Chinese).

Wu, C. AND Xue, S., 1992, Sedimentology of Petroliferous basins in China: The Petroleum Industry Press, 484 p. (in Chinese).

Yang, Y., 1990, Tectonic and Petroleum Geology of the West Thrust Belt of the Ordos Basin: Gansu Scientific Press, 160 p. (in Chinese).

Yang, Z., Cheng, Y., AND Wang, H., 1986, The Geology of China: Oxford Monographs on Geology and Geophysics, 303 p.

Yin, H., 1988, Paleobiogeography of China: China University of Geosciences Press, 329 p. (in Chinese).

Zhang, E., 1989, Regional Geology of Shaanxi Province: Beijing, Geological Publishing House, Geological Memoirs Series I, no. 13, 698 p. (in Chinese).

Zhang, K., 1989, Tectonic and Resources of Ordos Fault Block: Scientific Publishing House of Shanxi Province, 399 p. (in Chinese).

Xu, Z., Hou, L., AND Wang, Z., 1992, Orogenic processes of the Songpan-Garze Orogenic Belt of China: Geological Publishing House, 190 p. (in Chinese).

TECTONIC AND EUSTATIC CONTROLS ON THE STRATAL ARCHITECTURE OF MID-CRETACEOUS STRATIGRAPHIC SEQUENCES, CENTRAL WESTERN INTERIOR FORELAND BASIN OF NORTH AMERICA

MICHAEL H. GARDNER

Bureau of Economic Geology, The University of Texas at Austin, Austin, TX 78713-7508

ABSTRACT: Changes in stratal architecture defining a hierarchy of time-stratigraphic units record eustatic and tectonic controls on deposition in the central Western Interior foreland basin. Three temporal and spatial scales of stratigraphic cyclicity are recognized, each recording base-level changes of different periodicity. One long-term base-level cycle (600 m thick, 4.5 my) contains four intermediate-term base-level cycles (termed stratigraphic sequences, up to 300 m thick, 1 to 2 my). Each stratigraphic sequence consists of two to eight short-term cycles (up to 40 m thick, 0.3 my). A long-term stratigraphic cycle spanning Turonian through middle Coniacian stages consists of an upward-coarsening succession of marine and nonmarine deposits bounded by deposits formed during eustatic transgressions. Long-term base-level fall is recorded by the episodic eastward progradation of shoreface sandstones into the basin. The regressive maximum of the youngest stratigraphic sequence corresponds to a well-documented eustatic drop in the late Turonian.

Stratigraphic sequences of the Western Interior change as a function of local rates of sediment accommodation relative to supply in foreland subbasins comprising the western margin of the central Western Interior seaway. Turonian–Coniacian stratigraphic sequences were deposited under low accommodation and sediment supply conditions across Wyoming, Colorado, northeastern Utah, and northern New Mexico, and in high sediment accommodation to supply settings in western Wyoming, central Utah, and in northwestern New Mexico (upper part of the youngest sequence). Where sediment accommodation relative to supply rates are lower, stratigraphic sequences show: (1) more unconformities, (2) higher magnitude facies offsets across cycle boundaries, (3) seaward-stepping cycle stacking patterns, (4) vertically truncated facies tracts, (5) lower proportions of nonmarine strata, and (6) higher sandstone- to mudstone-ratios.

Along strike changes in sediment supply and accommodation are superimposed on a long-term pattern of westward-thickening and eastward-prograding basin fill. Regional variations in stratal architecture within stratigraphic sequences are related to southward migration of depocenters in successive foreland subbasins, and southward increases in accommodation recorded by more conformable stratal successions and higher proportions of deep-water carbonates and mudstones. Basinwide variations documented here show that a hierarchy of Turonian–Coniacian chronostratigraphic units may be resolved where stratal patterns are not consistent. Correlating base-level rise-to-fall turnarounds across foreland subbasins links dissimilar stratal patterns within chronostratigraphic units recording the same record of base-level change, but of varying magnitude and with changing contributions from primary controls (i.e., subsidence, sediment supply, and eustasy). For example, northward movement of subtropical (Atlantic-Tethyan) water masses and southward migration of depocenters appear to have locally combined in the late Turonian to produce high accommodation and sediment-supply conditions in the central Utah foreland basin. Here, short-term cycle stacking patterns closely resemble parasequence sets of a third-order depositional sequence. This suggests that this particular cycle stacking pattern and sequence model is most applicable to settings where accommodation and sediment supply are high.

INTRODUCTION

Sequence stratigraphy uses the geometric arrangement of strata, or "stratal architecture," to define chronostratigraphic units at a finer resolution than allowed by biostratigraphy or other temporal measures. Sequence stratigraphy concepts were popularized in seismic stratigraphy (e.g., Mitchum, and others, 1977; Vail and others, 1977), where terminations of seismic reflectors recording stratal interfaces were assumed to parallel time lines and bound chronostratigraphic units. Such units were associated with relative sea-level change to form the general sequence stratigraphic model of Vail and others (1977), Posamentier and others (1988, 1992), Posamentier and Vail (1988), Van Wagoner and others (1990), and Mitchum and Van Wagoner (1991). Sequence stratigraphic models emphasize the correlation of unconformities and changes in stratal geometry (e.g., lowstand, transgressive, and highstand systems tracts) to define and subdivide unconformity-bounded stratal packages.

Most sequence stratigraphic models are two-dimensional and focus on proximal to distal changes along a depositional profile. They do not address along-depositional-strike variations. These models generally assume that sea level controls base level. Stratal patterns of chronostratigraphic units are correlated to relative sea-level change. A test of whether sea level is a dominant control is to systematically examine along-strike variations in stratal architecture across marine foreland subbasins of an active orogenic belt. If sea level is the dominant control, then stratigraphic sequences should have unique characteristics that remain constant across foreland subbasins. This should also establish whether chronostratigraphic units containing potentially dissimilar stratal patterns can be correlated across foreland subbasins of varying subsidence and sediment supply.

This paper examines the stratal architecture of stratigraphic sequences from Cretaceous foreland subbasins bordering the western margin of the central Western Interior Cretaceous seaway (WICS). Four stratigraphic sequences are defined where the section is most complete in central Utah (see Gardner, this volume). Facies arrangements, cycle stacking patterns, and stratal geometries in stratigraphic sequences are documented along a 350-km depositional-strike transect of the Turonian–Coniacian shoreline of Utah. These four stratigraphic sequences are correlated for 1,500 km along depositional strike throughout the WICS spanning subbasins of higher and lower accommodation and sediment supply (Fig. 1). This time-stratigraphic framework encompasses a 1,500,000-km^2 region from Wyoming to central New Mexico north-south and from central Utah to Kansas east-west.

DATA AND APPROACH

This analysis builds on considerable previous sedimentologic, stratigraphic, biostratigraphic, and paleoecologic analysis of the WICS. Regional studies by Reeside (1944),

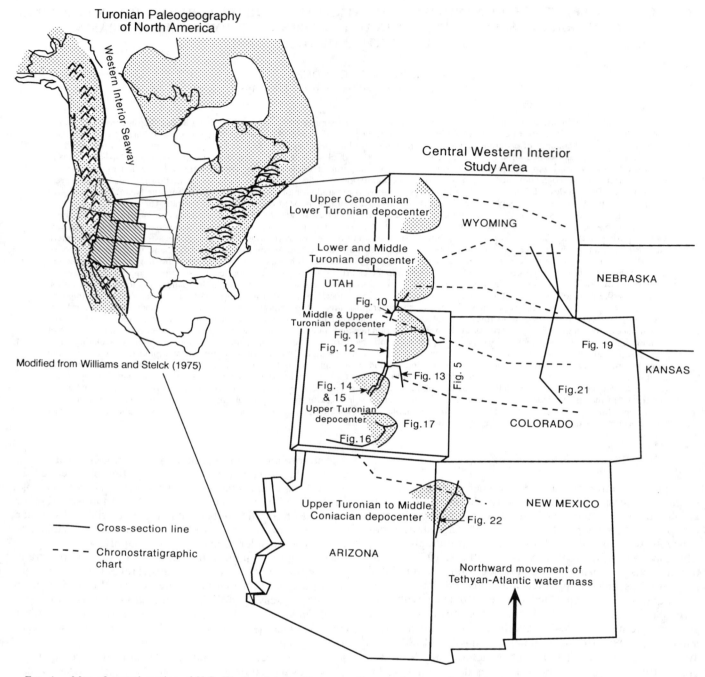

Fig. 1.—Map of central portion of U.S. Western Interior Cretaceous Seaway showing paleogeography and Utah study area. Note north-south migration of mid-Cretaceous depocenters across Utah and migration of Ferronensis sequence from southern to central Utah. Lines of cross sections and locations of major study areas discussed in text are shown.

Weimer (1960), McGookey and others (1972), Kauffman (1977), Molenaar (1983), Ryer and McPhillips (1983), Merewether and Cobban (1986), Ryer and Lovekin (1986), and Molenaar and Cobban (1991) provide a starting point for this research. Biostratigraphic studies on molluscan biozones (e.g., Kauffman and others, 1976; Kauffman, 1977; Cobban and Hook, 1979) help establish critical zonal boundaries across the basin. Major sources of data for documenting Turonian–Coniacian stratigraphic sequences are shown in Table 1.

Regional outcrop and subsurface stratigraphic data and a high resolution biostratigraphy are integrated to construct regional stratigraphic cross sections and facies maps documenting stratal architecture within a threefold hierarchy of time-stratigraphic units. This chronostratigraphic framework emphasizes two stratigraphic concepts. *Accommo-*

TABLE 1.—SOURCE OF STRATIGRAPHIC AND BIOSTRATIGRAPHIC DATA

Wyoming	Utah	Colorado	New Mexico
Cobban and Reeside (1952)	Gale (1910)	Gale (1910)	Cobban & Hook (1979, 1984, 1989)
Hunter (1952)	Lupton (1916)	Kent (1968)	Hook (1983)
Goodell (1962)	Walton (1944)	Kauffman (1969, 1977, 1985)	Molenaar (1974, 1977, 1983)
Barlow and Haun (1966)	Speiker (1949)	Weimer (1970, 1978, 1983)	Kirkland (1990)
Seimers (1975)	Hunt and others (1953)	Sonnenberg and Weimer (1981)	Valasek (1990)
Prescott (1975)	Davis (1954)	Weimer and Sonnenberg (1983)	Collom (1991)
Merewether and others (1975)	Katich (1954)	Pratt (1984)	
Merewether and others (1976)	Young (1955, 1957, 1960)	Elder (1985)	
Myers (1977)	Williams and Madsen (1959)	Pratt and others (1985)	
Weimer (1978, 1983)	Hale (1960, 1962, 1972)	Cobban (1988)	
Merewether and others (1979)	Hale and Van de Graaf (1964)	Collom (1991)	
Tilman and Almon (1979)	Hunt and Miller (1964)		
Merewether (1980)	Peterson (1969a, b)		
Tonnsen (1980)	Cotter (1971, 1975a, 1975b)		
Merewether (1983)	Maione (1971)		
Winn and others (1984)	Doelling (1972)		
Weimer and Flexer (1985)	Lessard (1973)		
Macdonald and Byers (1988)	Peterson and Ryder (1975)		
	Cobban (1976)		
	Kauffman (1977)		
	Ryer (1977, 1981, 1983), Ryer and others (1983, 1986)		
	Lawton (1985)		
	Eaton and others (1987, 1990)		
	Molenaar and others (1990, 1991)		
	Bobb (1991)		
	Reimersma and Chan (1991)		
	Shanley and McCabe (1992)		
	Leithold (1993)		
	Gardner, (1993)		

dation controls preservational trends in chronostratigraphic units, and it is the potential space available for sediment accumulation (Jervey, 1988). Accommodation is the sum of tectonic movement, lithospheric compensation to loads, compaction, and sea level. *Stratigraphic base level* (modified from Wheeler, 1964) measures the balance between accommodation and sediment transfer by surfical processes that erode, transport, and deposit sediment. Stratigraphic base level describes the energy required to move the Earth's physical surface up or down by deposition or erosion, respectively, to attain a condition of minimum energy expenditure where gradients, sediment supply, and accommodation are equilibrated. Stratigraphic base-level is not the same as geomorphic base-level (e.g., Schumm, 1993). Stratigraphic base level is neither sea level nor a geomorphic profile of equilibrium. These concepts provide the context in which variations in stratal architecture may be examined without an *a priori* assumption about specific relations between stratigraphic processes and responses. That is, it is unnecessary to know or assume what the contributions of primary forcing functions (e.g., tectonic movement, eustatic change, depositional topography, sediment supply, and lithosphere compensation to loads) were in creating the observed stratigraphy.

STRATIGRAPHIC FRAMEWORK

Three temporal and spatial scales of stratigraphic cycles are observed in Turonian–Coniacian strata (Fig. 2). A classification designated as long, intermediate, and short term is used to emphasize comparable scales (durations) of stratigraphic cyclicity rather than fixed cycle periods or assigning specific forcing mechanisms to each cycle frequency (i.e., second order = tectonic, third order = eustatic, etc.). Long-term cycles are about 4.5 my, intermediate-term cycles are 1 to 2 my, and short-term cycles are about 0.2 to 0.3 my.

Cycle durations were determined from overlapping sets of biozones within marine strata and from isotopic ages from volcanic ashes. Stratigraphic cycles are assumed to record synchronous base-level cycles and thus they can be used to extend the chronostratigraphic framework based on biochronozones and isotope data.

Regardless of scale, each stratigraphic cycle is the record of a complete base-level transit cycle *sensu* Wheeler (1964). During a base-level cycle, the accommodation/sediment supply ratio decreases to a limit (base-level fall), then increases to another limit (base-level rise). The limits, or "turnaround" points, of these accommodation/sediment supply ratio changes are synchronous throughout the spatial extent of each stratigraphic cycle. Thus, stratigraphic cycles of each scale are time-bounded rock units and comprise all strata accumulated during a base-level cycle. These base-level cycles are recorded by both rock and hiatal surfaces because sediment accumulation is not constant and continuous everywhere. Parts of the same base-level cycle can be preserved as rock, while other parts are preserved as surfaces of stratigraphic discontinuity.

Selection of the initiation point of a stratigraphic cycle is arbitrary but it must be consistent. In this study, the turnaround point from base-level rise to fall is used for stratigraphic cycles of all scales. This is the position of a maximum accommodation/sediment supply ratio. Selection of this turnaround point to define stratigraphic cycles has the potentially undesirable feature of placing major base-level

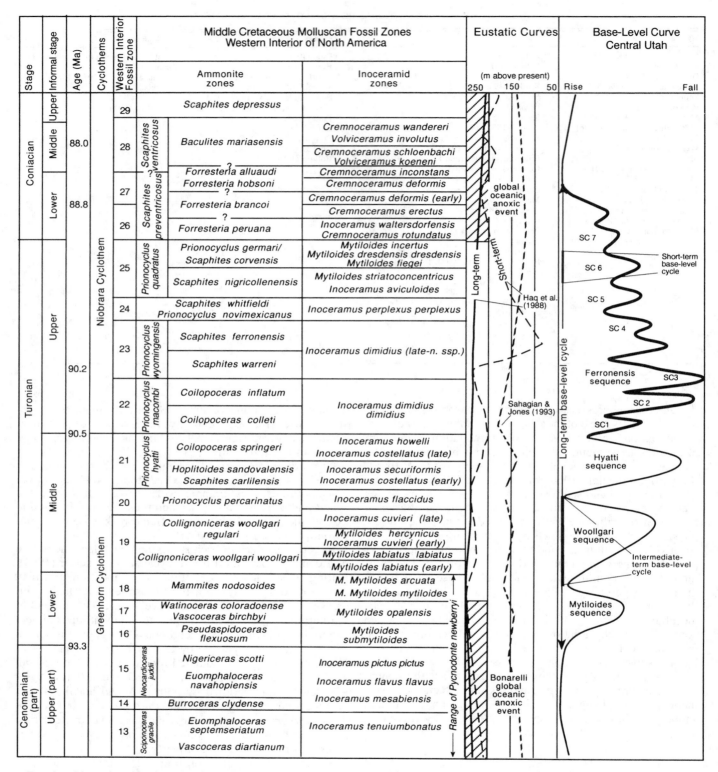

Fig. 2.—Biostratigraphic chart showing ammonite and inoceramid fossil zones for late Cenomanian through middle Coniacian Stages from the Western Interior of North America. Because of potential discrepancies associated with different absolute dates for stage boundaries, eustatic curves from Haq and others (1987) and Sahagian and Jones (1993) are calibrated to biozones. Base-level curves are based on stratigraphic relations in central Utah. Sources of data: Western Interior index fossils from Molenaar and Cobban (1991), upper Cenomanian and lower Turonian Substage biozones from Elder (1985); middle and upper Turonian Substage biozones from Kauffman and Collom (pers. commun., 1990), lower and middle Coniacian zones from Collom, (1991), argon-argon isotopic ages from Obradovich (pers. commun., 1991), and Obradovich (1991).

fall subaerial unconformities within stratigraphic cycles rather than at the cycle boundary. However, the base-level rise-to-fall turnaround point is most practical within this succession, as it is the position most easily recognized, documented, consistently picked, and physically traceable. In other deposits it might be preferable to select the other turnaround point (e.g., Van Wagoner and others, 1990).

Short-Term Stratigraphic Cycle

The smallest scale stratigraphic cycle comprises a progradational/aggradational unit that conforms to Walther's Law internally. Short-term cycles are generally conformable internally and include deposits and surfaces formed across all depositional environments during a short-term base-level cycle (Wheeler, 1964). Successive short-term cycles can be separated by surfaces of stratigraphic discontinuity marked by a facies offset. Such surfaces record erosional unconformities, sediment bypass surfaces, or nondepositional hiatal surfaces formed during either base-level rise or base-level fall time. Alternatively, successive short-term cycles can be separated by conformable strata that record the turnaround from base-level rise to fall. In such cases, a short-term cycle is marked by a progressive change in stratal geometry and facies architecture that records a history of first increasing and then decreasing accommodation/sediment supply ratio. For example, in the Ferronensis sequence of central Utah, short-term cycle boundaries occur at the top of condensed section deposits in marine-shelf strata, at the top of transgressive deposits that overlie a transgressive surface of erosion in shallow-marine strata, and at the change from sandstone-poor, organic-rich to sandstone-rich, organic-poor deposits in coastal-plain strata (Gardner, 1993).

Intermediate-Term Stratigraphic Cycle (Stratigraphic Sequence)

The boundaries of intermediate-term (1-to-2 my) stratigraphic sequences occur at the same base-level rise-to-fall turnaround of lower and higher frequency base-level cycles (e.g., long- and short-term stratigraphic cycles). Changes in accommodation during intermediate-term base-level cycles produce distinct stacking patterns of short-term cycles described geometrically as seaward stepping, landward stepping, and vertically stacked (Cross, 1988). Seaward-stepping cycles are deposited during intermediate-term base-level fall. Vertically stacked cycles can be deposited at the beginning of intermediate-term base-level rise. Landward-stepping cycles are deposited during intermediate-term base-level rise. Another succession of vertically stacked cycles may follow landward-stepping cycles and are deposited at the end of intermediate-term base-level rise and at the beginning of intermediate-term base-level fall (Fig. 3).

Four stratigraphic sequences are recognized in this study, each informally named after a molluscan biozone. In the basal Mytiloides sequence, the bivalve *Mytiloides* first occurs in the lower Turonian, providing a biostratigraphic calibration point. There is some overlap in the range of biozones in the successively younger Woollgari, Hyatti, and Ferronensis sequences (Fig. 4).

Boundaries of these stratigraphic sequences change character with position in the WICS and bounding strata type. In central portions of the WICS, stratigraphic sequences are commonly bounded by stratigraphic discontinuities within marine deposits. Such discontinuity surfaces formed either by nondeposition (sediment starvation) on top of landward-stepping cycles or by submarine erosion during the subsequent base-level fall. For example, hiatal (sediment starvation) surfaces can form in the basin center during base-level rise when all sediment is stored in coastal-plain and paralic facies tracts. Erosion surfaces bounding stratigraphic sequences in more proximal marine areas of the basin form during the subsequent base-level fall. Such erosion surfaces may record submarine channel scouring of base-level rise sediment and hiatal surfaces during base-level fall and sediment transported to the center of the seaway. Both types of surfaces (forming lacuna where merged; Wheeler, 1964) can be recognized paleontologically by missing molluscan biozones and stratigraphically by disconformity or paraconformity.

Along the western margin of the WICS, stratigraphic sequence boundaries are conformable in high-subsidence regions. Sequence boundaries are recognized here by progressive variations in the architecture of coastal-plain strata that indicates a change from increasing to decreasing accommodation. Landward of high subsidence regions in a foreland basin and in alluvial-plain strata, the sequence boundary is a presumably base-level fall unconformity on top of base-level rise strata. The change from unconformable sequence boundaries in the most landward position to conformable boundaries along the basin margin to unconformable boundaries in the center of the WICS is the reverse of the Exxon sequence model developed for passive continental margins (Vail and others, 1977).

Mid-Cretaceous Long-Term Stratigraphic Cycle

Turonian to early middle Coniacian strata of central Utah comprise one long-term stratigraphic cycle containing four intermediate-term stratigraphic sequences (Figs. 2–5). The long-term base-level cycle is a 500-m thick upward-coarsening succession of marine and nonmarine deposits bounded by deposits that formed during eustatic transgressions (Schlanger and Jenkyns, 1976; Arthur and Schlanger, 1979; Jenkyns, 1980; Schlanger and others, 1986), and is similar in duration (4.5 my) to cyclothems of the WICS (Hattin, 1964; Kauffman, 1969, 1977). Transgressions and associated landward shifts in sediment accumulation punctuate periods of long-term base-level fall and episodic eastward (basinward) shoreface progradation.

Base-level rise deposits are represented by pelagic carbonates in the Bridge Creek and Niobrara Limestones along the center of the WICS. In more proximal marine-shelf settings, base-level-rise time is represented by transgressive disconformities or horizons of calcareous concretions within marine mudstones. Along the western margin of the WICS adjacent to the Sevier thrust belt, vertically stacked coastal-plain and shoreface deposits record base-level rise. Long-term base-level fall is marked by a step-wise eastward shift of shoreface sandstones and laterally linked facies tracts into

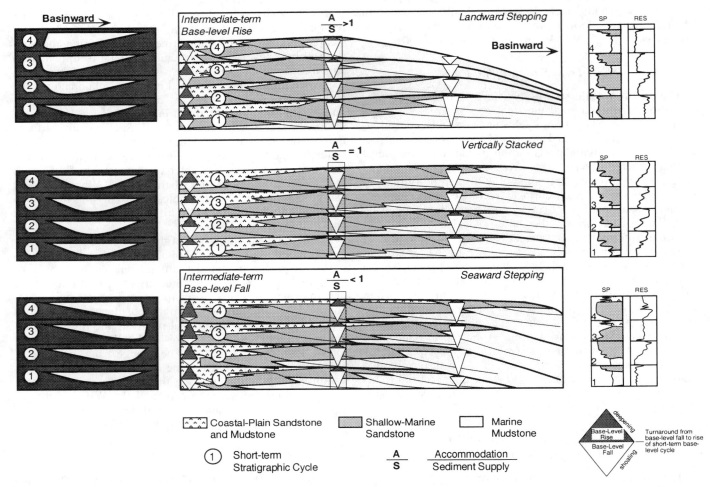

FIG. 3.—Diagram showing stacking patterns of short-term stratigraphic cycles. These are geometrically described as seaward stepping, landward stepping, and vertically stacked. Regularly occurring geographic changes in accommodation space during an intermediate-term base-level cycle produce distinct stacking patterns of short-term stratigraphic cycles. The accommodation profile shows how changes in accommodation are related to stratigraphic cycle stacking patterns. Schematic vertical profile shows how changes in facies arrangements are related to stratigraphic cycle stacking patterns. The accommodation-sediment supply ratio for different cycle stacking patterns is shown for a similar position along the longitudinal depositional profile. Diagram adapted from Van Wagoner and others (1991).

the basin, and basin-centered unconformities of stratigraphic sequences.

TURONIAN–CONIACIAN STRATIGRAPHIC SEQUENCES OF UTAH

Lithostratigraphic units of the four Turonian–Coniacian stratigraphic sequences of Utah and adjacent regions are shown in Table 2. The Utah study area encompasses the coastal-plain to marine-shelf segment of a longitudinal depositional profile (Fig. 3). Discussions about the stratigraphy of Turonian–Coniacian sequences in Utah are presented here in ascending stratigraphic order and from north to south across Utah. The three principal regions discussed are the Coalville area and Uinta basin of northern Utah, the San Rafael Uplift and flanking Castle Valley of central Utah, and the Henry Mountains region and Kaiparowits Plateau of southern Utah (Fig. 5). Across Utah these three regions coincide with middle Cretaceous foreland subbasins, each recording a unique accommodation and sediment supply history. A more detailed discussion of Turonian–Coniacian sequences from the central Utah foreland subbasin are given in Gardner (1994, this volume). Stratigraphic cross sections across Utah (Fig. 6), and facies maps of the Woollgari (Fig. 7), Hyatti (Fig. 8), and Ferronensis (Fig. 9) sequences provide a record of stratigraphic change in these foreland subbasins.

Mytiloides Sequence

In Utah, the lower boundary of the Mytiloides sequence coincides with the long-term stratigraphic cycle boundary.

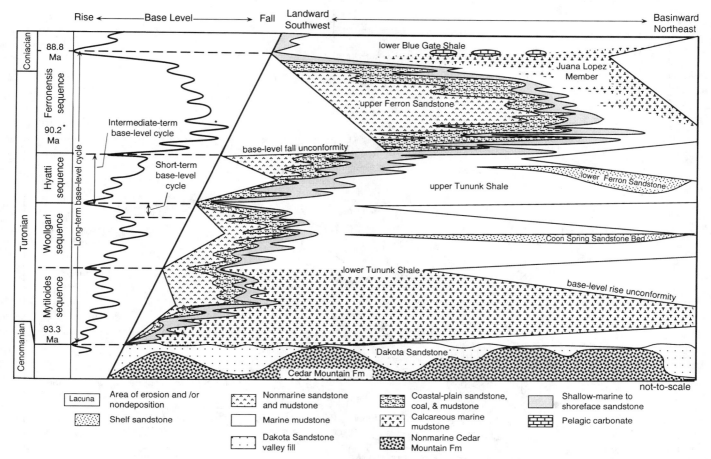

FIG. 4.—Schematic diagram showing stacking patterns of mid-Cretaceous strata in central Utah. Three temporal and spatial scales of stratigraphic cyclicity, each recording a base-level cycle of different periodicity, are recognized. Stratigraphic sequences show a step-wise progradation of more proximal deposits and a basinward migration of base-level fall unconformities in response to base-level fall of the long-term base-level cycle. Unconformities in the basin center are regionally expressed during intermediate-term base-level rise, when sediment starvation and nondeposition promoted hiatal surface development. Argon-argon isotopic data (Obradovich, pers. commun., 1993) are indicated by (*).

It is a disconformity separating the continental and shallow-marine Dakota Sandstone from the overlying marine shelf Tunuk Shale. This disconformity is regionally extensive and is overlain by an oyster coquina containing *Pycnodonte newberryi* (Stanton) formed during marine transgression and base-level rise. In central and southern Utah, the Dakota was deposited on a subaerial base-level fall unconformity during base-level rise. Thick Dakota sandstones are confined to incised paleovalleys. Outside of paleovalleys, the Dakota consists of thin and mudstone-dominated deposits with sandy concretions. Here, the base-level rise surface of disconformity is superimposed on, or has replaced, the previously formed base-level fall unconformity.

The Mytiloides sequence is an eastward-thinning, 50- to 200-m thick, calcareous mudstone wedge. Upward-shallowing marine-shelf mudstone facies successions define short-term cycles in this wedge. In central and southern Utah, short-term cycles are arranged in a seaward-stepping to vertically stacked to landward-stepping pattern. The stratigraphic sequence top is marked by the change from landward- (Mytiloides sequence) to seaward-stepping (Woollgari sequence) short-term cycles.

The most proximal facies in the Mytiloides sequence are exposed in northern Utah, near Coalville, where shallow-marine sandstones and mudstones contain the inoceramid *Mytiloides opalensis* (Fig. 10). From Coalville southward, the proportion of deeper outer-shelf strata (including calcareous shales and limestones) generally increases, and the frequency of unconformity surfaces decreases, reflecting deposition in deeper waters. Clastic supply to the WICS was from the north in Wyoming, and depositional areas in central and southern Utah were sediment starved.

The basal Mytiloides unconformity is most conspicuous and longest in duration east of Coalville in the Uinta basin, where strata are thinnest and accommodation was at a minimum. Here, short-term cycles of the Mytiloides sequence onlap a westward-dipping surface of unconformity with angular discordance (Fig. 11). This angular unconformity rises as much as 100 m to the west and separates the Albian Mowry Shale from the middle Turonian Tununk Shale (Weimer, 1960; Merewether, 1983).

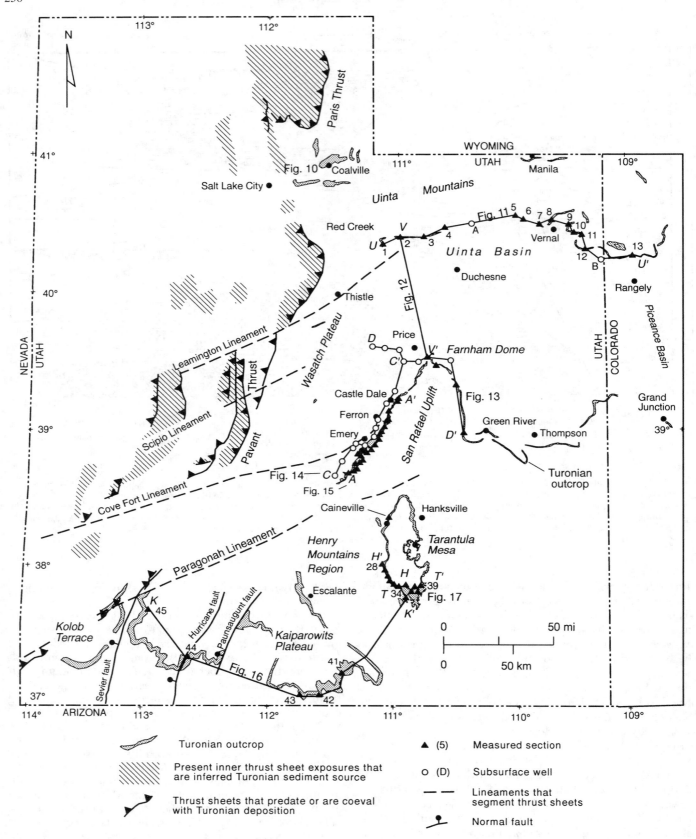

FIG. 5.—Map of Utah and western Colorado showing principal structural and paleogeomorphic elements, locations of Turonian outcrops, lines of cross sections, and locations of major study areas discussed in text.

TABLE 2.—CHRONOSTRATIGRAPHIC CHART FOR TURONIAN-CONIACIAN STRATIGRAPHIC SEQUENCES ACROSS CENTRAL WESTERN INTERIOR CRETACEOUS SEAWAY

(a)

Stage	Northern Wyoming				West Cody, Wyoming	Central Buffalo, Wyoming	East Newcastle, Wyoming
	Stratigraphic Sequence (1-2 Ma)	Western Interior Molluscan Fossil Zones	Fossil Number	Age (Ma)			
Coniacian (part) Lower (part)	Ferronensis sequence	Inoceramus deformis	27	88.8*	Cody Shale / Frontier Formation	Cody Shale / Sage Breaks Member (part)	Carlile Shale / Sage Breaks Shale Member
		Inoceramus erectus	26	89.3			
Turonian Upper		Prionocyclus quadratus	25			Frontier Formation Wall Creek Member	
		Scaphites whitfieldi	24				Turner Sandy Member
		Scaphites warreni 90.3*	23	90.3			
		Prionocylus macombi 90.5*	22				
Middle	Hyatti sequence	Prionocyclus hyatti	21	91.3	Frontier Formation Unnamed Member		
	Woollgari sequence	Prionocyclus percarinatus	20			Frontier Formation Billy Creek Sandstone	Pool Creek Member
		Collignoniceras woollgari	19	92.3			
Lower (part)	Mytiloides sequence	Mammites nodosoides	18				Greenhorn Formation
		Vascoceras birchbyi	17				
		Pseudaspidoceras flexuoxum	16	93.3*			

Legend: Marine mudstone | Shallow-marine to shelf sandstone | Pelagic carbonate | Area of erosion or nondeposition

\# Stratigraphic nomenclature arranged in ascending chronological order with highest rank at top.
Example:
Formation = Straight Cliffs Formation
(Older) Member or Bed = Tibbet Canyon Member
Member or Bed = Smokey Hollow Member
(Younger) Informal zone = Barren zone
Data for Table 2 shown in Table 1

Farther to the south in central and south-central Utah (San Rafael Uplift and northern Henry Mountains), seaward-stepping outer-shelf mudstones of the Mytiloides sequence downlap the basal sequence boundary (i.e., disconformity separating the Tunuk Shale from the underlying Dakota Sandstone and/or the Cedar Mountain Formation). In both regions, the Mytiloides sequence thins to the northeast. In the Henry Mountains, the sequence contains microfossil-rich calcareous shales (Hunt and others, 1953; Hunt and Miller, 1964; Smith and others, 1963; Lessard, 1973) that are correlative to fossiliferous calcareous shales that are capped by a bentonite cluster in the lower Tunuk and Tropic Shales of central and southern Utah, respectively (Table 2; Eaton and others, 1987). Locally along the east-flank of the San Rafael Uplift and farther southward in southern Utah (Kaiparowits Plateau), strata spanning the basal Mytiloides sequence boundary appear conformable, and in such cases the boundary is at the base of the most basinward positioned short-term cycle. Overlying thinly interbedded outer-shelf mudstone and sandstone are organized into vertically stacked cycles.

Woollgari Sequence

The Woollgari sequence is also dominated by outer- to inner-shelf facies organized into short-term cycles arranged in a seaward-stepping to vertically stacked to landward-stepping stacking pattern. Facies tracts of similar stratigraphic cycle position are displaced progressively farther east within the Woollgari sequence relative to those within the Mytiloides sequence. In the western part of the northern foreland subbasin, near Coalville, Utah, shallow-marine to shoreface sandstones cap short-term cycles and the proportion of sandstone in these cycles conspicuously increases to the east. Farther eastward, in the western Uinta basin, C. woollgari occurs at the base of the first marine sandstone

252   MICHAEL H. GARDNER

TABLE 2.—CONTINUED

(b) Central Wyoming Transect

Stage	Stratigraphic Sequence (1-2 Ma)	Western Interior Molluscan Fossil Zones	Fossil Number	Age (Ma)	West — Lander, Wyoming	Central — Southern Bighorn Basin, Wyoming	East — Casper, Wyoming	East — Douglas, Wyoming
Coniacian (part) / Lower (part)	Ferronensis sequence	Inoceramus deformis	27	88.8*	Cody Shale (part)	Cody Shale (part)	Cody Shale (part)	Area of erosion or nondeposition
Turonian / Upper	Ferronensis sequence	Inoceramus erectus	26	89.3	Frontier Formation Wall Creek Member	Frontier Formation Wall Creek Member	Frontier Formation Wall Creek Member	Frontier Formation Wall Creek Member
Turonian / Upper	Ferronensis sequence	Prionocyclus quadratus	25					
Turonian / Upper	Ferronensis sequence	Scaphites whitfieldi	24					
Turonian / Upper	Ferronensis sequence	Scaphites warreni 90.3*	23	90.3				
Turonian / Upper	Ferronensis sequence	Prionocylus macombi 90.5*	22					
Turonian / Middle	Hyatti sequence	Prionocyclus hyatti	21	91.3	Frontier Formation Emigrant Gap/Unnamed Member			
Turonian / Middle	Woollgari sequence	Prionocyclus percarinatus	20				Frontier Formation Emigrant Gap/Unnamed Member	
Turonian / Middle	Woollgari sequence	Collignoniceras woollgari	19					
Turonian / Lower (part)	Mytiloides sequence	Mammities nodosoides	18	92.3			Frontier Formation ?	Frontier Formation ?
Turonian / Lower (part)	Mytiloides sequence	Vascoceras birchbyi	17					
Turonian / Lower (part)	Mytiloides sequence	Pseudaspidoceras flexuoxum	16	93.3*				

(c) Southern Wyoming Transect

Stage	Stratigraphic Sequence (1-2 Ma)	Western Interior Molluscan Fossil Zones	Fossil Number	Age (Ma)	West — Kemmerer, Wyoming	Central — Rawlins, Wyoming	East — Cheyenne, Wyoming
Coniacian (part) / Lower (part)	Ferronensis sequence	Inoceramus deformis	27	88.8*	Hillard Shale	Cody Shale / Sage Breaks Member (part)	Niobrara Formation
Turonian / Upper	Ferronensis sequence	Inoceramus erectus	26	89.3	Frontier Formation	Frontier Formation Wall Creek Member	
Turonian / Upper	Ferronensis sequence	Prionocyclus quadratus	25		Dry Hollow Member		
Turonian / Upper	Ferronensis sequence	Scaphites whitfieldi	24				Frontier Formation Wall Creek Member
Turonian / Upper	Ferronensis sequence	Scaphites warreni 90.3*	23	90.3			
Turonian / Upper	Ferronensis sequence	Prionocylus macombi 90.5*	22		Oyster Ridge Member (upper part)		
Turonian / Middle	Hyatti sequence	Prionocyclus hyatti	21	91.3		Frontier Formation Unnamed Member	
Turonian / Middle	Woollgari sequence	Prionocyclus percarinatus	20		Oyster Ridge Member ? (lower part)		
Turonian / Middle	Woollgari sequence	Collignoniceras woollgari	19				
Turonian / Lower (part)	Mytiloides sequence	Mammities nodosoides	18	92.3	Allen Hollow Member / Coalville Member		
Turonian / Lower (part)	Mytiloides sequence	Vascoceras birchbyi	17				
Turonian / Lower (part)	Mytiloides sequence	Pseudaspidoceras flexuoxum	16	93.3*			

Legend: Marine mudstone; Shallow-marine to shelf sandstone; Pelagic carbonate; Coastal-plain sandstone and mudstone; Area of erosion or nondeposition.

TABLE 2.—CONTINUED

(d)

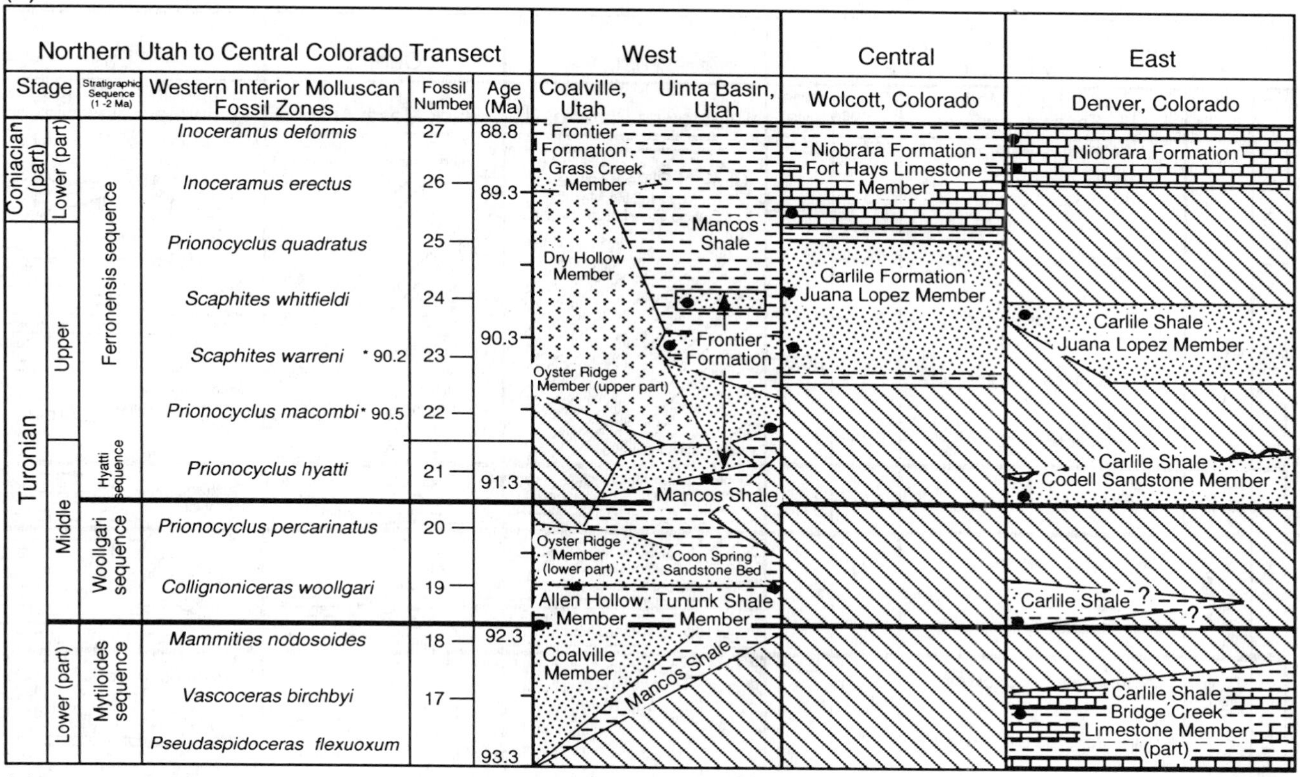

Northern Utah to Central Colorado Transect					West		Central	East
Stage		Stratigraphic Sequence (1-2 Ma)	Western Interior Molluscan Fossil Zones	Fossil Number	Age (Ma)	Coalville, Utah	Uinta Basin, Utah	Wolcott, Colorado	Denver, Colorado
Coniacian (part)	Lower (part)	Ferronensis sequence	Inoceramus deformis	27	88.8	Frontier Formation — Grass Creek Member		Niobrara Formation — Fort Hays Limestone Member	Niobrara Formation
Turonian	Upper	Ferronensis sequence	Inoceramus erectus	26	89.3	Dry Hollow Member	Mancos Shale	Carlile Formation Juana Lopez Member	
			Prionocyclus quadratus	25					
			Scaphites whitfieldi	24			Frontier Formation		Carlile Shale Juana Lopez Member
			Scaphites warreni *90.2	23	90.3	Oyster Ridge Member (upper part)			
			Prionocyclus macombi *90.5	22					
	Middle	Hyatti sequence	Prionocyclus hyatti	21	91.3		Mancos Shale		Carlile Shale Codell Sandstone Member
		Woollgari sequence	Prionocyclus percarinatus	20		Oyster Ridge Member (lower part)	Coon Spring Sandstone Bed		
			Collignoniceras woollgari	19		Allen Hollow Member	Tununk Shale Member		Carlile Shale ? ?
	Lower (part)	Mytiloides sequence	Mammites nodosoides	18	92.3	Coalville Member	Mancos Shale		
			Vascoceras birchbyi	17					Carlile Shale Bridge Creek Limestone Member (part)
			Pseudaspidoceras flexuoxum		93.3				

(e)

Central Utah to South-Central Colorado Transect					West	Central	East
Stage		Stratigraphic Sequence (1-2 Ma)	Western Interior Molluscan Fossil Zones	Fossil Number	Age (Ma)	San Rafael Swell, Utah	Delta, Colorado	Pueblo, Colorado
Coniacian (part)	Lower (part)	Ferronensis sequence	Inoceramus deformis	27	88.8		Niobrara Formation Fort Hays Limestone Member	Niobrara Formation Fort Hays Limestone Member
Turonian	Upper	Ferronensis sequence	Inoceramus erectus	26	89.3	Mancos Shale Blue Gate Member		
			Prionocyclus quadratus	25				
			Scaphites whitfieldi	24		Ferron Sandstone Member (upper part)	Carlile Formation Juana Lopez Member	Carlile Shale Juana Lopez Member
			Scaphites warreni 90.3*	23	90.3			
			Prionocyclus macombi 90.5*	22				
	Middle	Hyatti sequence	Prionocyclus hyatti	21	91.3	Mancos Shale Ferron Sandstone Member (lower part)	Mancos Shale	Carlile Shale Codell Sandstone Member Blue Hill Member
		Woollgari sequence	Prionocyclus percarinatus	20		Mancos Shale upper Tununk Member		Carlile Shale Fairport Member
			Collignoniceras woollgari	19		Coon Spring Sandstone Bed		Bridge Creek Limestone Member (upper part)
	Lower (part)	Mytiloides sequence	Mammites nodosoides	18	92.3	Mancos Shale lower Tununk Member	Mancos Shale	Carlile Shale Bridge Creek Limestone Member (part)
			Vascoceras birchbyi	17				
			Pseudaspidoceras flexuoxum	16	93.3			

Legend:
- Calcareous shale
- Marine mudstone
- Shallow-marine to shelf sandstone
- Pelagic carbonate
- Coastal-plain sandstone and mudstone
- Area of erosion or nondeposition
- ● Fossil collection

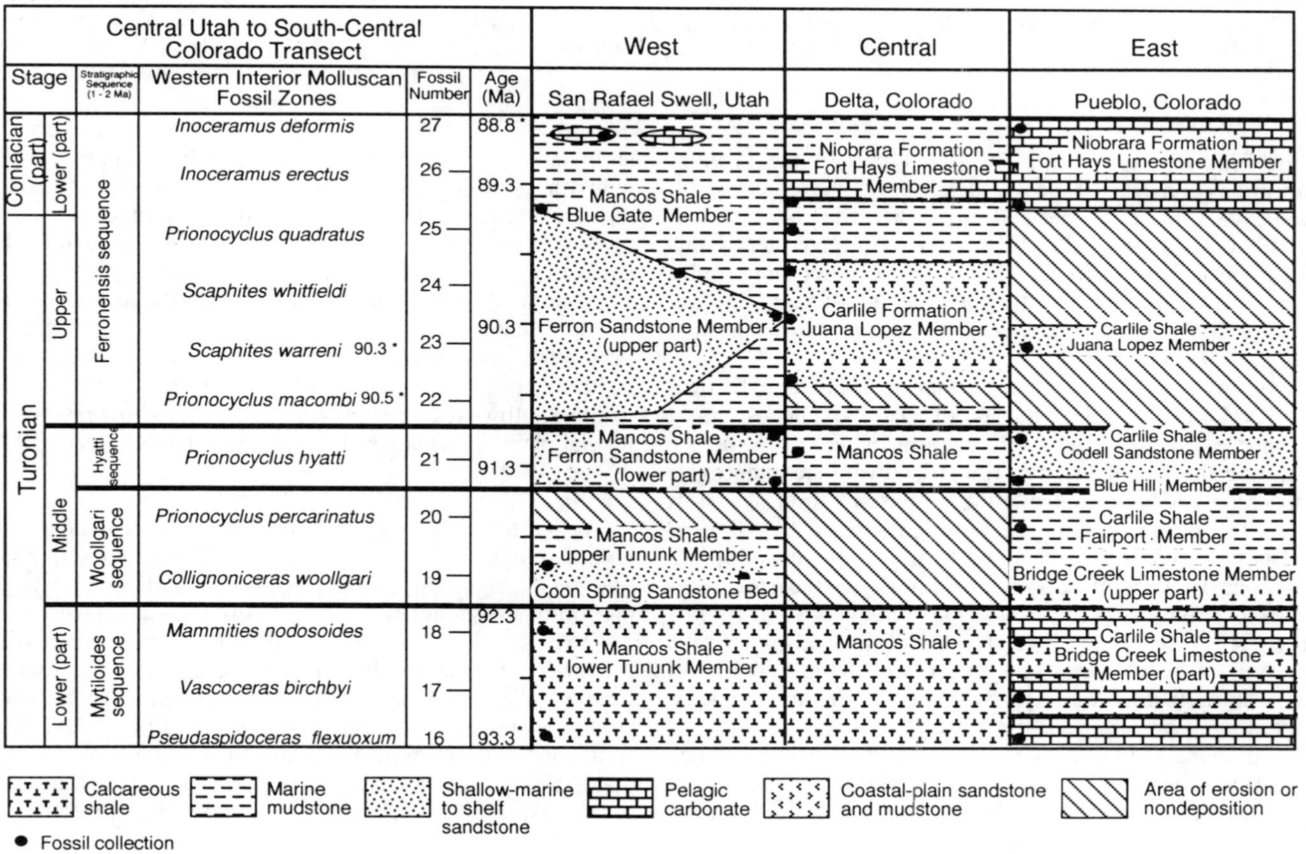

(f)

Southwestern Utah to Northern Arizona to West-Central New Mexico Transect					West	Central	East	
Stage		Stratigraphic Sequence (1 - 2 Ma)	Western Interior Molluscan Fossil Zones	Fossil Number	Age (Ma)	Kaiparowits Plateau, UT	Black Mesa, Arizona	Zuni Basin	San Juan Basin northern New Mexico
Turonian	Coniacian (part) / Lower (part)	Ferronensis sequence	Inoceramus deformis	27	88.8*	Straight Cliffs Formation Smokey Hollow Member Calico Sandstone bed	Toreva Formation (upper sandstone member)	Mancos Shale	Niobrara Formation
			Inoceramus erectus	26	89.3				Gallup Sandstone
	Upper		Prionocyclus quadratus	25					
			Scaphites whitfieldi	24					Mancos Shale
			Scaphites warreni 90.3*	23	90.3		Toreva Formation (middle carb. member)	Pescado Tongue	D-Cross Tongue
			Prionocyclus macombi 90.5*	22		Straight Cliffs Formation Smokey Hollow Member Tibbet Canyon Member		Tres Hermanos Formation Fite Ranch Member	Carlile Shale Juana Lopez Member
	Middle	Hyatti sequence	Prionocyclus hyatti	21	91.3		Toreva Formation (lower sandstone member)	Carthage Member	Semilla Sandstone Member
		Woollgari sequence	Prionocyclus percarinatus	20		Tropic Shale noncalcareous Member	Mancos Shale noncalcareous Member		Atarque Sandstone Member
			Collignoniceras woollgari	19		Hopi Sandstone Bed	Hopi Sandstone Bed		
	Lower (part)	Mytiloides sequence	Mammities nodosoides	18	92.3	Tropic Shale calcareous Member	Mancos Shale calcareous Member	Mancos shale Rio Salado Tongue	
			Vascoceras birchbyi	17					
			Pseudaspidoceras flexuoxum	16	93.3*				Greenhorn Limestone

Legend: Calcareous shale | Marine mudstone | Shallow-marine to shelf sndstn | Pelagic carbonate | Coastal-plain sandstone and mudstone | Area of erosion or nondeposition

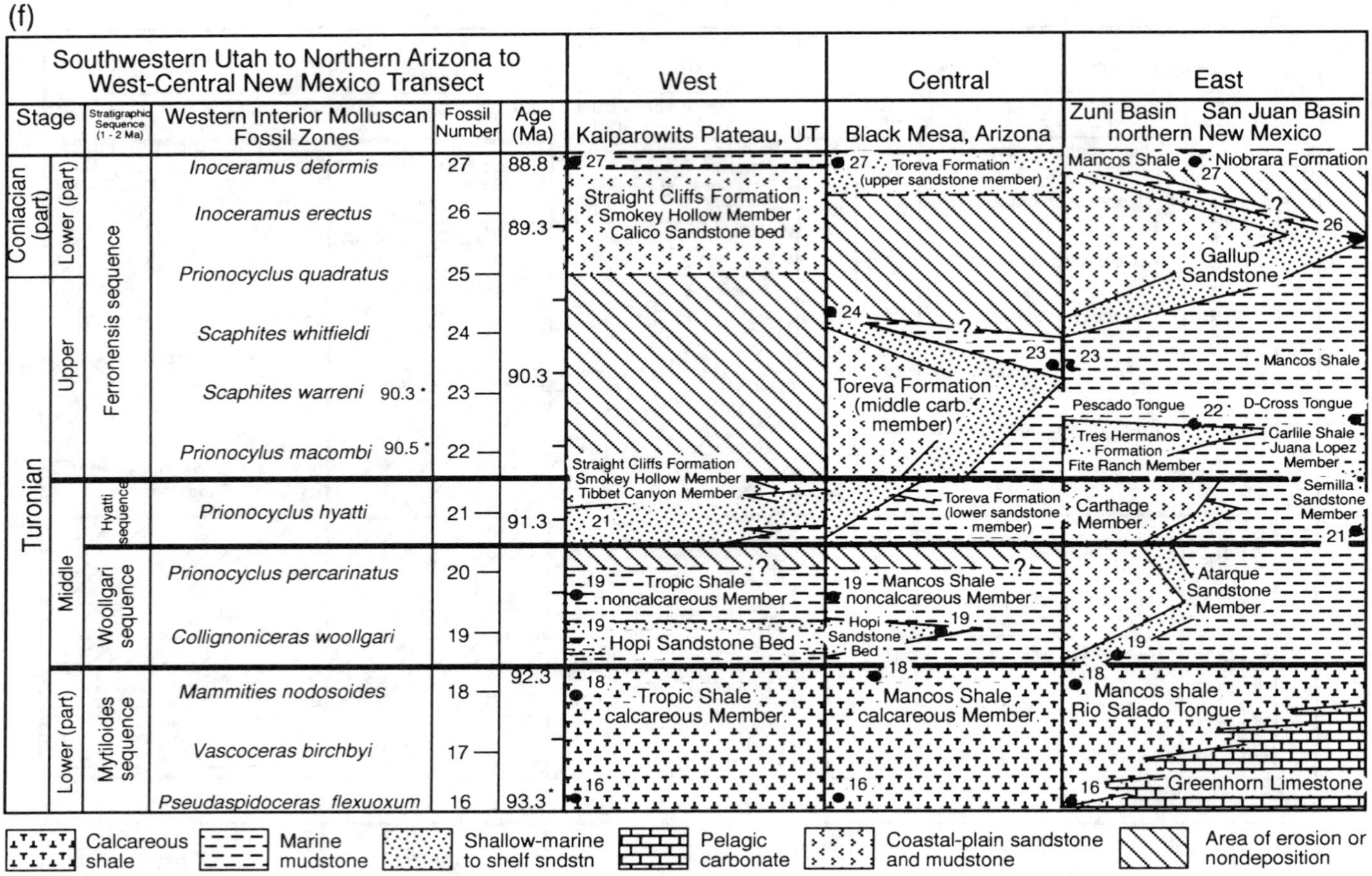

in the Frontier Formation. The lower part of the Frontier Formation consists of seaward-stepping shoreface sandstone cycles that lack coal and thin eastward (Fig. 12; Walton, 1944; Molenaar and Cobban, 1991). In the eastern Uinta basin, these sandstones are replaced by outer-shelf mudstones that onlap an unconformity, where strata of the underlying Mytiloides sequence are absent (Molenaar and Cobban, 1991).

A basinward shift in shallow-marine sandstone records base-level fall, and thus the Woollgari sequence starts at the base of the first seaward-stepping cycle (i.e., surface separating the top of the uppermost landward-stepping cycle of the Mytiloides sequence from the base of the lowermost seaward-stepping cycle of the Woollgari sequence). In central Utah, the Woollgari sequence starts with the Coon Spring Sandstone Bed locally containing poorly sorted, medium- to coarse-grained, fossiliferous sandstone with *C. woollgari* and chert-pebble lenses (Figs. 13–15). This sandstone forms a prominent bench along the east flank of the San Rafael Swell, thins to the south near Ferron, Utah, and thickens to the southwest along Castle Valley.

In the Henry Mountains region, the Woollgari sequence is a 60-m thick succession of outer-shelf mudstone and sandstone organized into seaward- then landward-stepping short-term cycles. Farther to the south (Kaiparowits Plateau) the Woollgari sequence coarsens upward from laminated calcareous shale to silty and sandy mudstones (Fig. 16; Peterson, 1969a). Here, Coon Spring Sandstone Bed equivalents (Hopi Sandy Member; Kirkland, 1990) recording base-level fall are about 60 m thick. The top of the Hopi Sandy Member marks the end of base-level fall and was previously recognized as a type-2 sequence boundary by Leithold (pers. commun., 1993). These sandstones thicken to the southwest and shows a conspicuous increase of terrestrial plant debris at Black Mesa, Arizona (Kirkland, 1990).

Hyatti Sequence

The vertical stacking pattern of short-term cycles in the Hyatti sequence is comparable to the two older sequences, but they prograded farther basinward and contain higher proportions of nonmarine and shallow-marine deposits (Figs. 13–15). The Hyatti sequence is sandstone rich in Utah, where it contains nonmarine deposits. In the Uinta basin, southern Castle Valley, and the Kaiparowits Plateau, shoreface sandstones occur in vertically stacked short-term cycles. These sandstones are overlain by a thin marine mudstone tongue that extends slightly landward in northern and southern Utah but is displaced much farther landward in central Utah.

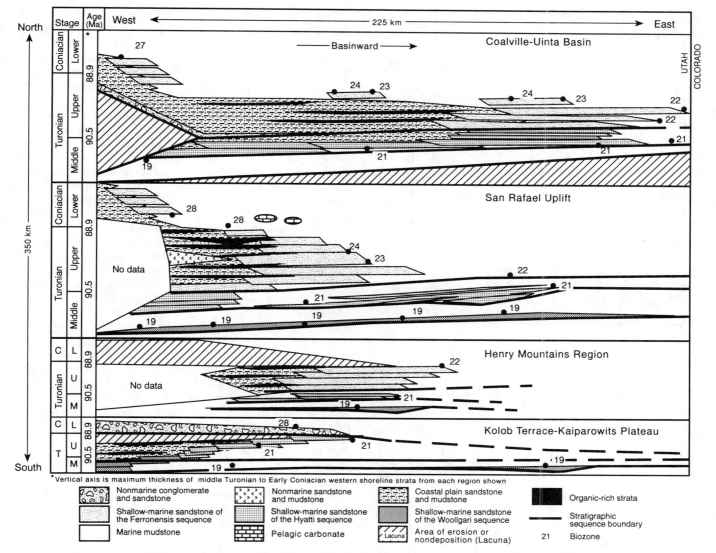

Fig. 6.—Diagram illustrating along-strike variations in stratal architecture of the upper three mid-Cretaceous stratigraphic sequences and component short-term stratigraphic cycles across Utah. Four stratigraphic cross sections along depositional dip illustrate cycle stacking patterns in stratigraphic sequences across northern, central, and southern Utah. Stratigraphic cross sections are modified from regional studies by Hale and Van De Graaf (1964), Peterson (1969a,b), Peterson and Ryder (1975), Ryer (1977), Ryer and McPhillips (1983), Eaton and others (1986), Molenaar and Cobban (1991), Reimersma and Chan (1993), and, Leithold (in press). Vertical axis of each cross section represents time scale to the average thicknesses of Turonian strata from each region. This representation of sediment accumulation rates illustrates along-strike variations in accommodation and sediment supply. Chronostratigraphic correlations are tied to biozones shown in Figure 2.

In northern Utah (near Coalville), the Hyatti sequence is cut out by an overlying unconformity (Hale, 1960, 1962; Hale and Van de Graaf, 1964). Farther to the east in the western Uinta basin, the Hyatti sequence records a pronounced basinward shift in shoreface sandstone (Fig. 11). Nearly 80 km basinward, these sandstones stack vertically producing a conspicuous stratal geometry attributed to relative sea-level rise (Ryer and Lovekin, 1986; Molenaar and Cobban, 1991).

In northern Castle Valley of central Utah, the Hyatti sequence contains five to seven short-term cycles arranged in a seaward-stepping, vertically stacked to landward-stepping stacking pattern. On outcrop, each short-term cycle grades upward from siltstone to hummocky cross-stratified sandstone to amalgamated hummocky and swaley cross-stratified sandstone. These deposits formed between storm- and fair-weather wave base on a storm-dominated shelf and were partly derived from the north (Figs. 12, 13). Coeval paralic and nonmarine strata occur in the subsurface near Price, Utah (Cotter, 1976; Ryer and McPhillips, 1983) and the Hyatti sequence thickens over 100 m farther landward northwest of Price (Fig. 13). Along depositional strike, in southern Castle Valley, south of Emery, Utah, these shelf sandstones pinch out at their depositional limit and are replaced through facies change by outer-shelf mudstones (Figs. 14, 15).

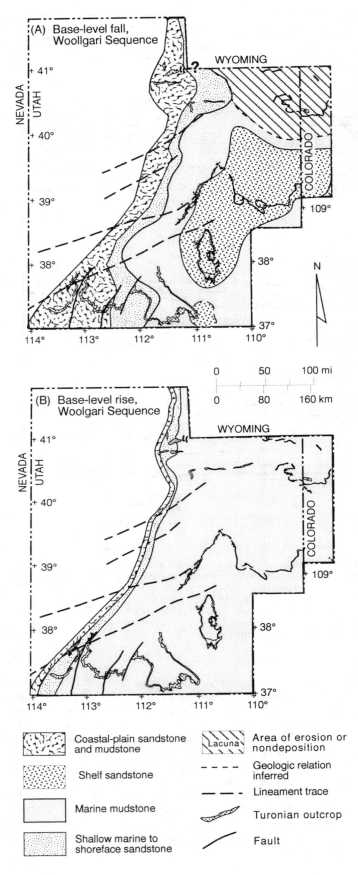

FIG. 7.—Map showing the distribution of facies and unconformities in the early middle Turonian Woollgari sequence of Utah. Each map shows a period of base-level fall and base-level rise for an intermediate-term base-level cycle about 1 to 2 my in duration. The lacuna in northeastern Utah spans 6 my, with strata from the underlying Mytiloides sequence absent.

The upper part of the Hyatti sequence in central Utah and the northern Henry Mountains is a distinctive black marine mudstone containing thin siltstone, sandstone, and bentonite interbeds. Organic matter in this southwestward-thinning mudstone is unusually rich in light carbon isotopes ($-26.19\ \partial^{13}C_{org}$ [‰ versus PDB]; Collom, unpubl. data, 1993). Similar carbon isotopic values (between -24 and $-26\ \partial^{13}C_{org}$ [‰ versus PDB]; Pratt and others, 1985) in organic-rich shale beds of the Bridge Creek Limestone have been interpreted to record a significant episode of global oceanic anoxia (Pratt and others, 1985). The organic-rich mudstone capping the Hyatti sequence in central Utah may also record a significant episode of oceanic anoxia. In southern Castle Valley, near Last Chance Canyon, this black mudstone is absent, and the Hyatti sequence consists of vertically stacked shoreface sandstones.

In the southern Henry Mountains region, the Hyatti sequence consists of northeast-thinning shallow-marine to shoreface sandstones. Shorelines trended north to northwest (Fig. 17). At least two seaward-stepping short-term cycles are recognized, which reach their depositional limits near Caineville Wash, Utah (Fig. 5). These are distal deposits of a northward-prograding shoreface complex that is thickest and broadest in the Kaiparowits Plateau region of southern Utah. In this region, the Hyatti sequence contains sandstone-dominated seaward-stepping to vertically stacked cycles that thin eastward, and *I. howelli* of the *P. hyatti* biozone occur (Eaton and others, 1987). Each short-term cycle shallows upward from wave-dominated shoreface sandstones that interfinger with and are overlain by nonmarine strata (Peterson, 1969a).

Ferronensis Sequence

In westernmost exposures of northern and southern Utah, the Ferronensis sequence is represented by a base-level fall unconformity overlain by a lag or bypass conglomerate formed during subsequent base-level rise. In northern Utah, this unconformity and overlying conglomerate is interpreted to correlate with locally conglomeratic, coarse arkosic fluvial sandstones in the upper Frontier Formation of the western Uinta basin (Walton, 1944; Williams and Madsen 1959; Ryer, 1977; Myers, 1977). In southern Utah, the Ferronensis sequence comprises a thin, locally conglomeratic sandstone interpreted as braided stream deposits locally confined to paleovalleys of the Calico Sandstone bed of the Smokey Hollow Member (Eaton and others, 1987; Bobb, 1991; Shanley and McCabe, 1992).

Where the base of the Ferronensis sequence is not an unconformity, it is placed at the base of the first seaward-stepping short-term cycle. In the Uinta basin, the Ferro-

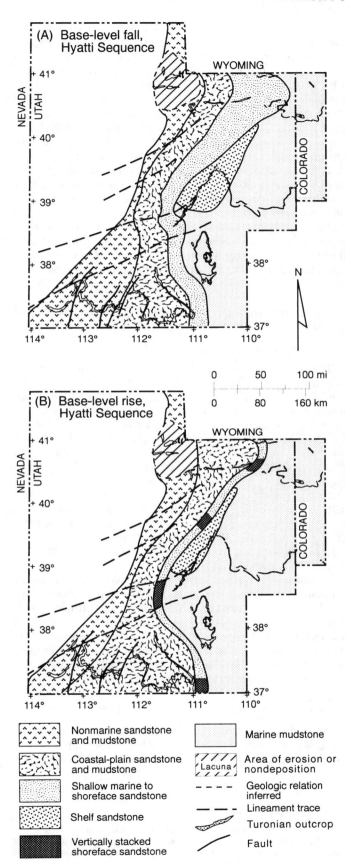

FIG. 8.—Map showing the distribution of facies and unconformities in the late middle Turonian Hyatti sequence of Utah. Each map shows a period of base-level fall and base-level rise for an intermediate-term base-level cycle about 1 to 2 my in duration. The Hyatti sequence is represented by an unconformity near Coalville, Utah.

nensis sequence begins with a series of seaward-stepping cycles. In the western Uinta basin, there is a vertical change from dominantly fine-grained floodplain strata to dominantly coarse fluvial sandstones containing conglomerate. This upsection change is inferred to record a basinward shift in alluvial deposition. Deposits coeval to these in the eastern Uinta basin contain seaward-stepping cycles consisting of the most basinward-positioned Turonian shoreface sandstones in Utah.

In central Utah and the Henry Mountains region, the Ferronensis sequence consists of northeast thinning clastic wedges sourced from the southwest (Katich, 1954). The Ferronensis sequence contains the greatest total sediment volume and most basinward-positioned Turonian fluvial deltaic deposits in central Utah (Figs. 11, 12). In central Utah, the Ferronensis sequence contains seaward-stepping, vertically stacked to landward-stepping short-term cycles that show an upward change from fluvial- to wave-dominated delta-front facies associations (Gardner, 1993). The Ferronensis sequence in the Henry Mountains is thinner, coarser-grained, and consists of seaward-stepping to vertically stacked cycles of coal-bearing fluvial-deltaic sandstone (Fig. 13). In southern Utah, conglomeratic fluvial sandstones in the Calico bed are inferred to represent the remnants of the fluvial system that fed these deltaic complexes. Unconformities have been recognized at the top of the Calico bed (Peterson, 1969a,b), at its base (sequence boundary of Shanley and McCabe, 1989, 1992), and at both lower and upper boundaries (Fig. 14; Eaton and others, 1987). The unconformity at the base of the Calico bed is inferred to be a bypass surface, with aggradation of fluvial deposits inferred to have occurred following deltaic deposition to the north. This unconformity may occur at the base of a similar fluvial sandstone succession in Black Mesa, Arizona (Franczyk, pers. commun., 1994).

A late Turonian transgression is recorded across Utah by base-level rise deposits in the Ferronensis sequence. Base-level rise is recorded by landward-stepping to vertically stacked cycle patterns (central Utah), mudstone-encased, shoreline-detached shoreface sandstones that were truncated by transgressive erosion surfaces (northern Utah), submarine disconformities (southern Utah, Henry Mountains), vertical accretion of coastal plain strata (central Utah), and fluvial aggradation in incised valleys (southern Utah, Kaiparowits Plateau).

Lower to middle Coniacian fauna (*Baculites mariasensis* and *Inoceramus deformis*) record maximum Ferronensis sequence water depths across Utah. These biozones occur in shallow-marine sandstones in northern Utah, near Coalville (Ryer, 1976, 1977), and in the Sanpete Valley of central Utah, (Ryer and McPhillips, 1983; Lawton, 1985). They occur in limestone concretions from southern Castle Val-

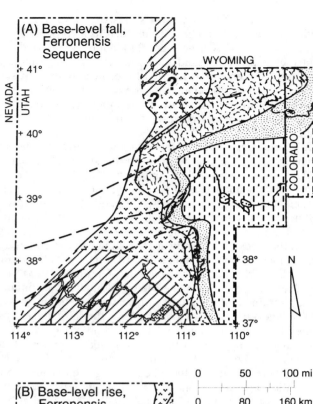
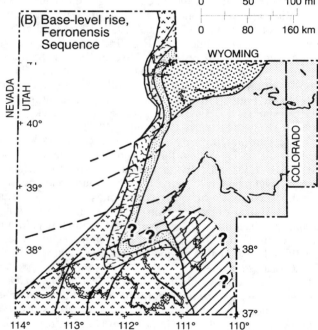

ley, and in strata (John Henry Formation) overlying the Calico bed in southern Utah (Eaton, 1990). In the southwestern Henry Mountains region, these deposits are replaced by an unconformity with up to 40 m of relief (Fig. 13; Peterson and Ryder, 1975). This unconformity is present over much of southern Utah and may be present in the Black Mesa region of northern Arizona, where it occurs at the base of a marine shale (Peterson and Ryder, 1975; Franczyk, pers. commun. 1994), but it rises stratigraphically to the north and becomes conformable in central Utah.

The unconformity is a sharp, low relief, laterally continuous surface overlain by an up to 0.5-m-thick, lenticular conglomerate containing shark's teeth and abraded oyster and inoceramid fragments (Hunt and others, 1953; Hunt and Miller, 1964; Peterson and Ryder 1975). Abraded marine macrofossils and overlying outer-shelf mudstones suggest the unconformity is submarine in origin. If so, this indicates the submarine unconformity formed during base-level rise.

CHARACTERISTICS OF TURONIAN–CONIACIAN STRATIGRAPHIC SEQUENCES IN WYOMING, COLORADO, AND NEW MEXICO

Mytiloides Sequence of the Central Western Interior

The initial Turonian eustatic transgression in the central WICS is recorded by an early Turonian disconformity overlain by a regionally extensive oyster coquina (*Pycnodonte newberryi*). This coquina occurs in shelf mudstones across central and southern Utah and in limestone beds of the lower Bridge Creek Limestone in west-central New Mexico and southern Colorado (Cobban and Hook, 1989). Deposition of the overlying Mytiloides sequence followed this long-term base-level rise, and it contains a high proportion of deeper water deposits relative to nonmarine deposits (Fig. 18). There is an eastward increase in the proportion of calcareous mudstone across Wyoming toward deeper basin areas (Fig. 19; Weimer, 1983). The best developed shoreface sandstones occur in southwestern Wyoming and northeastern Utah (Myers, 1977).

In Colorado and northern New Mexico, the Mytiloides sequence consists of outer-shelf deposits that change upward from calcareous to siliceous shales (Kauffman, 1977; Sonnenberg and Weimer, 1981; Weimer, 1983). In the San Juan and Zuni basins of north-central New Mexico, the Mytiloides sequence is a shale tongue (Rio Salado Tongue of the Mancos Shale) that becomes increasingly calcareous to the east and south (Molenaar, 1983).

Woollgari Sequence of the Central Western Interior

The Woollgari sequence is generally thinner and more sandstone-rich than the underlying Mytiloides sequence. The lateral persistence of shelf sandstones indicates widespread

FIG. 9.—Map showing the distribution of facies and unconformities in the late Turonian Ferronensis sequence of Utah. Each map shows a period of base-level fall and base-level rise for an intermediate-term base-level cycle about 1 to 2 my in duration.

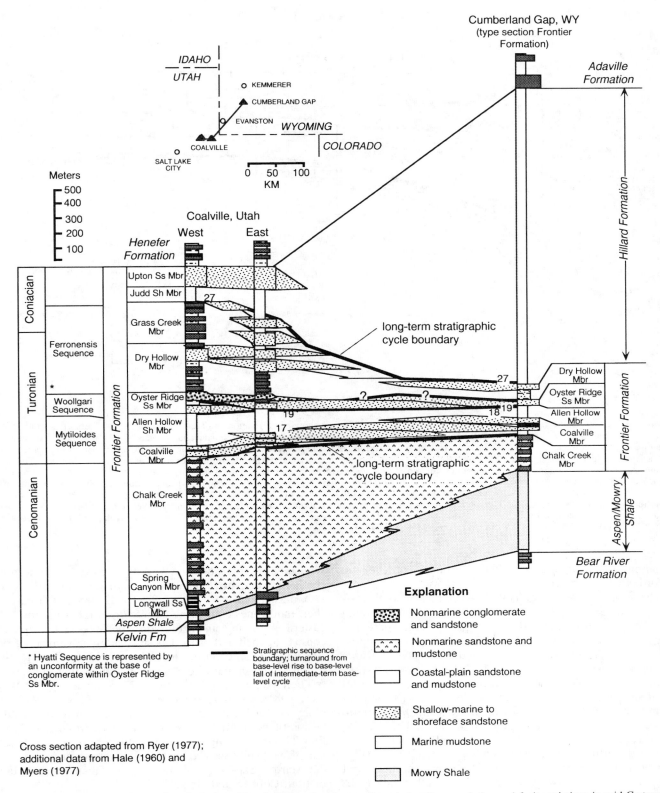

FIG. 10.—Stratigraphic cross section along depositional dip (southwest-to-northeast) showing correlation and facies relations in mid-Cretaceous stratigraphic sequences from Coalville, Utah, to Cumberland Gap, Wyoming. The measured section from Wyoming is from the type section of the Frontier Formation. Note that the Mytiloides sequence contains well-developed shoreline deposits and that the Hyatti sequence is absent in this region. This cross section is adapted from Ryer (1977) and shows stratigraphic framework outlined in text.

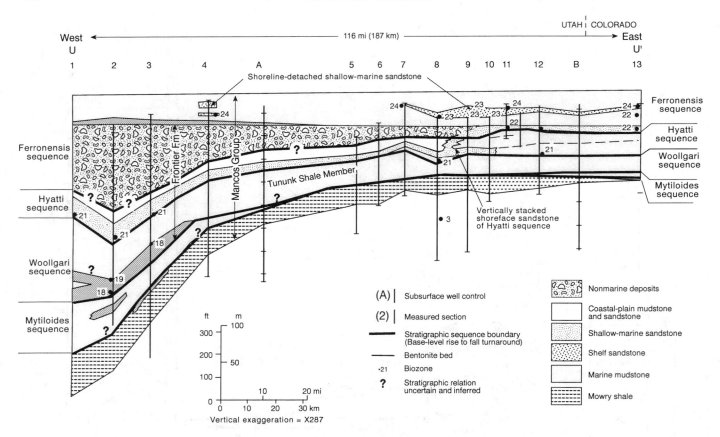

Fig. 11.—Stratigraphic cross section along depositional dip (west-to-east) showing correlation and facies relations in mid-Cretaceous stratigraphic sequences in the northern Uinta basin of northern Utah. This cross section is adapted from Molenaar and Cobban (1991) and shows stratigraphic framework outlined in text. Note pronounced vertical stack and stratigraphic rise of shoreface sandstones that comprise short-term stratigraphic cycles of the Hyatti sequence near section 8. The underlying, seaward-stepping stratigraphic cycles record base-level fall of the Hyatti sequence. The most basinward positioned, shallow-marine sandstones contain *P. macombi* (22) and comprise seaward-stepping stratigraphic cycles of the Ferronensis sequence. The Ferronensis sequence boundary is inferred to be near the base of these shallow-marine sandstones, whereas in nonmarine strata it is placed at the base of coarse, arkosic fluvial sandstones. Overlying shelf sandstones (23) are encased in shelf mudstones and were detached from the shoreline by ravinement during late Turonian relative sea-level rise. For location of cross section see Figure 5. Section 2 at Red Creek shown in Figure 12. Approximate distribution of nonmarine and coastal-plain deposits in Frontier Formation inferred from Red Creek section and descriptions from Walton (1944), Molenaar and Cobbam (1991), Molenaar (1994, pers. commun.).

shoaling related to an early middle Turonian sea-level fall recorded across the central Western Interior.

The Woollgari sequence is best developed in southwestern Wyoming, where two southwest-to-northeast-thinning delta lobes are recognized (Ryer, 1977; Fig. 10). These deltaic complexes each contain at least four seaward-stepping cycles of sharp-based shoreface sandstone (Oyster Ridge Sandstone Member; Myers, 1977). Across western Wyoming, the Woollgari sequence is a succession of seaward-stepping cycles of laterally discontinuous, unconformity-bounded, locally conglomeratic shallow-marine to shoreface sandstone (Merewether and others, 1975, 1979; Merewether, 1980, 1983; Merewether and Cobban, 1986). A broad north- to northwest-trending unconformity (Fig. 20) occupies the central third of Wyoming, north-central Colorado, and northwestern Utah (Merewether and Cobban, 1986). This unconformity is flanked by basin-restricted pebbly sandstone and mudstone deposits in central Wyoming, near Casper (Haun, 1953), and in north-central Colorado. In north-central Wyoming, a thick but restricted sandstone is replaced to the west by a broad paraconformity centered over central and northwestern Wyoming (Fig. 20).

A diastem in west-central Kansas occurs at the base of shelf mudstones containing *C. woollgari* (Hattin, 1964). In eastern Wyoming and Colorado, the Woollgari sequence consists of shelf and basin deposits that grade vertically from calcareous to siliceous shales (Weimer, 1978; Fig. 21). Siliceous shales contain ironstone concretions that increase in frequency upward (Kauffman, 1977; Sonnenberg and Weimer, 1981; Weimer, 1983; Weimer and Flexer, 1985). In more proximal deposits from the western San Juan and Zuni basins of northern New Mexico (R-1 of Molenaar, 1983; Fig. 22), the Woollgari sequence consists of seaward-stepping to vertically stacked cycles of shoreface sandstone (i.e., the Atarque Sandstone Member of the Tres Hermanos Formation; Cobban and Hook, 1979, 1989).

Hyatti Sequence of the Central Western Interior

In Wyoming, the Hyatti sequence contains twice the sandstone volume of underlying Mytiloides and Woollgari

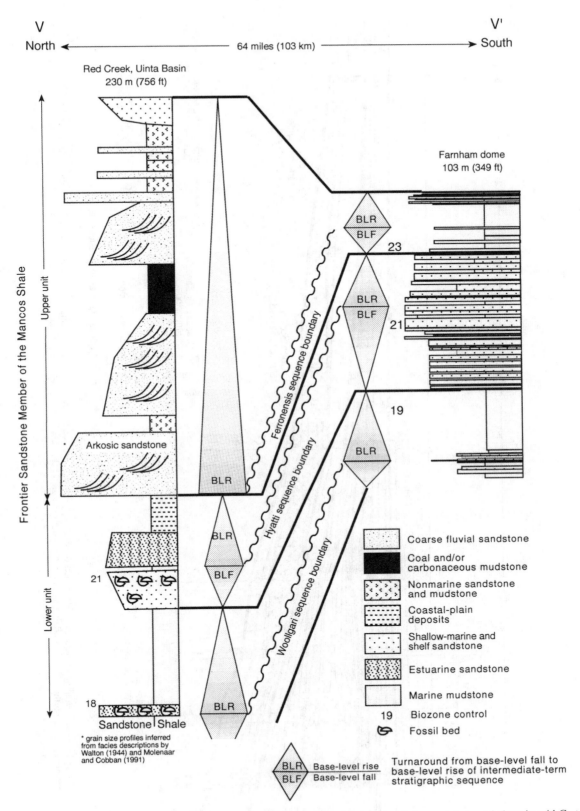

FIG. 12.—Stratigraphic cross section along depositional strike (north-to-south) showing correlation and facies relations in mid-Cretaceous stratigraphic sequences from the western Uinta basin in northern Utah, to Farnham dome in central Utah. Red Creek section is a composite profile from Walton (1944) and Molenaar and Cobban (1991; Fig. 15, p. 24). Stacking patterns of short-term stratigraphic cycles that comprise mid-Cretaceous stratigraphic sequences along depositional dip are shown in Figures 12 and 13. For location of cross section see Figure 5.

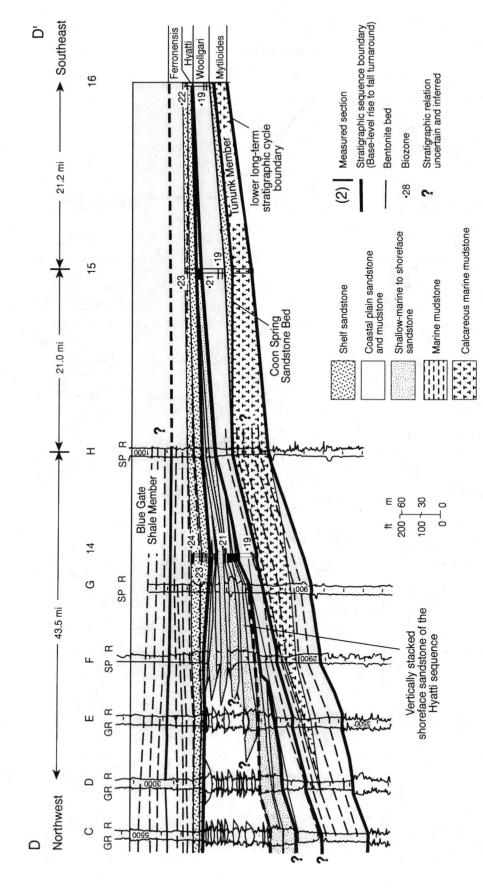

FIG. 13.—Stratigraphic cross section along depositional dip (northwest-to-southeast) of the Hyatti sequence showing correlation relationships in mid-Cretaceous strata of northern Castle Valley and along the eastern margin of the San Rafael Swell. For location of cross section D-D' see Figure 5.

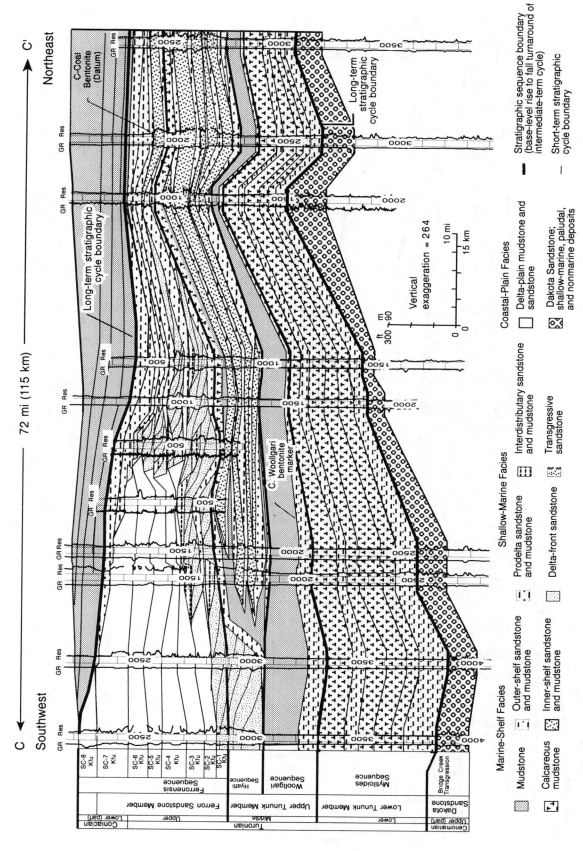

Fig. 14.—Subsurface stratigraphic cross section along depositional dip of the Ferronensis sequence showing correlation and facies relations in mid-Cretaceous stratigraphic sequences of central Utah. Cross section based on outcrops along the western margin of the San Rafael Swell in Castle Valley, Utah. For location of cross section C-C' see Figure 5.

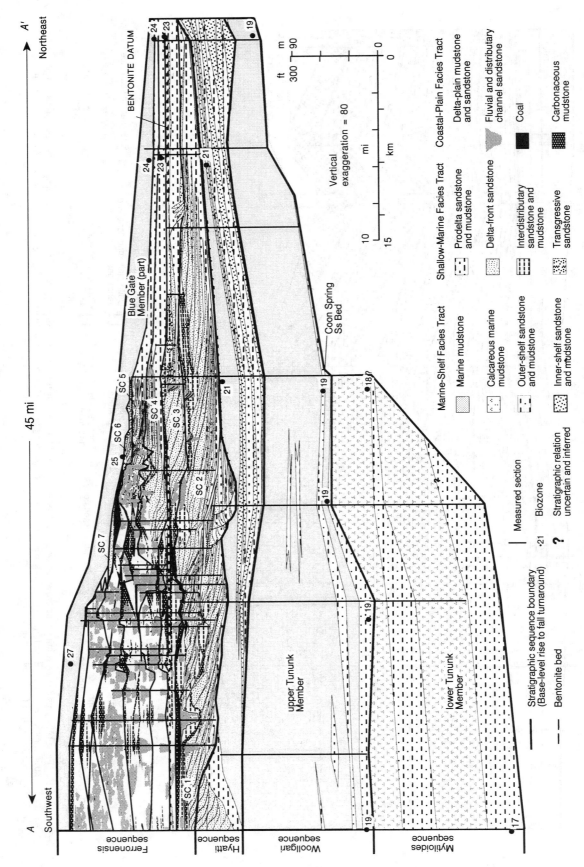

FIG. 15.—Stratigraphic cross section along depositional dip of the Ferronensis sequence showing correlation and facies relations in mid-Cretaceous stratigraphic sequences of central Utah. Cross section based on outcrops along the western margin of the San Rafael Swell in Castle Valley, Utah. For location of cross section A-A' see Figure 5.

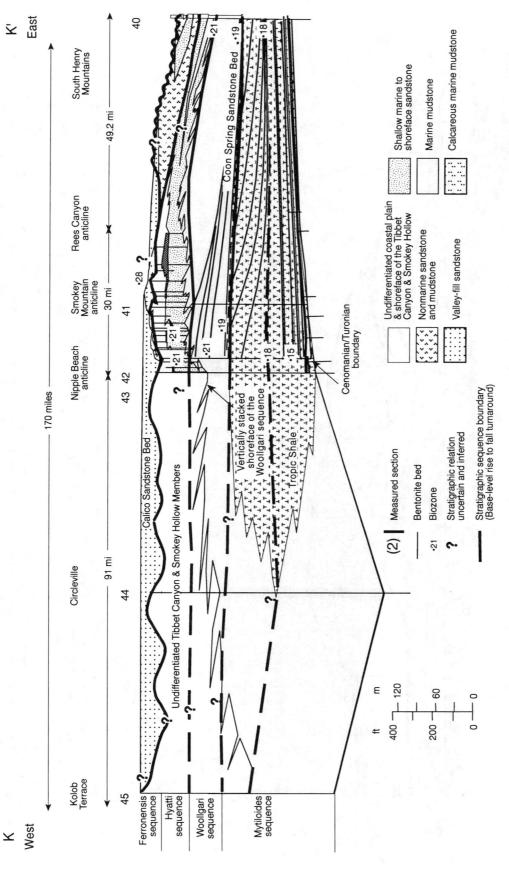

FIG. 16.—Stratigraphic cross section along depositional dip (west-to-east) showing correlation and facies relations in mid-Cretaceous stratigraphic sequences of southern Utah. Cross section is compiled from Peterson (1969a), Eaton and others (1987), and Leithold (in press). In the Kaiparowits region, the Tibbet Canyon Member contains *Prionocyclus hyatti* (21) of the Hyatti sequence. The Tibbet Canyon Member consists of wave-dominated shoreface sandstones that form a series of amalgamated, imbricated, seaward-dipping sandstone wedges. These sandstones show a change from seaward-stepping to vertically stacked to seaward-stepping stacking geometries that are inferred to record both the Woollgari and Hyatti sequences, as shown on cross section. Fluvial sandstones are confined to paleovalleys and overlie an unconformity at the base of the Calico Sandstone bed (Shanley and McCabe, 1992). The Calico sandstone bed is inferred to record base-level rise of the Ferronensis sequence. Aggradation of paleovalleys postdates landward-stepping short-term stratigraphic cycles in central Utah. For location of cross section see Figure 5.

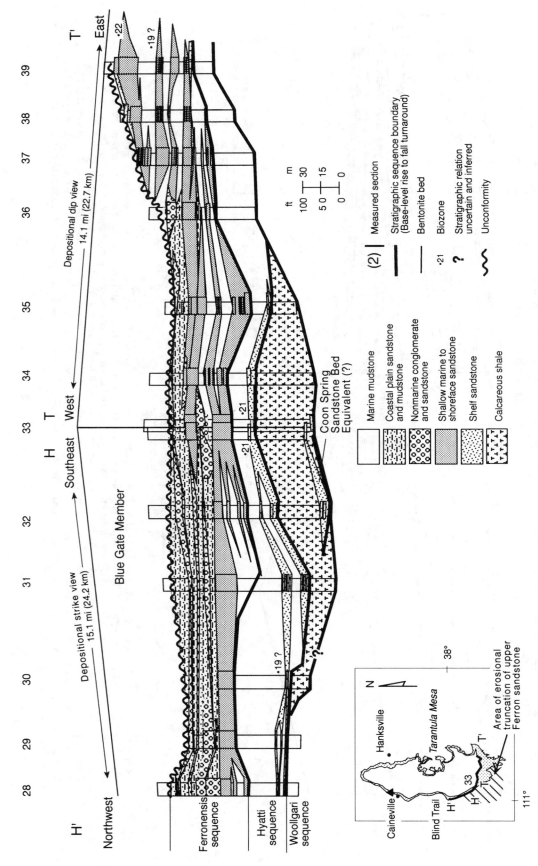

FIG. 17.—Stratigraphic fence diagram showing stratigraphic cross sections along depositional strike (southeast-to-northwest) and dip (west-to-east) of mid-Cretaceous stratigraphic sequences from Tarantula Mesa in the south-central Henry Mountains basin. Cross sections are modified from Peterson and Ryder (1975). Shallow-marine sandstones from the Post locality (33) contain *I. howelli* (21) of the Hyatti sequence. Correlations indicate that this sandstone bed and a stratigraphically higher sandstone bed both contain *I. cuveri* (19), an older inoceramid biozone. This biozone is five biozones below *I. howelli* and is part of the Woollgari sequence. Because *I. howelli* occurs in these sandstones (Leithold, pers. commun, 1993) and is consistent with collections of ammonites *P. macombi* listed above, the presence of *I. cuveri* remains problematic. The tongue of marine mudstone that separates shallow-marine sandstones containing *I. howelli* records base-level rise time of the Hyatti sequence. On the dip cross section, there is a basinward shift in the thickness of seaward-stepping cycles of shoreface sandstones containing *P. macombi* (22) of the Ferronensis sequence. The unconformity at the Ferron Sandstone–Blue Gate Shale contact shows a progressive increase in truncation to the southwest and is related to late Turonian tectonic movement and decreased sediment supply due to delta switching from southern to central Utah.

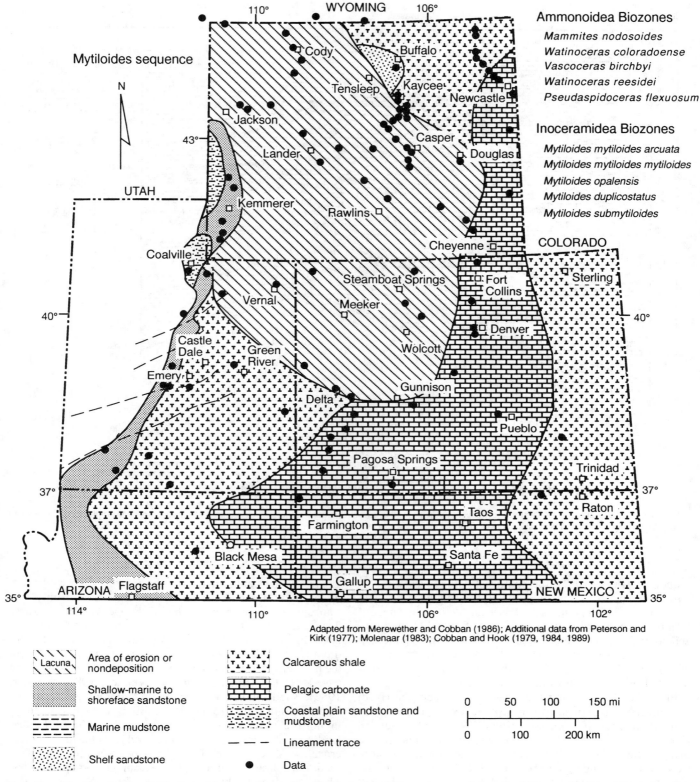

Fig. 18.—Map showing the distribution of facies and unconformities in the Mytiloides sequence of the central Western Interior of North America.

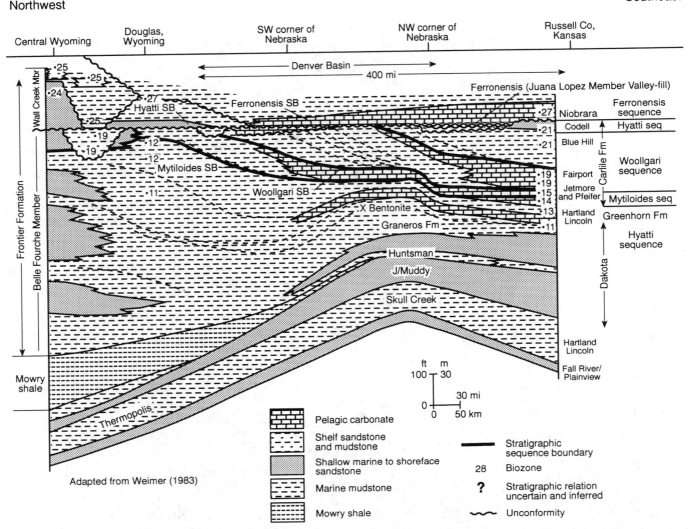

Fig. 19.—Stratigraphic cross section (northwest-southeast) showing correlation and facies relations in mid-Cretaceous stratigraphic sequences from central Wyoming to western Kansas. Cross section adapted from Weimer (1983).

sequences (Fig. 23). To the southeast across western Wyoming, nonmarine deposits in the Hyatti sequence occur in a clastic wedge that thins and is truncated by an unconformity to the northwest and east. It is flanked to the north and south by a broad area of shallow-marine sandstone that lacks nonmarine deposits (Fig. 23). In southwestern Wyoming and northwestern Utah, the Hyatti sequence is represented by an unconformity (Myers, 1977).

In southeastern Wyoming and eastern Colorado, the Hyatti sequence consists of seaward-stepping cycles of basin-restricted inner-shelf sandstone (Figs. 19, 21). In western Kansas, these shelf sandstones overlie an erosional surface that truncates progressively older deposits towards eastern Colorado (Sonnenberg and Weimer, 1981). Three facies are recognized: (1) sheetlike inner-shelf sandstone, (2) bioturbated sandstone, and (3) parallel- and ripple-laminated sandstone confined to paleovalleys (Sonnenberg and Weimer, 1981). In the southern Denver basin, sheetlike cycles are replaced by a broad northeast-trending unconformity that truncates strata over the center of the Denver basin (Table 2, Fig. 23). This unconformity joins a north-south-trending unconformity in north-central Colorado and isolates discontinuous shelf sandstones in the northern Denver basin. In the northern Denver basin deposits fill valleys incised during base-level fall (Sonnenberg and Weimer, 1981; Weimer, 1983). Shelf sandstones in southern Colorado are separated by an unconformity from thin, fossiliferous shelf sandstones in north-central New Mexico (Molenaar, 1983). Farther landward, in the western San Juan and Zuni basins of northern New Mexico, the Hyatti sequence contains shoreface sandstones arranged in a seaward-stepping short-term cycles. These are overlain by vertically stacked cycles of interfingering nonmarine deposits and shelf mudstones that lack preserved shoreface sandstone (Fig. 22; Molenaar, 1983).

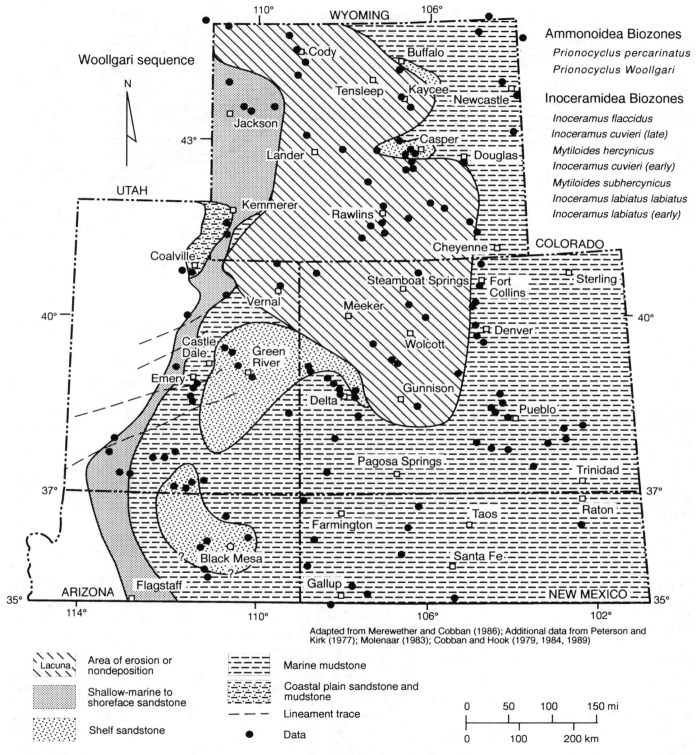

FIG. 20.—Map showing the distribution of facies and unconformities in the Woollgari sequence of the central Western Interior of North America.

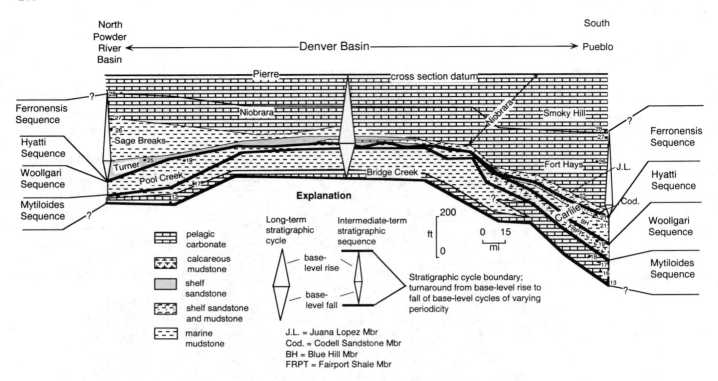

FIG. 21.—Stratigraphic cross section (north-south) along the seaway axis showing correlation and facies relations in mid-Cretaceous stratigraphic sequences from north-central Wyoming to Pueblo, Colorado. Cross section adapted from Weimer (1978).

Ferronensis Sequence of the Central Western Interior

Across Wyoming, short-term cycles comprising the Ferronensis sequence are characterized by high-energy, wave-dominated, sharp-based, shallow-marine to shoreface sandstones. Nonmarine deposits are sparse (Barlow and Haun, 1966; Prescott, 1975; Tilman and Almond, 1979; and Winn and others, 1983). These short-term cycles are arranged in a seaward-stepping stacking pattern and comprise the highest volume of basinward positioned sandstone in the central WICS (Figs. 24, 25). Near the seaward depositional limit of these short-term cycles, shallow-marine sandstones, commonly arranged in a landward-stepping stacking pattern, are truncated and detached from coeval shoreface deposits and are encased in marine mudstones. Interpreted as palimpsest deposits, these shoreline-detached sandstones were deposited during base-level fall but were reworked during the subsequent period of base-level rise, shoreline ravinement, and waning sediment supply.

Over much of Wyoming, the lower boundary of the Ferronensis sequence is an unconformity (Haun, 1953; Merewether and others, 1975, 1979; Merewether, 1980, 1983; Weimer, 1983; Merewether and Cobban, 1986). In southwestern and northern Wyoming, the Ferronensis sequence unconformably overlies the Woollgari sequence (Merewether and others, 1975). Northeast-trending shallow-marine sandstones east of Casper, Wyoming, overlie a disconformity present across the Powder River basin and are replaced to the west by a northwest-trending unconformity that extends southeastward to Pueblo, Colorado (Fig. 24; Haun, 1953; Merewether and others, 1979; Merewether and Cobban, 1986). In northeastern Wyoming, shelf sandstones in the Ferronensis sequence overlie a surface of unconformity that cuts more deeply towards the west, and is associated with northeast-trending paleovalleys (Weimer and Flexer, 1985).

In the center of the WICS, base-level fall in the Ferronensis sequence is recorded in part by thinly interbedded calcareous sandstone and mudstone shelf deposits (Juana Lopez and Turner Sandy Members). These shelf deposits are restricted to Colorado, southern and eastern Wyoming, and eastern Utah. Across Colorado, the Ferronensis sequence is mudstone-dominated and deposits are more calcareous upwards (Figs. 21, 24). To the east, these shelf deposits thin and overlie an erosional surface marked by a coarse sandstone lag with chert and phosphate pebbles (Hattin, 1975; Weimer, 1983; Witze and others, 1983; Merewether and Cobban, 1986). Argon-argon isotopic dates establish temporal equivalency of volcanic ash beds in the upper Ferron Sandstone (90.21 ± 0.72 Ma) and near the top of the Juana Lopez (90.5 ± 0.45 Ma; dates from Obradovich, unpubl. data, 1992) Members of the Mancos Shale in central Utah and northwestern New Mexico, respectively. Farther to the southwest, in the Zuni basin of western New Mexico, Juana Lopez Member shelf deposits are replaced by northeastward-thinning shallow-marine sandstones containing *P. macombi* (i.e., Fite Ranch Sandstone Member; Fig. 22; Cobban and Hook, 1989). These sandstones are interpreted as transgressive in origin (Molenaar, 1983). Overlying marine mudstones of the Pescado marine tongue record transgression, marine deepening, and base-level rise.

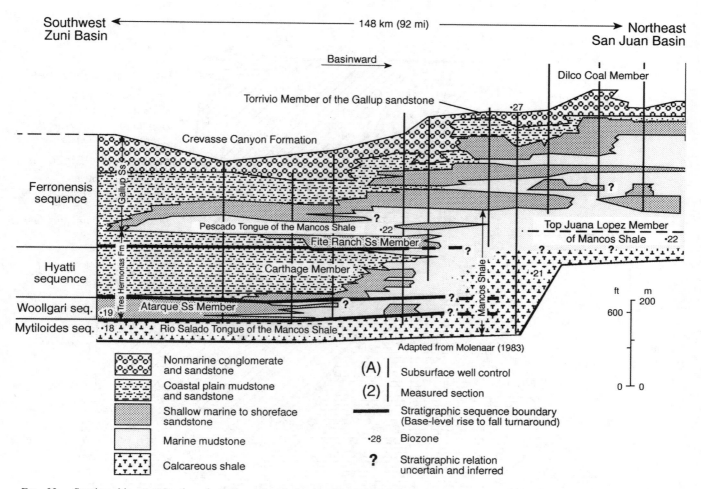

Fig. 22.—Stratigraphic cross section along depositional dip (southwest-northeast) showing correlation and facies relations in mid-Cretaceous stratigraphic sequences from the Zuni and western San Juan basins of north-central New Mexico. Cross section adapted from Molenaar (1983).

Across the central WICS, base-level rise in the Ferronensis sequence is recorded by pelagic carbonates and shelf mudstone deposits. In south-central Colorado, base-level rise is recorded by a submarine unconformity, where the Fort Hays Limestone, (containing limestone clasts of the Juana Lopez Member), disconformably overlies the Codell Sandstone (Sonnenberg and Weimer, 1981; Weimer, 1983; Weimer and Flexer, 1985; Cobban, 1988). This discontinuous submarine unconformity is present in Wyoming, southern Utah, and northern New Mexico, where it occurs at the top of the Juana Lopez Member.

Base-level rise in the Ferronensis sequence is also recorded by migration of sediment depocenters from the basin center to the basin margin, contraction of basin-centered unconformities, deposition of pelagic carbonates in the east and south, and deposition of marine mudstones that form a northward-extending tongue centered over western Colorado and eastern Utah (Fig. 25). In northern New Mexico, however, this period of base-level rise is recorded by wave-dominated and tidally influenced shoreface sandstones that prograded as seaward-stepping cycles northeastward into the basin (i.e., Gallup Sandstone; Molenaar, 1983; Valasek, 1990). The increased sandstone volumes recorded by these strata in southern reaches of the WICS records the large-scale north-to-south migration of depocenters through the Turonian to early middle Coniacian.

MIGRATION PATTERNS OF TURONIAN-CONIACIAN DEPOCENTERS

Six foreland subbasins comprise the western margin of the central WICS, each recording a unique accommodation and sediment supply history. A progressive southward increase in sediment volumes in progressively younger Turonian–Coniacian stratigraphic sequences records the migration of depocenters across these foreland subbasins (Fig. 26). Southward migration of depocenters and northward movement of Atlantic–Tethyan water masses promoted deeper water conditions for longer periods of time in southern reaches of the central WICS. Although eustasy should produce a diagnostic stratal architecture across this region, it does not because of variable tectonic subsidence and sediment supply rates in foreland subbasins. Intermediate-term base-level turnarounds of stratigraphic sequences are correlated across foreland subbasins, so that the role of tec-

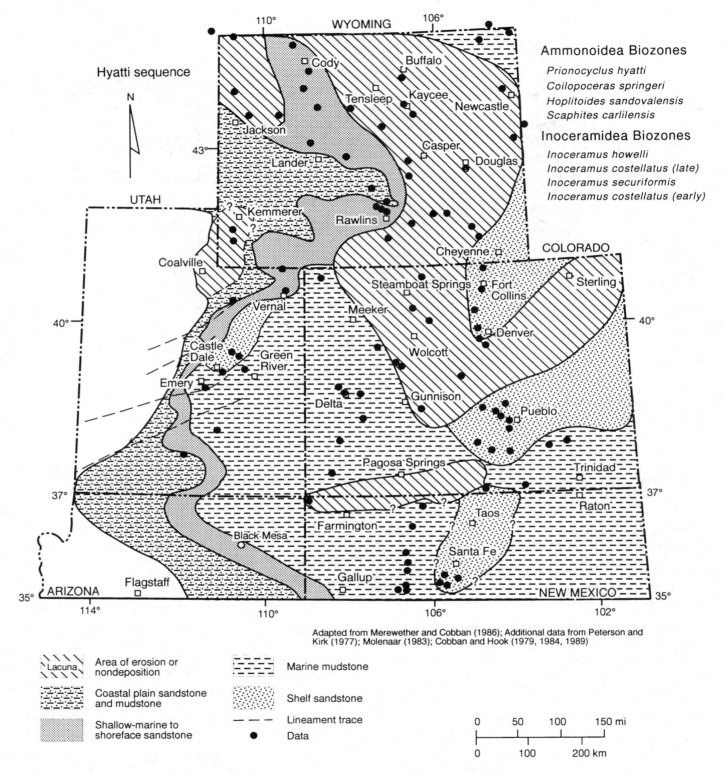

Fig. 23.—Map showing the distribution of facies and unconformities in the Hyatti sequence of the central Western Interior of North America.

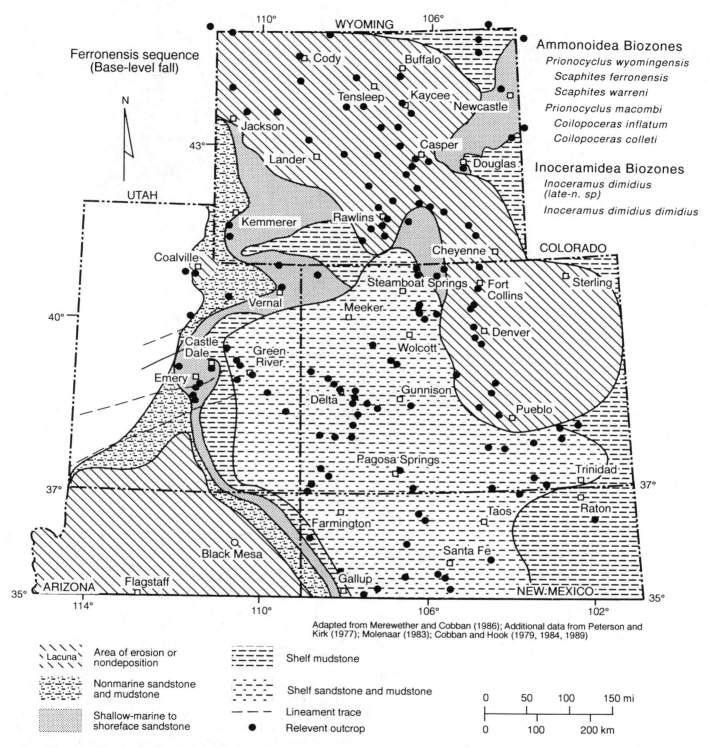

Fig. 24.—Map showing the distribution of facies and unconformities in the Ferronensis sequence of the central Western Interior of North America. This map shows only base-level fall for an intermediate-term base-level cycle about 1 to 2 my in duration.

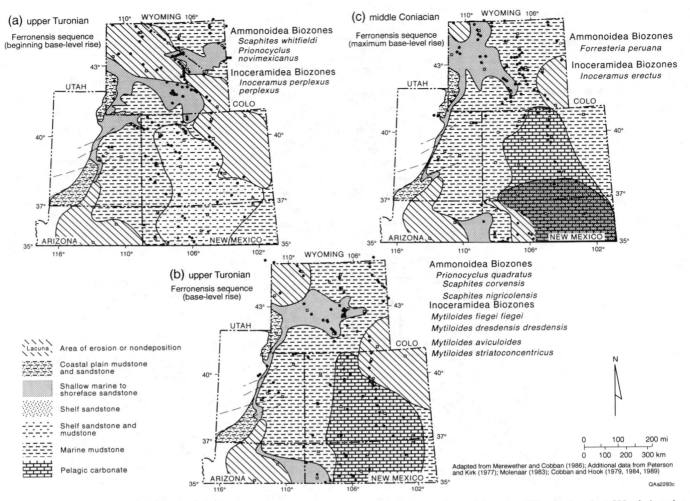

Fig. 25.—Maps showing the distribution of facies and unconformities in the Ferronensis sequence of the central Western Interior of North America. This map shows only base-level rise for an intermediate-term base-level cycle about 1 to 2 my in duration.

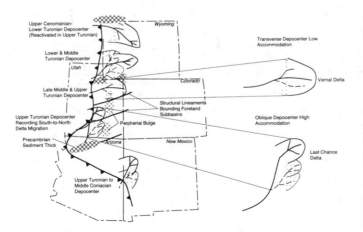

Fig. 26—Map of central portion of U. S. Western Interior Cretaceous Seaway showing the north-south migration of mid-Cretaceous depocenters across Utah, and the migration of Ferronensis sequence from southern to central Utah. The northward migration of Tethyan-Atlantic water mass is inferred to have affected the stratal architecture of stratigraphic sequences along the axis of the seaway.

tonic versus eustatic controls on three-dimensional variations in stratal architecture may be documented.

North-to-South Migration Patterns of Mid-Cretaceous Depocenters

Turonian–Coniacian strata of the central WICS show along-strike migration of depocenters. The major, long-term pattern is a north-to-south migration from Wyoming to northern New Mexico. North-to-south depocenter migration is related to emplacement of thrust sheets in the foreland fold-and-thrust belt earlier in the north than in the south. As successive thrust sheets were emplaced, rivers drained axially along the margin of the basin, and points of debouchment of river mouths migrated southward. A secondary intra-depocenter migration is south to north; the best documented example is in southern Utah during Ferronensis sequence time. South-to-north depocenter migration is related to abandonment of one major delta in the south, extension of the fluvial system to the north, and deposition of another delta to the north of the first. Facies tract maps of the Woollgari, Hyatti, and Ferronensis sequences (Figs. 7–9) show these important spatial changes across the basin.

The Mytiloides and Woollgari sequences are absent in the Bighorn basin of northwestern Wyoming, and an Upper Cenomanian clastic wedge represents the oldest and most northward positioned middle Cretaceous depocenter. The Mytiloides and Woollgari sequences are thickest in southwestern Wyoming. The maximum Hyatti shoreline regression extended into northeastern Utah (i.e., Vernal delta), but extended only into Sanpete Valley in central Utah, producing a pronounced northeast-to-southwest shoreline trend (Figs. 5, 8). The lower part of the Ferronensis sequence is thickest in central (i.e., Last Chance delta) and southern Utah, whereas the upper part of the Ferronensis sequence (Gallup Sandstone) is thickest, coarsest, and has the highest sandstone volumes in northern New Mexico.

In the San Juan basin of New Mexico, the inoceramid biozone *I. erectus* occurs near the seaward limit of the Gallup sandstone and is unconformably overlain by pebbly shelf sandstones containing *I. deformis* (i.e., Tocito Sandstone Lentil; Molenaar, 1983). Fossils of the inoceramid biozone *I. deformis* also occur in shallow-marine sandstones in west-central and northern Utah (Lawton, 1985) and in calcareous concretions in marine mudstone in central Utah. Shoreface sandstones in seaward-stepping short-term cycles of the Gallup Sandstone in the San Juan basin are at least 300 km east of their orogenic belt. By comparison, coeval shelf mudstones in the Ferronensis sequence are within 70 km of the central Utah thrust belt; shoreface sandstones occur within a narrow belt centered about 30 km from the thrust front in Sanpete Valley.

The Gallup Sandstone and Ferron Sandstone of the Ferronensis sequence record maximum progradation during the same intermediate-term base-level cycle. The Gallup Sandstone is temporally younger, however, and is equivalent to the upper Ferron Sandstone and Blue Gate Shale of the Ferronensis sequence. The difference is attributed to an increase in sediment-supply rates relative to accommodation southward in successively younger strata because of the southward migration of Turonian-Coniacian depocenters. The southward increase in Ferronensis sequence sediment supply, however, was not associated with southward increasing subsidence. For example, the site of maximum subsidence in the Ferronensis sequence is recorded by the Blue Gate Shale in Sanpete Valley of central Utah, but the highest sandstone volumes are contained within the coeval Gallup Sandstone of northern New Mexico. Although southward increasing sediment supply across foreland subbasins appears to be linked to tectonic activity in the orogenic belt, subsidence varies between foreland subbasins.

South-to-North Migration Patterns of Depocenters in the Ferronensis Sequence

Superimposed on the general north-to-south migration of Turonian-Coniacian depocenters is the systematic progradation of deltas within depocenters. These intra-depocenter variations record varying foreland subbasin subsidence and structurally controlled positioning of major fluvial systems along the western margin of the WICS (Fig. 5). As an example, the Ferronensis sequence records systematic south-to-north progradation of deltas from southern to central Utah.

In the southern Henry Mountains region, juvenile ammonites of *P. macombi* occur in the uppermost vertically stacked short-term cycle of the Ferronensis sequence (Peterson and Ryder, 1975). *P. macombi* occurs in outer-shelf mudstones in east-central Utah and shallow-marine sandstones in northeastern Utah, but it is not recorded in nonmarine or shallow-marine strata in central Utah. Thus, vertically stacked cycles in the Henry Mountains region are contemporaneous with seaward-stepping cycles in central and northern Utah. A base-level rise disconformity in the Henry Mountains is time equivalent to a succession of vertically stacked to landward-stepping cycles in central and northern Utah.

In the Henry Mountains, short-term cycles of the Ferronensis sequence step seaward, vertically stack, with uppermost cycles cut by a capping unconformity (Fig. 17). In central Utah, short-term cycles are arranged in a seaward-stepping, to vertically stacked, to landward-stepping stacking pattern (Fig. 13). In northern Utah, the short-term cycle stacking pattern is seaward- to landward-stepping (Fig. 11). Although the stacking patterns in the three foreland subbasins are different in a geometric sense, they record the same record of base-level change. Although eustasy is the common process recorded across foreland subbasins, changing accommodation and sediment supply within subbasins control stratal stacking patterns.

The along-strike change in the stratal geometry of the Ferronensis sequence is consistent with a north-northeast-oriented fluvial system derived from southern Utah. From southern to central Utah there is a northward decrease in grain size and a decrease in feldspar content in fluvial sandstones that show a persistent northeast transport direction (Katich, 1954; Peterson, 1969a,b; Peterson and Ryder, 1975; Eaton and others, 1987; Bobb, 1991). In the Kaiparowits Plateau, fluvial sandstones at the base of the Ferronensis sequence are arkosic (Peterson, 1969a; Bobb, 1991). In the Henry Mountains, the oldest delta-front sandstones and fluvial sandstones of the Ferronensis sequence are feldspathic. The delta in the Henry Mountains was abandoned at the top of vertically stacked short-term cycles, as indicated by the capping transgressive disconformity. Following abandonment of the southern delta, the same river initiated a new delta about 50 km to the north in Castle Valley. Fluvial and delta-front sandstones in Castle Valley are feldspathic. These petrologic, facies, and geometric attributes show that progressively younger fluvial-deltaic complexes in the Ferronensis sequence prograded from southwest to northeast across Utah. Southern sourced, fluvial-deltaic complexes terminated in central Utah. Intra-depocenter variations are also indicated in the Mytiloides and Woollgari sequences by progradation of progressively younger deltas from southwest to northeast and northwest to southeast across southwestern Wyoming, respectively.

SOUTHWARD-INCREASING ACCOMMODATION ACROSS THE CENTRAL WESTERN INTERIOR SEAWAY

Across the WICS, subtropical macrofauna and microfauna in the Juana Lopez Member show a northward migration and radiation, and the overlying Niobrara Formation records a northward increase in water temperature and decrease in salinity (Eicher, 1969; Kauffman, 1977; Eicher and Diner, 1985; Collom, 1991). Similar depositional patterns are reported for the Bridge Creek Limestone (Elder, 1985; Pratt, 1985). These paleoecologic and geochemical trends indicate a south-to-north movement of subtropical Atlantic–Tethyan water masses. These trends also appear to be associated with geographic changes in accommodation along the Turonian–Coniacian seaway axis. In southern reaches of the WICS, Turonian–Coniacian stratigraphic sequences record intermediate-term base-level cycles under relatively deeper water conditions for longer periods of time.

Eustatic sea-level changes are considered the primary driving force on the northward movement of Atlantic–Tethyan marine waters along the Turonian–Coniacian seaway axis. Associated changes in accommodation are illustrated by the stratal architecture of Turonian–Coniacian stratigraphic sequences in Wyoming compared with those from northern New Mexico and southern Colorado. These regions occupy similar positions on west-to-east transverse basin profiles but different positions along the north-south seaway axis.

Stratigraphic sequences in New Mexico and southern Colorado contain higher proportions of deep-water carbonates and more frequent transgressive disconformities formed during base-level rise time. They are relatively thin, having fewer (base-level fall) unconformities and containing seaward-stepping to vertically stacked short-term cycles. These short-term cycles contain attenuated facies tracts that show an increased proportion of carbonate and nonmarine deposits.

In contrast, stratigraphic sequences in Wyoming contain more basin-centered shallow-marine sandstones. Short-term cycles have seaward-stepping stacking patterns in these deposits, each cycle commonly truncated by unconformities developed during base-level fall time. These sandstones are detached from contemporaneous shoreface deposits by extensive transgressive erosion surfaces. Thus, shoreface sandstones were deposited during base-level fall, but were reworked during base-level rise.

Along-basin changes in facies architecture within stratigraphic sequences suggest higher sediment supply relative to accommodation rates in Wyoming compared with southern Colorado and northern New Mexico. In part, these facies architecture changes appear to record persistent northward movement of Atlantic–Tethyan water mass up the WICS axis producing a southward deepening along the seaway axis. Alternatively, this depositional pattern may record northward decreasing tectonic subsidence from New Mexico to Wyoming. If true, this basin-scale subsidence trend may record southward initiation of progressively younger thrusts loading successive foreland subbasins flanking the western margin of the WICS.

FORELAND SUBBASINS

Flanking the western margin of the central WICS are at least six regions recording unique accommodation and sediment supply conditions and inferred to represent foreland subbasins. Foreland subbasins are 100 km^2 structural depressions recording load-induced flexure and asymmetric subsidence between a proximal thrust sheet and distal peripheral bulge. Across Utah these subbasins are offset by northeast-trending tectonic lineaments segmenting thrust sheets in the orogenic belt (Picha, 1986). Differential subsidence between foreland subbasins is recorded by transverse (low accommodation) or oblique (high accommodation) depocenter patterns. For example, foreland subbasins in northern and central Utah contain transverse (Vernal delta) and oblique (Last Chance delta) depocenters, respectively (Gardner and Cross, 1994).

Sediment depocenters generally conform to foreland subbasins, but are not restricted to them (i.e., northerly-derived Hyatti and southerly-derived Ferronensis sequences of central Utah). For example, the southward migration of Cenomanian-Coniacian depocenters occurs in progressively more southerly positioned foreland subbasins. The stratigraphic hierarchy and tectonic history of the central Utah foreland subbasin is discussed in Gardner (1994, this volume).

CONTROLS ON THE STRATAL ARCHITECTURE OF MID-CRETACEOUS STRATIGRAPHIC SEQUENCES

Across a 1,500-km strike-oriented segment of the western shoreline, consistent changes in stratigraphic sequence architecture suggest a common driving force produced synchronous stratigraphic responses over a 100,000-km^2 area and over periods of 1 to 2 my. Three driving forces are considered: (1) synchronous basin-scale tectonic subsidence, (2) synchronous sediment supply variations, and (3) eustasy.

Given local variations in tectonic movement documented across the central WICS, synchronous tectonism across the central WICS seems implausible (e.g., nonsynchronous tectonic movement along inferred peripheral bulges located in the San Rafael uplift, Henry Mountains, and Uinta basins of Utah). Synchronous basin-scale tectonic movement is not consistent with New Mexico to Wyoming accommodation trends discussed above. Nonsynchronous tectonic subsidence is recorded by changing depocenter patterns between Utah foreland subbasins (Gardner and Cross, 1994).

The Ferronensis sequence records the greatest basinward shift in deposition and contains the highest sandstone volumes and the first occurrence of feldspar-bearing sandstone in Turonian strata of Utah. This association of increased sediment volume and provenance change may suggest an increase in sediment supplied from the mountain belt. However, along-strike variations in sandstone sediment volumes and shoreface stacking patterns in short-term cycles of the Ferronensis sequence indicate that sediment supply increases were not uniform across foreland subbasins. Along-strike variations in sediment supply are reflected in migration patterns of Turonian-Coniacian depocenters.

If stratigraphic cycles of the central WICS were driven by eustasy, Turonian–Coniacian sea-level changes must be shown to be global. There is strong evidence for eustatic sea-level rise at the Cenomanian–Turonian boundary and in lower to middle Coniacian strata (Rona, 1973; Schlanger and Jenkyns, 1976; Kauffman, 1977; Arthur and Schlanger, 1979; Hancock and Kauffman, 1979; Jenkyns, 1980;

Weimer, 1983; Schlanger and others, 1986). These eustatic transgressions are considered a primary control on deposition of base-level rise deposits bounding the long-term Turonian-Coniacian stratigraphic cycle in the WICS.

The eustatic sea-level chart of Haq and others (1988) for the Turonian-Coniacian time shows three short-term eustatic cycles superimposed on a longer term (4.5 my) eustatic cycle (Fig. 2). A eustatic sea-level chart constructed from a high accommodation basin in West Siberia shows four high-frequency eustatic cycles superimposed on a lower frequency eustatic cycle (Sahagian and Jones, 1993). These correlate surprisingly well to base-level cycles documented here. Isotopic ages from volcanic ash beds support correlation of a late Turonian eustatic sea-level fall and unconformity on the Russian Platform (Sahagian and Jones, 1993), with a base-level fall disconformity in the Ferronensis sequence of the central Western Interior.

Stratigraphic Responses to Late Turonian Eustatic Sea-Level Changes

Because the Ferronensis sequence includes both the base-level fall and rise maxima of the long-term stratigraphic cycle, it records dramatic changes in stratal architecture that may be related to eustasy. The following section examines stratigraphic responses in the Ferronensis sequence to late Turonian eustasy.

The maximum shoreline regression and the highest proportion of nonmarine facies in Turonian–Coniacian stratigraphic sequences of the central WICS is recorded by the Ferronensis sequence. In Wyoming, Utah, and New Mexico, the lower part of the Ferronensis sequence consists of short-term cycles of shoreface sandstone that are arranged in a seaward-stepping stacking pattern. In eastern Utah, northern New Mexico, across Colorado, and eastern Wyoming, coeval shelf and basin deposits are represented by the Juana Lopez and Turner Sandy Members.

Eustatic sea-level fall during deposition of the Ferronensis sequence was superimposed on the base-level fall turnaround of the long-term Turonian-Coniacian stratigraphic cycle. Late Turonian base-level fall unconformities of the Ferronensis sequence increase in duration and frequency to the north and along the eastern margin of the WICS (Table 2, Fig. 25), where the Juana Lopez Member forms an unique basin-centered deposit (Molenaar and Cobban, 1991). The disjunct interbedding of coarser-grained sandstone and finer-grained mudstone and volcanic ash beds led early workers to interpret Juana Lopez facies as transgressive in origin. Instead, these sandstone-rich basinal deposits are inferred to record intermediate-term base-level fall of the Ferronensis sequence. High volumes of basin-centered sandstones record decreased accommodation and higher energy deposition during short-term cycles superimposed on intermediate-term base-level fall (Ferronensis sequence). The broad areal distribution of these basinal deposits records deposition in an intracratonic basin that lacks a shelf-slope break, and reflects lowstand conditions in a ramp setting.

Basinal sedimentation patterns and synchronous shoreline responses across the central WICS suggest subsequent late Turonian base-level rise was eustatically controlled. Despite a decipherable sedimentological signature of eustasy, local accommodation and sediment-supply changes in foreland subbasins produced variable cycle stacking patterns and architectural styles in the Ferronensis sequence. Stratigraphic responses to a late Turonian eustatic transgression include submarine erosion and shoreface ravinement, resulting in (1) shoreline-detached, shallow-marine sandstones in northern Utah, across Wyoming, and "transgressive sands" in northern New Mexico, (2) vertically stacked to landward-stepping cycle stacking patterns in central Utah, and (3) submarine unconformities in southern Utah, southern Colorado, northern New Mexico, and eastern Wyoming.

Base-level fall deposits exhibit a more consistent stratal architecture spatially. Stratigraphic cycles have seaward-stepping cycle stacking patterns recording increased sediment flux to the basin and regression during base-level fall. By contrast, base-level rise deposits record more variable cycle stacking patterns, stratal architecture, and diverse facies. Base-level rise deposits appear to be more sensitive to local changes in sediment supply reflecting variable efficiencies of updip sediment trapping along a bathymetric profile. For example, shoreface sandstones in seaward-stepping cycles in the upper part of the Ferronensis sequence of northern New Mexico are coeval to landward-stepping cycles of outer-shelf mudstone in central Utah.

CONCLUSIONS

A threefold stratigraphic hierarchy is recognized in Turonian–Coniacian strata of the central WICS. One long-term Turonian to early middle Coniacian stratigraphic cycle (4.5 my) contains four intermediate-term stratigraphic sequences (1 to 2 my) consisting of 2 to 8 short-term cycles (0.3 my). Each stratigraphic cycle records a complete base-level cycle of varying periodicity, and the base-level rise-to-fall turnaround can be correlated, at all scales, across regions showing variations in stratal architecture. Variations in the four stratigraphic sequences are related to (1) unique tectonic subsidence and sediment supply histories in six subbasins comprising the western margin of central WICS, (2) southward migration of depocenters along the western margin, and (3) southward increasing accommodation related to northward movement of eustatically driven marine water masses.

Stratigraphic sequences do not conform to "standard" sequence models. For example, sequence boundaries are unconformable near the basin center but are conformable landward, the inverse from published sequence models. Stratal architecture cannot be used to correlate stratigraphic sequences to a specific temporal duration or standard stratigraphic hierarchy. Therefore, regional correlation of chronostratigraphic units containing variable stratal patterns is achieved by linking stratal variations to a record of base-level change. For example, systematic variations in stratal architecture recording base-level rise of the Ferronensis sequence include (1) submarine erosion and shoreface ravinement producing shoreline-detached, shallow-marine sandstones in northern Utah, across Wyoming, and "transgressive sands" in northern New Mexico; (2) vertically stacked to landward-stepping cycle stacking patterns in central Utah; and (3) submarine unconformities in southern Utah, southern Colorado, northern New Mexico, and eastern Wyoming.

Regional variations in stratigraphic sequences can be related to eustatic and tectonic controls. Eustatic controls on architecture are related to the persistent northward movement of marine-water masses along the WICS axis. Eustatic controls on facies architecture appear to have promoted deep-water carbonate and mudstone deposition for longer durations in southern reaches of the WICS. Coeval strata in northern reaches of the WICS contain a higher volume of widespread and basin-centered shallow-marine sandstone; these are commonly truncated by base-level fall unconformities. Turonian—Coniaciean strata record higher sediment supply but lower accommodation in Wyoming relative to southern Colorado and northern New Mexico. Eustasy also appears to have been the "pulse" modulating the periodicity of intermediate-term base-level cycles.

Tectonic controls on stratigraphic sequence architecture are related to: (1) structural lineaments segmenting thrust sheets, producing foreland subbasins, and organizing paleodrainage of a few moderate to large rivers; (2) southward migration of depocenters recording emplacement of thrust sheets of the foreland fold-and-thrust belt earlier in the north than in the south; and (3) unique foreland subbasin subsidence recorded by deltas oriented oblique or transverse to the axis of high subsidence. Transverse depocenters record low accommodation (i.e., Vernal delta in the Uinta basin of northern Utah). These intra-depocenter patterns record the migration of deltas (i.e., south-to-north oblique migration of deltas in the Ferronensis sequence of central and southern Utah) within and across foreland subbasins.

ACKNOWLEDGMENTS

This paper represents part of a Ph.D. study conducted at the Colorado School of Mines. Financial support provided through various grants from the American Association of Petroleum Geologists grants-in-aid program, American Chemical Society Petroleum Research Fund, SEPM Donald Smith Research grant, Sigma Xi grants-in-aid program, U.S. Geological Survey Branch of Coal Resources, and the Utah Geological and Mineralogical Survey. Additional support provided from Industrial Associates of the Genetic Stratigraphy Research Program administered by T. A. Cross at the Colorado School of Mines. Figures drafted by Jana S. Robinson under the supervision of Richard L. Dillon, Chief Cartographer, Bureau of Economic Geology. Technical editing by Tucker Hentz, additional editing by Lana Dieterich and comments by Tim Cross and Brian Willis significantly improved manuscript quality. Critical reviews by Robert Weimer and Terry Jordan greatly enhanced focus and clarity, and their efforts are gratefully acknowledged.

Published with permission, Bureau of Economic Geology, The University of Texas at Austin.

REFERENCES

ARTHUR, M. A. AND SCHLANGER, S. O., 1979, Cretaceous "oceanic anoxic events" as causal factors in development of reef reservoired giant oil fields: American Association of Petroleum Geologists Bulletin, v. 63, p. 870–885.

BARLOW, J. A. AND HAUN, J. D., 1966, Regional restratigraphy of the Frontier and relation to Salt Creek field, Wyoming: American Association of Petroleum Geologists Bulletin, v. 50, p. 2185–2196.

BOBB, M. C., 1991, The Calico Bed, Upper Cretaceous, Southern Utah: a fluvial sheet deposit in the Western Interior foreland basin and its relationship to eustasy and tectonics: Unpublished M.S. Thesis, University of Colorado, Boulder, 166 p.

COBBAN, W. A., 1976, Ammonite record from the Mancos Shale of the Castle Valley–Price–Woodside area, east-central Utah: Provo, Brigham Young University Geological Studies, v. 22, p. 117–126.

COBBAN, W. A., 1988, Ammonites in Clasts of the Juana Lopez Member of the Carlile Shale (Upper Cretaceous) near Pueblo, Colorado: United States Geological Survey Bulletin 1837, 5 p.

COBBAN, W. A. AND HOOK, S. C., 1979, *Collignoniceras woollgari woollgari* (Mantell) Ammonite Fauna from Upper Cretaceous of Western Interior, United States: Socorro, New Mexico Bureau of Mines and Mineral Resources Memoir 37, 51 p.

COBBAN, W. A. AND HOOK, S. C., 1984, Mid-Cretaceous molluscan biostratigraphy and paleogeography of southwestern part of Western Interior, United States, in Westermann, G. E. G., ed., Jurassic-Cretaceous Biochronology and Paleogeography of North America: Geological Association of Canada Special Paper 27, p. 257–271.

COBBAN, W. A. AND HOOK, S. C., 1989, Mid-Cretaceous molluscan record from west-central New Mexico, in Southeastern Colorado Plateau: Socorro, New Mexico Geological Society Guidebook, 40th Field Conference, p. 247–264.

COBBAN, W. A. AND REESIDE, J. B., JR., 1952, Frontier Formation: Wyoming and adjacent areas: American Association of Petroleum Geologists Bulletin, v. 36, p. 1913–1961.

COLLOM, C. J., 1991, High-resolution stratigraphic and paleoenvironmental analysis of the Turonian–Coniacian Stage boundary interval (Late Cretaceous) in the lower Fort Hays Member, Niobrara Formation, Colorado and New Mexico: Unpublished M. S. Thesis, Brigham Young University, Provo, 371 p.

COTTER, E., 1971, Paleoflow characteristics of a Late Cretaceous river in Utah from analysis of sedimentary structures in the Ferron Sandstone: Journal of Sedimentary Petrology, v. 41, p. 129–138.

COTTER, E., 1975a, Deltaic deposits in the Upper Cretaceous Ferron Sandstone, Utah, in Broussard, M. L. S., ed., Deltas, Models for Exploration: Houston, Houston Geological Society, p. 471–484.

COTTER, E., 1975b, Late Cretaceous sedimentation in a low-energy coastal zone: the Ferron Sandstone of Utah: Journal of Sedimentary Petrology, v. 45, p. 15–41.

COTTER, E., 1976, The role of deltas in the evolution of the Ferron Sandstone and its coals: Provo, Brigham Young University Studies, v. 22, pt. 3, p.15–41.

CROSS, T. A., 1988, Controls on coal distribution in transgressive-regressive cycles, Upper Cretaceous, Western Interior, U.S.A., in Wilgus, C. K., Hasting, B. S., Kendall, C. G. St. C., Posamentier, H. W., Ross, C. A., and Van Wagoner, J. C., eds., Sea-level Changes: An Integrated Approach: Tulsa, Society of Economic Paleontologists and Mineralogists Special Publication 42, p. 371–380.

DAVIS, L. J., 1954, Stratigraphy of the Ferron Sandstone: Salt Lake City, Intermountain Association of Petroleum Geology Fifth Annual Field Conference Guidebook, p. 55–58.

DOELLING, H. H., 1972, Central Utah coal fields: Sevier–Sanpete, Wasatch Plateau, Book Cliffs and Emery: Salt Lake City, Utah Geological and Mineralogical Survey Monograph Series 3, p. 418–496.

EATON, J. G., KIRKLAND, J. I., GUSTASON, E. R., NATIONS, J. D., FRANCZYK, K. J., RYER, T. A., AND CARR, D. A., 1987, Stratigraphy, correlation and tectonic settings of Late Cretaceous rocks in the Kaiparowits and Black Mesa Basins, in Davis, G. H., and Vanden Dolder, E. M., eds, Geologic Diversity of Arizona and its margins: Excursions to choice areas: Arizona Bureau of Geology and Mineral Technology Special Paper 5, p. 113–125.

EATON, J. G., KIRKLAND, J. I., AND KAUFFMAN, E. G., 1990, Evidence and dating of Mid-Cretaceous tectonic activity in the San Rafael Swell, Emery County, Utah: The Mountain Geologist, v. 27, p. 39–45.

EICHER, D. L., 1969, Cenomanian and Turonian planktonic foraminifera from the Western Interior of the United States: Geneva, Proceedings of the First International Conference of Planktonic Microfossils, v. 2, p. 163–174.

EICHER, D. L. AND DINER, R., 1985, Foraminifera as indicators of water mass in the Cretaceous Greenhorn sea, Western Interior, in Pratt, L. M., Kaufman, E. G., and Zelt, F. B., eds., Fine-Grained Deposits and Biofacies of the Cretaceous Western Interior Seaway: Evidence of Cyclic Sedimentary Processes: Denver, Society of Economic Paleontologists and Mineralogists Field Trip Guidebook 4, p. 60–71.

ELDER, W. P., 1985, Biotic patterns across the Cenomanian Turonian extinction boundary near Pueblo, Colorado, in Pratt, L. M., Kauffman, E. G., and Zelt, F. B., eds., Fine-Grained Deposits and Biofacies of the Cretaceous Western Interior Seaway: Evidence of Cyclic Sedimentary Processes: Denver, Society of Economic Paleontologists and Mineralogists Field Trip Guidebook 4, p. 157–169.

GALE, H. S., 1910, Coal Fields of Northwestern Colorado and Northeastern Utah: United States Geological Survey Bulletin 415, 265 p.

GARDNER, M. H., 1993, Sequence stratigraphy and facies architecture of the Upper Cretaceous Ferron Sandstone Member of the Mancos Shale, East-Central Utah: Unpublished Ph.D. Dissertation, Colorado School of Mines, Golden, 528 p.

GARDNER, M. H. AND CROSS, T. A., 1994, Middle Cretaceous Paleogeography of Utah, in Caputo, M., Peterson, J., and Franczyk, K., eds., Mesozoic Systems of the Rocky Mountains Region, USA: Denver, Rocky Mountain Section, Society of Economic Paleontologists and Mineralogists, p. 471–502.

GOODELL, H. G., 1962, The stratigraphy and petrology of the Frontier Formation of Wyoming, in Enyert, R. L. and Curry, W. M. III, eds., Symposium on Early Cretaceous Rocks of Wyoming and Adjacent Areas: Casper, Wyoming Geological Association 17th Annual Field Conference Guidebook, p. 173–210.

HALE, L. A., 1960, Frontier Formation—Coalville, Utah, and nearby areas of Wyoming and Colorado, in McGookey, D. M. and Miller, D. N. Jr., eds., Overthrust Belt of SW Wyoming and Adjacent Areas: Casper, Wyoming Geological Association 15th Annual Field Conference Guidebook, p. 137–146.

HALE, L. A., 1962, Frontier Formation—Coalville, Utah, and nearby areas of Wyoming and Colorado (revised), in Enyert, R. L. and Curry, W. M. III, eds., Symposium on Early Cretaceous Rocks of Wyoming and Adjacent Areas: Casper, Wyoming Geological Association 17th Annual Field Conference Guidebook, p. 211–220.

HALE, L. A., 1972, Depositional history of the Ferron Formation, Central Utah, in Plateau Basin and Range Transition Zone: Salt Lake City, Utah Geological Association, p. 115–138.

HALE, L. A. AND VAN DE GRAAF, F. R., 1964, Cretaceous stratigraphy and facies patterns—northeastern Utah and adjacent areas, in Geology and Mineral Resources of the Uinta Basin: Salt Lake City, Intermountain Association of Petroleum Geologists Thirteenth Annual Field Guidebook, p. 115–138.

HANCOCK, J. M. AND KAUFFMAN, E. G., 1979, The great transgressions of the Late Cretaceous: Geological society of London Journal, v. 136, p. 175–186.

HAQ, B. U., HARDENBOL, J., AND VAIL, P. R., 1987, Mesozoic and Cenozoic chronostratigraphy and eustatic cycles, in Wilgus, C. K., Hasting, B. S., Kendall, C. G. St. C., Posamentier, H. W., Ross, C. A., and Van Wagoner, J. C., eds., Sea-level Changes: An Integrated Approach: Tulsa, Society of Economic Paleontologists and Mineralogists Special Publication 42, p. 71–109.

HATTIN, D. E., 1964, Cyclic sedimentation in the Colorado Group of west-central Kansas, in Merriam, D. F., ed., Symposium on Cyclic Sedimentation: Kansas Geological Survey Bulletin 169, p. 205–217.

HATTIN, D. E., 1975, Stratigraphic study of the Carlile-Niobrara (Upper Cretaceous) unconformity in Kansas and northeastern Nebraska: Calgary, Geological Association of Canada Special Paper 13, p. 195–210.

HAUN, J. D., 1953, Stratigraphy of the Frontier Formation, Powder River Basin: Unpublished Ph.D. Dissertation, University of Wyoming, Laramie.

HOOK, S. C., 1983, Stratigraphy, paleontology, depositional framework, and nomenclature of marine Upper Cretaceous rocks, Socorro County, New Mexico, in Chapin, C. E., ed., Socorro Region II: Socorro, New Mexico Geological Society Guidebook, 34th Field Conference, p.165–172.

HUNT, C. B., AVERITT, P., AND MILLER, R. L., 1953, Geology and Geography of the Henry Mountains Region, Utah: Washington, D. C., United States Geological Survey Professional Paper 228, 234 p.

HUNT, C. B. AND MILLER, R. L., 1964, General geology of the region-stratigraphy, in Hunt, C. B., ed., Guidebook to the Geology of the Henry Mountain Region: Salt Lake City, Utah Geological Society Guidebook No. 1, p. 6–10.

HUNTER, L. D., 1952, Frontier Formation along the eastern margin of the Big Horn basin, Wyoming, in Southern Bighorn Basin: Casper, Wyoming Geological Association 7th Annual Field Conference Guidebook.

JENKYNS, H. C., 1980, Cretaceous anoxic events: from continents to oceans: Journal Geological Society of London, v. 137, p. 171–188.

JERVEY, M. T., 1988, Quantitative geological modeling of siliciclastic rock sequences and their seismic expression, in Wilgus, C. K., Hasting, B. S., Kendall, C. G. St. C., Posamentier, H. W., Ross, C. A., and Van Wagoner, J. C., eds., Sea-level Changes: An integrated Approach: Tulsa, Society of Economic Paleontologists and Mineralogists Special Publication 42, p. 109–125.

KATICH, P. J., JR., 1954, Cretaceous and early Tertiary stratigraphy of central and south-central Utah with emphasis on the Wasatch Plateau area: Salt Lake City, Intermountain Association of Petroleum Geologists Fifth Annual Field Guidebook, p. 42–54.

KAUFFMAN, E. G., 1969, Cretaceous marine cycles of the Western Interior: The Mountain Geologist, v. 6, p. 227–245.

KAUFFMAN, E. G., 1977, Geological and biological overview: Western Interior Cretaceous basin, in Cretaceous Facies, Faunas and Paleoenvironments across the Western Interior Basin: Mountain Geologist, v. 14, p. 75–99.

KAUFFMAN, E. G., 1985, Cretaceous evolution of the Western Interior Basin of the United States, in Pratt, L.M., Kauffman, E. G., and Zelt, F. B., eds., Fine-Grained Deposits and Biofacies of the Cretaceous Western Interior Seaway: Evidence of Cyclic Sedimentary Processes: Denver, Society of Economic Paleontologists and Mineralogists Field Trip Guidebook 4, p. iv–xiii.

KAUFFMAN, E. G., COBBAN, W. A., AND EICHER, D. L., 1976, Albian through Lower Coniacian Strata, Biostratigraphy and Principal Events, Western Interior United States: Annales Du Museum D'Histoire Naturelle De Nice-Tome IV, 51 p.

KENT, H. C., 1968, Biostratigraphy of Niobrara-equivalent part of Mancos Shale (Cretaceous) in northwestern Colorado: American Association of Petroleum Geologists Bulletin, v. 52, p. 2098–2115.

KIRKLAND, J. I., 1990, The paleontology and paleoenvironments of the Middle Cretaceous (Late Cenomanian–Middle Turonian) Greenhorn Cyclothem at Black Mesa, northeastern Arizona: Unpublished Ph.D. Dissertation, University of Colorado, Boulder, 1,350 p.

LAWTON, T. F., 1985, Style and timing of frontal thrust structures, thrust belt, central Utah: American Association of Petroleum Geologists Bulletin, v. 69, p. 1145–1159.

LEITHOLD, E. L., in press, Stratigraphic architecture of an ancient muddy basin margin, southern Utah: Sedimentology.

LESSARD, R. H., 1973, Micropaleontology and paleoecology of the Tununk Member of the Mancos Shale: Salt Lake City, Utah Geological and Mineral Survey, Special Studies 45, 28 p.

LUPTON, C. T., 1916, Geology and Coal Resources of Castle Valley in Carbon, Emery, and Sevier Counties, Utah: United States Geological Survey Bulletin 628, 84 p.

MAIONE, S. J., 1971, Stratigraphy of the Frontier Sandstone Member of the Mancos Shale (Upper Cretaceous) on the south flank of the eastern Uinta Mountains, Utah and Colorado: Earth Science Bulletin, Wyoming Geological Association, v. 4, p. 27–58.

MACDONALD, R. H. AND BYERS, C. W., 1988, Depositional history of the Greenhorn Formation (Upper Cretaceous) Northwestern Black Hills: The Mountain Geologist, v. 25, no. 3, p. 71–85.

MCGOOKEY, D. P., HAUN, J. D., HALE, L. A., MCCUBBIN, D. G., WEIMER, R. J., AND WULF, G. R., 1972, Cretaceous System, in Mallory, W. W., ed., Geologic Atlas of the Rocky Mountain Region: Denver, Rocky Mountain Association of Geologists, p. 190–228.

MEREWETHER, E. A., 1980, Stratigraphy of Mid-Cretaceous Formations at Drilling Sites in Weston and Johnson Counties, Northeastern Wyoming: Washington, D. C., United States Geological Survey Professional Paper 1186-A, 25 p.

MEREWETHER, E. A., 1983, The Frontier Formation and Mid-Cretaceous orogeny in the foreland of southwestern Wyoming: The Mountain Geologist, v. 20, p. 121–138.

MEREWETHER, E. A. AND COBBAN, W. A., 1986, Biostratigraphic units and tectonism in the mid-Cretaceous foreland of Wyoming, Colorado, and adjoining areas, in Peterson, J. A., ed., Paleotectonics and Sedimentation in the Rocky Mountain region, United States, part III, middle Rocky Mountains: Tulsa, American Association of Petroleum Geologists Memoir 41, p. 443–468.

MEREWETHER, E. A., COBBAN, W. A., AND CAVANAUGH, E. T., 1979, Frontier Formation and equivalent rocks in eastern Wyoming: The Mountain Geologist, v. 16, p. 67–102.

MEREWETHER, E. A., COBBAN, W. A., AND RYDER, R. T., 1975, Lower Upper Cretaceous strata, Bighorn Basin, Wyoming and Montana, in Crum, F. A. and George, G. R., eds., Geology and Mineral Resources of the Bighorn Basin: Casper, Wyoming Geological Association 27th Annual Field Conference Guidebook, p. 73–84.

MEREWETHER, E. A., COBBAN, W. A., AND SPENCER, C. W., 1976, The Upper Cretaceous Frontier Formation in the Kaycee-Tisdale Mountain area, Johnson County, Wyoming, in Laudon, R. B., Curry, W. H., III, and Runge, J. S., eds., Geology and Mineral Resources of the Powder River: Casper, Wyoming Geological Association 28th Annual Field Conference Guidebook, 33–45.

MITCHUM, R. M., JR., VAIL, P. R., AND THOMPSON, S., III, 1977, Seismic stratigraphy and global changes in sea level, part 2: the depositional sequence as a basic unit for stratigraphic analysis, in Payton, C. E., ed., Seismic Stratigraphy—Applications to Hydrocarbon Exploration: Tulsa, American Association of Petroleum Geologists Memoir 26, p. 53–62.

MITCHUM, R. M. AND VAN WAGONER, J. C., 1991, High frequency sequences and their stacking patterns: sequence stratigraphic evidence of high frequency eustatic cycles: Sedimentary Geology, v. 70, p. 131–160.

MOLENAAR, C. M., 1974, Correlation of the Gallup Sandstone and associated formations, Upper Cretaceous, eastern San Juan and Acoma Basins, New Mexico, in Seimers, C. T., ed., Ghost Ranch (central-northern New Mexico): Socorro, New Mexico Geological Society 25th Field Conference Guidebook, p. 251–258.

MOLENAAR, C. M., 1977, Stratigraphy and depositional history of Upper Cretaceous rocks of the San Juan Basin area, New Mexico and Colorado, in Fassett, J. E., ed., San Juan Basin III: Socorro, New Mexico Geological Society 28th Field Conference Guidebook, p. 159–166.

MOLENAAR, C. M., 1983, Major depositional cycles and regional correlations of Upper Cretaceous rocks, southern Colorado Plateau and adjacent areas, in Reynolds, M. W. and Dolly, E. D., eds., Mesozoic Paleogeography of West-Central United States: Denver, Rocky Mountain Section, Society of Economic Paleontologists and Mineralogists, Rocky Mountain Paleogeography Symposium 2, p. 201–224.

MOLENAAR, C. M. AND COBBAN, W. A., 1991, Middle Cretaceous stratigraphy on south and east sides of the Uinta Basin, northeastern Utah and northwestern Colorado: United States Geological Survey Bulletin 1787-P, 34 p.

MOLENAAR, C. M. AND WILSON, B. W., 1990, The Frontier Formation and Associated Rocks of Northeastern Utah and Northwestern Colorado: United States Geological Survey Bulletin 1787-M, 21 p.

MYERS, R. G., 1977, Stratigraphy of the Frontier Formation (Upper Cretaceous), Kemmerer Area, Lincoln County, Wyoming, in Heisey, E. L., Lawson, D. E., Norwood, E. R., Wach, P. H., and Hale, L. A., eds., Rocky Mountain Thrust Belt Geology and Resources: Casper, Wyoming Geological Association 29th Annual Field Conference Guidebook, p. 271–311.

OBRADOVICH, J. D., 1991, A revised Cenomanian–Turonian time scale based on studies from the Western Interior United States (abs.): Geological Society of America, Abstracts with Programs, 1991 Annual Meeting, A296p.

PETERSON, F., 1969a, Cretaceous Sedimentation and Tectonism in the Southeastern Kaiparowits Region, Utah: Washington, D. C., United States Geological Survey Open-File Report, 259 p.

PETERSON, F., 1969b, Four new members of the Upper Cretaceous Straight Cliffs Formation in the Southeastern Kaiparowits Region Kane County, Utah: United States Geological Survey Bulletin 1274-J, p. J1–J28.

PETERSON, F. AND KIRK, A. R., 1977, Correlation of the Cretaceous rocks in the San Juan, Black Mesa, Kaiparowits, and Henry Basins, Southern Colorado Plateau, in Fassett, J. E., ed., San Juan Basin III: Socorro, New Mexico Geological Society Guidebook, 28th Field Conference, p. 167–178.

PETERSON, F. AND RYDER, R. T., 1975, Cretaceous rocks in the Henry Mountains region, Utah and their relation to neighboring regions, in Fassett, J. E. and Wengerd, S. A., eds., Canyonlands Country: Durango, Four Corners Geological Society Conference, 8th Guidebook, p. 167–189.

PICHA, F., 1986, The influence of preexisting tectonic trends on geometries of the Sevier Orogenic Belt and its foreland in Utah, in Peterson, J. A., ed., Paleotectonics and Sedimentation: Tulsa, American Association of Petroleum Geologists Memoir 41, p. 309–320.

POSAMENTIER, H. W., ALLEN, G. P., JAMES, D. P., AND TESSON, M., 1992, Forced regressions in a sequence stratigraphic framework: Concepts, examples and exploration significance: American Association of Petroleum Geologists Bulletin, v. 76, p. 1687–1710.

POSAMENTIER, H. W., JERVEY, M. T., AND VAIL, P. R., 1988, Eustatic controls on clastic deposition I—conceptual framework, in Wilgus, C. K., Hasting, B. S., Kendall, C. G. St. C., Posamentier, H. W., Ross, C. A., Van Wagoner, J. C., eds., Sea-level Changes: An Integrated Approach: Tulsa, Society of Economic Paleontologists and Mineralogists Special Publication 42, p. 109–125.

POSAMENTIER, H. W. AND VAIL, P. R., 1988, Eustatic controls on clastic deposition II—sequence and systems tract models, in Wilgus, C. K., Hasting, B. S., Kendall, C. G. St. C., Posamentier, H. W., Ross, C. A., Van Wagoner, J. C., eds., Sea-level Changes: An Integrated Approach: Tulsa, Society of Economic Paleontologists and Mineralogists, Special Publication 42, p. 125–155.

PRATT, L. M., 1984, Influence of paleoenvironmental factors on the preservation of organic matter in middle Cretaceous Greenhorn Formation, Pueblo, Colorado: American Association of Petroleum Geologists Bulletin, v. 68, p. 1146–1159.

PRATT, L. M., 1985, Isotopic studies of organic matter and carbonate in rocks of the Greenhorn marine cycle, in Pratt, L. M., Kauffman, E. G., and Zelt, F. B., eds., 1985, Fine-Grained Deposits and Biofacies of the Cretaceous Western Interior Seaway: Evidence of Cyclic Sedimentary Processes: Denver, Society of Economic Paleontologists and Mineralogists Field Trip Guidebook 4, p. 38–48.

PRATT, L. M., KAUFFMAN, E. G., AND ZELT, F. B., eds., 1985, Fine-Grained Deposits and Biofacies of the Cretaceous Western Interior Seaway: Evidence of Cyclic Sedimentary Processes: Denver, Society of Economic Paleontologists and Mineralogists Field Trip Guidebook 4, 249 p.

PRESCOTT, M. W., 1975, Spearhead Ranch, Converse County, Wyoming, in A Symposium on Deep Drilling Frontiers in the Central Rocky Mountains: Rocky Mountain Association Geologists, p. 239–244.

REESIDE, J. B., JR., 1944, Map showing thickness and general character of the Cretaceous deposits in the Western Interior of the United States: Washington, D. C., United States Geological Survey, Oil and Gas Investigations Map 0M-10.

REPENNING, C. A. AND PAGE, H. G., 1956, Late Cretaceous stratigraphy of Black Mesa, Navajo and Hopi Indian Reservations: American Association of Petroleum Geologists Bulletin, v. 40, p. 255–294.

RIEMERSMA, P. AND CHAN, M., 1991, Facies of the lower Ferron Sandstone and Blue Gate Shale Members of the Mancos Shale: lowstand and early transgressive facies architecture, in Swift, D. J. P., Oertel, G. F., Tilman, R. W., and Thorne, J. A., eds., Shelf Sands and Sandstone Bodies, Geometry, Facies and Sequence Stratigraphy: Oxford, International Association of Sedimentologists Special Publication 14, p. 489–510.

RONA, P. A., 1973, Relations between rates of sediment accumulation on continental shelves, sea-floor spreading, and eustasy inferred from the central North Atlantic: Geological Society of America Bulletin, v. 84, p. 2851–2872.

RYER, T. A., 1976, Cretaceous invertebrate faunal assemblages of the Frontier and Aspen Formations, Coalville and Rockport areas, North-Central Utah: The Mountain Geologist, v. 13, p. 101–114.

RYER, T. A., 1977, Patterns of Cretaceous shallow-marine sedimentation, Coalville and Rockport areas, Utah: Geological Society of America Bulletin, v. 88, p. 177–188.

RYER, T. A., 1981, Deltaic coals of Ferron Sandstone Member of Mancos Shale: predictive model for Cretaceous coal-bearing strata of western interior: American Association of Petroleum Geologists Bulletin, v. 65, p. 2323–2340.

RYER, T. A., 1982, Possible eustatic control on the location of Utah Cretaceous coal fields: Salt Lake City Proceedings—5th ROMOCO Symposium, Utah Geological and Mineral Survey Bulletin 118, p. 89–93.

RYER, T. A., 1983, Transgressive-regressive cycles and the occurrence of coal in some Upper Cretaceous strata of Utah: Geology, v. 11, p. 207–210.
RYER, T. A. AND LOVEKIN, J. R., 1986, The Upper Cretaceous vernal delta of Utah—depositional or paleotectonic feature? in Peterson, J. A., ed., Paleotectonics and Sedimentation: Tulsa, American Association of Petroleum Geologists Memoir 41, p. 497–509.
RYER, T. A. AND MCPHILLIPS, M., 1983, Early Late Cretaceous paleogeography of east-central Utah, in Reynolds, M. W. and Dolly, E. D., eds., Mesozoic Paleogeography of West-Central United States: Denver, Rocky Mountain Section, Society of Economic Paleontologists and Mineralogists, Rocky Mountain Paleogeography Symposium 2, p. 253–271.
SAHAGIAN, D. AND JONES, M., 1993, Quantified Middle Jurassic to Paleocene eustatic variations based on Russian Platform stratigraphy: stage level resolution: Geological Society of America Bulletin, v. 105, p. 1109–1118.
SCHLANGER, S. O., ARTHUR, M. A., JENKYNS, H. C., AND SCHOLLE, P. A., 1986, The Cenomanian–Turonian oceanic anoxic event I. Stratigraphy and distribution of organic carbon-rich beds and the marine excursion, in Brooks, J. and Flebt, A. J., eds., Marine Petroleum Sources Rocks: London, Geological Society Special Publication 24, p. 347–375.
SCHLANGER, S. O. AND JENKYNS, H. C., 1976, Cretaceous Oceanic anoxic events: causes and consequences: Geologie En Mijnbouw, v. 55, p. 179–184.
SCHUMM, S. A., 1993, River response to base-level change: Implications for sequence stratigraphy: Journal of Geology, v. 101, p. 279–294.
SEIMERS, C. T., 1975, Paleoenvironmental analysis of the Upper Cretaceous Frontier Formation, northwestern Bighorn Basin, Wyoming, in Crum, F. A. and George, G. R., eds., Geology and Mineral Resources of the Bighorn Basin: Casper, Wyoming Geological Association 27th Annual Field Conference Guidebook, p. 85–101.
SHANLEY, K. W. AND MCCABE, P. J., 1989, Sequence stratigraphic relationships and facies, architecture of Turonian–Campanian strata, Kaiparowits Plateau, south-central Utah (abs.): American Association of Petroleum Geologist Bulletin, v. 73, p. 410–411.
SHANLEY, K. W. AND MCCABE, P. J., 1992, Predicting facies architecture through sequence stratigraphy—an example from the Kaiparowits Plateau, Utah: Geology, v. 19, p. 742–745.
SMITH, J. F., JR., HUFF, L. C., HINRICHS, E. N., AND LUEDKE, R. G., 1963, Geology of the Capital Reef area, Wayne and Garfield Counties, Utah: Washington, D. C.,United States Geological Survey Professional Paper 363, 102 p.
SONNENBERG, S. A. AND WEIMER, R. J., 1981, Tectonics, sedimentation and petroleum potential, northern Denver basin, Colorado, Wyoming, and Nebraska: Colorado School of Mines Quarterly, v. 76, p. 7–45.
SPEIKER, E. M., 1949, Sedimentary facies and associated diastrophism in the Upper Cretaceous of Central and Eastern Utah: Geological Society of America Memoir 39, p. 55–82.
TILMAN, R. W. AND ALMON, W. R., 1979, Diagenesis of Frontier Formation offshore bar sandstones, Spearhead Ranch field, Wyoming, in Scholle P. A. and Schluger, P. R., eds., Aspects of Diagenesis: Tulsa, Society of Economic Paleontologists and Mineralogists Special Publication 26, p. 337–378.
THOMPSON, S. L., 1985, Ferron Sandstone Member of the Mancos Shale: a Turonian mixed-energy deltaic system: Unpublished M.S. Thesis, The University of Texas at Austin, Austin, 165 p.
TONNSEN, J. J., 1980, The Frontier Formation in northwestern Wyoming and adjacent areas, in Hollis, S., ed., Stratigraphy of Wyoming: Casper, Wyoming Geological Association 31st Annual Field Conference Guidebook, p. 173–184.
VAIL, P. R, MITCHUM, R. M., JR., AND THOMPSON, S., 111, 1977, Seismic stratigraphy and global changes in sea level: Relative changes of sea level from coastal onlap, in Payton, C. E., ed., Seismic Stratigraphy—Applications to Hydrocarbon Exploration: Tulsa, American Association of Petroleum Geologists Memoir 26, p. 63–81.
VALASEK, D. W., 1990, Compartmentalization of shoreface sequences in the Cretaceous Gallup Sandstone west of Shiprock, New Mexico: implications for the reservoir quality and continuity (abs.): American Association of Petroleum Geologists Bulletin, v. 74, p. 784.
VAN WAGONER, J. C., MITCHUM, R. M., CAMPION, K. M., AND RAHMANIAN, V. D., 1990, Siliciclastic Sequence Stratigraphy: Tulsa, American Association of Petroleum Geologists Methods in Exploration Series 7, 55 p.
WALTON, P. T., 1944, Geology of the Cretaceous of the Uinta Basin, Utah: Geological Society of America Bulletin, v. 55, p. 91–130.
WEIMER, R. J., 1960, Upper Cretaceous stratigraphy Rocky Mountain area: American Association of Petroleum Geologists Bulletin, v. 44, p. 1–20.
WEIMER, R. J., 1970, Rates of deltaic sedimentation and intrabasin deformation, Upper Cretaceous of Rocky Mountain region, in Morgan, J. P., ed., Deltaic Sedimentation, Modern and Ancient: Tulsa, Society of Economic Paleontologists and Mineralogists Special Publication 15, p. 270–292.
WEIMER, R. J., 1978, Influence of transcontinental arch on Cretaceous marine sedimentation: a preliminary report, in Pruitt, J. D. and Coffin, P. E., eds., Energy Resources of the Denver Basin: Denver, Rocky Mountain Association of Geologists, p. 211–222.
WEIMER, R. J., 1983, Relation of unconformities, tectonics, and sea-level changes in the Cretaceous of the Denver Basin and adjacent areas, in Reynolds, M. W. and Dolly, E. D., eds., Mesozoic Paleogeography of West-Central United States: Denver, Rocky Mountain Section, Society of Economic Paleontologists and Mineralogists, Rocky Mountain Paleogeography Symposium 2, p. 359–377.
WEIMER, R. J. AND FLEXER, A., 1985, Depositional patterns and unconformities, eastern Powder River Basin, Wyoming: Casper, Wyoming Geological Association 36th Annual Field Conference Guidebook, p. 131–147.
WEIMER, R. J. AND SONNENBERG, S. A., 1983, Codell Sandstone, new exploration play, Denver basin, in Mid-Cretaceous Codell Sandstone Member of the Carlile Shale, Eastern Colorado: Denver, Society of Economic Paleontologists and Mineralogists, Rocky Mountain Section Guidebook, p. 26–48.
WHEELER, H. E., 1964, Base level, lithosphere surface, and time stratigraphy: Geological Society of America, v. 75, p. 599–610.
WILLIAMS, N. C. AND MADSEN, J. H., 1959, Late Cretaceous stratigraphy of the Coalville area, Utah, in Williams, N. C., ed., Guidebook to the Geology of the Wasatch and Uinta Mountains Transition Area: Salt Lake City, Intermountain Association of Petroleum Geologists, 10th Annual Field Conference, p. 122–126.
WILLIAMS, G. D. AND STELCK, C. R., 1975, Speculations on the Cretaceous paleogeography of North America, in Caldwell, W. G. E., ed., The Cretaceous System in the Western Interior of North America: Calgary, The Geological Association of Canada Special Paper 13, p. 1–20.
WINN, R. D., STONECIPHER, S. A., AND BISHOP, M. G., 1983, Depositional controls on diagenesis in offshore sand ridges, Frontier Formation, Spearhead Ranch field, Wyoming: The Mountain Geologist, v. 20, p. 41–58.
WINN, R. D., STONECIPHER, S. A., AND BISHOP, M. G., 1984, Sorting and wave abrasion: controls on the composition and diagenesis in lower Frontier sandstones, southwestern Wyoming: American Association of Petroleum Geologists Bulletin, v. 68, p. 268–284.
WITZE, B. J., LUDVIGSON, G. A., POPPE, J. R., AND RAVN, R. L., 1983, Cretaceous paleogeography along the eastern margin of the Western Interior Seaway, Iowa, Southern Minnesota, and eastern Nebraska and South Dakota, in Reynolds, M. W. and Dolly, E. D., eds., Mesozoic Paleogeography of West-Central United States: Denver, Rocky Mountain Section, Society of Economic Paleontologists and Mineralogists, Rocky Mountain Paleogeography Symposium 2, p. 225–253.
YOUNG, R. G., 1955, Sedimentary facies and intertonguing in the Upper Cretaceous of the Book Cliffs, Utah–Colorado: Geological Society of America Bulletin, v. 66, p. 177–202.
YOUNG, R. G., 1957, Late Cretaceous cyclic deposits, Book Cliffs, eastern Utah: American Association of Petroleum Geologists Bulletin, v. 41, p. 1760–1774.
YOUNG, R. G., 1960, Dakota Group of the Colorado Plateau: American Association of Petroleum Geologists Bulletin, v. 44, p. 156–194.

THE STRATIGRAPHIC HIERARCHY AND TECTONIC HISTORY OF THE MID-CRETACEOUS FORELAND BASIN OF CENTRAL UTAH

MICHAEL H. GARDNER
Bureau of Economic Geology, The University of Texas at Austin, Austin, TX 78713-7508

ABSTRACT: A high-resolution stratigraphic framework links nonmarine through marine-shelf deposits across the Turonian–Coniacian foredeep basin exposed in central Utah. This record of cyclic lithologic change and biostratigraphic and isotopic age data document basin-scale patterns of foreland basin deposition. Three temporal and spatial scales of stratigraphic cyclicity are recognized in the middle Cretaceous foreland basin of central Utah, each recording base-level changes of different periodicity. A long-term stratigraphic cycle spans early Turonian through middle Coniacian Stages (93.25 to 88.8 Ma; 4.5 my in duration) and is a 500-m-thick upward-coarsening succession of marine and nonmarine deposits bounded by deposits formed during eustatic transgressions. Four intermediate-scale stratigraphic cycles (300 m thick, 1 to 2 my in duration) punctuate this succession, each recording an episode of eastward shoreline progradation followed by transgression. Intermediate-scale successions are defined by the stacking pattern of small-scale stratigraphic cycles (40 m thick, 0.25 my in duration) defined by progressive upward change in stratal geometry and facies.

In proximal portions of the marine foreland basin, the lowest two intermediate-term stratigraphic cycles (Mytiloides and Woollgari sequences) are westward-thickening mudstone-dominated clastic wedges. Seventy kilometers into the basin, these stratigraphic sequences show depositional thinning in successive short-term stratigraphic cycles over a westward migrating peripheral bulge. Deposition of the third intermediate-term cycle (Hyatti sequence) records the development of two spatially discrete depositional systems in central Utah. This sequence is broadly tabular and probably records a 1–2-my period of tectonic quiescence and foreland basin expansion. The final intermediate-term cycle (Ferronensis sequence) records an increase in sediment supply from the southwest related to structurally controlled longitudinal drainage of a moderate to large fluvial system.

Changes in stratal geometry and facies architecture across these stratigraphic sequences record accelerated subsidence, basin contraction, and westward migration of a peripheral bulge, followed by basin expansion and widespread deposition of coarse clastics. In other foreland basins, such patterns of basin filling have been related to episodic tectonism, but a coeval tectonic event is not recognized in the adjacent orogenic belt. Because Turonian marine mudstones overlap the hanging wall of the initial thrust and constrain its cessation, these foreland basin depositional patterns appear to record viscoelastic responses to thrust deformation. There is, however, strong evidence for important eustatic sea-level changes in the middle Cretaceous. Accordingly, criteria used to distinguish tectonic versus eustatic controls on foreland basin deposition may be overly simplistic.

INTRODUCTION

The cyclic arrangement of strata reflects a repetitive sequence of processes and conditions. Traditionally it has been assumed that stratigraphic cycles begin and end under the same conditions and record a similar series of events repeated at regular intervals. It is now recognized that stratigraphic cycles of different scales are superimposed in a hierarchy that records changes in accommodation and sediment supply of different magnitude and duration (Fig. 1). This paper documents the hierarchy of stratal cyclicity within the Mid-Cretaceous foreland basin of central Utah (Fig. 2).

Sedimentary responses to structural development of foreland basins are generally assumed to represent a temporally and spatially related process-response system (Blair and Bilodeau, 1988; Heller and others, 1988; Heller and Paola, 1989; Fleming and Jordan, 1990). The loading of thrust sheets produces uplift and asymmetric flexural subsidence and associated changes in basin slope, accommodation, sediment supply, and lithofacies distributions in an adjacent basin. Dating inferred synorogenic conglomerates suggested uplift and erosion of the source area was directly related to deposition of a prograding clastic wedge (Armstrong and Oriel, 1965; Wiltschko and Dorr, 1983). An episode of thrusting, however, increases both subsidence and sediment supply to the basin restricting coarser-grained sediment to proximal areas where accommodation is high (i.e., typically within 70 km of active thrusting; e.g., Heller and Paola, 1989). Following the cessation of thrusting, a progressive decrease in subsidence and sediment supply and progradation of coarse-grained clastics across the basin records the gradual adjustment of basin slopes to the topographic decay and isostatic adjustment of thrust sheets. This depositional pattern requires sediment supply to exceed subsidence rates following thrust cessation. Episodic tectonism thus produces large-scale, upward-coarsening successions interpreted to correspond to these "keep-up" then "fill-up," long-term base-level cycles (Beck and Vondra, 1985; Heller and others, 1988; Blair and Bilodeau, 1988; Fleming and Jordan, 1990).

The western margin of the Western Interior Cretaceous seaway (WICS) of North America bordered an east-vergent, retro-arc foreland fold and thrust belt for about 53 my (Armstrong and Oriel, 1965; Armstrong, 1968; Dickinson, 1974; Burchfiel and Davis, 1975; Jordan, 1981). Load-induced flexural subsidence associated with the emplacement of thrust sheets produced an asymmetric foreland basin (Dickinson, 1974; Beaumont, 1981; Jordan, 1981). In central Utah the initial marine incursion and development of the Cretaceous seaway began in the early Turonian. A sea-level maximum in the early Turonian (Hancock and Kauffman, 1979; Haq and others, 1988) was followed by deposition of five to six eastward-thinning clastic wedges. These clastic wedges record long-term base-level cycles over 25 my of foreland basin evolution (Spieker, 1949; Young, 1957; Weimer, 1960, 1983; Molenaar, 1983). This paper examines the initial 4.5-my period of this record.

Turonian strata of central Utah coarsens upward with the coarsest sediments post-dating the inferred cessation of initial thrusting by over 3 my (Pavant 1 thrust; Villien and Kligfield, 1986). Similar stratigraphic successions have been interpreted to record tectonic cycles, but no Turonian tectonic event is recognized in the central Utah orogenic belt. There is, however, strong evidence for a global late Turonian sea-level drop. A high-resolution stratigraphic

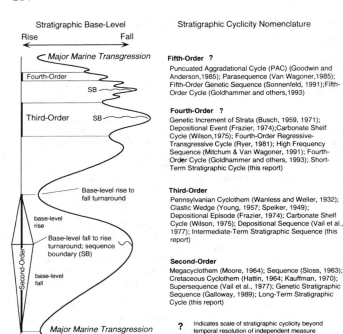

Fig. 1.—Nomenclature applied to different scales of stratigraphic cyclicity.

framework constructed from sequential changes in the stratal geometry and facies is used here to interpret the record of tectonism and eustasy on foreland basin deposition.

Biostratigraphic and lithologic data are integrated to develop a chronostratigraphic framework utilizing two stratigraphic concepts. *Accommodation*, the potential space available for sediment accumulation, controls preservational trends in chronostratigraphic units (Jervey, 1988). Accommodation is the sum of tectonic movement, lithospheric compensation to loads, compaction, and sea level. *Stratigraphic base level* (modified from Wheeler, 1964) reflects the balance between accommodation and the transfer of sediment mass across the Earth's surface by surficial processes that erode, transport, and deposit sediment. Stratigraphic base level is an imaginary potentiometric surface that describes the energy required to move the Earth's physical surface up or down by deposition or erosion, respectively, to attain a condition of minimum energy expenditure where gradients, sediment supply, and accommodation are equilibrated.

HIERARCHY OF MID-CRETACEOUS STRATA

Three temporal and spatial scales of stratigraphic cycles are observed in middle Cretaceous strata of central Utah (Fig. 3). High subsidence rates in central Utah combined with a Cretaceous sea level highstand to accommodate high volumes of sediment shed from the adjacent thrust belt. Because these high-accommodation and sediment-supply conditions promote sediment burial and preservation, stratigraphic cycles here contain the most complete record of deposition.

Durations of stratigraphic cycles were estimated from overlapping age ranges of fossils in marine strata and isotopic ages of volcanic ashes (Fig. 4). Stratigraphic cycles record synchronous base-level cycles and thus they can be used to extend biochronozones into strata where key fossils are absent. A classification designated as long, intermediate, and short term, is used to emphasize comparable scales and durations of stratigraphic cyclicity rather than emphasizing fixed cycle periods or assigning specific forcing mechanisms to each cycle frequency (i.e., second order = tectonic, third order = eustatic, etc.). The long-term stratigraphic cycle is about 4.5 my in duration, the intermediate-term stratigraphic cycles are 1 to 2 my in duration, and the short-term stratigraphic cycles are about 0.2 to 0.3 my in duration (Fig. 3).

In central Utah, a long-term stratigraphic cycle encompasses Turonian and early Middle Coniacian strata and includes the 500-m thick interval from the base of the Tununk Shale through the Ferron Sandstone and lower part of the Blue Gate Shale Members of the Mancos Shale. Argon-argon isotopic ages from volcanic ash beds tied to ammonite biozones indicate this long-term cycle extends from 93.25 ± 0.55 Ma to 88.8 ± 0.8 Ma (J. D. Obradovich, unpubl. data, 1992) and thus is 4.45 my in duration (Fig. 4). The 12 biozones comprising this long-term stratigraphic cycle are each about 0.37 my in duration. These dates suggest the average duration of the four intermediate-term stratigraphic sequences is 1 my. In the Ferron Sandstone Member, an argon-argon isotopic date of 90.25 ± 0.45 Ma was obtained from a volcanic ash bed in the C-coal bed (J. D. Obradovich, unpubl. data, 1992). Argon-argon isotopic ages from this and an overlying middle Coniacian volcanic ash bed indicate that part of the late Turonian intermediate-term stratigraphic sequence (*P. macombi* and *S. warreni* to *I. deformis* of the Ferronensis sequence) extends from 90.25 Ma to 88.8 Ma, thereby establishing a minimum age of 1.45 my for this sequence. Together these dates suggest that intermediate-term stratigraphic sequences are 1 to 2 my in duration.

Linear division of the five to six ammonite biozones that span the 1.45-my Ferronensis sequence indicates each biozone represents between 0.24 and 0.29 my. Inasmuch as the 0.13 my discrepancy between the two isotopically dated biozone intervals is within the error range of the analyses, each biozone is assumed to approximate a 0.24 to 0.37 my duration. Because the five to six ammonite biozones are contained within six short-term stratigraphic cycles of the Ferronensis sequence, each short-term cycle is a minumum of 0.24 my in duration. Below is a brief summary of this stratigraphic hierarchy; see Gardner (this volume) for a more complete discussion.

Short-Term Stratigraphic Cycle

The shortest-period stratigraphic cycle, or short-term cycle, is a progradational/aggradational unit that conforms to Walther's Law (Fig. 3; Gardner, this volume, Fig. 3). Short-term stratigraphic cycles are generally conformable internally and include deposits and surfaces formed across all depositional environments during a short-term base-level

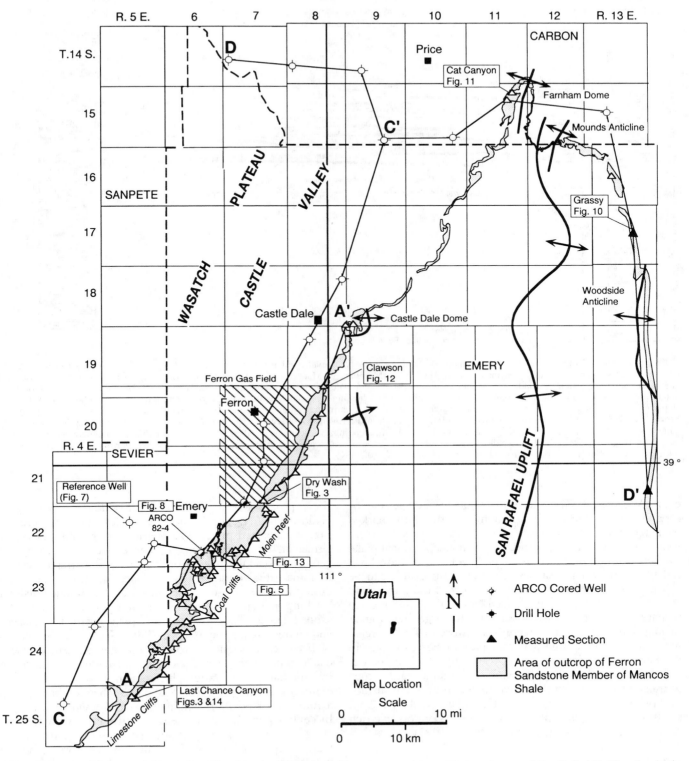

FIG. 2.—Map of central Utah showing study area and cross section and outcrop locations. Flank structures of San Rafael Uplift active in the middle Cretaceous shown.

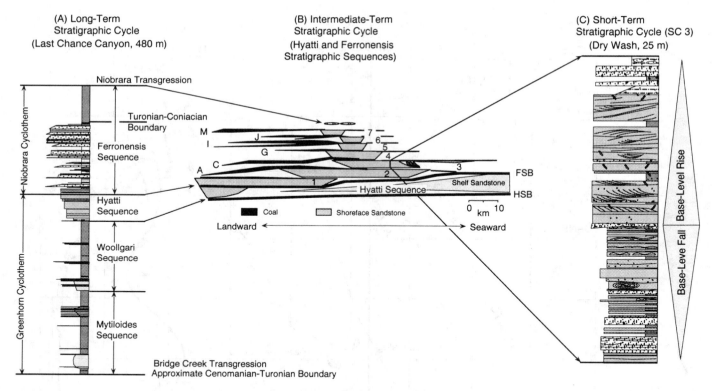

Fig. 3.—Diagram illustrating stratigraphic hierarchy and scales of base-level cyclicity for Turonian and Coniacian (part) strata of central Utah. (A) Measured section of Tununk, Ferron and Blue Gate (part) Members of Mancos Shale at Last Chance Canyon, Utah. This locality is thickest and most complete section of middle Cretaceous strata in central Utah. The long-term stratigraphic cycle records a 4.5-my base-level cycle initiated and terminated by eustatic transgressions. The stratal architecture of the long-term stratigraphic cycle records long-term base-level fall as a progressive step-wise progradation described by the stacking pattern of intermediate-term stratigraphic sequences (1 to 2 my). These are shown as the Mytiloides, Woollgari, Hyatti, and Ferronensis sequences. (B) Intermediate-term Ferronensis and Hyatti sequences may be characterized by the internal stacking pattern short-term stratigraphic cycles, here shown in a seaward-stepping, vertically stacked and landward-stepping arrangement for Ferronensis sequence. (C) Vertical profile from shallow-marine facies tract of seaward-stepping stratigraphic cycle (SC 3) in the Ferronensis sequence. Short-term stratigraphic cycles record high-frequency base-level cycles (0.3 my).

cycle (Wheeler, 1964). Successive short-term cycles can be separated by surfaces of stratigraphic discontinuity marked by a facies offset. Such surfaces record erosional unconformities, sediment bypass surfaces, or nondepositional hiatal surfaces formed during either base-level rise or base-level fall time. Elsewhere, successive short-term cycles can be separated by conformable strata that record the turnaround from base-level rise to fall. In such cases, a short-term cycle is marked by a progressive change in stratal geometry and/or facies architecture that records a history of first increasing and then decreasing accommodation. For example, in the Ferronensis sequence, short-term cycle boundaries occur at condensed sections in marine-shelf strata, at the tops of transgressive deposits that overlie a transgressive surface of erosion in shallow-marine strata, and at the change from sandstone-rich, organic-poor to sandstone-poor, organic-rich deposits in coastal-plain strata (Gardner, 1993).

Intermediate-Term Stratigraphic Cycle (Stratigraphic Sequence)

Four intermediate-term stratigraphic sequences are recognized, each informally named after a molluscan biozone (i.e., progressively upwards the Mytiloides, Woollgari, Hyatti, and Ferronensis sequences; Gardner, this volume, Fig. 4, enclosed plate). Geometries and boundaries of intermediate-term stratigraphic sequences are described by the geometric stacking pattern of short-term stratigraphic cycles (Gardner, this volume, Fig. 3). Changes in accommodation during intermediate-term base-level cycles produce distinct stacking patterns of short-term stratigraphic cycles, described geometrically as seaward-stepping, vertically stacked, and landward-stepping (Cross, 1988). See below discussion of Ferronensis sequence for stratigraphic tendencies associated with short-term cycle stacking patterns comprising intermediate-term stratigraphic sequences.

Intermediate-term cycle boundaries change character as a function of position within the seaway and facies type. In central portions of the seaway, stratigraphic sequences are generally bounded by stratigraphic discontinuities within marine deposits. Such discontinuity surfaces form either by nondeposition on top of short-term landward-stepping stratigraphic cycles during base-level rise or by submarine erosion during subsequent base-level fall. For example, hiatal (sediment starvation) surfaces can form during base-level rise time when all sediment is stored in coastal plain and

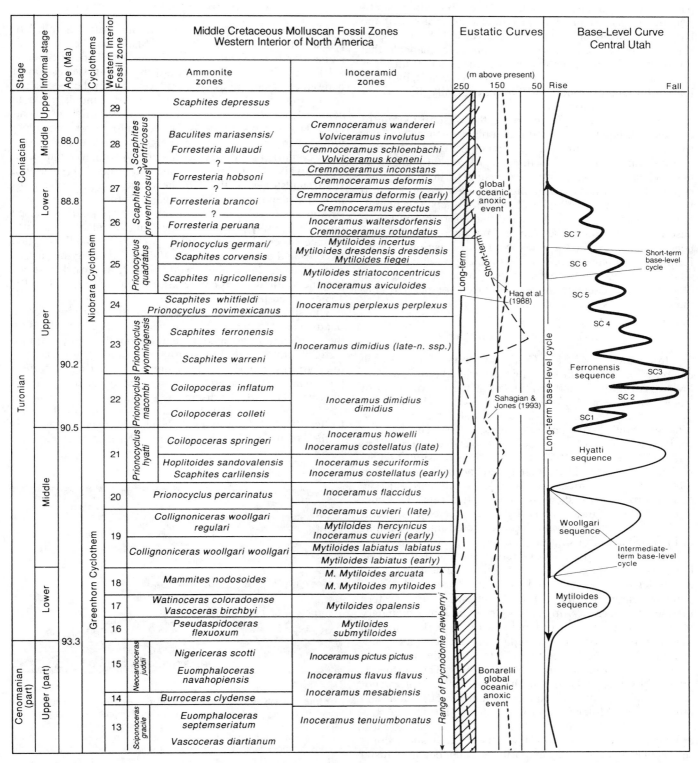

FIG. 4.—Biostratigraphic chart showing ammonite and inoceramid fossil zones for late Cenomanian through middle Coniacian Stages from the Western Interior of North America. Because of potential discrepancies associated with different stage boundary dates, eustatic curves from Haq and others (1988) and Sahagian and Jones (1993) are calibrated to biozones. Base-level curves based on stratigraphic relations in central Utah. Sources of data: Western Interior index fossils from Molenaar and Cobban (1991); upper Cenomanian and lower Turonian Substage biozones from Elder (1985); middle and upper Turonian Substage biozones from Kauffman and Collom (unpubl. data, 1990); lower and middle Coniacian zones from Collom (1991); argon-argon isotopic ages from Obradovich (unpubl. data, 1991).

paralic facies tracts. Erosion surfaces bounding stratigraphic sequences in more proximal marine areas of the basin form during subsequent base-level fall. Such erosion surfaces may record scouring of submarine channels that transport sediment to the center of the seaway. Both types of surfaces are recognized both paleontologically by missing molluscan biozones and stratigraphically by disconformity or paraconformity.

In high-subsidence regions along the western margin of the seaway, boundaries of stratigraphic sequences are conformable. These boundaries are recognized here by progressive variations in the architecture of coastal-plain strata that indicate a change from increasing to decreasing accommodation. Landward of high subsidence regions in a foreland basin and in alluvial-plain facies tracts, this boundary is presumably a base-level fall unconformity on top of base-level rise strata. The change from unconformable sequence boundaries in the most landward position to conformable boundaries along the basin margin to unconformable boundaries in the center of the seaward is the reverse of the Exxon sequence model developed for passive continental margins (Vail and others, 1977).

Long-Term Stratigraphic Cycle

Turonian to early middle Coniacian strata of central Utah comprise one long-term stratigraphic cycle recording a base-level cycle initiated and terminated at the top of strata of base-level rise and eustatic transgression (Figs. 3, 4; Schlanger and Jenkyns, 1976; Arthur and Schlanger, 1979; Jenkyns, 1980; Schlanger and others, 1986). This cycle is similar in duration (4.5 my) to Cretaceous cyclothems (Hattin, 1964; Kauffman, 1969, 1977a). Transgressions and associated landward shifts in sediment accumulation punctuate a period of long-term base-level fall and episodic eastward (basinward) shoreface progradation (Fig. 3). Long-term base-level fall is marked by a stepwise eastward shift of shoreface sandstones and basin-centered unconformities of intermediate-term stratigraphic sequences into the basin.

Base-level rise deposits are represented by pelagic carbonates in the Bridge Creek and Niobrara Limestones along the center of the seaway and by vertically stacked coastal plain and shoreface facies tracts along the active western margin of the seaway. In central Utah, the lower boundary of this stratigraphic cycle is marked by an early Turonian transgressive disconformity separating the Dakota Sandstone from the overlying marine Tununk Shale (Eaton and others, 1990; Molenaar and Cobban, 1991). This surface is overlain by a regional lag of chert pebbles and/or *Pycnodonte newberryi* oyster coquina (Young, 1960; Eaton and others, 1990; Molenaar and Cobban, 1991). The disconformity records two missing faunal zones, marks the turnaround from long-term base-level rise to fall, and records sediment starvation and submarine ravinement in a distal marine-shelf setting. The upper boundary of the long-term stratigraphic cycle is represented by bioturbated, fossiliferous, concretionary calcarenites in the Blue Gate Shale approximately 60 m above the top of the Ferron Sandstone. These concretions contain middle Coniacian fauna that are also present in the Niobrara Limestone.

STRATAL ARCHITECTURE OF STRATIGRAPHIC SEQUENCES

Along the western margin of the San Rafael Swell, Turonian strata include the Tununk Shale and Ferron Sandstone Members of the Mancos Shale. The Tununk Shale is a 100- to 250-m-thick, westward-thickening succession of bentonite, shale, mudstone, and very thinly bedded sandstone (Fig. 5). An isopach map of the Tununk Shale in central Utah (Fig. 6) shows this westward-thickening pattern. The Tununk Shale is divided into lower and upper units, at the base of the Coon Spring Sandstone Bed, the parts comprising the Mytiloides and Woollgari sequences, respectively. Near Emery, Utah, two minor ledge-forming shelf sandstones in the Tununk Shale comprise the most proximal deposits in these sequences (Fig. 5). Short-term stratigraphic cycles in the Tununk Shale are poorly exposed on weathered slopes and are best documented in the subsurface by conductivity well logs (Fig. 7). Well logs tied to cores and outcrops in Castle Valley and along the eastern margin of the San Rafael Swell (Figs. 7, 8) provide a stratigraphic framework (Gardner, this volume, Figs. 13, 14, 15, enclosed plate).

The stratigraphic nomenclature of the overlying Ferron Sandstone Member of the Mancos Shale is complicated by the grouping of two spatially and temporally distinct clastic wedges into a single lithostratigraphic unit. These partially overlapping clastic wedges have contrasting sediment provenance and dispersal patterns. Each clastic wedge is interpreted to reflect a discrete depositional system (Katich, 1951, 1954; Davis, 1954; Hale and Van De Graaf, 1964; Cotter, 1971, 1975a, b, 1976; Hale, 1972) and corresponds to the Hyatti and lower part of the Ferronensis sequences.

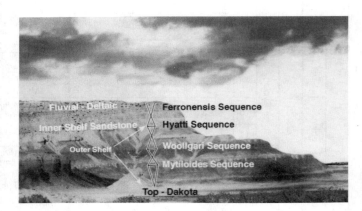

Fig. 5.—View looking northeast of the Tununk and Ferron Members near Emery, Utah. Valley floored by Dakota Sandstone. The mutual lower boundary of long- and intermediate-term stratigraphic cycles is placed at Dakota Sandstone and overlying Tununk Member contact. The Tununk Member is about 200 m thick here. Two minor resistant benches in the Tununk Member contain outer-shelf sandstones in Mytiloides and Woollgari sequences. The intermediate-term base-level fall to rise turnaround of the Mytiloides sequence occurs at the lower white resistant bench. The upper resistant bench is formed by shelf sandstones of the Coon Spring Sandstone Bed. The base of the Coon Spring Sandstone Bed marks intermediate-term base-level rise to fall turnaround and lower boundary of Woollgari sequence. The Hyatti sequence forms the lowermost cliff-forming sandstone. Overlying cuesta forming sandstones comprise the Ferronensis sequence.

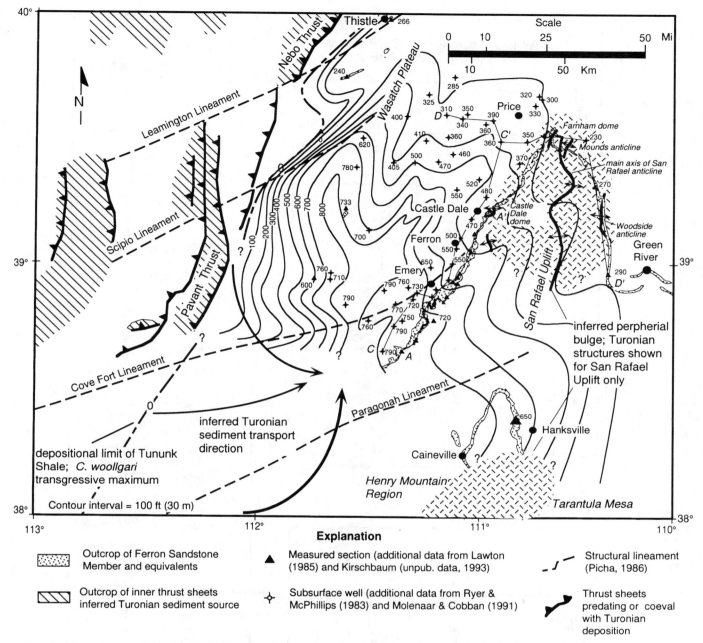

Fig. 6.—Isopach map of Tununk Member in central Utah. Westward thickening of Tununk Member records southwest increasing tectonic subsidence. Note location of inferred perpherial bulge in vicinity of San Rafael Uplift. Map compiled from Peterson and Ryder (1975), Ryer and McPhillips (1983), Lawton (1985), Picha (1986), Villien and Kligfield (1986), Kirschbaum (unpubl. data 1989), and Molenaar and Cobban (1991).

An isopach map of the Ferron Sandstone Member in central Utah (Fig. 9) shows the thickness of both the Hyatti and Ferronensis sequences.

Mytiloides Sequence

The Mytiloides sequence is a 50- to 200-m-thick westward-thickening calcareous mudstone succession (Figs. 3, 5, 6; Gardner, this volume, enclosed plate). It is based by a major transgressive disconformity and is capped by the overlying Coon Spring Sandstone Bed (as discussed above). This stratigraphic sequence is dominated by deep-water calcareous mudstones deposited below storm wave base. The position of the Mytiloides shoreline in central Utah is unknown, but coeval marine mudstones (Allen Valley Shale) are exposed 70 km west of Castle Valley (Gardner, this volume, Fig. 5; Villien and Kligfield, 1986).

Short-term stratigraphic cycles in the Mytiloides sequence are 10- to 20-m thick mudstone-dominated successions that coarsen upward from laminated shale to thin

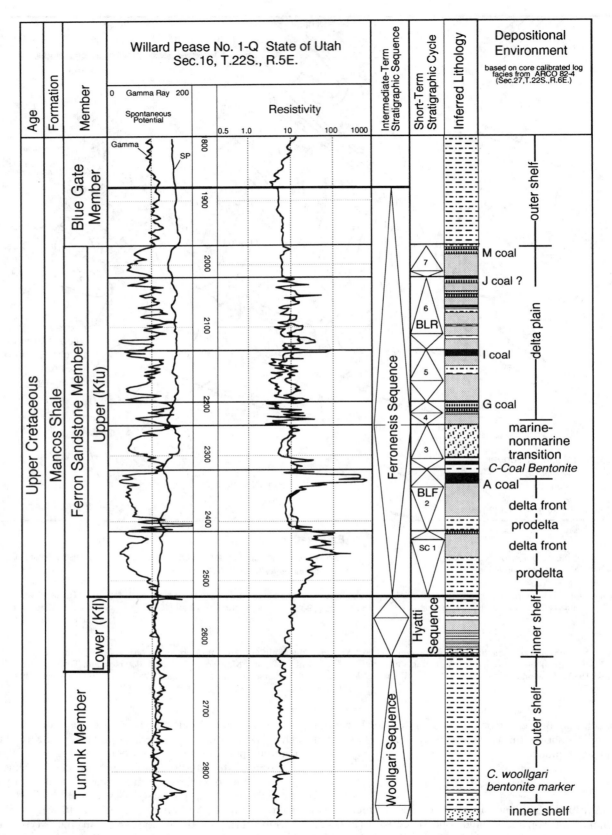

Fig. 7.—Subsurface well log showing middle Cretaceous stratigraphic sequences in central Utah. Well location shown in Figure 2. Core-calibrated log facies based on ARCO 82-4 cored well shown in Figure 8.

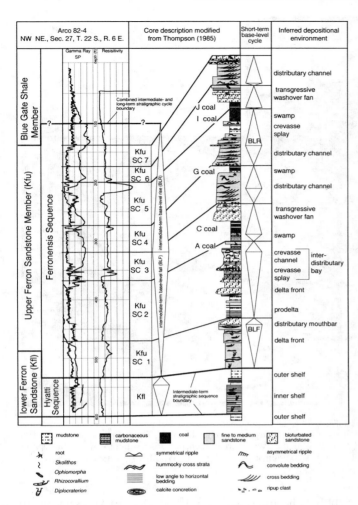

FIG. 8.—Core-calibrated well log facies in the Ferron Member in southern Castle Valley. Depositional environments are based on nearby outcrops in Miller Canyon. ARCO 82–4 core description adapted from Thompson (1985).

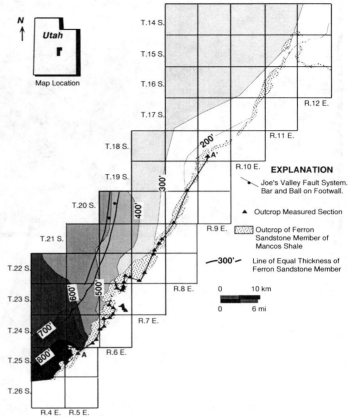

FIG. 9.—Isopach map of upper Cretaceous Ferron Sandstone Member in Castle Valley, central Utah. Map compiled from Ryer and McPhillips (1983), Lines and Morrissey (1983), and data from this study.

sandstone and siltstone interbeds. Sandstone-rich cycle caps form minor benches on outcrop slopes, commonly overlain by bentonites. Thin beds of well-sorted, fine-grained to very fine-grained sandstone are massive, or they display horizontal to wavy laminations and microhummocky cross-stratification internally. Sandstone beds are fossiliferous, contain burrow tracks and trails, and are capped by formsets of combined-flow ripple marks. These upward-shoaling, substorm wave base facies are inferred to record low sedimentation rates on a distal shelf.

Five to seven short-term stratigraphic cycles comprise the Mytiloides sequence. The lower five cycles step seaward and form a northeastward thinning, mudstone wedge. These cycles show a progressive increase in the sandstone to mudstone ratio and in the proportion of higher energy sedimentary structures within sandstones. The most basinward positioned and most sandstone-rich short-term cycle marks the change from intermediate-term base-level fall to rise. This cycle marks a pronounced low-conductivity zone on subsurface logs (Gardner, this volume, Fig. 14, enclosed plate) and contains microhummocky cross-stratified sandstones with abundant bioclastic debris in outcrop. Along the Molen Reef escarpment, this cycle forms a prominent white bench capped by bentonites (Figs. 5, 13). The overlying base-level rise deposits are thinner, vertically stacked to landward-stepping cycles, each progressively more mudstone-rich. These deposits are recognized regionally in the subsurface by increased intensities on gamma ray logs (Gardner, this volume, Fig. 14).

Woollgari Sequence

The Woollgari sequence contains the oldest middle Turonian strata in central Utah (Fig. 4). Along Molen Reef in southern Castle Valley, the basal Woollgari sequence deposits form a laterally continuous bench of brown siltstone and sandstone in the upper Tununk Shale (Fig. 5, 13), and in east-central Utah, they are the Coon Spring Sandstone Bed (Fig. 10). The overlying middle Turonian index fossil *Prionocyclus percarinatus* is absent in central Utah (Molenaar and Cobban, 1991, p. 7) suggesting a paraconformity caps the sequence. Shallow-marine sandstones containing *C. Woollgari* east of Salina, Utah document a more basinward position for the Woollgari shoreline than the Mytiloides shoreline (Ryer, 1983).

Fig. 10.—Outcrop photo of the Coon Spring Sandstone Bed of the Tununk Member and upper part of the Mytiloides sequence and lower part of the Woollgari sequence along the northeast flank of the San Rafael Uplift at Grassy. Sharp lower contact of the Coon Spring Sandstone Bed records intermediate-term base-level rise to fall turnaround forming lower boundary of the Woollgari sequence. Closeup photo of cuesta capping sandstone. Planar tabular, cross-bedded and burrowed inner-shelf sandstone records maximum base-level fall in Woollgari sequence.

From southwest to northeast along Castle Valley, the Woollgari sequence is a 10- to 60-m thick, eastward-thinning mudstone wedge consisting of at least five 3- to 20-m thick short-term cycles (Fig. 5, 10, 13). Woollgari sequence facies are similar to, but coarser grained than, those in the underlying Mytiloides sequence. The Coon Spring Sandstone Bed is distinctive and forms an upward-coarsening succession capped locally by chert pebbles and fossils. This fine- to medium-grained sandstone bed is thickest along the eastern San Rafael Swell (Fig. 10) but is coarsest grained where the Woollgari sequence is at its thinnest point near Ferron, Utah. The Coon Spring Sandstone Bed extends eastward almost as far as Grand Junction, Colorado (Molenaar and Cobban, 1991) and to the north onlaps a regional unconformity in the Uinta Basin (Molenaar and Wilson, 1990).

There is no evidence of subaerial erosion associated with the Coon Spring Sandstone Bed, but there is a regional basinward shift in facies. These deposits appear to have accumulated mostly below storm wave base, at least 20 km offshore. If pebbles of the Coon Spring Sandstone Bed are derived from a landward source, then they have been transported significant distances across the shelf. The mechanism(s) that could accomplish this are unclear.

Pebble-bearing shelf sandstones in the Ferronensis sequence occur at the inferred base-level fall maximum. These pebbly sandstones are present at least 70 km seaward of their coeval shoreline. By analogy, pebble-bearing sandstones in the Coon Spring Sandstone Bed may record the efficiency of sediment bypass across the shelf during base-level fall.

The turnaround from intermediate-term base-level fall to rise occurs near the top of the Coon Spring Sandstone Bed. Two thin landward-stepping cycles in the upper Tununk Shale progressively fine upward and record marine deepening. A thick volcanic ash bed occurs approximately 7 to 10 m above the Coon Spring Sandstone Bed and can be correlated throughout central Utah (Gardner, this volume, Figs. 13, 14, 15, enclosed plate).

Hyatti Sequence

The Hyatti sequence (Fig. 3, 4, 5, 11, 12, 13, 14) is temporally equivalent to the lower Ferron Sandstone Member and a thin interval of overlying Mancos Shale in northern Castle Valley and to vertically stacked shoreface sandstones in the lower part of the Ferron Sandstone in southern Castle Valley. These along-strike variations reflect two distinct shelf and shoreface depositional systems comprising the Hyatti sequence here. Minor erosional relief at the paraconformity at the base of this sequence is indicated by truncation of sandstone beds in the upper Tununk Shale in northern Castle Valley (Fig. 11).

Fig. 11.—Outcrop photo of outer- to inner-shelf facies of lower Ferron Member in northern Castle Valley. Short-term stratigraphic cycles in Hyatti sequence are shown. The turnaround from intermediate-term base-level fall to rise of Hyatti sequence is placed at erosional lower contact of capping sandstone bench. Inner-shelf sandstones form highest energy facies exposed in the northerly derived depositional system of the Hyatti sequence. Note hummocky cross stratified sandstone (HCS). Approximately 20 m of roadcut shown.

Fig. 12.—Outcrop photo showing short-term stratigraphic cycles of the Hyatti sequence (lower Ferron Member) in central Castle Valley near Clawson, Utah. Middle of photo shows meterwide spheroidal, botyroidal, calcite-cemented sandstone concretions occurring in Clawson Sandstone Bed. The top of the sandstone concretions mark turnaround from intermediate-term base-level fall to base-level rise of the Hyatti sequence. Overlying sandstone beds thin and mudstone beds thicken upward and record base-level rise of the Hyatti sequence. These deposits comprise short-term stratigraphic cycles arranged in a landward-stepping stacking pattern. Cliff face approximately 15 m high.

A southeastward prograding and thinning clastic wedge in northern Castle Valley is recorded by five to seven, short-term cycles arranged in a seaward- then landward-stepping stacking pattern (Gardner, this volume, Fig. 13). Facies in cycles grade upwards from siltstones to hummocky cross-stratified sandstones (Fig. 11). Intensity and diversity of burrowing increase to the south. These storm-dominated shelf sandstones are replaced along depositional-strike to the south by marine-shelf mudstones. Vertically stacked shoreface sandstones occur farther south near Last Chance Canyon (Fig. 14; Gardner, this volume, enclosed plate). Across this transect the thickness of mudstone between a volcanic ash bed in the upper Woollgari sequence and the base of the Hyatti sequence almost doubles. This thickness change reflects southwestward increasing subsidence and suggests sandstone deposition began earlier in the north, making inner-shelf sandstones in the north older.

In northern Castle Valley, the most seaward-stepped short-term cycle in the Hyatti sequence contains the most sandstone-rich, highest energy, and shallowest water deposits. These deposits are amalgamated, swaley cross-stratified sandstones with swaley depressions lined with inoceramid and oyster shell lags (Fig. 11). A type 2 sequence boundary is postulated at the base of this cycle (Riemersma and Chan, 1991) marking the turnaround from intermediate-term base-level fall to rise. Along depositional strike, to the south this cycle contains distinctive meterwide, spheroidal, botyroidal, calcite-cemented sandstone concretions (Fig. 12). These seaward-stepping cycles are overlain by three, vertically stacked to landward-stepping cycles of upward-shallowing outer- to inner-shelf deposits that become progressively thinner and more mudstone rich and are bounded by bioturbated transgressive deposits (Fig. 12, 13; Gardner, this volume Fig. 15). Hence, changes in stratal geometry, li-

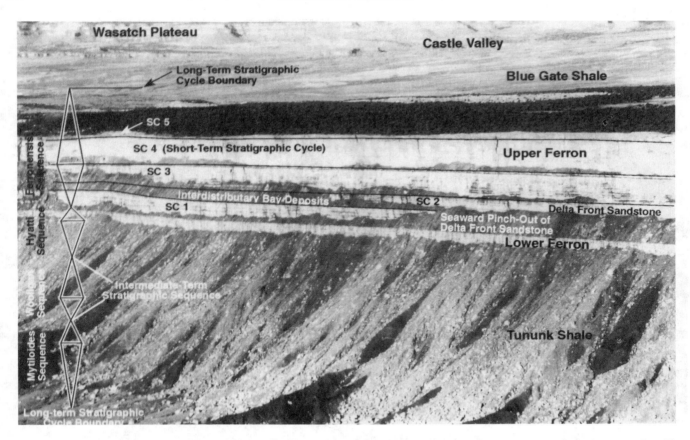

Fig. 13.—View looking west toward Emery, Utah, of the Tununk and Ferron Members along the Molen Reef in southern Castle Valley. Note the seaward pinch-out of shallow-marine sandstone in SC 1 of the Ferronensis sequence near center of photo. From valley floor to cuesta top is approximately 330 m.

thology ratio, and facies architecture within these short-term cycles are recognized. Seaward-stepping cycles are thicker, broader, more sandstone rich, and contain higher energy deposits than landward-stepping cycles.

The Hyatti sequence is capped by a thin, dark, organic-rich siltstone and thinly bedded sandstone succession that divides the Ferron Sandstone Member into two units. This siltstone contains a volcanic ash bed extending over most of Castle Valley and records the cessation of northerly-derived sediment (Gardner, this volume, Figs. 13, 14, 15, enclosed plate). Soft-sediment deformation features in this bed include chevron and recumbent folds and small-scale reverse faults and record sediment loading of the overlying Ferronensis sequence. To the southwest, this siltstone bed interfingers with and is downlapped by successive seaward-stepping cycles of the Ferronensis sequence (Fig. 13). Farther to the southwest, near Last Chance Canyon, this siltstone bed is replaced by a 60-m thick shoreface sandstone (Fig. 14).

The northeast to southwest trend of the Hyatti shoreline can be related to an increase in subsidence rates from northern to central Utah (Ryer and McPhillips, 1983; Lawton, 1985; Ryer and Lovekin, 1986; Molenaar and Cobban, 1991; Gardner, this volume). In northern Utah, middle to late Turonian strata are conspicuously thin and exhibit a laterally expanded deltaic depositional pattern that occupies much of the present Uinta Basin (Ryer and Lovekin, 1986). This broad depositional feature is the inferred source for sediment transported southward (alongshore) to central Utah. Southward sediment transport is supported by paleocurrent data from symmetrical ripple crests and theoretical circulation patterns modeled for the Cretaceous seaway (Ericksen and Slingerland, 1990). Erosion of a fluvial-fed delta to the north by southward-directed, shore-parallel currents would promote sediment transport to the southern shallow, storm- and wave-dominated inner shelf.

Other workers have interpreted the southward thinning of sandstones as evidence for elongate, shoreline-detached shelf bars (Hale and Van De Graaff, 1964; Hale, 1972; Cotter, 1976; Thompson, 1985; Molenaar and Cobban, 1991; Riemersma and Chan, 1991). Coal deposits capping shoreface sandstones in the subsurface near Price, Utah, however, establish the proximity of the shoreline to northern Castle Valley outcrops. Regional sedimentation patterns in the Hyatti sequence suggest that there was no single paleogeomorphological offshore bar (Ryer and McPhillips, 1983). Rather, increased sediment input from a northern source may simply have increased the supply of sand to the inner shelf of southeastward prograding deltas.

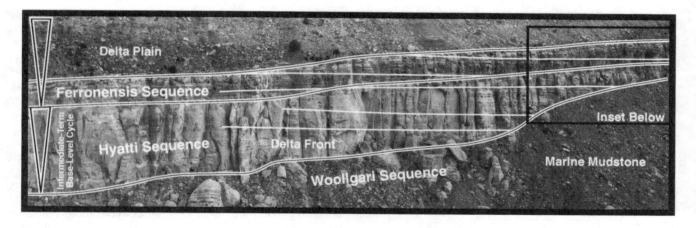

Fig. 14.—Outcrop photos of more than 60-m-thick, cliff-forming sandstone in the Ferron Member at Last Chance Canyon, Utah. (A) The lower two-thirds of cliff contain clinoforms that dip toward right margin of photo and toplap flat continuous surface in upper third of cliff. Clinoforms in sandstone forming upper one-third of cliff downlap this surface. This discontinuity surface separates the Hyatti sequence from the overlying Ferronensis sequence and marks intermediate-term base-level rise to fall turnaround. The lower two-thirds of cliff comprise shoreface sandstones arranged in three vertically stacked short-term cycles. Upper one-third of cliff is first seaward-stepping cycle of the Ferronensis sequence. (B) Enlarged photo showing right margin of photo in (A). Vertically stacked shoreface sandstones in Hyatti sequence pinch-out and are replaced by marine mudstones. Overlying shoreface sandstones forming the lowermost seaward-stepping cycle of the Ferronensis sequence prograde 30 km basinward from this point to their depositional pinch-out near Emery, Utah, shown in Figure 13.

Shoreface sandstones exposed in southern Castle Valley are arranged in a 50-m thick succession of at least three vertically stacked short-term cycles (Fig. 14). Vertical stacking and oblique clinoforms in these shoreface sandstones suggest sedimentation rates were equal to subsidence rates here. Shelf and shoreface sandstones in northern and southern Castle Valley are spatially and temporally distinct sediment bodies. Consequently two distinct and isolated sandstone wedges in the Hyatti sequence comprise the laterally continuous sandstone distribution inferred by Ryer and McPhillips (1983).

Ferronensis Sequence

The Ferronensis sequence records northeastward progradation of Turonian fluvial-deltaic deposits in central Utah (Figs. 3, 4, 5, 13, 14, 16). The Ferronensis sequence has a greater volume than other stratigraphic sequences discussed here (Gardner, 1993). Exceptionally continuous exposures allow correlation of short-term cycles across marine-shelf, shallow-marine, and coastal-plain facies tracts.

In Castle Valley, the Ferronensis sequence contains at least eight, short-term stratigraphic cycles arranged in a seaward-stepping to vertically stacked to landward-stepping pattern (Gardner, this volume, enclosed plate). Facies tracts in successive short-term cycles (SC 1–3) are displaced seaward and progressively increase in width. Sediment volumes in more landward facies tracts decrease, whereas sediment volumes in more seaward facies tracts increase. Such seaward-stepping stratigraphic cycles deposited during intermediate-term base-level fall record progressively reduced accommodation. Coastal-plain facies tracts thin up-

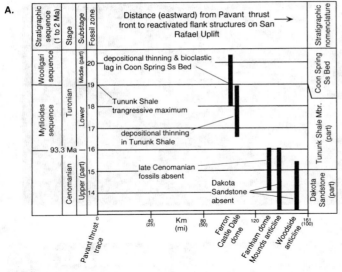

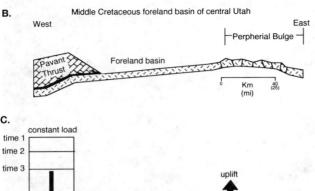

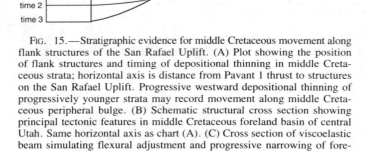

Fig. 15.—Stratigraphic evidence for middle Cretaceous movement along flank structures of the San Rafael Uplift. (A) Plot showing the position of flank structures and timing of depositional thinning in middle Cretaceous strata; horizontal axis is distance from Pavant 1 thrust to structures on the San Rafael Uplift. Progressive westward depositional thinning of progressively younger strata may record movement along middle Cretaceous peripheral bulge. (B) Schematic structural cross section showing principal tectonic features in middle Cretaceous foreland basin of central Utah. Same horizontal axis as chart (A). (C) Cross section of viscoelastic beam simulating flexural adjustment and progressive narrowing of foreland basin in response to a constant load applied over time. Diagram from Beaumont (1981).

erally straight and wave-dominated initially but became progressively more irregular with lobate and elongate deltaic promontories (Gardner, 1993).

Vertically stacked cycles (undifferentiated SC 4) following this succession of seaward-stepping cycles record the beginning of intermediate-term base-level rise. Vertically stacked cycles have little or no offset of facies tracts across cycle boundaries and little shift in the depositional limits of successive stratigraphic cycles.

Landward stepping short-term cycles (SC 5–7) record decreased facies tract widths. Facies tracts are displaced landward across successive landward-stepping cycle boundaries. Sediment volumes in more landward facies tracts increase, whereas volumes of more seaward facies tracts decrease. Coastlines became progressively more embayed with or without local barrier bars and increased tidal influence. Other attributes are the reverse tendencies of those described for successive seaward-stepping cycles. Changes in the facies architecture and sediment volume partitioning in these short-term cycles are discussed in Gardner and others (1992) and Gardner (1993).

Coastal-plain deposits in the Ferronensis sequence progressively shift basinward in successive seaward-stepping cycles recording a profound Turonian base-level fall. High coastal-plain sediment volumes reflect, in part, close proximity to sediment source. It seems unlikely that numerous small deltas prograded eastward, in concert, to form the depositional pattern displayed by the Ferronensis sequence here (*sensu* Ryer and McPhillips, 1983). Rather, the stacking pattern of the Ferronensis sequence in central Utah is inferred to record longitudinal axial drainage from a large fluvial system flowing parallel (northeastward) to the thrust belt from southern to central Utah.

Near Last Chance Canyon, the lower boundary of the Ferronensis sequence marks the initiation of northward shoreface progradation of a succession of seaward-stepping cycles that extend more than 70 km basinward from underlying, vertically stacked shoreface sandstones in the Hyatti sequence (Fig. 14). Farther northward along Castle Valley, successive seaward-stepping cycles downlap onto the top of a black organic-rich siltstone (Fig. 13; Gardner, this volume, enclosed plate). Along the eastern margin of the San Rafael Swell and in northern Castle Valley, the top of this siltstone forms the lower boundary of the Ferronensis sequence. Locally, this siltstone is cut by a coarse-grained shelf sandstone deposit that is up to 1 m thick and locally contains 2.5-cm black chert pebbles (Fig. 15). This sandstone records sediment bypass during base-level fall. The geometry of this pebbly sandstone has not been established, but these deposits locally occur as far east as Thompson, Utah (Molenaar and Cobban, 1991).

The upper part of the Ferronensis sequence is the lower Blue Gate Shale Member in Castle Valley and the upper part of the Funk Valley Formation in Sanpete Valley 70 km to the west. Early to middle Coniacian *Baculites mariasensis* and *Inoceramus deformis* occur in limestone concretions about 40 m above the Ferron Sandstone Member, and *I. deformis* occur in the Funk Valley Formation. These deposits record maximum marine deepening in the Ferronensis sequence.

ward and display increased sandstone to shale ratios, increased facies diversity, deeply incised channels, laterally extensive erosion surfaces, and thin aggradational paleosols. Shallow-marine facies tracts become increasingly heterolithic, having decreased sandstone to mudstone ratios, increased sediment volume, more diverse facies associations, and more deformational sedimentary structures produced by slumping and dewatering. Coastlines were gen-

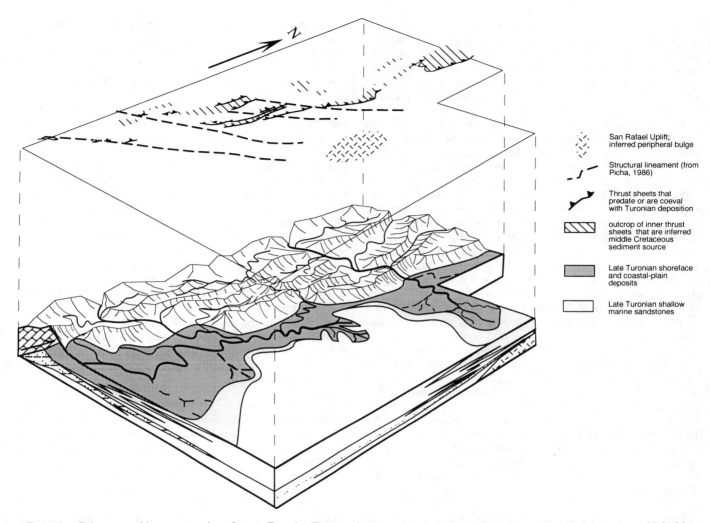

FIG. 16.—Paleogeographic reconstruction of upper Turonian Ferronensis sequence at base-level fall maximum (*P. macombi* biozone, 90.21 Ma). The position of structural features thought to control the position of major middle Cretaceous fluvial systems are shown above block diagram. Note the northeastward orientation of the inferred trunk stream that transported sediment to the Ferronensis shoreline of southern and central Utah. This fluvial system originated from southern Utah and transported feldspar-bearing sandstones that record the south-to-north migration of deltaic complexes along the mid-Cretaceous shoreline of Utah. The terminal margin of these deltaic complexes in the Ferronensis sequence occurs in central Utah.

FORELAND TECTONISM AND STRATIGRAPHIC SEQUENCE ARCHITECTURE

Tectonic control on deposition in the Mid-Cretaceous foreland basin of central Utah is shown by (1) westward-thickening stratigraphic sequences recording load-induced flexural subsidence (Figs. 6, 9), (2) depositional thinning of successive short-term cycles over a westward-migrating peripheral bulge (Figs. 2, 15), and (3) depocenter and paleodrainage patterns controlled by lineaments segmenting thrust sheets and creating foreland subbasins flanking the western margin of the Cretaceous seaway (Fig. 16). These tectonic controls are addressed below.

The Mid-Cretaceous Record of Foreland Tectonics in Central Utah

Mid-Cretaceous stratigraphic sequences in central Utah thicken westward toward the east-vergent Pavant 1 thrust. This trend is most distinct in the Mytiloides and Ferronensis sequences (Figs. 6, 9) and can be related to westward-increasing flexural subsidence. About 70 km farther into the basin, westward migration of a peripheral bulge is inferred by depositional thinning in successively younger mid-Cretaceous strata over more progressively westward positioned flank structures on the San Rafael uplift. Along the eastern margin of the San Rafael uplift, depositional thinning is suggested by (1) the absence of the Dakota Sandstone at Woodside and Mounds anticline, (2) depositional thinning in successive short-term cycles of the Mytiloides and Woollgari sequences over Mounds anticlines, and (3) the absence of late Cenomanian fossils and north-south depositional thinning in the Mytiloides sequence over Farnham dome (Fig. 7; Molenaar and Cobban, 1991). Along the northwestern margin of the San Rafael uplift, depositional thinning is indicated by northeast-southwest thinning of short-term cycles in the Woollgari and lower Hyatti sequences over Castle Dale dome and thinning of the Woollgari sequence near Ferron, Utah (Fig. 15). A poorly sorted, fossiliferous chert pebble lag deposit caps the Coon Springs Sandstone Bed at the thinnest point of the Woollgari sequence near Ferron, Utah. Thinning of short-term cycles in the Woollgari sequence containing this locally distributed lag suggests the region was a paleotopographic high.

Marine mudstones containing Turonian palynomorphs unconformably overlap the hanging wall of the Pavant 1 thrust near Gunnison, Utah (70 km west of outcrops in Castle Valley), suggesting that thrust movement ceased at the Cenomanian–Turonian boundary (Villien and Kligfield, 1986, p. 298). However, thinning of successive short-term stratigraphic cycles over a westward migrating peripheral bulge suggests crustal adjustments to thrusting and tectonic subsidence continued throughout the Turonian. Thus, it would appear that rapid basin subsidence continued for a period of at least 2 my after thrusting and tectonic loading in the orogenic belt ceased. Continued rapid basin subsidence for such a long period after crustal loading has stopped does not support models of foreland basin development that assume an elastic lithosphere rheology (e.g., Fleming and Jordan, 1989, 1990).

Rather, foredeep contraction, westward migration, and rise of a peripheral bulge may reflect viscoelastic relaxation of the crust following thrust loading (Beaumont and others, 1984, 1987). Uniform viscoelastic crust models predict that thrust-imposed loading would be followed by deepening of the proximal foredeep and migration and steepening of the peripheral bulge (Dorobek, this volume). These models suggest that thrust loading along the central Utah thrust salient could have ceased during the early Turonian, with tectonic subsidence in the foreland extending into the middle Turonian.

Sedimentary Response to Foreland Tectonics in Central Utah

The tectonic history of increasing then decreasing subsidence and migration of a peripheral bulge is superimposed on a progressive increase in sediment supply recorded by the Woollgari through Ferronensis sequences in central Utah. Because of the time lag between thrust cessation and increased sedimentation, it is not clear how the basin fill relates to a tectonic cycle. Neither elastic nor viscoelastic lithosphere models predict a significant time lag between thrust cessation and increased rates of sediment delivery to the basin. Viscoelastic models are consistent with the Turonian history of central Utah tectonic activity, but suggest a 3-million-year time lag between thrust cessation and sediment delivery. Elastic models are incompatible with this tectonic history, requiring undetected tectonic activity, but suggest only a 1 to 2 million-year time lag. An explanation of sediment volume changes in the most sandstone-rich (Hyatti and Ferronensis) sequences suggests possible causes for this time lag. These are addressed below.

The Hyatti sequence records regression of shorelines from both the thrust belt and peripheral bulge into the basin (Fig. 6, 15). The sequence is relatively constant in thickness despite southward and northward depositional thinning of shelf and shoreface sandstones, respectively. These broad shelf sandstones indicate decreased tectonic subsidence, expansion of sites of increasing accommodation in response to a cessation of tectonic activity, and widespread shoaling.

A pronounced increase in sediment supply rates during deposition of the Ferronensis sequence is inferred from proximal foredeep deposits exposed near Last Chance Canyon. Here, the Hyatti sequence consists of a 50-m thick succession of three, vertically stacked cycles of shoreface sandstone showing an upward decrease in burrowing and increase in multidirectional trough cross bedding. Internal clinoforms in vertically stacked sandstones toplap a low-angle disconformity surface separating them from capping seaward-stepping shoreface sandstones (Ferronensis sequence). These clinoforms downlap a sharp basal surface with underlying marine mudstones forming the lower boundary of the Hyatti sequence (Fig. 14).

The first short-term cycle of the Ferronensis sequence at Last Chance Canyon steps seaward and is a 20-m thick, upward-coarsening succession of amalgamated, multidirectional, trough cross-stratified sandstone. Internal clinoforms downlap the lower boundary of the Ferronensis sequence. Across this boundary, there is a change in cycle stacking pattern from vertically stacked (i.e., Hyatti sequence) to seaward stepping (i.e., Ferronensis sequence), and internal clinoforms toplap and downlap it. These stratal geometry changes are associated with a 30 km basinward shift in the Ferronensis shoreline and record decreased accommodation and/or increased sediment supply. Because these strata thicken, indicating increasing accommodation at Last Chance Canyon, this shoreline regression is inferred to record an increase in sediment supply. The addition of a sediment source from the south during Ferronensis deposition would increase sediment supply rates and may be related to delta lobe switching and northward migration of depocenters from southern to central Utah (Gardner, this volume).

In southern Utah, an unconformity at the base of the Calico sandstone bed in the Straight Cliffs Formation is overlain by northeast transported, feldspathic pebble-bearing fluvial sandstones (Peterson, 1969; Eaton and others, 1987; Bobb, 1991). In central Utah, the Ferron sandstone is a feldspathic arenite with granitic rock fragments. Synorogenic conglomerates adjacent to the Pavant 1 thrust sheet contain only quartzite, chert, and limestone, with no feldspar clasts. The closest feldspar source is southern Utah, where fluvial sandstones of the Calico bed most likely represent the remnants of the fluvial system that was the source of the Ferronensis sequence in central Utah. The position of this fluvial system may have been controlled by northeast-trending tectonic lineaments (Fig. 16), which segmented thrust sheets along the thrust belt (Picha, 1986). Analogous thrust segmentation controlling point-source sediment dispersal is reported from the Miocene Himalayan orogenic belt by Mulder and Burbank (1993) and Willis (1993).

Late Turonian deposition of the Ferronensis sequence records a change in sediment provenance inferred to be synorogenic. The time lag between thrusting and deposition of coarser-grained sediment in central Utah may indicate uplift between the thrust belt and foreland basin. Such patterns have been documented in other forelands, where "piggy back" basins delay sediment transported to the foreland (Mulder and Burbank, 1993). Depositional patterns in the Ferronensis sequence are oriented oblique to the foreland axis, suggesting high subsidence rates induced longitudinal drainage.

Longitudinal drainage along the foreland basin axis may also have contributed to sediment dispersal across central Utah. Such patterns are particularly prevalent in foreland basins where subsidence rates exceed sedimentation rates (underfilled; e.g., Fuchtbaur, 1967; Eisbacher and others, 1974; van Houten, 1977; Sevon, 1985; Raynolds and Johnson, 1985; Mulder and Burbank, 1993; Willis, 1993). A structural reentrant in southern Utah (Paragonah lineament of Picha, 1986), may have determined the entrance of a major river system to the basin. This river appears to have flowed longitudinally northward along the foreland basin axis and deposited thick deltaic deposits of the Ferronensis sequence in central Utah (Fig. 16).

The rate of sediment supply appears to have increased significantly at the beginning of Ferronensis deposition from the addition of a source of sediment from the south. This sediment supply increase does not support models that predict a decrease in sediment supply following thrusting (Heller and others, 1988; Fleming and Jordan, 1989, 1990). This depositional pattern does support, however, the assumption that the supply of sediment outpaces subsidence, and facies prograde into the basin as the basin widens during thrust cessation (Fleming and Jordan, 1990). Marine mudstones overlying the hanging wall of the Pavant 1 thrust indicate movement ceased up to three million years prior to Ferronensis deposition. This time lag between thrust cessation and increased deposition rates in the basin reflects sediment supply rates outpacing tectonic subsidence rates and may be related to (1) duration of 100-km west-to-east progradation into the basin, (2) undetected Turonian tectonism along a blind thrust (required for elastic model), (3) rates of isostatic uplift in the source area greater than tectonic subsidence rates in the basin, (4) temperature-dependent viscoelastic flexural response to thrust deformation, and (5) delayed sediment delivery related to longitudinal drainage and sediment storage in a piggy-back basin.

Inferred Subsidence and Sea-Level History

Tectonic and eustatic movements produced accommodation changes in mid-Cretaceous strata from central Utah. Tectonic subsidence rates appear to have been always greater than rates of eustatic fall. In such a high accommodation setting, unconformities are absent, and physical stratigraphic criteria used to define cycle boundaries are more subjective. Sedimentological indicators of stratigraphic cyclicity in a high accommodation setting are outlined below.

The conformable and interfingering relationship between the Tunuk Shale, Ferron Sandstone, and Blue Gate Shale Members of the Mancos Shale indicates continuous accommodation (Figs. 5, 13; Gardner, this volume, enclosed plate). Nine volcanic ash beds in the Tunuk Shale provide traceable markers and show that strata of the Mytiloides, Woollgari, and Hyatti sequences thicken to the southwest along the west flank of the San Rafael Swell. Sandstones in the Hyatti sequence thin to the southwest (Gardner, this volume, Figs. 14, 15). All these sequences thicken to the west toward positions of increasing subsidence rates. To the southwest these stratigraphic sequences preserve more complete and continuous biostratigraphic zones than in the northeast, where strata are punctuated by condensed sections. These relationships suggest that relative sea level was rising prior to Ferronensis deposition.

Relative sea level also appears to have been rising during deposition of the Ferronensis sequence. Delta lobes in this sequence lack incised valleys or regionally extensive erosional surfaces that would indicate negative accumulation during times of relative sea-level fall. Short-term cycles of the Ferronensis sequence have a sigmoidal geometry of topset aggradation with clinoforms rising during progradation (Fig. 17; Ryer, 1981). Facies within successive sigmoidal strata rise stratigraphically, such that delta-front deposits thicken seaward by progressive increase of mudstone at their bases.

Stratigraphic rise within prograding delta front or shoreface strata requires accumulation under conditions of relative sea-level rise. Delta fronts prograde across and downlap onto a low-relief, gently seaward-dipping surface, which is formed by a combination of deposition and ravinement of the underlying progradational unit. Accommodation space available reflects compaction or subsidence. The prograding delta fills space by building a seaward-thickening, wedge-shaped platform of mud overlain by delta-front sands of uniform thickness. The thickness of delta-front sands is controlled by the shoreface slope and storm wave-base. Space below storm wave-base must be filled before the upper delta-front sands may prograde across the shelf. When relative sea-level is rising, the thickness of the mud platform wedge must increase more before shoreface progradation than times when relative sea level is static or falling. Progradation of the shoreface during a relative sea-level rise is associated with vertical aggradation of the coastal plain and steeper shoreface slopes. The deposits of an episode of progradation would thus have more a pronounced sigmoidal ge-

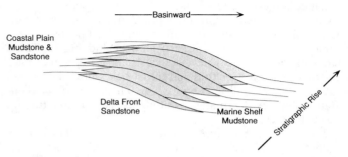

FIG. 17.—Schematic diagram illustrating concept of stratigraphic rise in shallow-marine strata. Stratigraphic rise is demonstrated by the progressive seaward offlap of thinner sandstone beds onlapped by thicker mudstone beds. By contrast, the upper shoreface of the delta front contains thicker sandstone beds that offlap seaward and are onlapped by progressively thinner mudstone interbeds. This produces a sigmoidal geometry during delta progradation–aggradation. See text for discussion.

ometry. This geometry is generally best expressed near the seaward depositional limits of delta-front sandstones, where they interfinger with prodelta mudstones. As shown schematically in Figure 17, in this setting the lower shoreface of the delta is marked by the progressive seaward offlap of thinner sandstone beds that are onlapped by thicker mudstone beds. By contrast, the upper shoreface of the delta front contains thicker sandstone beds that offlap seaward and are onlapped by progressively thinner mudstone interbeds. A similar interfingering relationship occurs between delta-plain and delta-front facies within the shingled delta lobes.

Preservation of shallow-marine strata deposited during transgressions also indicate an increase in accommodation and relative sea-level rise during Ferronensis time. Well-developed transgressive deposits up to 5-m thick cap shallow-marine and coastal-plain strata of each cycle, some associated with an intervening ravinement surface. These transgressive deposits are the erosional products of shoreline retreat, ravinement, and cannibalization of shoreface and coastal-plain sediments. Preservation of shelf and lower shoreface sediments during transgression suggests long-term relative sea-level rise during Ferron time. When relative sea level is falling, such transgressive sediments are generally eroded during progradation of the subsequent delta front.

The accumulation of peats and their preservation as coals is another indicator of relative sea-level rise. Ryer and others (1980) documented low-lying peat mires (as opposed to raised mires) bidirectional expansion, and aggradation during delta progradation. They showed that ash beds within coals terminate in progressively more seaward and landward positions as peat-forming swamps expanded through time, encroaching upon drier land and burying underlying sediment. Multiple coal beds within coastal-plain strata of the Ferronensis sequence indicate a ground-water table near the surface. As facies tracts of the Ferronensis sequence rise stratigraphically within short-term cycles (see above), expansion of peat-forming mires indicates that the water table was rising during delta progradation, probably reflecting relative sea-level rise.

SUMMARY

Three temporal and spatial scales of stratigraphic cyclicity are recognized in Turonian and Coniacian strata of central Utah. This stratigraphic hierarchy documents the tectonic history of the central Utah foreland basin. These stratigraphic sequences indicate that tectonic subsidence rates in front of the Sevier fold and thrust belt were greater than rates of eustatic fall in central Utah. Thus, tectonic and eustatic changes facilitated continuous accommodation at variable rates.

Intermediate-term stratigraphic sequences in central Utah record a period of accelerated tectonic subsidence and westward migration of a peripheral bulge followed by a period of foreland basin expansion and coarse-grained sediment transport into the basin. Expansion of the foreland basin produced a broad accommodation profile and promoted shelf sandstone deposition. Widespread Late Turonian deposition of coarse-grained clastics postdates tectonism and records the northward migration of deltaic complexes from southern to central Utah. These coarse clastics record a change in sediment provenance that is inferred to be synorogenic in origin.

The middle Cretaceous basin-fill pattern is consistent with a tectonic cycle, but Turonian tectonism is not recognized in the central Utah thrust belt. If there was Turonian tectonism (viscoelastic flexural response or undetected thrust), the basin fill records a 1- to 2-my time lag between thrust cessation and coarse-grained sediment transport to the basin. If tectonism ended at the Cenomanian–Turonian boundary, as has been previously suggested, there is a 4-my time lag. In either case, the time lag between tectonism and coarser grained sediment transport to the basin may represent a complex feedback response. Such sedimentary feedback responses can be related to (1) point-source sediment dispersal, (2) longitudinal drainage along the foredeep axis, (3) structural lineaments producing segmented thrust sheets and foreland subbasins with unique sediment supply and subsidence records, (4) structural uplift basinward of the leading thrust sheet producing piggy-back basins.

The mid-Cretaceous tectonic and stratigraphic (long-term stratigraphic cycle) record is consistent with viscoelastic flexural models for foreland basin subsidence. If an elastic flexural model is assumed, an undetected blind thrust in the thrust belt would be indicated. Neither flexural model adequately explains the time lag recorded by sedimentation patterns in the long-term stratigraphic cycle. There is strong evidence of a Late Turonian, eustatic sea-level fall, and regional correlations (Gardner, this volume) document its affect on sedimentation patterns in the central Western Interior. The long-term stratigraphic cycle is bounded by base-level rise deposits that record well-documented eustatic transgressions. If the long-term stratigraphic cycle is eustatic in origin, then this sedimentation pattern mimics a tectonic depositional cycle suggesting that slope adjustments and sediment supply rates are directly linked to eustasy. Regardless of the mechanism, the stratigraphic record from the foreland basin appears to be a more sensitive recorder of tectonism than structural relationships in the thrust belt.

ACKNOWLEDGMENTS

This paper is part of a Ph.D. study conducted at the Colorado School of Mines. The American Association of Petroleum Geologists grants-in-aid program, American Chemical Society Petroleum Research Fund, SEPM Donald Smith Research grant, Sigma Xi grants-in-aid program, U.S. Geological Survey Branch of Coal Resources, and the Utah Geological and Mineralogical Survey provided financial support for this project. Industrial Associates of the Genetic Stratigraphy Research Program administered by T. A. Cross at the Colorado School of Mines provided additional financial support. The seminal work on the Ferron by T. A. Ryer provided the stratigraphic framework that dictated the conduct of this study, and the author appreciates discussions and geologic insights Tom Ryer graciously shared. The Bureau of Economic Geology, The University of Texas at Austin, provided numerous opportunities for the author to guide geologists through Ferron outcrops in central Utah and this manuscript benefits from their questions, observations, and comments. Figures drafted by Jana Robinson under the supervision of Richard L. Dillon, Chief Cartographer, Bureau of Economic Geology. Technical editing by Tucker Hentz and editing by Lana Dieterich provided by the Bureau of Economic Geology. Helpful reviews and discussions on various parts of this manuscript by Tim Cross, Brian Willis, and Mark Sonnenfeld significantly improved the manuscript. Critical reviews of an early version of this manuscript by Dag Nummedal and Tom Ryer greatly enhanced manuscript quality and clarity, and their efforts are gratefully acknowledged.

Published with permission, Bureau of Economic Geology, The University of Texas at Austin.

REFERENCES

ARMSTRONG, R. L., 1968, Sevier orogenic belt in Nevada and Utah: Geological Society of America Bulletin, v. 79, p. 429–458.

ARMSTRONG, R. L. AND ORIEL, S. S., 1965, Tectonic development of Idaho-Wyoming thrust belt: American Association of Petroleum Geologists Bulletin, v. 49, p. 1847–1866.

ARTHUR, M. A. AND SCHLANGER, S. O., 1979, Cretaceous "oceanic anoxic events" as causal factors in development of reef reservoired giant oil fields: American Association of Petroleum Geologists Bulletin, v. 63, p. 870–885.

BEAUMONT, C., 1981, Foreland basins: Geophysical Journal of the Royal Astronomical Society, v. 65, p. 291–329.

BECK, R. H. AND VONDRA, C. A., JR., 1985, Sedimentary-tectonic model for Laramide basement-thrusting: Geological Society of America Abstracts with Programs, v. 17, p. 521.

BLAIR, T. C. AND BILODEAU, W. L., 1988, Development of tectonic cyclothems in rift, pull-apart, and foreland basins: sedimentary responses to episodic tectonism: Geology, v. 16, p. 517–520.

BOBB, M. C., 1991, The Calico Bed, Upper Cretaceous, Southern Utah: a fluvial sheet deposit in the Western Interior foreland basin and its relationship to eustasy and tectonics: Unpublished M.S. Thesis, University of Colorado, Boulder, 166 p.

BURCHFIEL, B. C. AND DAVIS, G. A., 1975, Nature and controls of Cordilleran orogenesis, western United States, extensions of an earlier synthesis: American Journal of Science, v. 275A, p. 363–396.

BUSCH, D. A., 1959, Prospecting for stratigraphic traps: American Association of Petroleum Geologists Bulletin, v. 43, p. 2829–2843.

BUSCH, D. A., 1971, Genetic units in delta prospecting: American Association of Petroleum Geologists Bulletin, v. 55, p. 1137–1154.

COBBAN, W. A., 1976, Ammonite record from the Mancos Shale of the Castle Valley–Price–Woodside area, east-central Utah: Brigham Young University Geological Studies, v. 22, pt. 3, p. 117–126.

COLLOM, C. J., 1991, High-resolution stratigraphic and paleoenvironmental analysis of the Turonian-Coniacian Stage boundary interval (Late Cretaceous) in the lower Fort Hays Member, Niobrara Formation, Colorado and New Mexico: Unpublished M.S. Thesis, Brigham Young University, Provo, 371 p.

COTTER, E., 1971, Paleoflow characteristics of a Late Cretaceous river in Utah from analysis of sedimentary structures in the Ferron Sandstone: Journal of Sedimentary Petrology, v. 41, p. 129–138.

COTTER, E., 1975a, Deltaic deposits in the Upper Cretaceous Ferron Sandstone, Utah, in Broussard, M. L. S., ed., Deltas, Models for Exploration: Houston, Houston Geological Society, p. 471–484.

COTTER, E., 1975b, Late Cretaceous sedimentation in a low-energy coastal zone: the Ferron Sandstone of Utah: Journal of Sedimentary Petrology, v. 45, p. 15–41.

COTTER, E., 1976, The role of deltas in the evolution of the Ferron Sandstone and its coals: Brigham Young University Geological Studies, v. 22, p. 15–41.

CROSS, T. A., 1988, Controls on coal distribution in transgressive-regressive cycles, Upper Cretaceous, Western Interior, U.S.A., in Wilgus, C. K., Hasting, B. S., Kendall, C. G. St. C., Posamentier, H. W., Ross, C. A., and Van Wagoner, J. C., eds., Sea-level Changes: An Integrated Approach: Tulsa, Society of Economic Paleontologists and Mineralogists Special Publication 42, p. 371–380.

DAVIS, L. J., 1954, Stratigraphy of the Ferron Sandstone, in Grier, B., ed., Geology of Portions of the High Plateaus and adjacent Canyonlands, Central and South-Central Utah: Intermountain Association of Petroleum Geology Fifth Annual Field Guidebook, p. 55–58.

DICKINSON, 1974, Plate tectonics and sedimentation, in Dickinson, W. D., ed., Tectonics and Sedimentation: Tulsa, Society of Economic Paleontologists and Mineralogists Special Publication 22, p. 1–27.

EATON, J. G., KIRKLAND, J. I., GUSTASON, E. R., NATIONS, J. D., FRANCZYK, K. J., RYER, T. A., AND CARR, D. A., 1987, Stratigraphy, correlation and tectonic settings of Late Cretaceous rocks in the Kaiparowits and Black Mesa Basins, in Davis, G. H. and Vanden Dolder, E. M., eds, Geologic Diversity of Arizona and its Margins: Excursions to Choice Areas: Flagstaff, Arizona Bureau of Geology and Mineral Technology, Special Paper 5, p. 113–125.

EATON, J. G., KIRKLAND, J. I., AND KAUFFMAN, E. G., 1990, Evidence and dating of Mid-Cretaceous tectonic activity in the San Rafael Swell, Emery County, Utah: The Mountain Geologist, v. 27, p. 39–45.

EISBACHER, G. H., CARRIGY, M. A., AND CAMPBELL, R. B., 1974, Paleodrainage pattern and late-orogenic basins of the Canadian Cordillera, in Dickinson, W. D., ed., Tectonics and Sedimentation: Tulsa, Society of Economic Paleontologists and Mineralogists Special Publication 22, p. 143–166.

ELDER, W. P., 1985, Biotic patterns across the Cenomanian Turonian extinction boundary near Pueblo, Colorado, in Pratt, L. M., Kauffman, E. G., and Zelt, F. B., eds., 1985, Fine-Grained Deposits and Biofacies of the Cretaceous Western Interior Seaway: Evidence of Cyclic Sedimentary Processes: Denver, Society of Economic Paleontologists and Mineralogists Field Trip Guidebook 4, p. 157–169.

ERICKSEN, M. C. AND SLINGERLAND, R., 1990, Numerical simulations of tidal and wind-driven circulation in the Cretaceous Interior Seaway of North America: Geological Society of America Bulletin, v. 102, p. 1499–1516.

FLEMING, P. B. AND JORDAN, T. E., 1989, A synthetic stratigraphic model of foreland basin development: Journal of Geophysical Research, v. 94, p. 3851–3866.

FLEMING, P. B. AND JORDAN, T. E., 1990, Stratigraphic modeling of foreland basins: interpreting thrust deformation and lithosphere rheology: Geology, v. 18, p. 430–434.

FRAZIER, D. E., 1974, Depositional episodes: their relationship to the Quaternary stratigraphic framework in the northwestern portion of the Gulf basin: Austin, Bureau of Economic Geology Geological Circular 74-1, The University of Texas at Austin, 28 p.

FUCHTBAUR, V. H., 1967, Die sandsteine in der molasse Nordlich der Alpin: Geological Rundshau, v. 56, p. 266–300.

GALLOWAY, W. E., 1989, Genetic stratigraphic sequences in basin analysis I: architecture and genesis of flooding-surface bounded depositional units: American Association of Petroleum Geologists Bulletin, v. 73, p. 125–142.

GARDNER, M. H., 1993, Sequence stratigraphy and facies architecture of the Upper Cretaceous Ferron Sandstone Member of the Mancos Shale, East-Central Utah: Unpublished Ph.D. Dissertation, Colorado School of Mines, Golden, 528 p.

GARDNER, M. H., BARTON, M. D., TYLER, N., AND FISHER, R. S., 1992, Architecture and permeability structure of fluvial-deltaic sandstones, Ferron Sandstone, east-central Utah, in Flores, R. M., ed., Mesozoic of the Western Interior: Fort Collins, Society of Economic Paleontologists and Mineralogists, Theme Meeting Field Guidebook, p. 5–21.

GOLDHAMMER, R. K., OSWALD, E. J., AND DUNN, P. A., 1991, Hierarchy of stratigraphic forcing: example from Middle Pennsylvanian shelf carbonates of the Paradox Basin, in Franseen, E. K., Watney, W. L., Kendall, C. C. G. St. C., and Ross, W., eds., Sedimentary Modeling: Computer Simulations and Methods for Improved Parameter Definition: Kansas Geological Survey Bulletin 233, p. 361–415.

GOODWIN, P. W. AND ANDERSON, E. J., 1985, Punctuated aggradational cycles: a general hypothesis of episodic stratigraphic accumulation: Journal of Geology, v. 93, p. 515–533.

HALE, L. A., 1972, Depositional history of the Ferron Formation, central Utah, in Plateau–Basin and Range Transition Zone: Salt Lake City, Utah Geological Association, p. 115–138.

HALE, L. A. AND VAN DE GRAAFF, F. R., 1964, Cretaceous stratigraphy and facies patterns—northeastern Utah and adjacent areas: Salt Lake

City, Intermountain Association of Petroleum Geologists Thirteenth Annual Field Guidebook, p. 115–138.
HANCOCK, J. M. AND KAUFFMAN, E. G., 1979, The great transgressions of the Late Cretaceous: Geological Society of London Journal, v. 136, p. 175–186.
HAQ, B. U., HARDENBOL, J., AND VAIL, P. R., 1988, Mesozoic and Cenozoic chronostratigraphy and eustatic cycles, in Wilgus, C. K., Hasting, B. S., Kendall, C. G. St. C., Posamentier, H. W., Ross, C. A., and Van Wagoner, J. C., eds., Sea-level Changes: An Integrated Approach: Tulsa, Society of Economic Paleontologists and Mineralogists Special Publication 42, p. 71–109.
HATTIN, D. E., 1964, Cyclic sedimentation in the Colorado Group of west-central Kansas, in Merriam, D. F., ed., Symposium on Cyclic Sedimentation: Kansas Geological Survey Bulletin 169, v. 1, p. 205–217.
HELLER, P. L., ANGEVINE, C. L., AND WINSLOW, N. S., 1988, Two-phase stratigraphic model of foreland-basin sequences: Geology, v. 14, p. 501–504.
HELLER, P. L. AND PAOLA, C., 1989, The paradox of Lower Cretaceous gravels and the initiation of thrusting in the Sevier orogenic belt, United States Western Interior: Geological Society of America Bulletin, v. 101, p. 864–875.
JENKYNS, H. C., 1980, Cretaceous anoxic events: from continents to oceans: Journal of the Geological Society of London, v. 137, p. 171–188.
JERVEY, M. T., 1988, Quantitative geological modeling of siliciclastic rock sequences and their seismic expression, in Wilgus, C. K., Hasting, B. S., Kendall, C. G. St. C., Posamentier, H. W., Ross, C. A., and Van Wagoner, J. C., eds., Sea-Level Changes: An Integrated Approach: Tulsa, Society of Economic Paleontologists and Mineralogists Special Publication 42, p. 109–125.
JORDAN, T. E., 1981, Thrust loads and foreland basin evolution, Cretaceous, western United States: American Association of Petroleum Geologists Bulletin, v. 65, p. 2506–2520.
KATICH, P. J., 1951, The stratigraphy and paleontology of pre-Niobrara upper Cretaceous rocks of Castle Valley, Utah: Unpublished Ph.D. Dissertation, Ohio State University, 208 p.
KATICH, P. J., JR., 1954, Cretaceous and early Tertiary stratigraphy of central and south-central Utah with emphasis on the Wasatch Plateau area: Salt Lake City, Intermountain Association of Petroleum Geologists Fifth Annual Field Guidebook, p. 42–54.
KAUFFMAN, E. G., 1969, Cretaceous marine cycles of the Western Interior: The Mountain Geologist, v. 6, p. 227–245.
KAUFFMAN, E. G., 1977, Geological and biological overview: Western Interior Cretaceous basin, in Cretaceous Facies, Faunas and Paleoenvironments across the Western Interior Basin: The Mountain Geologist, v. 14, p. 75–99.
KAUFFMAN, E. G., COBBAN, W. A., AND EICHER, D. L., 1976, Albian through Lower Coniacian Strata, Biostratigraphy and Principal Events, Western Interior United States: Annales Du Museum D'Histoire Naturelle De Nice-Tome IV, 51 p.
LAWTON, T. F., 1985, Style and timing of frontal thrust structures, thrust belt, central Utah: American Association of Petroleum Geologists Bulletin, v. 69, p. 1145–1159.
LINES, G. C. AND MORRISSEY, D. J., 1983, Hydrology of the Ferron Sandstone Aquifer and Effects of Proposed Surface-Coal Mining in Castle Valley, Utah: Washington, D. C., United States Geological Survey Water-Supply Paper 2195, 40 p.
MEREWETHER, E. A. AND COBBAN, W. A., 1986, Biostratigraphic units and tectonism in the mid-Cretaceous foreland of Wyoming, Colorado, and adjoining areas, in Peterson, J. A., ed., Paleotectonics and Sedimentation: Tulsa, American Association of Petroleum Geologists Memoir 41, p. 443–468.
MITCHUM, R. M. AND VAN WAGONER, J. C., 1991, High frequency sequences and their stacking patterns: sequence stratigraphic evidence of high frequency eustatic cycles: Sedimentary Geology, v. 70, p. 131–160.
MOLENAAR, C. M., 1983, Major depositional cycles and regional correlations of Upper Cretaceous rocks, southern Colorado Plateau and adjacent areas, in Reynolds, M. W. and Dolly, E. D., eds., Mesozoic Paleogeography of West-Central United States: Denver, Rocky Mountain Section, Society of Economic Paleontologists and Mineralogists, Rocky Mountain Paleogeography Symposium 2, p. 201–224.
MOLENAAR, C. M. AND COBBAN, W. A., 1991, Middle Cretaceous stratigraphy on south and east sides of the Uinta Basin, northeastern Utah and northwestern Colorado: United States Geological Survey Bulletin 1787-P, 34 p.
MOLENAAR, C. M. AND WILSON, B. W., 1990, The Frontier Formation and Associated Rocks of Northeastern Utah and Northwestern Colorado: United States Geological Survey Bulletin 1787-M, 21 p.
MOORE, R. C., 1964, Pennsylvanian and Permian Cyclothems, Kansas, in Merriam, D. F., ed., Symposium on Cyclic Sedimentation: Kansas Geological Survey Bulletin 169, v. 1, p. 287–380.
MULDER, T. L. AND BURBANK, D. W., 1993, The impact of incipient uplift on patterns of fluvial deposition: an example from the Salt Range, Northwest Himalayan Foreland Pakistan, in Alluvial Sedimentation: Special Publications of the International Association of Sedimentologists 17, p. 521–539.
PETERSON, F., 1969, Cretaceous Sedimentation and Tectonism in the Southeastern Kaiparowits Region, Utah: Denver, United States Geological Survey Open-File Report, 259 p.
PETERSON, F. AND RYDER, R. T., 1975, Cretaceous rocks in the Henry Mountains region, Utah and their relation to neighboring regions, in Fassett, J. E., and Wengerd, S. A., eds., Canyonlands Country: Four Corners Geological Society Guidebook, 8th Field Conference, p.167–189.
PICHA, F., 1986, The influence of preexisting tectonic trends on geometries of the Sevier Orogenic Belt and its foreland in Utah, in Peterson, J. A., ed., Paleotectonics and Sedimentation: Tulsa, American Association of Petroleum Geologists Memoir 41, p. 309–320.
RAYNOLDS, R. G. H. AND JOHNSON, G. D., 1985, Rates of Neogene depositional and deformational processes, northwest Himalayan foredeep margin, Pakistan, in Snelling, N. J., ed., The Chronology of the Geological Record: The Geological Society Memoir 10, p. 297–311.
RIEMERSMA, P. AND CHAN, M., 1991, Facies of the lower Ferron Sandstone and Blue Gate Shale Members of the Mancos Shale: lowstand and early transgressive facies architecture, in Swift, D. J. P., Oertel, G. F., Tilman, R. W., and Thorne, J. A., eds., Shelf Sands and Sandstone Bodies, Geometry, Facies and Sequence Stratigraphy: International Association of Sedimentologists Special Publication 14, p 489–510.
RYER, T. A., 1981, Deltaic coals of Ferron sandstone member of Mancos Shale: predictive model for Cretaceous coal-bearing strata of Western Interior: American Association of Petroleum Geologists Bulletin, v. 65, p. 2323–2340.
RYER, T. A., 1983, Transgressive-regressive cycles and the occurrence of coal in some Upper Cretaceous strata of Utah: Geology, v. 11, p. 207–210.
RYER, T. A. AND LOVEKIN, J. R., 1986, The Upper Cretaceous vernal delta of Utah—depositional or paleotectonic feature? in Peterson, J. A., ed., Paleotectonics and Sedimentation: Tulsa, American Association of Petroleum Geologists Memoir 41, p. 497–509.
RYER, T. A. AND MCPHILLIPS, M., 1983, Early Late Cretaceous paleogeography of east-central Utah, in Reynolds, M. W., and Dolly, E. D., eds., Mesozoic Paleogeography of West-Central United States: Denver, Rocky Mountain Section, Society of Economic Paleontologists and Mineralogists, Rocky Mountain Paleogeography Symposium 2, p. 253–271.
RYER, T. A., PHILLIPS, R. E., BOHOR, B. F., AND POLLASTRO, R. M., 1980, Use of altered volcanic ash falls in stratigraphic studies of coal-bearing sequences: an example from the Upper Cretaceous Ferron Sandstone Member of the Mancos Shale in central Utah: Geological Society of America Bulletin, v. 91, p. 579–586.
SAHAGIAN, D. AND JONES, M., 1993, Quantified Middle Jurassic to Paleocene eustatic variations based on Russian Platform stratigraphy: Stage level resolution: Geological Society of America Bulletin, v. 105, p. 1109–1118.
SCHLANGER, S. O., ARTHUR, M. A., JENKYNS, H. C., AND SCHOLLE, P. A., 1986, The Cenomanian–Turonian oceanic anoxic event I. Stratigraphy and distribution of organic carbon-rich beds and the marine $d^{13}C$ excursion, in Brooks, J. and Flebt, A. J., eds., Marine Petroleum Sources Rocks: Geological Society Special Publication 24, p. 347–375.
SCHLANGER, S. O. AND JENKYNS, H. C., 1976, Cretaceous oceanic anoxic events: causes and consequences: Geologie En Mijnbouw, v. 55, p. 179–184.

SEVON, W. D., 1985, Nonmarine facies of the middle and later Devonian Catskill coastal alluvial plain, *in* Woodrow, D. L. and Sevon, W. D., eds., The Catskill Delta: Boulder, Geological Society of America Special Paper 201, p. 79–90.

SLOSS, L. L., 1963, Sequences in the Cratonic Interior of North America: Geological Society of America Bulletin, v. 71, p. 93–114.

SONNENFELD, M. D., 1991, Anatomy of offlap in a shelf-margin depositional sequence: Upper San Andres Formation (Permian, Guadalupian), Last Chance Canyon, Guadalupe Mountains, New Mexico: Unpublished M.S. Thesis, Colorado School of Mines, Golden, 297 p.

SPEIKER, E. M., 1949, Sedimentary facies and associated diastrophism in the Upper Cretaceous of central and eastern Utah, *in* Sedimentary Facies in Geologic History: New York, Geological Society of America Memoir 39, p. 55–82.

THOMPSON, S. L., 1985, Ferron Sandstone Member of the Mancos Shale: a Turonian mixed-energy deltaic system: Unpublished M.S. Thesis, The University of Texas at Austin, Austin, 165 p.

VAIL, P. R, MITCHUM, R. M., JR., AND THOMPSON, S., III, 1977, Seismic stratigraphy and global changes in sea level: Relative changes of sea level from coastal onlap, *in* Payton C. E., ed., Seismic Stratigraphy—Applications to Hydrocarbon Exploration: Tulsa, American Association of Petroleum Geologists Memoir 26, p. 63–81.

VAN HOUTEN, F. B., 1977, Northern Alpine mollasic and similar Cenozoic sequences of southern Europe, *in* Dott, R. H., Jr. and Shaver, R. H., eds., Modern and Ancient Geosynclinal Sedimentation: Tulsa, Society of Economic Paleontologists and Mineralogists Special Publication 19, p. 260–273.

VAN WAGONER, J. C., 1985, Reservoir facies distribution as controlled by sea-level change (abs.): Golden, Society of Economic Paleontologists and Mineralogists Mid-Year Meeting, p. 91–92.

VAN WAGONER, J. C., MITCHUM, R. M., CAMPION, K. M., AND RAHMANIAN, V. D., 1990, Siliciclastic Sequence Stratigraphy: Tulsa, American Association of Petroleum Geologists Methods in Exploration Series 7, 55 p.

VILLIEN, A. AND KLIGFIELD, R. M., 1986, Thrusting and synorogenic sedimentation in central Utah, *in* Peterson, J. A., ed., Paleotectonics and Sedimentation: Tulsa, American Association of Petroleum Geologists Memoir 41, p. 281–307.

WANLESS, H. R. AND WELLER, J. M., 1932, Correlation and extent of Pennsylvanian cyclothems: Geological Society of America Bulletin, v. 43, p. 1003–1016.

WEIMER, R. J., 1960, Upper Cretaceous stratigraphy Rocky Mountain area: American Association of Petroleum Geologists Bulletin, v. 44, p. 1–20.

WEIMER, R. J., 1983, Relation of unconformities, tectonics, and sea-level changes in the Cretaceous of the Denver Basin and adjacent areas, *in* Reynolds, M. W. and Dolly, E. D., eds., Mesozoic Paleogeography of West-Central United States: Denver, Rocky Mountain Section, Society of Economic Paleontologists and Mineralogists, Rocky Mountain Paleogeography Symposium 2, p. 359–377.

WHEELER, H. E., 1964, Base level, lithosphere surface, and time stratigraphy: Geological Society of America, v. 75, p. 599–610.

WILTSCHKO, D. B. AND DORR, J. A., 1983, Timing of deformation in overthrust belt and foreland of Idaho, Wyoming and Utah: American Association of Petroleum Geologists Bulletin, v. 67, p. 1304–1322.

WILLIS, B., 1993, Evolution of Miocene fluvial systems in the Himalayan foredeep through a 2-kilometer-thick succession in northern Pakistan: Sedimentary Geology, v. 87, p. 1–45.

WILSON, J. L., 1986, Carbonate Facies in Geologic Time: New York, Springer-Verlag, 471 p.

YOUNG, R. G., 1957, Late Cretaceous cyclic deposits, Book Cliffs, eastern Utah: American Association of Petroleum Geologists Bulletin, v. 41, p. 1760–1774.

YOUNG, R. G., 1960, Dakota Group of the Colorado Plateau: American Association of Petroleum Geologists Bulletin, v. 44, p. 156–194.

INDEX

INDEX

A

Accommodation, 9, 15, 97, 133–134, 141–142, 183, 185, 187–208, 243–249, 271, 275–278, 283–284, 286, 288, 296, 298–300
Aggradation, 15, 17, 21–22, 28–29, 131, 133, 143, 181, 195–196, 257
Alabama promontory, 111, 113–115, 123
Alabama structural recess, 111
Alberta, iii, 21, 84, 135, 213–215, 218, 221–222, 227
Alberta Basin, 216, 222
Anaerobic, 198, 200, 205, 218, 223
Anchor Limestone, 177–182, 184
Ancient Wall reef complex, 216–217, 224
Anoxic, 208–209, 216, 222–223
Antler Orogen, 175
Antler orogeny,
    in California, 184–185
    in Idaho and Montana, 135–138, 213, 225
    in Nevada, 184–185, 188, 208, 225
    in Western Canada, 135–136, 225–227
Arc-continent collision, 53, 61, 111, 123, 188, 225
Arcs Member, 217, 220
Arctic Devonian clastic wedge, 77
Arrowhead Limestone, 177

B

Back-bulge basin, 129, 133, 187–188, 194, 196–198, 200, 203, 208
Backward reconstructions, 31
Bakken Formation, 225
Banff Formation, 214–215, 219–225, 226
Bangor Limestone, 117, 119–122
Basement fault system, 111–113, 115, 117, 121–123
Bashaw-Duhamel reef trend, 216
Beaverhill Lake Group, 220
Beaverhill Lake megacycle, 216, 220
Bessemer transverse zone, 117, 122
Big Valley Formation, 218
Biostratigraphy, 190, 200, 213, 216, 243
Bird Fiord Formation, 78–80, 82–83, 88, 90
Birmingham basement fault system, 111, 113, 115, 121–123
Black Warrior foreland basin, 111–124
Black shale, 198–199, 216, 218–223
Blind thrust anticlines, 140–141
Blountian, 53–54, 57–61
Brachiopods, 93, 177–178, 182–183, 198, 202, 205, 216
British Columbia, 65, 72–74, 188, 213–215, 225–227
Bullion Limestone, 177, 181–182

C

Cahaba synclinorium, 116–117, 120–123
Cairn, 216, 222
Caledonian, 77, 79, 82, 87–90
Campanian, 65–68, 73
Canadian Rocky Mountains, iv, 213–214, 220–227
Canyon Formation, 141, 167
Carbonate bank, 196
Carbonate platforms,
    backstepping of, 130–131, 170
    drowning of, 53, 132–133, 185, 223
    growth rates of, 133
    progradation of, 117, 120, 123, 142, 181–182, 185, 196–197, 205, 207, 216, 283, 295–296, 300
Carbonate ramps, 129–131, 141–143
Carboniferous, 77, 82, 213, 226–227, 227–229
Cariboo Mountains, 225
Cariboo orogeny, 225
Catskill Delta, iii, 226
Cenomanian, 66, 274, 276, 287, 300
Central Basin Platform, 37–40, 42–49, 136, 139, 150–165, 167–168, 170–172
Chainman Shale, 204–205, 207–208
Chert, 69–70, 79, 81, 93, 97–103, 105–106, 109, 114, 120, 175, 179–180, 182–185, 192–194, 197–201, 203, 218, 220, 288, 292, 296, 298–299
Chronostratigraphic charts, 31
Cisco Formation, 141, 167
Clastic wedge, iii, 77–79, 82, 84–89, 111, 113–124, 215, 222, 224–226, 235, 238, 240, 257, 268, 288, 283, 293
Clastic-wedge progradation, 12, 15
Clay mineral composition, 53–55, 57–61
Cline Channel, 216
Coast Belt, 65–70, 72–74
Columbia Icefield, 214, 222
Compaction, 25, 27, 46, 160, 243, 284, 300
Conglomerates, 60, 68, 97–109, 121, 192–193, 195, 197, 205, 235–236, 283, 298
Conodonts,
    anchoralis-latus Zone, 220
    Bispathodid conodont assemblage, 183
    Cavusgnathus Zone, 185
    crepida Zone, 220
    expansa Zone, 218, 222–223
    Hindeodid biofacies, 180
    isosticha-Upper *crenulata* Zone, 185
    linguiformis Zone, 217
    rhomboidea Zone, 220
    triangularis Zone, 217
    typicus Zone, 184–185
Continental margin, 10, 17, 53, 113, 124, 213–214, 225–227
Contraction, 271, 283, 298
Convergent margin, 123
Cooking Lake platform, 216, 220
Corals, 175, 177–178, 181–183, 202, 205, 216
Costigan Member, 218
Cratonic margin, 213, 222, 226
Cretaceous, iv, 21, 65–68, 70, 72–74, 97–109, 132, 226, 235, 236, 240, 243–278, 285–286, 290–291, 296–297, 299–300
Crustal loading, 224, 226, 298
Cycle stacking patterns, 243, 277, 286

D

Dawn Limestone, 177–178, 180–182
Deformation, ii–iv, 4, 6, 9, 37–39, 45, 53, 65, 77, 97, 123, 127–129, 134–136, 139–141, 149–170, 195, 213, 224–226, 233, 294, 296, 299
Delaware Basin, 37–39, 43–49, 150–170
Deltaic, 27, 65, 68–69, 81, 111, 117, 119–120, 122, 185, 207–209, 238, 257, 260, 294–297, 299–300
Depth-dependent rheology, 4
Deseret Limestone, 204–205
Detrital zircons, iii, 65–74, 79, 82–85, 87–90
Devonian, iii, 68, 71, 77–79, 84, 86–90, 135–138, 175, 177, 184, 188–201, 203, 213, 215, 220–227
Diamond Peak Formation, 207–208
Diapirs, 139–140
Diffusion, 3–4, 6–7, 15, 21
Drowning sequences, 25
Duvernay Formation, 216, 222
Dysaerobic, 180, 182, 200, 205, 218

E

Early Mississippian, 79, 88, 188–189, 200–201, 203, 205, 218, 220, 224–225, 227
Earn Group, 213, 225
East Greenland Caledonides, 89
Eastern California, iv, 175, 182–185

Elk Point Basin, 216
Ellesmere Island, 77, 79–82, 84–86, 87, 93
Ellesmerian fold belt, 216, 225
Ely Limestone, 188, 207
Episodic thrusting, 31
Erosion, iii, 5–6, 9, 22, 25–27, 42, 46, 48, 53, 60, 65, 68, 73–74, 77, 84–85, 87–90, 108, 133–134, 143, 155, 159, 164–165, 167–168, 170, 183, 188, 190, 193–194, 197–201, 203, 205, 207–209, 213, 225, 235, 245, 247, 257, 276–277, 283–284, 286, 288, 292–294, 296, 299–300
Eustasy, iii–iv, 4, 10, 25, 28, 187–188, 207–209, 272, 275–278, 284, 300
Eustatic sea-level changes, 25, 28, 129, 132–134, 185, 187, 208, 245, 247–248, 272–273, 276–278, 283
Exshaw Formation, 213–216, 218–227
Exshaw-Banff megacycle(s), 218–219
Extension Formation, 66–70, 72

F

Fairholme Group, 222
Fairholme reef complex, 216–217, 222
Famennian, 77–79, 138, 189–190, 194, 197, 199, 201, 214–218, 220–227
Fenfanghe Formation, 235, 238, 240
Fernie, British Columbia, 217
Ferron Sandstone, 271, 275, 284, 288, 294, 297–299
Filled foreland basin, definition, 9
Fitchville Limestone, 200
Flexure,
  of an elastic plate, 8
  flexural rigidity, 25
  flexural folding, 25
  visco-elastic relaxation, 31, 133–134
Floyd Shale, 119–120
Flume Formation, 216, 220
Forebulge,
  see also peripheral bulge
Foredeep, 37–41, 78, 108, 127–131, 133–136, 139–144, 149–153, 166–167, 169–171, 224, 235, 240, 278, 298, 300
Foreland, iii–iv, 3–5, 8–10, 12, 15, 17, 21–23, 25–26, 28, 37–39, 48, 53–61, 65–66, 71–72, 77–78, 87–90, 97, 111–124, 127–144, 149–150, 158, 168–169, 171, 187–209, 213, 223, 225–226, 233–240, 243, 283–284, 288, 297–300
Foreland basin, iii–iv, 3–5, 8–10, 12, 15, 17, 21–23, 25, 28, 37–38, 53–61, 65–66, 71–72, 77–78, 97, 108–109, 111–124, 127–135, 141–144, 149, 165, 187–188, 190–209, 213, 225–226, 233, 240, 247, 283–284, 288, 297–300
Foreland basin subsidence, iii, 129, 131–133, 137, 223, 299
Fram Formation, 79, 81–82, 93
Franklinian, 77–78, 82, 86
Frasnian, 77–79, 82, 138, 188–190, 195, 200–201, 214–218, 220–227

G

Gabriola Formation, 68–69, 71–74
Geochronology, 60, 85, 87–89, 105
Georgia Basin, 65, 67, 74
Givetian, 79, 195, 214–217, 220–221, 227
Graded river, concept of, 9
Grantland Uplift, 80, 87
Greenland, 77–78, 82, 84–90
Grenville Orogen, 88
Guilmette Limestone, 190, 192, 200

H

Hartselle Sandstone, 117, 119–120, 123
Hecla Bay Formation, 78–79, 81, 93
Helan orogenic belt, 233
Hemicycles, 222
Homoclinal ramp, 129, 178, 182–183, 185

I

Indian Springs Formation, 176, 182
Indosinian Orogeny, 233, 236, 240
In-plane stress, 25–34, 128–129, 134–136
Intraforeland uplifts, 37, 136, 149
Intraplate stress,
  see also in-plane stress,
Iowa baseline, 226
Isostatic rebound, 25, 31, 207

J

Jasper Basin, 217
Jasper, Alberta, 214–215, 217, 220
Jinsha-Menglian suture, 233
Joana Limestone, 202–204
Jura Creek, Alberta, 218–219

K

Keystone Thrust, 181
Kinderhookian, 181, 182–185, 225
Kootenay arc, 213, 224–225

L

Late Devonian, 77–79, 135–138, 175, 184, 188–190, 194, 197, 199–200, 202, 208, 213, 220, 223–227
Leaning Rock Formation, 182–184
Leduc Woodbend megacycle, 222
Lee Canyon thrust, 181
Lithofacies, 40, 97–98, 106, 134, 149–150, 160–161, 163, 168, 201, 204, 215–216, 218–220, 223, 226, 283
Lithosphere,
  deformation of, 170
  flexure of, 12, 15, 25, 37–49, 127–134, 149, 187–188, 194, 208
  heterogeneities in the, iii, 47–48
  properties of, iii, 25, 27, 39–42, 45–48
Loading,
  distributed, 5, 37, 39–41, 298
  redistribution of, iii, 3–4, 6, 149
Lodgepole Formation, 131
Lower Mississippian, 111, 114, 188, 205, 213, 216
Lowstand, 143, 194, 196–198, 200, 205, 216
Lussier syncline, British Columbia, 220, 243, 277

M

Maastrichtian, 65, 67–68, 74
Magmatic arc, 53, 60, 224
Maligne Formation, 216
Marathon-Ouachita foreland, 37–38, 136, 139, 165–171
Marathon-Ouachita orogenic belt, 37, 47, 136, 139, 149–150, 165–171
Marble Canyon thrust, 183
Mass fluxes,
  competition among tectonic, surface, and isostatic, 12–21
Mechanical coupling, 26
Meekwap reefs, 220
Meramecian, 122–123, 175, 181–185, 207
Mexican Spring Formation, 182–185
Middle Devonian, 79, 188, 213
Midland Basin, 37–39, 41–49, 151, 163, 165–168, 170–171
Miette reef complex, 216, 222, 224
Mississippian, iv, 38, 79, 88, 111, 113, 116–117, 120, 122, 131, 135–136, 160, 164–170, 175, 181–185, 188–190, 200–201, 205, 207–209, 213, 215–216, 218, 220, 223–227
Model,
  coupled diffusive/advective surface processes, 6
  doubly-vergent critical Coulomb orogenic wedge, 6
  feedback among tectonic, surface, and isostatic model processes, 12–21
  orographically controlled precipitation (Climate), 8
  parameter values for planform orogen and foreland basin model, 6
  planform flexural isostatic compensation, 8

planform kinematic orogen and clastic foreland basin, 4–6
Model results,
 axial drainage network, 22
 longitudinal transport and progradational basin filling, 17–22
 orographically controlled precipitation, effects of, 17, 22
 planform synchronous continent/continent collision, 12–17
 planform diachronous continent/continent collision, 17–22
 river power facies, 12, 17, 21
Montana, 72, 131, 135, 184–185, 213–214, 224
Monte Cristo Group, 176, 182
Monteagle Limestone, 117, 119–120
Morro Member, 218
Mount Hawk Formation, 217
Mudstone, 53–54, 59–60, 65, 97–109, 114, 117, 119–121, 177, 179–180, 182–183, 193, 196–198, 200–205, 217–220, 236, 247, 249, 251, 254–258, 260, 270–271, 275, 277, 283, 288–289, 291–294, 296, 298–300

N

Nanaimo Group, iii, 65–74
Nevada, iv, 175–178, 181, 183–185, 188, 200, 203, 207–209, 225
Nisku-Winterburn megacycle, 222
Nutrient recharge, 218

O

Oblique convergence, 129, 135
Offlap, 299–300
Okse Bay Formation, 79–80, 82, 87, 93
Onlap, 25, 41, 55, 136, 150, 155, 158, 160–161, 163, 168, 170, 188–189, 196, 249, 254
Opposing relative sea-level trends, 187
Ordos basin, iv, 233–240
Ordos Lake, 236–238
Ordovician, iii, 53–54, 60–61, 77, 82, 84–85, 88–90, 114–115, 123, 131–132, 158, 160–161, 163–164, 170, 233
Orogenic wedge,
 advancement of, 130–131
 creep rate within, 26
 critical, 3–6, 23, 26–27, 207
 growth of, 3–5, 15, 17, 21, 25, 27
 rheological properties of, iii, 26–27
Osagean, 175, 181, 183–185, 189, 225
Ouachita orogen, 115
Outboard terranes, 53, 60
Overfilled foreland basin, definition, 9
Oxygen-minimum zone (OMZ), 222–223

P

Paleocurrents, 69, 79, 121
Paleogeography, 73, 213
Palliser Formation, 214–216, 218–223, 225–227
Parkwood Formation, 117, 119–122
Passive-margin shelf, 114
Peace River Arch, 216, 225
Pearya, 77, 88
Peechee Member, 222
Perdido Formation, 182
Perdrix Formation, 216, 222
Peripheral bulge, iv, 9–10, 12, 15, 17, 21–22, 25, 28, 53, 128–131, 133–135, 141, 187–188, 194, 196–197, 200, 203, 205, 207–208, 276, 283, 297–298, 300
Permian Basin, iii–iv, 37–49, 136, 149–172
Persian Gulf, 140, 143
Pilot Shale, 192, 194–195, 197–200, 222
Pinecone Sequence, 192, 194, 197, 200
Pinyon Peak Limestone, 193, 196–198
Platforms, iv, 53, 127–144, 163, 175–178, 180–185, 216–217, 220, 226
Pottsville Formation, 120–123
Pride Mountain Formation, 120–122

Pro-foreland basin, definition, 5
Progradation, 12, 15, 21, 25, 29, 79, 119–121, 131, 134–135, 141, 181–182, 184–185, 187–188, 196–197, 205, 207–208, 216, 247, 275, 283–284, 288, 295–296, 299–300
Prophet Trough, 224
Protection Formation, 67–70, 72–74
Provenance, iii, 25, 53–61, 65–74, 77, 82–89, 106–108, 117, 207, 225–226, 276, 288, 299–300
Purcell Mountains, 213, 225

Q

Qinling-Qilian orogenic belt, 233
Quartzofeldspathic clastic rocks, 219–220, 223–224, 226

R

Reactivation of basement structures, 38, 129, 134–136
Reefs, 129, 131–133, 140–141, 216–217, 222
Regression, 122, 134, 222, 240, 274, 277, 298
Retro-foreland basin, definition, 5
Rimbey-Meadowbrook reef trend, 216
Rimmed platform, 185
Roberts Mountains allochthon, 185, 188–190, 194, 201, 207, 225
Roberts Mountains thrust, 190, 225
Rocky Mountains, iv, 213–227
Ronde Member, 217

S

Sahul Shelf, 132, 143
San Juan thrust system, 65–69, 72
Santa Rosa Hills Limestone, 182–183
Santonian, 66
Sappington Member, 222
Sassenach Formation, 214–215, 217, 220–224
Sea-level changes, 25, 28, 127, 129, 132–134, 187, 196, 220–223, 276–277
Sediment dispersal, 111, 117, 121, 142–143, 298–300
Sediment provenance, iii, 25, 31, 288, 299–300
Selywn Basin, 225
Sequence stratigraphy, 216
Shale-conglomerate-facies belt, 175
Shallowing upward cycles, 216, 220
Shelf-slope break, 218, 277
Sichuan foredeep, 240
Simla Member, 217
Sino-Korean massif, 233
Slave Point Platform, 216
Southern Great Basin, 175
Southesk Formation, 216, 222
Southesk-Cairn reef complex, 222
Stone Canyon limestone, 182–185
Strawn Formation, 159, 163, 165, 167–168, 170
Strike-slip deformation, 47, 135–136, 151, 153, 155, 157, 160–161
Subduction, 5, 8, 10, 17, 23, 53, 74, 111, 115, 141, 143, 149, 207, 225–226
Submarine fan, 66, 68–69, 74, 205, 207–208
Subsidence,
 analyses, iv, 37, 135, 137, 168–169, 213, 223–227
 differential, 97, 127, 129, 135–136, 150, 160, 166–168, 208, 225–226
 flexural, iii, 3–4, 8, 12, 15, 17, 25, 37–49, 78, 111, 127–141, 150, 170–171, 187–188, 205, 207–209, 225, 283, 297, 299–300
 rate, iii, 6–7, 9–10, 15, 17, 21–23, 117, 122, 131–133, 184–185, 187, 223, 226, 240, 283–284, 295, 299–300
 regional, iii–iv, 9, 12, 37–49, 53, 65, 78, 87, 97–109, 116, 122, 135, 137, 184–185, 213, 223–224, 226–227, 233, 276–278, 288, 292, 294, 300
Swan Hills Formation,
 platform and reefs, 216–217, 220, 222
 reef stage, 214
Swan Hills-Beaverhill megacycle, 220, 222

Synsedimentary basement fault, 106–109
Synsedimentary fault reactivation, 117

T

Taconic, 53–54, 56–61, 114
Tectonostratigraphic model, 53, 69
Tethys tectonic domain, 227
Three Forks Formation, 222
Timor Trough, 143
Tin Mountain Limestone, 182–184
Topographic inversion, 208
Tournaisian, 77–78, 88
Transgression, 98, 108, 113–114, 119–120, 120, 181–182, 184–185, 189, 194, 196–197, 200, 202, 205, 218, 220–223, 247, 249, 257–258, 271, 276–277, 283, 288, 300
Transportation coefficients, 27
Transpression, 227, 234
Transtension, 213, 226
Turonian, 65, 67, 244–278, 283–284, 288, 291–292, 294–296, 298–300

U

U-Pb analyses, 65, 70
U-Pb geochronology, 71, 82–83
Underfilled foreland basin, definition, 9
Upper Devonian, 77, 88, 136, 177, 192–194, 203, 205, 213, 220, 225–227
Upwelling, 218, 222
Utah, iv, 97, 184–185, 188, 194, 200, 202, 207–209, 245–278, 283–284, 288–289, 291–292, 294–300

V

Val Verde Basin, 37–39, 41–43, 46–49, 139, 141, 151–152, 158, 164–165, 167–171
Vancouver Island, 65
Volcanic ash, 224, 245, 270, 277, 284, 292–294, 299

W

Wabamun Group, 214, 216–218, 225
Waterways Formation, 216
West Alberta Arch, 216
West Pembina, 216, 222
Western Interior, iv, 97, 243, 258, 260, 268, 270, 275, 278
Winterburn megacycle, 216–217
Woodbend megacycle, 216, 222
Woodruff Formation, 192, 194, 197
Wrangellia terrain, 65, 67–73
Wyoming, iv, 97–98, 109, 185, 243, 249, 258, 260, 268, 270–271, 274–278

X

X-ray diffraction, 55, 57–59
X-ray fluorescence, 55

Y

Yanan Formation, 235, 237–238
Yanchang Group, 235–236, 240
Yellow River, 233
Yellowpine Limestone, 177, 181–182
Yinshan-Tianshan orogenic belt, 233